Analyzing the Role of Citizen Science in Modern Research

Luigi Ceccaroni
1000001 Labs, Spain

Jaume Piera
ICM-CSIC, Spain

A volume in the Advances in Knowledge Acquisition, Transfer, and Management (AKATM) Book Series

IGI GLOBAL
DISSEMINATOR OF KNOWLEDGE

www.igi-global.com

Published in the United States of America by
IGI Global
Information Science Reference (an imprint of IGI Global)
701 E. Chocolate Avenue
Hershey PA, USA 17033
Tel: 717-533-8845
Fax: 717-533-8661
E-mail: cust@igi-global.com
Web site: http://www.igi-global.com

Library of Congress Cataloging-in-Publication Data

Names: Ceccaroni, Luigi, 1969- author. | Piera, Jaume, 1966- author.
Title: Analyzing the role of citizen science in modern research / Luigi Ceccaroni and Jaume Piera, editors.
Description: Hershey PA : Information Science Reference, [2017] | Series: Advances in knowledge acquisition, transfer, and management | Includes bibliographical references and index.
Identifiers: LCCN 2016033137| ISBN 9781522509622 (hardcover) | ISBN 9781522509639 (ebook)
Subjects: LCSH: Science--Social aspects. | Science--Study and teaching--Social aspects. | Science and state--Citizen participation.
Classification: LCC Q175.5 .C44 2017 | DDC 507.2--dc23 LC record available at https://lccn.loc.gov/2016033137

This book is published in the IGI Global book series Advances in Knowledge Acquisition, Transfer, and Management (AKATM) (ISSN: 2326-7607; eISSN: 2326-7615)

British Cataloguing in Publication Data
A Cataloguing in Publication record for this book is available from the British Library.

Advances in Knowledge Acquisition, Transfer, and Management (AKATM) Book Series

Murray E. Jennex
San Diego State University, USA

ISSN:2326-7607
EISSN:2326-7615

Mission

Organizations and businesses continue to utilize knowledge management practices in order to streamline processes and procedures. The emergence of web technologies has provided new methods of information usage and knowledge sharing.

The Advances in Knowledge Acquisition, Transfer, and Management (AKATM) Book Series brings together research on emerging technologies and their effect on information systems as well as the knowledge society.AKATM will provide researchers, students, practitioners, and industry leaders with research highlights surrounding the knowledge management discipline, including technology support issues and knowledge representation.

Coverage

- Cognitive Theories
- Cultural Impacts
- Information and Communication Systems
- Knowledge acquisition and transfer processes
- Knowledge management strategy
- Knowledge Sharing
- Organizational Learning
- Organizational Memory
- Small and Medium Enterprises
- Virtual Communities

The Advances in Knowledge Acquisition, Transfer, and Management (AKATM) Book Series (ISSN 2326-7607) is published by IGI Global, 701 E. Chocolate Avenue, Hershey, PA 17033-1240, USA, www.igi-global.com. This series is composed of titles available for purchase individually; each title is edited to be contextually exclusive from any other title within the series. For pricing and ordering information please visit http://www.igi-global.com/book-series/advances-knowledge-acquisition-transfer-management/37159. Postmaster: Send all address changes to above address.

Titles in this Series

For a list of additional titles in this series, please visit: www.igi-global.com

Handbook of Research on Social, Cultural, and Educational Considerations of Indigenous Knowledge in Developing Countries
Patrick Ngulube (University of South Africa, South Africa)
Information Science Reference • copyright 2017 • 462pp • H/C (ISBN: 9781522508380) • US $265.00 (our price)

Research 2.0 and the Impact of Digital Technologies on Scholarly Inquiry
Antonella Esposito (University of Milan, Italy)
Information Science Reference • copyright 2017 • 343pp • H/C (ISBN: 9781522508304) • US $185.00 (our price)

Handbook of Research on Theoretical Perspectives on Indigenous Knowledge Systems in Developing Countries
Patrick Ngulube (University of South Africa, South Africa)
Information Science Reference • copyright 2017 • 516pp • H/C (ISBN: 9781522508335) • US $275.00 (our price)

Harnessing Social Media as a Knowledge Management Tool
Ritesh Chugh (Central Queensland University, Australia)
Information Science Reference • copyright 2017 • 393pp • H/C (ISBN: 9781522504955) • US $185.00 (our price)

Contemporary Approaches to Dissertation Development and Research Methods
Valerie A. Storey (University of Central Florida, USA) and Kristina A. Hesbol (University of Denver, USA)
Information Science Reference • copyright 2016 • 360pp • H/C (ISBN: 9781522504450) • US $195.00 (our price)

Organizational Knowledge Facilitation through Communities of Practice in Emerging Markets
Sheryl Buckley (University of South Africa, South Africa) Grzegorz Majewski (University of the West of Scotland, UK) and Apostolos Giannakopoulos (University of South Africa, South Africa)
Business Science Reference • copyright 2016 • 326pp • H/C (ISBN: 9781522500131) • US $175.00 (our price)

Mixed Methods Research for Improved Scientific Study
Mette L. Baran (Cardinal Stritch University, USA) and Janice E. Jones (Cardinal Stritch University, USA)
Information Science Reference • copyright 2016 • 335pp • H/C (ISBN: 9781522500070) • US $195.00 (our price)

Handbook of Research on Innovations in Information Retrieval, Analysis, and Management
Jorge Tiago Martins (The University of Sheffield, UK) and Andreea Molnar (University of Portsmouth, UK)
Information Science Reference • copyright 2016 • 580pp • H/C (ISBN: 9781466688339) • US $325.00 (our price)

Editorial Advisory Board

List of Reviewers

Table of Contents

Detailed Table of Contents

The first goal of this chapter is to propose a slight re-framing of citizen science, which will contextualize the information presented in the rest of the book. The authors propose a perspective on and a definition for citizen science (which is alternative to the numerous previously documented definitions) as: "work undertaken by civic educators together with citizen communities to advance science, foster a broad scientific mentality, and/or encourage democratic engagement, which allows society to deal rationally with complex modern problems". By explaining the rationale behind this definition, the authors also hope to raise awareness of the role that the meaning of words and phrases (semantics) plays in understanding and supporting citizen science. A second goal of this chapter is to explain how different organizations already use certain software solutions to organize knowledge about citizen science, how these systems can be classified and how they can facilitate or impede interoperability – the ability of humans and machines to pass information between each other.

Citizen science activity is growing rapidly around the world and diversifies into new disciplines with recent advances in technology. This expansion is accompanied by the formation of associations and networks dedicated to citizen science practitioners, which aim at supporting citizen science as a research

This chapter focuses on the risk that, when citizen science is introduced in social environments different from those in the Global North where it originated, it may be subject to the error of providing the right answer to the wrong question. To avoid this type of errors, it is necessary to train those who participate in citizen-science studies: citizens as well as researchers. Otherwise, we may encounter new forms of scientific dependence that benefit knowledge accumulation and policy decision-making in the Global North, without contributing to the quality of life of those who carry out the studies. This chapter analyzes the relationship between civic development, citizen science and ways of implementing research conclusions through public policies, given the characteristics of political and citizen participation in the Global South. Here, the introduction of citizen science is seen as an opportunity to construct a more inclusive and participatory society, and to reduce the risk of returning to paternalistic, passivity-inducing and purely instrumental approaches to development.

This chapter shows how citizen-science initiatives have been known to exist for a long time, but only recently they were further enhanced thanks to technological and societal developments, such as the availability of mobile devices, the widespread use of the internet and the low cost of location devices. These developments shaped the geographic information system (GIS) world as it is known today: a group of technologies that allows retrieving, storing, analyzing and sharing spatial information, by people who are not necessarily GIS professionals. This chapter starts with a general background about GIS, adding then more detail in topics of particular relevance in the context of citizen science. The rest of the chapter is focused on reviewing and classifying the use of GIS in citizen-science initiatives; and some use cases are described in order to provide practical examples of the use of these technologies for solving specific spatial problems. The chapter closes with a brief discussion of the future of GIS in citizen science, in the light of current technological trends.

The chapter aims to analyse the role of citizen science in sustainable development, including case studies implementation, with specific focus on its suitability of citizen science in environmental sustainability. The authors structured this chapter in five sections: Background; Main focus; Solutions and recommendations for designing and executing citizen science initiatives; Future research directions with thoughts on the future role of citizen science; and Conclusion. In section of main focus, first, the authors reviewed the state of citizen science in sustainable development and explored the potential of citizen science for environmental research and governance. Second, authors identified and elaborated the core components that support the role of citizen science and demonstrated the practical approach to realize its objective. Third, using several citizens' observatories studies from various regions in Europe and within diverse environmental fields, authors highlighted the lessons learned, and reflected on major outcomes, challenges and opportunities.

This chapter provides a brief history of citizen science in our societies, identifies the main stakeholders involved in projects of this topic, and analyzes the main points to take into consideration, from a social perspective, when designing a citizen-science project: communicating; recruiting and motivating participants; fostering innovation, interdisciplinarity and group dynamics; promoting cultural changes, healthy habits, inclusion, awareness and education; and guiding policy goals and decisions. Different governance structures, and a coexistence of different approaches, are analyzed together with how they suit different communities and scientific studies.

Citizen Observatories are the technological platforms where a diverse range of tools are developed, such as web portals, smartphone apps, electronic devices, that allow the development of citizen science projects, particularly those with the principal objective of large scale participation of the people, covering large geographical areas and long periods of time. These new observatories integrate the latest Information and Communication Technologies (ICT) to connect the citizens digitally, improve their observational capabilities and provide information flows. The concept of Citizen Observatories offers great possibilities as an educational experience, precisely due to the opportunities offered by the participation of the people, with different levels and roles and therefore, it is assumed in terms of active collaboration of the citizens, in shared processes of knowledge creation. This is especially clear when we pay attention to the complexity of the challenges education must face today, within the framework of a society of knowledge like ours.

Citizen Science is increasing in popularity and used by many academics, community groups and Non-Governmental Organizations in scientific data collection. Despite this, little is known about the motivations and experiences of those who contribute to citizen science projects, nor about the impacts of involvement in citizen science upon the individual. Moreover, few have considered the pedagogic

Foreword

The following chapters underline the fact that citizen science is now a significant international phenomenon. No longer simply a broad concept, but instead firmly taking shape in a range of specific actions, citizen science has emerged on a scale that few could have anticipated. As the contributors to this book discuss, professional and practitioner organizations have sprung up across the globe, tens – even hundreds - of thousands of participants are engaging in citizen science world-wide, and thousands of citizen science projects have been developed. And, just in case there should still be any doubt about the emergence of citizen science, even the Oxford English Dictionary has found it a place.

However, and taking again an important lesson from this book, it seems impossible to contemplate the dramatic growth of citizen science over the last decade in particular without asking a number of questions. Is this solely a matter of data gathering or are citizen science projects raising larger issues with regard to science? What could be the function and contribution of citizen science within moves towards global sustainability? Should this really be seen as a new movement or is it simply a new way of describing what practical enthusiasts having been doing for a very long time? And, as an important focus for many of the contributors, what is the role of *information and communication technologies* (ICTs) in all this? One also becomes curious about the citizen scientists themselves. What motivates them? Does the experience of citizen science have some kind of transformative effect upon them? What barriers do they face?

Now that citizen science has become a recognized scientific and social activity, it is essential that we take the time to analyze, explore, document and, very importantly, learn from accumulated experience. One significant implication of the following chapters is that the challenge is no longer simply to advocate citizen science but instead to consider what it has actually become – and to reflect seriously upon where it might go from here. This is especially important when several commentators point to the diversity in meaning (or, put less politely, the differences and even contradictions) within citizen science.

In discussing the dramatic rise of citizen science, we should be alert to the negative as well as positive future possibilities. Of course, one's only response to citizen science's emergence and growth should not be to look for problems. Nevertheless, it is undeniable that the rapid expansion of citizen science raises questions about its further development. Could one imagine a situation where differences in the definition and practice of citizen science undermine the current sense of shared experience? Equally, we cannot ignore the question of how to network, build and even institutionalize citizen science without losing its contextualized and citizen-generated appeal. Sometimes, what thrives on the margins can struggle once it becomes 'mainstream'.

Considering the chapters that follow, and with special regard for this book's overall focus on 'analyzing the role of citizen science in modern research', three closely-linked questions come to mind – although

I am sure there are also many others. The first addresses the relationship between what we can call *the 'democratic' and the 'data-gathering'* aspects of citizen science. Often, these are presented as fundamentally different activities or even as alternatives. On the one side, we have the idea of citizen science as a matter of 'remote sensing' or 'crowdsourcing' (in the broadest sense of that term). This suggests that citizen science is basically concerned with providing support for mainstream science in a large-scale and 'participatory' fashion: with 'participation' here being defined in strictly non-political terms. On the other side, we have the notion that citizen science should be seen as a matter of 'opening up' science, asking questions about the form and direction of contemporary research, and giving citizens a voice within scientific institutions. This suggests a form of 'participation' with directly political consequences: how can 'scientific citizenship' be defined, developed and implemented?

My first question then is whether we should continue to think of these as two separate paths within citizen science. Is there really a distinction between democracy and data-gathering - or should we not see the two as closely inter-connected? If we think for example of local engagement with issues of sustainability, it seems to me that seeking further knowledge of one's environment can also be understood as a practical expression of citizenship. Certainly, the distinction between the tracking of neighbourhood pollution by citizen scientists and drawing political attention to local environmental problems can at times be so slender as to be almost non-existent. One can make a good case that the different meanings of citizen science should not be kept apart – especially when the boundaries between 'science' and 'society' become ever more fluid in contemporary society. Another way of putting this is to ask whether citizen science is also an emergent expression of citizenship in a world where traditional forms of politics are under widespread challenge (for further discussion of this, see Irwin, 2015; Irwin & Horst, 2016).

That takes me very nicely to my second question. This one concerns the relationship between *citizen science and larger changes in science itself.* Sometimes, we talk as if 'science' is a fixed, heavily-guarded and unchanging institution which occasionally opens its doors to 'citizens' - only to politely dismiss them after accepting their kind gifts and compliments. But what if we also look at contemporary science in more fluid and open terms, and consider it as a field of changing and diverse practice – from nuclear physics to classroom biology, from corporate R&D laboratories to government environment inspectors? Citizen science seems then much less like an outlier and much more as (to steal a phrase from one of the following chapters) an 'advanced learning environment'. Rather than citizen science needing to evolve in order to fit with the requirements of science, it could just be that science (or at least some parts of it) is also shifting in the direction of more flexible and dynamic relations with larger society – as increasingly-used expressions such as 'open science' (or indeed 'open innovation') suggest. This might mean that citizen science will in future be seen as an important and integral aspect of science (like big data or interdisciplinarity) rather than as an unusual activity or 'add-on'.

The third question flows directly from one strong theme of this book and was already hinted at above. What is the relationship between *changing ways of networking and sharing knowledge – notably in the area of information and communication technologies – and the nature of citizen science*? Are digital platforms, mobile devices and Geographical Information Systems best seen as enablers of citizen science or do they change the relationship between citizens and science in a more fundamental way?

When I was first writing about citizen science over twenty years ago, I had no real understanding or imagination of the changes ahead. Back in the 1990s, how did one even find out about the existence of related projects, especially beyond one's own region? Data collection was a much slower business and it was simply impossible to think of citizen science projects communicating simultaneously with thousands of participants across the globe. Of course, and as again several of the following chapters discuss, this raises important questions concerning the motivation for citizens to get involved in ICT-based citizen science. Equally, one wonders about what is lost as well as gained in this situation – or is that simply the misplaced nostalgia of someone who grew up in a pre-digital era? More specifically, does ICT-based citizen science allow the same opportunities for mutual learning and engagement as 'traditional' (i.e. face-to-face) interaction? Do we inevitably lose the depth of citizen participation as connectivity increases? I do not know the answer to that, but I am very glad to see such issues being raised and opened up to larger reflection.

One can say then that citizen science has come of age. With this increased maturity and new standing, come further questions, fresh possibilities and, inevitably, challenges. I am very glad that international colleagues are taking up these issues enthusiastically and boldly. I very much hope this book will provide just the foundation we need for the next level of dialogue, research, learning and practical action.

Alan Irwin
Copenhagen Business School, Denmark

REFERENCES

Irwin, A. (2015). On the local constitution of global futures. *Nordic Journal of Science and Technology Studies*, *3*(2), 24–33.

Irwin, A., & Horst, M. (2016). Engaging in a decentred world: overflows, ambiguities and the governance of climate change. In *Remaking participation: Science, environment and emergent publics*. Routledge.

Preface

THE VALUE OF INFORMATION

The goal of this book is to help educators and researchers to increase the value of the information they deliver to others and to open up opportunities for others to learn and engage in the domain of citizen science. Some of the chapters of the book explain how to teach people about a scientific subject in ways that make what is learned more likely to influence subsequent decisions which involves citizens. In other chapters, the authors show that relatively simple principles can help educators and researchers to deliver the kinds of information that make desired learning and engaging outcomes, often in terms of citizen activities, more likely.

Typical activities included in citizen science range from data collection and analysis to information access and delivery. It is then important to define *data*, *information*, and also the related concepts of *knowledge* and *competence;* and their relationships to one another. Understanding their differences is a key to increasing knowledge and competence. Data are collected and analyzed to create information suitable for making decisions, while knowledge is derived from extensive amounts of experience dealing with information on a subject. Information is what educators and researchers can convey to others directly. The same is not true for knowledge or competence. Knowledge is memories of how concepts and objects are related to one another. Knowledge requires information. Conveying information is the means by which educators and researchers can increase others' knowledge. Competence is the ability to perform a task in a particular way. A competent choice is one that is consistent with a relevant set of facts (say, about water quality) and that is in sufficient alignment with a chosen criterion (e.g., how one feels about tradeoffs between environmental sustainability and economic growth). Competence is always with respect to a task and an evaluative criterion. and requires knowledge. Educators' and researchers' information can increase citizens' competence only if the citizens think about the information in ways that transform it into applicable types of knowledge. Educators and researchers can achieve their objectives more effectively and efficiently by understanding what kinds of information are more relevant to increasing specific competences.

When *collecting data*, activities typically involved are voluntary, active and conscious with respect to citizen science, even if some activities do not include all these characteristics, for example:

- Some school activities are not voluntary for the participant students;
- In some cases, citizens are passively wearing/carrying/using a sensor device and might be collecting data in a form which is not active;

- When your tweets are mined, you are not conscious of your participation in citizen-science projects.

Often, data collection means citizens using lightweight and accessible sensor technologies to gather, and then share, data in order to collectively monitor the environment. Data-collection technologies range from specific sensors or applications that augment mobile phones to increase their functionality, to dedicated, smart and connected devices. Although there has been and there is a vast number of citizen-sensing projects (e.g., instigated by civic educators), modern, large-scale research, in which citizen science has a role, demands new applications (software solutions or systems), and better integration within and among different organizations collecting and using scientific information.

An important aspect in citizen-science *data analysis, processing and interpretation* is quality control. The data are coming from various sources and are collected under different conditions, which influence the quality of the observations and their use. For wider use it is required to validate the observations by means of data quality-control procedures including giving quality flags as is normal practice in data management.

Part of the quality control is also related to the fact that people provide (in a more or less voluntary, active or conscious way) sufficient metadata and context data, such as time, location, name, instrument, when uploading their observations to project databases. This meta-documentation provides essential information next to the data themselves. Sensing technologies can sometimes automatically provide contextual information, which is an essential requirement for personalization and for real-time data-quality validation. Mobile applications can check in real time if a measurement is taken correctly (taking into account, for example, position, orientation and temperature). Another aspect to take into account in quality control is the potential knowledge transfer among participants. Participants with more expertise may help in validating or providing additional information on observations reported by new users.

For a citizen-centered project to bring about benefits to society a critical aspect is *information access*. For example, by sharing information about environmental factors, citizens can become aware of how their lifestyles affect the ecosystem, identify local issues such as air pollution, and learn more and act on their environment. However, there are challenges associated to information delivery and information access.

It is important to focus on how to *deliver information* more effectively. The starting premise is that a person's ability to pay attention to information is extremely limited. Unchangeable aspects of human biology lead people to ignore almost all of the information to which they are, or could be, exposed. Since learning and engaging require attention, educators and researchers who want to increase citizens' knowledge or competence have to find ways to get their attention.

A common requirement for obtaining attention is citizens perceiving information as being highly relevant to their immediate needs. The ineffectiveness of many educational and engagement strategies can be linked to mistaken beliefs about how citizens perceive the value of different kinds of information. Citizens often perceive as abstract or uncertain the benefits of learning and engaging about many of the things about which educators and researchers are. Educators and researchers who fail to recognize these perceptions tend to overwhelm citizens with information that they do not want and will not use.

With the goal of helping educators and researchers in mind, it is important to recognize that many educators and researchers channel their energies into ineffective strategies that do little for the improvement of the knowledge, or the increase of the competences, that motivated them to develop educational or research strategies in the first place. Educational and research ineffectiveness occur because many educators and researchers are mistaken about how people learn and make decisions. Mistaken beliefs lead many educators and researchers to offer information that others do not value or cannot use.

In the book, the authors examine relationships among the information to which people are exposed, the knowledge that such information produces, and knowledge's effect on important competences. The book's central proposition is that educators and researchers can be much more effective if they know more about how people think and learn about science, about ways to differentiate between information to which prospective learners and citizens in general will pay attention and information that these same people will ignore.

It is important to focus particular attention on learning dynamics that are common to citizen science; and offer ways for educators and researchers to increase the value and effectiveness of their educational and engagement strategies. It is important to show educators and researchers how to develop more effective ways of increasing useful kinds of knowledge in a wide range of citizen-science contexts, and relative to three components: *issue complexity*, *stakeholder roles* and *learning/engaging costs*.

By *complexity*, we mean that the domain of citizen science has diverse and occasionally contradictory components, experienced by an equally diverse range of communities; and, when a domain is complex, educators and researchers have to choose how to frame it (i.e., they have to choose how to formally represent it and what parts of the topic to emphasize). Choices about how to frame issues in citizen science will impact the effectiveness of the communication and exchange of information, and determine whether or not that information has any subsequent effect on others' knowledge and competence. In this respect, we offer a vision in which, even if citizen science moves towards more and more formal knowledge-representation and the ability to carry out automatic reasoning by machines, humans are responsible for checking whether any given knowledge representation is still an accurate reflection of reality.

If we focus on *stakeholder roles*, we can find a variety of situations to be taken into account. A member of the board of directors has different roles than do individual members of an association. Legislators have different roles than citizens who hold no elective offices. These role differences mean that information that increases a policy maker's competence at his most important tasks may have little value to people with different roles (say, a member of a school board), and vice versa. Understanding stakeholder roles can help educators and researchers to direct information to more valuable ends.

If we focus on *learning and engaging costs*, it is important to note that increasing knowledge or competence can take a lot of work. Educators and researchers have to put effort into developing and implementing their strategies. Citizens have to devote resources to learning and engaging. Everyone involved in educational and engagement efforts pays costs of one kind or another. For some, the costs are sacrificed time and effort. For others, money is the main expense. These costs alter the net benefit of different types of information to different kinds of people. Understanding different types of learning and engaging costs reveals an important implication for educators and researchers: even if people share important values and stakeholder roles, they may disagree about whether the costs of an educational or engagement effort are worth paying *because some people face greater costs than others*. Educators and researchers can use this kind of knowledge to elicit broader participation in important educational endeavors and to accomplish important educational and engagement goals more efficiently.

PLAN OF THE BOOK

In Chapter 1, the authors begin by proposing a slight re-framing of citizen science, which will contextualize the information presented in the rest of the book. The authors propose a perspective on and a definition for citizen science (which is alternative to the numerous previously documented definitions) as: "work undertaken by civic educators together with the general public to advance science, foster a broad scientific mentality or encourage democratic engagement in local concerns, which allows society to deal rationally with complex modern problems". By explaining the rationale behind this definition, the authors also hope to raise awareness of the role that the meaning of words and phrases (semantics) plays in understanding and supporting citizen science. This chapter also explains how different organizations already use certain software solutions to organize knowledge about citizen science, how these systems can be classified and how they can facilitate or impede interoperability – the ability of machines to pass information between each other.

In Chapter 2, the authors show how, with recent advances in technology, citizen-science activity is growing rapidly around the world and diversifies into new disciplines. This expansion is accompanied by the formation of associations and networks dedicated to citizen-science practitioners, which aim to support citizen science as a research approach. This chapter examines how four such organizations in the United States, Europe, Australia, and China have begun to take shape, and are working with citizen-science communities and stakeholders in their respective regions and globally. Challenges and future plans of these groups are also discussed. This chapter identifies three core roles of citizen-science practitioner organization: (1) establishing communities of practitioners; (2) building expertise through the sharing of existing knowledge and the development of new knowledge; and (3) representing community interests. The authors aim then to stimulate further research, discussion and critical reflection on these associations and networks in the emerging citizen-science landscape.

In Chapter 3, the authors focus on how SciStarter has developed a new digital infrastructure to support sustained engagement in citizen science, and research into the behaviors and motivations of participants. The new digital infrastructure of SciStarter includes integrated registration and contribution tracking tools to make it easier to participate in multiple projects, enhanced GIS information to promote locally relevant projects, an online personal dashboard to keep track of contributions, and the use of these tools (integrated registration, GIS, dashboard) by project owners and researchers to better understand and respond to the needs and interests of citizen-science participants. In this chapter, the authors explore how these new tools build pathways to participatory policymaking, expand access to informal STEM experiences, and lower barriers to citizen science. The chapter concludes with a design for a citizen-science future with increased access to tools, trackable participation, and integrated competencies.

In Chapter 4, the authors show that in order for citizen-science initiatives to pan out well, various actors need to be willing to engage in citizen-science activities. The authors' particular interest lies with the citizens and their motivations to participate in technology-enabled citizen science since without citizen participation there is no citizen-science activity. The authors examine in detail a particular case: citizens' willingness to collect weather-related data using personal weather stations and share them via online amateur weather networks. To better understand what determines citizens' interest to participate in such online networks and how their activities could be up-scaled to address hydro-meteorological data gaps, the authors use the lens of a decision-making theory to guide their empirical research. The authors find that there are no regional differences in the main drivers and incentives for citizens to share their personal-weather-stations data; they also confirm that weather observation is a male-dominated activity, which has implications for up-scaling it as a citizen-science initiative.

Chapter 5 focuses on the risk that, when introduced in social environments different from those in the Global North where it originated, citizen science may be subject to the error of providing the right answer to the wrong question. To avoid this type of errors, it is necessary to train those who participate in citizen-science studies: citizens as well as researchers. Otherwise, we may encounter new forms of scientific dependence that benefit knowledge accumulation and policy decision-making in the Global North, without contributing to the quality of life of those who carry out the studies. This chapter analyzes the relationship between civic development, citizen science and ways of implementing research conclusions through public policies, given the characteristics of political and citizen participation in the Global South. Here, the introduction of citizen science is seen as an opportunity to construct a more inclusive and participatory society, and to reduce the risk of returning to paternalistic, passivity-inducing and purely instrumental approaches to development.

Chapter 6 shows how citizen-science initiatives have been known to exist for a long time, but only recently they were further enhanced thanks to technological and societal developments, such as the availability of mobile devices, the widespread use of the internet and the low cost of location devices. These developments shaped the *geographic information system* (GIS) world as it is known today: a group of technologies that allows retrieving, storing, analyzing and sharing spatial information, by people who are not necessarily GIS professionals. This chapter starts with a general background about GIS, adding then more detail in topics of particular relevance in the context of citizen science. The rest of the chapter is focused on reviewing and classifying the use of GIS in citizen-science initiatives; and some use cases are described in order to provide practical examples of the use of these technologies for solving specific spatial problems. The chapter closes with a brief discussion of the future of GIS in citizen science, in the light of current technological trends.

Chapter 7 aims to analyze the role of citizen science in sustainable development, including case studies implementation, with specific focus on the suitability of citizen science in environmental sustainability. The authors present solutions and recommendations for designing and executing citizen-science initiatives, and thoughts on the future role of citizen science. Firstly, the authors review the state of citizen science in sustainable development and explore the potential of citizen science for environmental research and governance. Secondly, the authors identify and elaborate on the core components that support the role of citizen science. Thirdly, using several citizens' observatories studies from various regions in Europe and within diverse environmental fields, the authors highlight the lessons learned, and reflect on major outcomes, challenges and opportunities.

Chapter 8 provides a brief history of citizen science in our societies, identifies the main stakeholders involved in projects of this topic, and analyzes the main points to take into consideration, from a social perspective, when designing a citizen-science project: communicating; recruiting and motivating participants; fostering innovation, interdisciplinarity and group dynamics; promoting cultural changes, healthy habits, inclusion, awareness and education; and guiding policy goals and decisions. Different governance structures, and a coexistence of different approaches, are analyzed together with how they suit different communities and scientific studies.

Chapter 9 focuses particular attention on learning dynamics that are common to citizen observatories, the technology-driven environments where a diverse range of tools is developed, such as web portals, smartphone apps, monitoring devices, and allows the growth of citizen-science projects, particularly those with the principal objective of a large-scale participation of people, covering large geographical areas and long periods of time. These observatories integrate the latest *information and communication technologies* (ICT) to digitally connect the citizens, improve their observational capabilities and provide information flows. The concept of *citizen observatories* offers great possibilities as an educational experience, precisely due to the opportunities offered by the participation of citizens, with different levels and roles, in terms of active collaboration, and in shared processes of knowledge creation. This is especially clear when we pay attention to the complexity of the challenges education must face today, within the framework of a society of knowledge.

Chapter 10 focuses on the role of citizen science in environmental education. Citizen science is increasing in popularity and is used by academics, communities and a wide range of non-governmental organizations. Many within the field of environmental education see the convergence of citizen science, science education and environmental education as a major opportunity for enhancing sustainable thinking and behaviors. Through synthesizing existing literature, policy documents and educational materials, this chapter critically reflects upon the pedagogic potential of the convergence of citizen science and environmental education. It also challenges the notion that environmentally concerned citizen-science projects enhance awareness of ecological issues and encourage the adoption of more sustainable behaviors. The author draws upon insights from practical projects to explore the motivations and experiences of citizen scientists; and discusses the apparent impacts of involvement in citizen science upon the individual in the development of environmental citizenship.

Chapter 11 shows how global environmental changes put society in front of new challenges, and how immediate and intense actions have to be undertaken in order to foster necessary progress in global sustainability research. The technological infrastructure has reached a status of ubiquitous computing and virtually unlimited data availability. Yet, the dynamic nature of the global environment makes continuous and in-situ monitoring challenging. Citizen-driven *geographic information science* can bridge this gap by building on inputs, observations, and the wisdom of the crowd, represented by the citizens themselves. This chapter argues for the important role of citizen science in geographic information science, presents its position in current research, and discusses future potential research streams, based on the participation by and collaboration with citizens. In particular, the chapter sheds light on three major pillars of the future of citizen-driven geographic information science, namely: big geo-data; education; and open science.

Chapter 12 considers the potential for citizen science to contribute to policy development. A background to evidence-based policy making is given, and the requirement for data to be robust, reliable and, increasingly, cost-effective is noted. The potential for the use of 'co-design' strategies with stakeholders, to add value to their engagement as well as provide more meaningful data that can contribute to policy development, is presented and discussed. Barriers to uptake can be institutional and the quality of data used in evidence-based policy making will always need to be fully assured. Data must be appropriate to the decision making process at hand and there is potential for citizen science to fill important, existing data-gaps.

Chapter 13 focuses on how, by "activating" the citizen's engagement in the research process, the scientific community has a smart way to benefit from the wisdom of the "crowd". There are countless success stories in which citizens participate, contributing with their knowledge, cognitive capacity, creativity, opinion, and skills. However, for many scientists, the lack of familiarity with the particular nature of citizen participation, which is usually anonymous and volatile, turns into a barrier for its adoption. This chapter presents a problem-based typology for citizen-science projects that aims to help scientists to choose the best strategy for engaging and counting on citizen participation based on the scientific problem at hand; and some examples are included. Moreover, the chapter discusses the main challenges for researchers who intend to start involving the citizens in order to solve their specific scientific needs.

Chapter 14's starting premise is that for decades the United States Environmental Protection Agency region subsuming most of the Great Lakes watershed has been partially monitored by private citizens, but collected data have been underutilized by water managers, scientists, and policymakers. Today, citizens with only a smartphone can dramatically increase our understanding of surface water, help managers and policymakers, and educate the general public about the quality of water. The US Clean Water Act and National Strategy for Civil Earth Observations have helped to coordinate citizen scientists and direct funds to surface-water monitoring. And more contributors are being solicited and trained to help with the enormous task of monitoring lakes and streams. At the same time, technology allows citizens with a smartphone to accomplish what previously required experts in a lab: to act for clean water!

CONCLUSION

In *Analyzing the Role of Citizen Science in Modern Research*, we argue that we are living at the dawn of a dramatic change in science. This change is being driven by powerful new cognitive tools, enabled by the internet, which are greatly accelerating scientific research. There are many books about how the internet is changing business or the workplace or the government. This is the first book about something different and fundamental: how the internet is empowering citizens in transforming the nature of science.

Citizen science draws from different fields, such as environmental sciences, biological sciences, Earth observation, crowdsourcing, do-it-yourself approaches, participatory science, environmental mapping, intelligent data-analysis, social sciences and artificial intelligence. Initiatives and projects based on citizen-science are being developed at local, national and global levels, and are reaching out to ordinary citizens and decision makers for them to engage and take part in science together with researchers. Prominent examples of citizen-science projects are the European citizens' observatories [http://www.citizen-obs.eu/]. Citizens are now valued as a key component in the global transition towards a sustainable development, and citizen science aims to: enhance capacities with regard to citizens' initiatives; collect and analyze data from citizens; identify good practices and challenges, such as data accessibility and interoperability; and deliver information to decision makers and, importantly, back to the citizens.

Citizen science can support decision makers by intensifying the dialogue at different scales: from improving early-warning systems, to assisting grassroots activities to protect an endangered species, to supporting environment-related policy targets. It provides the methodology and tools to listen to and engage with citizens on issues such as environmental sustainability, and thus acquire essential local knowledge to determine if national and international environmental programs are working.

This book looks to:

- Discuss how to formalize the new discipline of citizen science in its early stages;
- Allow everyone in the research community to find out what everyone else has been doing in citizen science;
- Allow greater cooperation among citizen-science initiatives.

It also addresses topics which are not often explored; specifically it will:

- Address how citizen science relates with other sciences;
- Provide a classification of all the definitions of this domain;
- Address data harmonization, describing how citizen science can be integrated with existing standards.

Based on the analysis of existing experiences, it defines best practices in the methodologies to set up and implement citizen-science initiatives, especially at the international level. Citizen science is formally defined through the description of its relation with the following areas of knowledge: Earth observation, environment quality, education, socioeconomic benefits, information acquisition, information processing and interpretation, information delivery, data interoperability, standards, novel low-cost technology, collaboration and teamwork, security, and privacy.

This book aims to be an essential reference source about how novel, low-cost technology might be used in citizen science, providing landmarks and guidelines to decision makers and researchers exploring this new territory in knowledge and working at the cutting edge of this field. It will provide inspiration to researchers who designed a tool with one specific use in mind, to generalize it to other uses and to realize the social innovation potential when this tool is used in citizen-science projects.

The book will help to relax the constraints of governing metaphors in science. Society is failing to see the citizen-science revolution coming, in part because science's governing metaphor, especially in the Global North, is drawn from the idea of the scientific document in highly specialized journals: expensive materials and methods, not people. Consequently, most science is not considering low-cost instruments, alternative ways of publishing and social networks.

Even if citizen science is not a plug-and-play solution to sustainability, this book helps to perceive important developments and possibilities, such as that individuals are taking an active role in the transition to a sustainable society, and in helping to protect and improve health and the environment; and does this not assuming that traditional trends will continue to follow their trajectory. Sciences, in the last century, have been characterized by their increasing complexity: an obvious and indisputable trend. Despite this fact, this book documents the rise of citizen science, whose simplicity, openness and ability to empower and engage citizens make it perfect for a society which is already been changed by Facebook and Alphabet.

Finally, this book considers the risks, especially in the Global South, of innovating our way out of strong dependency on traditional science only to be plunged into chaos when the world suddenly finds itself dealing with different approaches to science at the same time. Because every new solution often hides its own set of problems, one objective of this book is to define paths to conciliate and mutually strengthen traditional and citizen science.

We hope that academicians, researchers, policy makers, technology developers and government officials aiming at developing a citizen-science initiative will find this text useful in furthering their exposure to pertinent topics and research efforts in this field.

Acknowledgment

The editors would like to thank the members of the Editorial Advisory Board for their assistance and suggestions at several stages in the preparation of this book, all the reviewers who shared their expertise in the review process of this publication, 1000001 Labs and ICM-CSIC which supported this project, and all our friends who gave us the inspiration to launch this book successfully.

The wellspring for *Analyzing the Role of Citizen Science in Modern Research* are 1000001 Labs' and ICM-CSIC's citizen-science projects. Since 2012, these organizations have conducted several citizen-science projects and related activities, involving thousands of citizens, and numerous authorities and organizations representing many different vantage points. This book has truly been a collaborative effort, with many individuals who have contributed very much in special ways, above and beyond the call. Every aspect of the book has benefited from their knowledge, experience and technical expertise, and their vital guidance and encouragement made this work a reality.

Chapter 1
Civic Education and Citizen Science:
Definitions, Categories, Knowledge Representation

Luigi Ceccaroni
1000001 Labs, Spain

Anne Bowser
Woodrow Wilson International Center for Scholars, USA

Peter Brenton
Atlas of Living Australia, Australia

ABSTRACT

The first goal of this chapter is to propose a slight re-framing of citizen science, which will contextualize the information presented in the rest of the book. The authors propose a perspective on and a definition for citizen science (which is alternative to the numerous previously documented definitions) as: "work undertaken by civic educators together with citizen communities to advance science, foster a broad scientific mentality, and/or encourage democratic engagement, which allows society to deal rationally with complex modern problems". By explaining the rationale behind this definition, the authors also hope to raise awareness of the role that the meaning of words and phrases (semantics) plays in understanding and supporting citizen science. A second goal of this chapter is to explain how different organizations already use certain software solutions to organize knowledge about citizen science, how these systems can be classified and how they can facilitate or impede interoperability – the ability of humans and machines to pass information between each other.

INTRODUCTION

According to numerous surveys and news reports (e.g., in the US: Dimock, Kiley, Keeter, Doherty, & Tyson, 2014; Annenberg Public Policy Center, 2014), the mass public appears to know very little about politics, government, policy and the environment. When pollsters ask even simple questions on any

DOI: 10.4018/978-1-5225-0962-2.ch001

of these topics, many people fail to give correct answers. In response to such evidence, the question is "What can be done about it?"

Some people are very critical and frustrated about citizens' inability to answer basic questions on policy and the environment. In democratic countries, they ask how we can expect ignorant citizens to choose qualified candidates for office or offer defensible views on social or environmental topics. In response to what can be done about this, some seek a constructive approach to the evidence of civic ignorance (Lupia, 2015), and try to open new avenues for civic education. Former US Supreme Court Justice Sandra Day O'Connor, for example, argues that "we have to ensure that citizens are well informed and prepared to face tough challenges. If there is a single child not learning about civics or not being exposed to what they must do as citizens, then all our lives are poorer for that." (Terhune, 2013)

The extraordinary amount of work undertaken by volunteers in all areas of society across the world is testament to the fact that many people are answering calls for this kind of education. They include researchers, teachers, scientists, issue advocates, journalists, reporters and political campaigners - to simplify, the authors consider all of these individuals and groups as *civic educators:*

Civic educators are people who believe that providing information to others, and/or creating opportunities for others to learn, are paths to greater civic competence and a better future.

Civic educators are principal instigators for the engagement of citizens (e.g., the public) in affairs of public interest (including, but not limited to, scientific research). They develop and implement educational strategies; design plans to provide certain kinds of information to certain people; or, open up opportunities for others to learn in certain ways.

Civic educators' strategies are diverse. Some write articles. Others teach students. Some seek to draw attention to important facts and causes while working for widely recognized and highly reputable organizations. Others seek educational innovation through startups. Some seek to educate at places of work. Others operate in settings like high schools, colleges, fabrication laboratories and universities. Some educators do many of these things. Educators differ in their ambitions. In the context of this book (citizen science, which the authors define and explore in detail later), civic educators often wish to educate a specific audience (e.g., young adults, fishermen, inhabitants of particular neighborhoods, residents of a particular city) about a specific or a general topic.

Civic educators also have diverse identities. Some consider themselves *advocates*. These advocate educators (e.g. Greenpeace, the Sierra Club, etc.) are motivated by a desire to achieve policy outcomes for social and environmental equity such as a safe and healthy community in which to live; securing appropriate regulatory standards for air and water pollution; and many more.

Others identify themselves as *experts* on a topic rather than as policy-driven advocates. Many expert educators are motivated by ideas from science and do not make explicit appeals for or against specific policies. Instead, these experts seek to involve the public in their scientific endeavors, sometimes to educate audiences about how things or the environment work, and sometimes to collect data for their public interest research. Many academics think of themselves in this manner.

Still other civic educators identify as *advocates and experts*. They not only want to teach audiences about how things or the environment work, but they also wish to enlighten others about how things or the environment could be if certain options were chosen. These educators often provide information or create learning opportunities for the purpose of bringing policy outcomes in line with the lessons of their expertise and their own points of view.

Civic educators see themselves as knowing things that can help others, and believe that greater knowledge will benefit others. Issues for which civic educators are active include world hunger, causes related to health, general environmental topics, and specific environmental issues such as whether land should be preserved, how to monitor natural-waters quality, the early detection and reporting of weeds, and myriad other causes. Many people are civic educators of one kind or another, and in this chapter the authors explore the role of civic educators and "the public" in the emerging domain of citizen science.

So what is "citizen science"? The authors begin with a few illustrative examples, returning to formal definitions and approaches in a later section.

PIGEON SCIENCE

In London, in 2016, you could see pigeons wearing backpacks. As part of an experiment, French startup Plume Labs outfitted ten pigeons with tiny sensors designed to offer a real-time snapshot of the city's ozone and nitrogen dioxide levels. But the ultimate goal was to get humans to wear similar devices, that researchers, in this case from Imperial College London, could use to collect and analyze data for shareable insights, like which route people should walk if they want to breathe the cleanest air. It is presently harder to get insights on air pollution from fixed monitoring stations, which is how researchers and public administrators measure pollution right now (Waxman, 2016).

In Barcelona, later in 2016, you could see people wearing a sensor device. As part of a much publicized citizen-science experiment, the multimillion-euro European project CITI-SENSE provided several citizens with tiny sensors designed to offer a snapshot of the city's ozone and nitrogen dioxide levels.

Are we witnessing a transition from "pigeon science" to citizen science?

We might be, if we consider citizens just as data collectors, data carriers or a crowd that can be used as a new and expanded resource for data or information gathering. There is no doubt that, by engaging large numbers of people over large geographic areas in specific aspects of scientific projects, we can collect larger volumes of data and cover significantly more area than is physically or financially possible when only using more traditional scientific approaches. Researchers cannot physically be everywhere, and enabling non-researchers (i.e. the general public) to participate in a specific resource-intensive aspect of a project, such as data collection, can be time- and cost-efficient, as well as rewarding to both parties.

CITIZEN SCIENCE

Citizen science has many benefits over other research methods and should be seen as a powerful enabler and augmenter for the scientific process. Also, when humans are involved, there is the possibility of achieving significant positive social outcomes through civic education and participation. Individual contributors benefit from enhanced topical knowledge, or knowledge of the scientific research process. Social networks expand; and communities become more resilient, with enhanced capacity to influence research agendas and contribute to public policy dialogues (Haklay, 2015; Bonney et al., 2009; Irwin, 1995). Citizen science can also provide a form of "workplace experience" which can provide pathways to new or even first employment opportunities. In 2008, Stuart Harris, a vineyard worker in Canberra, Australia, discovered a new species of peacock spider. This experience created opportunities for him to work closely with practicing scientists and set him on a path towards a new career in which he has developed many new skills and found a new personal sense of purpose and contribution to society (Vyver, 2014).

Today, "citizen science" is often just a convenient label for certain types of projects. There is no single, agreed-upon definition and typology by all parties involved, despite efforts from numerous researchers over the past 20 or so years.

Existing Classifications of Citizen: Science Projects

Here the authors will present select classification efforts of citizen-science projects to date. These are offered with the caveat that we are living at the dawn of dramatic changes in science, enabled by the internet, which are greatly accelerating scientific research, and empowering civic educators and citizens in transforming the nature of science.

Perhaps the most elusive problem—and also the most important—in describing citizen science originates from the multiple meanings of the concept itself. On a qualitative level this is evident by observing how two distinct meanings have developed in the social and natural sciences respectively since the mid-1990s (Kullenberg, & Kasperowski, 2016). Researchers often distinguish between:

1. Citizen science primarily conducted with goals including democratization, public engagement, equity, and justice in the discourse of science and in setting the research agenda (e.g., Irwin, 1995; Irwin, & Horst, 2015);
2. Citizen science that is focused on something else, usually on public involvement in scientific research, with members of the public partnering with professional scientists to collectively gather, submit, or analyze large quantities of data (e.g., Bonney, 1996; Dickinson, Zuckerberg, & Bonter, 2010).

While the second approach has often dominated scholarly dialogues over the past 20 years, the dramatic changes in technology that we are experiencing and the maturation of citizen-science communities could favor an increase in significance of the first one, at least if citizen science's stakeholders recognize values in the discipline which extend beyond the value of "pigeon science". Fully appreciating this trend of balancing of purpose within citizen-science communities, as they evolve and mature, requires exploring existing classifications in more detail, to better understand the history of the field and the ongoing discussion.

The examples provided in this chapter are mainly meant to facilitate the comparison among typologies. Most of these classifications are not mutually exclusive; for example, a project could be classified in terms of: governance model; goals and tasks; or intellectual property concern. Additionally, in some cases, the same project may be classified according to a number of classes within a single typology. For example, a project may involve, at the same time and with equal priority, data collection and data processing as the nature of the activities participants engage in. Other times, classifications are designed as exhaustive and mutually exclusive. While project governance models may change over time, no single project will employ two distinct governance models simultaneously (e.g., Shirk et al., 2012).

Citizen-science projects are often classified by the nature of the activities participants engage in (Bonney et al., 2015):

- **Data-Collection Projects:** (the National Audubon Society's Christmas Bird Count, numerous projects hosted by the Atlas of Living Australia, scientist-lead ecological projects, etc.), For which contributors who may or may not have any formal training as scientists collect data that can be used in organized scientific research;
- **Data-Processing Projects:** (those hosted by the Zooniverse suite, Australia's DigiVol digitization project (Ellwood et al., 2015), etc.), Focused on categorization, transcription and interpretation, enabled by the Internet, and sometimes referred to as "crowdsourcing" or "crowd science";
- **Curriculum-Based Projects:** (BirdSleuth, the Basin Champions program in Australia, etc.), Take place in schools or in "informal" youth-development settings, collecting and submitting data to a larger, "parent" citizen-science project;
- **Community-Science Projects:** (the West Oakland Environmental Indicators project (California Energy Commission, 2012), The highly successful Waterwatch program in south eastern Australia (Chalkley, Brendan, & Gowland, 1999), etc.), which place local or regional issues at the heart of the research, and typically seek to affect policy or local decision-making for public health, environmental health, or conservation.

Citizen-science projects can also be classified by governance model, or the extent to which the public participates in different parts of the scientific process (Shirk et al., 2012):

- **Contractual Projects:** (exemplified by European Science Shops (Jorgensen et al., 2004)), Where communities ask professional researchers to conduct a specific scientific investigation and report on the results;
- **Contributory Projects:** (the Christmas Bird Count, Western Australia's MicroBlitz project (Gruber, 2015), Australia's Waterwatch program, etc.), Generally designed by scientists and for which members of the public primarily contribute data;
- **Collaborative Projects:** (e.g., community-based monitoring of wetlands in Madagascar (Andrianandrasana, Randriamahefasoa, Durbin, Lewis, & Ratsimbazafy, 2005), Generally designed by scientists and for which members of the public contribute data, but also help to refine project design, analyze data or disseminate findings;
- **Co-Created Projects**: (e.g., the West Oakland Environmental Indicators project), Designed by scientists and members of the public working together and for which at least some of the public participants are actively involved in most or all steps of the scientific process;
- **Collegial Contributions:** (as exemplified by amateur astronomers, archaeologists, and taxonomists, who often work on their own (Hopkins, & Freckleton, 2002)), Non-credentialed individuals conduct research independently with varying degrees of expected recognition by institutionalized science and/or professionals.

Another classification of citizen science according to governance models, framed as the level of participation and collaboration between professional and non-professional scientists, is offered by Haklay (2013):

- **Crowdsourcing Projects:** (the Christmas Bird Count, the Australian DigiVol project, etc.), In which participation is limited to the provision of resources, and the cognitive engagement is minimal;

- **Distributed-Intelligence Projects:** (e.g., Galaxy Zoo), In which the cognitive ability of the participants is the resource that is being used;
- **Community-Science or Participatory-Science Projects:** (e.g., the West Oakland Environmental Indicators project), In which the problem definition is set by the participants, and in consultation with scientists and experts, a data collection method is devised;
- **Collaborative-Science or Extreme Citizen-Science Projects**: Completely integrated activities, where professional and non-professional scientists are involved in deciding on which scientific problems to work and the nature of the data collection, so it answers the needs of scientific protocols while matching the motivations and interests of the participants.

Other definitions are specific to public participation in certain domains. For example, citizen-science projects have been defined by the degree of local participation in the domain of natural resource monitoring (Danielsen et al., 2009):

- **Externally-Driven, Professionally Executed Monitoring Projects**: Do not involve local stakeholders;
- **Externally-Driven Monitoring Projects with Local Data Collectors:** (e.g., the Citclops project on natural-water monitoring (Wernand, Ceccaroni, Piera, & Zielinski, 2012)), Involve local stakeholders mainly in data collection;
- **Collaborative Monitoring Projects with External Data Interpretation:** (e.g., community-based monitoring of wetlands in Madagascar), Involve local people in data collection and management-oriented decision making, but in which the design of the scheme and the data analysis are undertaken by external scientists;
- **Collaborative Monitoring Projects with Local Data Interpretation:** (e.g., ranger and community-based monitoring of resource use and wildlife in China (Van Rijsoort & Jinfeng, 2005)), Involve local stakeholders in data collection, interpretation or analysis, and management decision making, although external scientists may provide advice and training;
- **Autonomous Local Monitoring Projects:** (e.g., the West Oakland Environmental Indicators project), In which the whole monitoring process -from design, to data collection, to analysis, and finally to use of data for management decisions- is carried out autonomously by local stakeholders; there is no direct involvement of external agencies.

In addition, numerous typologies extend beyond examining citizen science through the degree of public participation. For example, citizen science projects may be defined in terms of project goals and tasks (Wiggins, & Crowston, 2011):

- **Action-Oriented Projects**: (e.g., the West Oakland Environmental Indicators project), Encourage participant intervention in local concerns, using scientific research as a tool to support civic agendas;
- **Conservation Projects:** (e.g., the Missouri Stream Team program on river conservation), Support stewardship and natural resource management goals, primarily in the area of ecology; they engage citizens as a matter of practicality and outreach;
- **Investigation Projects**: (e.g., Citclops) Focused on scientific research goals requiring data collection from the physical environment;

- **Science-Oriented Virtual Projects:** (e.g., Galaxy Zoo), In which all project activities are ICT-mediated with no physical elements whatsoever, differentiating them from the investigation projects in which the physical places of contributor participation was also important;
- **Education Projects:** (e.g., BirdSleuth), Make education and outreach primary goals, all of which include relevant aspects of place.

Citizen-science projects are also defined by the different ways that scientific inquiry can permeate the management of natural resources and collaboration between professional and non-professional scientists (Cooper, Dickinson, Phillips, & Bonney, 2007):

- **Scientific Consulting Research Projects:** (e.g., ranger and community-based monitoring of resource use and wildlife in China), In which knowledge-producing institutions (e.g., universities) function as consultants to community groups to answer questions raised by the community groups;
- **Citizen Science Research Projects:** (e.g., Citclops), Engage a dispersed network of contributors to assist in professional research using methodologies that have been developed by or in collaboration with professional researchers;
- **Adaptive Citizen Science Research Projects**: Involve providing a centralized organizational infrastructure that is specifically designed to promote individual, community, and regional science-based management via an interactive feedback loop;
- **Adaptive Co-Management Research Projects:** Community groups, individuals, and professional land-managers and urban planners work together such that management objectives are carried out and evaluated as "experiments" tailored to specific locations;
- **Participatory Action Research Projects:** (e.g., community-based monitoring of wetlands in Madagascar), Begin with the interests of participants, who work collaboratively with professional researchers through all steps of the scientific process to find solutions to problems of community relevance.

Citizen science may also be classified in terms of issues including *intellectual property* (IP) concerns (Scassa, & Chung, 2015), and many more topics. The existence and use of different classifications suggests that researchers take alternative views regarding what is and is not important to pay attention to in the field of citizen science, and how to structure their vocabularies in accordance with different values.

Existing Definitions of Citizen Science

By highlighting important aspects of the citizen science experience, these typologies lead to the related question of "What is *citizen science*?" Various definitions have been proposed, including:

- "The participation of nonscientists in the process of gathering data according to specific scientific protocols and in the process of using and interpreting that data; the engagement of nonscientists in true decision-making about policy issues that have technical or scientific components; and the engagement of research scientists in the democratic and policy process" (Lewenstein, 2004).
- "The systematic collection and analysis of data; development of technology; testing of natural phenomena; and the dissemination of these activities by researchers on a primarily avocational basis" (i.e., done regularly for enjoyment rather than as a job; OpenScientist, 2011).

- "The scientific activities in which non-professional scientists volunteer to participate in data collection, analysis and dissemination of a scientific project" (Haklay, 2013; based on Cohn (2008) and Silvertown (2009))
- "A contribution by the public to research, actively undertaken and requiring thoughtful action" (Simpson, 2013)
- "Scientific work undertaken by members of the general public, often in collaboration with or under the direction of professional scientists and scientific institutions" (Oxford English Dictionary, 2014)
- "The collection and analysis of data relating to the natural world by members of the general public, typically as part of a collaborative project with professional scientists" (Oxford Dictionaries, 2014).
- A paradigm where "people who are not professional scientists take part in one or more aspects of science—systematic collection and analysis of data, development of technology, testing of natural phenomena and dissemination of the results of activities. They mainly participate on a voluntary basis." (Park, 2014)
- "The general public engagement in scientific research activities when citizens actively contribute to science either with their intellectual effort or surrounding knowledge or with their tools and resources." (Serrano Sanz, Holocher-Ertl, Kieslinger, Sanz García, & Silva, 2014)
- "The public involvement in inquiry and discovery of new scientific knowledge. A citizen science project can involve one person or millions of people collaborating towards a common goal. Typically, public involvement is in data collection, analysis, or reporting." (SciStarter, 2016)

In seeking to understand who contributes to citizen science, it is also important to consider that neither civic educators nor citizen scientists are homogeneous groups. Social scientists made the important argument that "the public" as a single entity does not exist. Instead, we have to acknowledge the presence of a plurality of "publics" (Irwin, & Horst, 2015). In this sense, citizen scientists can be characterized as members of "communities." Such communities are thought to be at the opposite end of the spectrum of the larger "crowd" that is referred to in discussions on crowdsourcing. These communities are subsets of the public with specific and shared interests, whereas the crowd usually refers to a broader citizenry. Citizen-science community members may have some training and expertise; thus, they can be considered "expert amateurs" and not representative of the full suite of potential participants in citizen-science projects (Lukyanenko, Parsons, & Wiersma, 2016).

While existing conceptualizations of citizen science and citizen scientists are helpful points of departure, many of these major understandings and definitions do not exhaust all forms of citizen science that are of relevance for researchers interested in this phenomenon. Some leave out activities not related to the "natural world", such as activities conducted in the domain areas of health, medical science, and social science. Most of them focus on data and information, and leave knowledge and competence out of the equation. And there are still questions related to citizen science with no easy answer:

- Must citizen science generate data used in science, policymaking, or management planning? Or can experiential learning activities be conducted without an impact on science, management or policy also be citizen science?
- Does participation need to be opt-in, meaning that a project mining citizen Twitter feeds about water and flooding would be out of scope?

- What degree of community participation is required? Is a project involving the use of camera-traps out of scope if members of the general public participate only in the deployment of the instruments? And what about if they just change the batteries of the cameras once a year? And by the way, who owns the data collected: those who built the (possibly do-it-yourself) camera trap, those who deployed it, those who changed the batteries and retrieved the data, those who reviewed and interpreted the images, or the researchers who designed the experiment?

In addition to the above, there is the question of whether "citizen science" is even the best or most accurate term to use. *Citizenship* is the status of a person recognized under the custom or law as being a member of a country and this status plays no role in "citizen science". Perhaps "community science", "public science" or "participatory science" are better expressions. This is precisely the point made by a group of researchers in the *United States* (USA) who re-branded "citizen science" as "*public participation in scientific research* (PPSR)" in the early 2000s (Bonney et al., 2009). Notably, in the USA some organizations, including the *National Science Foundation* (NSF), still use the expressions PPSR to mean *citizen science* and also *public participation in science, technology, engineering, and mathematic research.* However other leading organizations, like the recently formed *US Citizen Science Association*, the *European Citizen Science Association* (ECSA) and the *Australian Citizen Science Association* (ACSA), use the now *de facto* standard term "citizen science," even while recognizing this nomenclature as problematic.

Finally, there is the issue of defining the relationships between citizen science and related types of open innovation activities such as: participatory mapping; *volunteered geographic information* (VGI); participatory health monitoring; social studies; and bio-medical studies, just to name some of the research and activity areas included in this book.

An Updated Distinction of Meaning

The authors believe that achieving a practical working definition of *citizen science* is less important than communicating and understanding the general characteristics of citizen-science projects. Perhaps it would be more constructive to consider the *role* of these projects, in terms of supporting research, education, and/or public policy.

By synthesizing a number of the above definitions in a context of civic education, citizen science, and *public engagement in science* (PES), the authors suggest that scientific projects in which citizens are engaged in matters of public interest, or in driving social learning, scientific endeavor or policy development, can be categorized into one of two forms:

1. **Instrumental:** Projects which involve the public in specific and limited parts of a process, for example data collection. These projects usually take place in a traditional social and political structure with discrete and fixed actors who "engage" with one another in a specific context for a particular period of time, and then resume their separate, business-as-usual existences; and
2. **Capacity Building:** Projects of a scientific nature undertaken by groups of citizens with a common goal or interest, either independently or in collaboration with professional scientists. These projects are not necessarily established exclusively to answer specific scientific questions, but rather are conducted to reach a range of social, scientific, learning, and/or environmental outcomes.

This new distinction of meaning expands and builds upon the approaches presented earlier, which framed the goal of citizen science as encouraging a more informed and active citizenry (developed in the social sciences) or for scaling data collection (developed in the natural sciences) (Kullenberg, & Kasperowski, 2016). Through this new categorization, and slight re-framing, the authors recognize that citizen-science projects are conducted in any domain of interest to society. This categorization also expands the previous distinction of meaning by acknowledging additional benefits of citizen science, including knowledge gains, which were not at the forefront of scholarly discourse when the previous approaches emerged in 1995. Finally, the authors suggest that citizen-science projects of the capacity-building type are generally initiated by civic educators and involve the public as active participants (as opposed to passive data-collectors) in one or more aspects of the project activities.

While this re-framing is similar to the presentation of goals and tasks advanced by other researchers (Wiggins & Crowston, 2011), in this case the authors are not looking at the primary purpose of a particular project; rather, they are exploring how the paradigm of citizen science itself is framed. In other words, the authors are seeking to explain the ultimate and overarching value of public participation in scientific research:

Citizen science is work undertaken by civic educators together with citizen communities to advance science, foster a broad scientific mentality, and/or encourage democratic engagement, which allows society to deal rationally with complex modern problems.

This definition shifts the focus from the action-oriented, data-centered point of view of *collect*, *participate* and *contribute* (e.g., the instrumentalist point of view) towards a re-framing, based on civic education, of how science and society should respond to a call for *openness*, *inclusiveness*, *responsiveness*, *democratic engagement*, *consultation*, *dialogue* and *commons* e.g., the capacity-building point of view). The definition reflects the values civic educators see in citizen science, which usually include some of the following: supporting and advancing scientific research; public engagement in scientific discourse; public engagement in informing policy at various levels, from local to international; desire to achieve a particular environmental, social or policy outcome; increased capacity to respond to community needs, such as concerns about water quality or access to scientific information; and enhancing lifelong learning/education about the scientific process, and the world around us.

By explaining the rationale behind this definition, the authors hope to raise awareness of the role that semantics, or the meaning of words and phrases, plays in understanding and supporting citizen science and civic education. Semantics is important in human conversations, when diverse speakers and listeners must rely on shared or interoperable vocabularies to get their points across. Semantics is even more important in conversations between humans and machines, or between machines.

In the rest of this chapter the authors will show how different organizations use software solutions to organize knowledge about citizen science, how these systems can be classified, and how they can facilitate or impede interoperability. Following this discussion, the chapter explores different processes for representing knowledge. Attention is devoted to knowledge representation through human consensus-building, and to the introduction of semantic technologies to a non-technical audience. Finally the authors will examine how developing a formal knowledge representation can be used as a strong basis for:

1. Developing better designed, more robust, more interoperable software solutions for citizen science;
2. Making the work of civic educators more functional towards a contribution to research.

CITIZEN SCIENCE, KNOWLEDGE REPRESENTATION, AND INTEROPERABILITY

Modern research, in which citizen science has a role, demands new *applications* (software solutions or systems for citizen science), and better integration within and among different organizations collecting and using scientific information. Concurrently, we have seen a significant increase in individuals, groups of individuals, and formal organizations related to citizen science projects (mainly small ones) who provide modern services and solutions to today's citizen scientists.

Coordination among these services and solutions is important because there can be a large number of local research projects that emerge around community concerns. Communities usually need a local anchor for participation to be meaningful, as well as means of being able to choose which project or solution is most appropriate for lodging their data or for them to participate in. Some citizen scientists working in these projects hope to collaborate with other organizations to achieve complementary goals, and thus require clear and open communication channels to share information. Civic educators dealing with *software solutions for citizen science*, which are a fundamental part of communication for most projects, are realizing they must satisfy the needs of a wide range of organizations and develop applications tailored to each niche. This involves the daunting task of understanding differences among existing software solutions (David, McCarthy, & Sommer, 2003), and also identifying the differences resulting from various scientific research segments, and associated disciplinary standards.

The types of systems that will be described below are characterized by their *knowledge organization*, or a consistent way of describing and organizing knowledge within the software domain. These systems also require standardized *knowledge representation* of concepts and relations in the domain of citizen science. Knowledge representations refer to taxonomies and other classifications that are understandable by both humans and machines, and aim to support communication among them (for example, by ensuring that two humans in dialogue understand important words to mean the same thing; and, by enabling two artificial intelligences to exchange and understand data sets between themselves without human facilitation or intervention).

Both knowledge organization and knowledge representation are valuable to citizen science for a number of reasons.

Knowledge Organization

The authors categorize citizen-science software solutions—in other words, the variety of technical tools for supporting work processes, and achieving project goals—into:

- Those with no overall organizing rationale;
- Those with inward organization;
- Those with outward organization, or the ability to communicate and interact as communities.

Systems with No Overall Organizing Rationale

These systems do not incorporate any organizing principle for their data, information and knowledge. Instead, information is collected in an irregular or ad-hoc manner. These systems may work when an individual or community is not concerned with sharing data with external parties, such as researchers working to support scientific research or drive public policy. Some curriculum-based projects, which may take place in schools or in "informal" youth-development settings, are examples of systems with no overall organizing rationale. These systems are irrelevant to conversations of interoperability, and outside of the scope of this analysis of knowledge organization in citizen science.

Systems with Inward Organization

More advanced software solutions for citizen science, such as iNaturalist (Pimm et al., 2015) or Citclops (Wernand et al., 2012), incorporate an organizing principle (such as one that builds upon a standard-based metadata schema) to bolster their categorization and processing capabilities. They standardize data-collection procedures and knowledge representation, and provide methods to access data in these standardized formats. By enhancing communications, these systems have facilitated the growth of international models for citizen-science data and information exchange. However, these systems impose constraints on adopting organizations: many of these systems are inflexible, and organizations seeking to use them must adapt to them, rather than the other way around. As a result, some organizations are struggling to implement these more advanced solutions to achieve better functionality, but integration using this approach may be difficult and expensive. If certain systems -such as iNaturalist or CitSci.org (Wang et al., 2015)- become widely adopted (because projects cannot always build their own software solutions, even if they may wish to, and thus need systems such as these), these systems with inward organization will likely become, at least temporarily, the *status quo* as software for citizen-science projects' data and metadata management. It should be noted that systems with inward organization can transition, and in some cases are already transitioning, to become systems with outward organization.

Systems with Outward Organization

These systems are based on standards that are already accepted or in use. Outwardly organized software solutions for citizen science, such as the databases and knowledge bases of citizen science projects being developed by ECSA, CSA, ACSA, Atlas of Living Australia, CitSci.org, and Woodrow Wilson International Center for Scholars, support not only the work of a single organization and immediate collaborators, but also facilitate future interaction between diverse organizations by providing data to other participants in predictable and mutually agreed upon formats. In some cases, systems with outward organization are based on the specifications of a single system (with inward organization) that became accepted over time. For example, the way that information is structured in the SciStarter database has influenced how shared standards for citizen-science project metadata have evolved. Newer project databases (including the US Federal Crowdsourcing and Citizen Science Catalog developed by the Woodrow Wilson International Center for Scholars; and the Atlas of Living Australia), have taken SciStarter's structures as model to build upon and expand.

At the most basic level, projects with shared outward organization can share data through a set of custom-designed APIs. One benefit of systems with outward organization, and a more explicit specification of a shared conceptualization, is that they are readable by a computer and can enhance inter-organizational communication by focusing on a standard set of definitions based on a single, machine-readable format like RDF/XML OWL (a family of knowledge-representation languages for authoring ontologies in a Web environment) or on the *JavaScript object notation for Linked Data* (JSON-LD, a method of encoding Linked Data using JSON) allowing uniform integration of data elements.

Future software solutions for citizen science in a semantically-rich context may go even further by utilizing common, independently-viewed definitions. For example, a system that captures water-quality data could give the citizen, via an app, information for the improvement of scuba-diving activities (*citizen-app view*), while the research organization involved could use the same data to contribute to ocean-color research (*research-organization view*). These systems increase storage efficiency and this is especially valuable in resource-poor environments. If a particular citizen-science project experiences a lapse in funding, data storage through integrated systems will ensure that data collected by this project do not become inaccessible to other researchers. These solutions with outward organization are typically based on existing ones that capture detailed semantics about, for example, environment biodiversity, resources, characteristics and events, as envisioned in a Web services context.

Knowledge Representation

In order for humans and machines to reason and communicate with others in a semantically-rich way, a formal representation of important concepts and relations is required. For example, many individuals and organizations can act as civic educators, or conduct citizen-science activities, without explicitly labeling their actions. But, in order for these people to find each other, enter into dialogue, share information and learn from one another, it is necessary to utilize common forms of knowledge representation. The difficulty of representing knowledge is evident in the struggle to find a single definition of "citizen science" in a way that resonates with the experience of diverse researchers, practitioners, and contributors in the field. In the following, the reasons for which knowledge representation is valuable to citizen science are presented.

Shared Knowledge Representation Indicates a Shared Ontological Commitment

While there may never be absolute agreement, a high degree of consensus regarding how knowledge is represented within a domain—or, a shared *ontological commitment* among citizen science researchers, practitioners, and communities—shows cohesion and solidarity. People who are committed to reaching a shared understanding of domain knowledge are not only making it easier for people within the domain to communicate, but are also distinguishing themselves as a broad citizen-science community of practice distinct from groups of citizens cohered around other topics.

Along these same lines, a shared ontological commitment, e.g. a shared vocabulary, can help newcomers to the citizen-science domain understand what is important in this paradigm. If parts of the vocabulary are developed in more details, for example by one group having spent years developing a set of scales to articulate and measure the educational benefits of citizen science (Phillips, Ferguson, Minarchek, Porticella, & Bonney, 2014), a new researcher can see that these scales are broadly used and cited, and understand that the citizen-science community considers their work valuable not just for the scientific benefits, but for social and educational benefits as well.

A shared ontological commitment helps researchers studying the field of citizen science understand each other's contributions, and pose new research questions that build upon and extend previous work. As described earlier, researchers already categorize key aspects of citizen science, such as the range of participation goals and tasks (Wiggins, & Crowston, 2011), governance models (Shirk et al., 2012; Haklay, 2013), and intellectual property concerns (Scassa, & Chung, 2015). When the same categorizations are widely accepted and understood, they provide a shared point of reference for exploring new research questions. For example, one group of researchers used an established typology of governance models to ask and answer a research question about how different outcomes relate to projects of different "types" (Bonney et al., 2015).

A Common Knowledge-Representation Helps Citizen-Science Researchers and Practitioners Explain Their Work to Domain Outsiders

Data quality is an important and frequently debated topic in citizen science, as in many other domains (Bonney et al., 2014). Developing a way to communicate about data quality that is specific to citizen science (covering mechanisms to support data quality *before* research, like through contributor training; *during* research, like point-of-capture standardization and validation; and *after* research, like through human-based or automated validation strategies) can help different citizen-science researchers understand each other's work. Some mechanisms for supporting data quality, like contributor training, derivations of new data from combinations of existing data, and automated real-time validation, are common to a number of sub-domains. Other mechanisms, such as using existing species distribution maps to assess the plausibility of a new sighting, may be specific to application sub-domains such as biodiversity monitoring. Additionally, developing a vocabulary to represent data quality in citizen science that matches (or is compatible with) the vocabularies used by researchers outside of citizen science can help to integrate citizen-science contributions with other research contributions, and increase the likelihood that citizen-science data are incorporated into other formal research and policymaking systems.

Certainly, not all citizen-science activities align with the goals of policymaking systems or potential downstream applications for their data. For example, if one community wants to secure new, more appropriate regulatory standards for air and water pollution, this group might prefer to develop their own metrics for what constitutes high-quality air and water, and propose alternative ways for measuring and communicating about air and water quality and contamination (e.g., Ottinger, 2011). In some cases, the decision to reject an existing knowledge representation can be associated with a suggestion for a re-framing of policymaking and engagement. Groups who suggest alternatives to existing standards and frameworks for science-related activities still need to explain their work to outsiders, and thus should provide a clear and complete documentation of their choices and activities that shows where they depart from existing knowledge representations (e.g., in terms of metadata), and why.

Knowledge Organization and Knowledge Representation Can Offer Value Through Classification

There is an overwhelming number of software solutions for citizen science that support activities such as contributor recruitment and management, data collection, data analysis, data storage, and data retrieval e.g., GEO BON's "BON in a Box". Formal knowledge organization and representation allow either humans or machines to better survey existing software solutions, identifying the most promising ones to suite a project's needs. If organizations and project managers can better understand available citizen-science systems, they can quickly narrow their search to a category that supports their needs, either by adopting a single software application that supports all processes, or by compiling a range of applications into a systems assemblage (Prestopnik, & Crowston, 2012) where different solutions are combined to serve a set of goals.

Knowledge Representation Can Support Increased Adoption of Citizen Science

One route to increased adoption of existing technologies is for developers of citizen-science software solutions to identify closely related research niches and modify their system to become compatible with these niches. For example, citizen science has been traditionally focused on species observations; thus, models such as iNaturalist have been *limited* to manage biodiversity-related questions. However, there are projects (e.g. Citclops, CITI-SENSE, Air Quality Egg, among others) that use sensors to measure a variety of environmental parameters, to monitor, for example, water quality, land cover and disease-vector species; there are projects that collect and manage non-environmental variables (e.g., Agent Exoplanet, Galaxy Zoo, ARTigo, among others); and there are projects managing just metadata, including citizen-science project repositories. Therefore, robust and proven architectures and representations of existing software solutions focused on biodiversity could be adapted to include data about projects, instruments, devices and new types of variables.

Knowledge Representation Can Support Coordination Among Web-Based Portals

By creating software based on semantically-rich technologies and formal knowledge representations, and operating it through their portals, citizen-science portal providers (iNaturalist, SciStarter, Atlas of Living Australia, among others (Beaman, & Cellinese, 2012)) can offer additional services such as co-ordinating relationships among related sites and organizations. Taken to the extreme, this model could result in a global community of citizen scientists working together.

Citizen-science software-providing portals may debate whether to develop complete information solutions, which span the full range of project activities, or focus on core competencies such as supporting a specific type of data collection. If the latter occurs, the need for inter-portal communication standards will be critical to a portal's success. If different portals streamline inter-organizational communication, the data could be stored once and then made available to all interested parties (e.g., citizen-science communities, researchers, and policymakers). For example, records of natural-water quality could be used by: citizens, focusing on users of the beach ranking the best beaches; researchers, who could use these same records to perform analyses on water quality in different lakes; and policymakers, who could use this information to manage lake-water demand.

Paths Towards Knowledge Representation in Citizen Science

Citizen science is a specific domain. While there is some overlap of citizen-science knowledge representation with other domain ontologies and general, top-level ontologies, many key concepts are articulated in a way that is unique to the citizen-science domain. This section briefly explores different methods for categorizing and representing knowledge that are currently being used in citizen science, before exploring the opportunity for additional work.

Acquiring and representing knowledge in citizen science are difficult tasks, which have to balance two elements. On the one hand, because domains like citizen science are constantly evolving to reflect real-world changes, knowledge representation must be considered an ongoing task (for example, contributors' gender was once classified as binary but no longer is). On the other hand, in order for knowledge representation to successfully support communication between machines, a certain level of stability and agreement is required.

Knowledge acquisition is a process where domain information is collected for use in knowledge representation (e.g., Sowa, 1999; Jakus, Milutinović, Omerović, & Tomažić, 2013). Methods of knowledge acquisition rely on the capabilities of human domain-experts, and also machines. Different methods of knowledge acquisition, which comprise elements that are peculiar to citizen science, include:

- **Knowledge Elicitation:** Human domain-experts pool their individual knowledge to reach a shared understanding of what is important to model in a particular domain. Knowledge elicitation may take place through: brainstorming sessions, including workshops; the ongoing activities of working groups; real-world application, learning, hypothesis testing and validation; and/or the use of folksonomy web-platforms where the general public is encouraged to submit and vote on different terms. It can often be difficult for domain experts to say what they know, or to make tacit knowledge explicit. Techniques designed to elicit tacit or procedural knowledge, such as observation or card sorting, may be used to augment the declarative knowledge that is more easily shared. In citizen science, knowledge elicitation has typically occurred through workshops (e.g., Bowser, McMonagle, & Tyson, 2015) and the activities of working groups (e.g., the Data and Metadata Standardization Working Group of the Citizen Science Association). Web and mobile technologies have expanded social knowledge-elicitation, thanks to the way they link people together and facilitate collaboration. An interesting form of knowledge elicitation is being explored, for example, through YAMZ and other folksonomies (e.g., Hotho, Jäschke, Schmitz, & Stumme, 2006), platforms that use internet-based collaboration on information architecture to elicit and evaluate different metadata terms within the domain of ontology and taxonomy construction.
- **Knowledge Discovery:** Information is automatically extracted from digital sources, for example through different types of data mining. To date, knowledge discovery is an under-utilized tool for knowledge representation in citizen science. To demonstrate the value of knowledge discovery, a database of citizen-science tools—compiled by SciStarter and the Woodrow Wilson International Center for Scholars—utilizes a knowledge-discovery process for determining relevant metadata used in the description of citizen-science tools and technologies.

Knowledge representation is a process where, typically, an individual or committee drafts up a general guideline for the construction of an ontology (or an equivalent construct; see description of related technologies below), which acts as a skeleton that plans the shape of the ontology. It takes into account the goal of the ontology and the data that are to be within its scope, and attempts to represent it within a tree. Rigorous rules are laid down for fleshing out that skeleton, to ensure that the initial population and subsequent addenda are internally consistent. If this is done properly, the result is a coherent and consistent ontology. A well-constructed ontology is not trivial to create. The Cyc top-level ontology (Reed, & Lenat, 2002), for instance, has been factored and re-factored over the years as its creators have learned through hard experience the way of organizing commonsense knowledge.

The classic ontology has a steep learning curve for construction and maintenance. Guidelines have to be both well-planned and rigorously adhered to. For example, The Atlas of Living Australia tried to implement RDF OWL vocabularies and ontologies between 2009 and 2011, and found it to be extremely difficult to build them in the biodiversity domain. Implementing RDF OWL representations in citizen science has proven to be a major hurdle due largely to the resources required and complexity in building comprehensive vocabularies and ontologies, and in mapping and defining the relationships between the concepts, making machine inference unrealistic in most cases.

Newer and perhaps more implementable web technologies, such as:

1. JSON-LD (Lanthaler, Sporny, & Kellogg, 2014) with numerous major players such as Alphabet, the BBC, HealthData.gov, Yahoo!, and Microsoft already deploying its specification in production,
2. Internet-of-things technologies, and
3. Authoring technologies empowering users to become contributors, appear to be getting traction and could represent the basis for the next step for knowledge representation in citizen science: working with data that are important to stakeholders and have to interoperate across the Web.

NEXT STEPS FOR CITIZEN SCIENCE

With civic educators and citizen science's identities established, the question becomes "What kinds of information should be conveyed and in which form?" Decision-makers working with small and large projects depend on the provision of the right information, and the right representations of information. By mapping the terms used in existing active datasets and relevant legacy datasets to a standardized ontological framework, the authors seek to help civic educators develop more effective and efficient educational strategies. This can have consequences in re-education, cultural-change management, and systems which handle terms translation for information exchange.

Providing better-represented information will:

- Increase people's knowledge of complex modern problems, and how these problems may be addressed through science and policy;
- Provide pathways for related paradigms, such as the do-it-yourself/maker/hacker movement, to become more integrated into citizen science thanks to a better interoperability with their own "language";
- Offer ways to formalize citizen science information, to support better decision-making;
- Help scholars of the citizen-science topic to improve the accuracy and value of their research;
- Uncover common errors in representation in citizen science, for example errors arising from outdated conceptualizations of the paradigm, and help to fix them; enabling these corrections to make subsequent action and scholarship more useful to educators of all kinds.

Information restructuring is also the consequence of a paradigm shift. Previous definitions of citizen science focused on how the public participated in the research and policy-making processes; the authors' definition shifted the focus from an earlier distinction between scientific activities and public policy, to emphasize citizen science as a comprehensive strategy towards a shared social and scientific mentality and new level of competence. This shift is related to what civic educators consider important, and how they model key features of the citizen-science domain; and any ongoing re-framing of the domain will trickle down to impact a wide range of knowledge-organization and knowledge-representation terms.

CONCLUSION

This chapter's central proposition is that greater knowledge of the topics dealt with in citizen science can empower civic educators and researchers to more effectively:

1. Lead prospective learners to knowledge that matters;
2. Increase the value of the information they deliver to others; and
3. Open up opportunities for others to learn.

In the first part of the chapter, the authors analyzed the numerous existing classifications and definitions of citizen science, and proposed a slight re-framing of the topic and a definition for citizen science that is alternative to existing ones. The second part of the chapter showed the role that the meaning of words and phrases (semantics) plays in understanding and supporting citizen science. Here, the goal was to explain how different organizations already use certain software solutions to organize knowledge about citizen science, how these systems can be classified and how they can facilitate or impede interoperability – the ability of humans and machines to pass information between each other. The third and final part of the chapter introduced the idea that providing better-represented information will:

1. Increase people's knowledge of complex modern problems, and how these problems may be addressed through science and policy;
2. Provide pathways for related paradigms, such as the do-it-yourself/maker/hacker movement, to become more integrated into citizen science thanks to a better Interoperability with their own "language";
3. Offer ways to formalize citizen science information, to support better decision-making;
4. Help scholars of the citizen-science topic to improve the accuracy and value of their research; and
5. Uncover common errors in representation in citizen science, for example errors arising from outdated conceptualizations of the paradigm, and help to fix them, enabling these corrections to make subsequent action and scholarship more useful to educators of all kinds.

This final part of the chapter focused on knowledge-representation complexity. By complexity, the authors mean that the domain of citizen science has diverse and occasionally contradictory components, experienced by an equally diverse range of communities; and, when a domain is complex, educators and researchers have to choose how to frame it (i.e., they have to choose how to formally represent it and what parts of the topic to emphasize). Choices about how to frame issues in citizen science will impact the effectiveness of the communication and exchange of information, and determine whether or not that information has any subsequent effect on others' knowledge and competence. In this respect, the authors offer a vision in which, even if citizen science moves towards more and more formal knowledge-representation and the ability to carry out automatic reasoning by machines, humans are responsible for checking whether any given knowledge representation is still an accurate reflection of reality.

REFERENCES

Andrianandrasana, H. T., Randriamahefasoa, J., Durbin, J., Lewis, R. E., & Ratsimbazafy, J. H. (2005). Participatory ecological monitoring of the Alaotra wetlands in Madagascar. *Biodiversity and Conservation, 14*(11), 2757–2774. doi:10.1007/s10531-005-8413-y

Annenberg Public Policy Center. (2014). *Americans know surprisingly little about their government, survey finds*. Retrieved from http://www.annenbergpublicpolicycenter.org/americans-know-surprisingly-little-about-their-government-survey-finds/

Beaman, R. S., & Cellinese, N. (2012). Mass digitization of scientific collections: New opportunities to transform the use of biological specimens and underwrite biodiversity science. *ZooKeys*, (209): 7. PMID:22859875

Bonney, R. (1996). Citizen Science: A lab tradition. *Living Bird, 15*(4), 7–15.

Bonney, R., Ballard, H., Jordan, R., McCallie, E., Phillips, T., Shirk, J., & Wilderman, C. C. (2009). *Public Participation in Scientific Research: Defining the Field and Assessing Its Potential for Informal Science Education. A CAISE Inquiry Group Report*. Online Submission.

Bonney, R., Phillips, T. B., Ballard, H. L., & Enck, J. W. (2015). Can citizen science enhance public understanding of science? *Public Understanding of Science (Bristol, England)*. PMID:26445860

Bonney, R., Shirk, J., Phillips, T., Wiggins, A., Ballard, H., Miller-Rushing, A., & Parrish, J. (2014). Next steps for citizen science. *Science, 343*(6178), 1436–1437. doi:10.1126/science.1251554 PMID:24675940

Bowser, A., McMonagle, R., & Tyson, E. (2015). Citizen Science and Crowdsourcing Metadata Workshop Summary. Washington, DC: Woodrow Wilson International Center for Scholars. Retrieved from https://www.wilsoncenter.org/publication/citizen-science-and-crowdsourcing-metadata-workshop-summary

California Energy Commission. (2012). *Community-based climate adaptation planning: case study of Oakland*. Author.

Chalkley, S., Brendan, E., & Gowland, K. (1999). *Who is watching our water? Waterwatch Australia. In The Challenge of Rehabilitating Australia's Streams* (pp. 163–168). Melbourne: Co-operative Research Centre for Catchment Hydrology.

Cohn, J. P. (2008). Citizen science: Can volunteers do real research? *Bioscience, 58*(3), 192–197. doi:10.1641/B580303

Cooper, C. B., Dickinson, J., Phillips, T., & Bonney, R. (2007). Citizen science as a tool for conservation in residential ecosystems. *Ecology and Society, 12*(2), 11.

Danielsen, F., Burgess, N. D., Balmford, A., Donald, P. F., Funder, M., Jones, J. P., & Yonten, D. et al. (2009). Local participation in natural resource monitoring: A characterization of approaches. *Conservation Biology, 23*(1), 31–42. doi:10.1111/j.1523-1739.2008.01063.x PMID:18798859

David, J. S., McCarthy, W. E., & Sommer, B. S. (2003). Agility---: The key to survival of the fittest in the software market. *Communications of the ACM, 46*(5), 65–69. doi:10.1145/769800.769803

Davis, R., Shrobe, H., & Szolovits, P. (1993). What is a Knowledge Representation? *AI Magazine*, *14*(1), 17–33.

Dickinson, J. L., Zuckerberg, B., & Bonter, D. N. (2010). Citizen science as an ecological research tool: Challenges and benefits. *Annual Review of Ecology Evolution and Systematics*, *41*(1), 149–172. doi:10.1146/annurev-ecolsys-102209-144636

Dimock, M., Kiley, J., Keeter, S., Doherty, C., & Tyson, A. (2014). Beyond Red vs. Blue: The Political Typology. *The Pew Research Center for the People & the Press*. Retrieved from http://www.people-press.org/files/2014/06/6-26-14-Political-Typology-release1.pdf

Ellwood, E. R., Dunckel, B. A., Flemons, P., Guralnick, R., Nelson, G., Newman, G., & Mast, A. R. et al. (2015). Accelerating the digitization of biodiversity research specimens through online public participation. *Bioscience*, *65*(4), 383–396. doi:10.1093/biosci/biv005

Gruber, K. (2015). Deep influence of soil microbes. *Nature Plants*, *1*(12), 15194. doi:10.1038/nplants.2015.194 PMID:27251726

Haklay, M. (2013). Citizen science and volunteered geographic information: Overview and typology of participation. In *Crowdsourcing geographic knowledge* (pp. 105–122). Springer Netherlands. doi:10.1007/978-94-007-4587-2_7

Haklay, M. (2015). Citizen Science and Policy: A European Perspective. Washington, DC: Woodrow Wilson International Center for Scholars. Retrieved from https://www.wilsoncenter.org/sites/default/files/Citizen_Science_Policy_European_Perspective_Haklay.pdf

Hopkins, G. W., & Freckleton, R. P. (2002). Declines in the numbers of amateur and professional taxonomists: Implications for conservation. *Animal Conservation*, *5*(3), 245–249. doi:10.1017/S1367943002002299

Hotho, A., Jäschke, R., Schmitz, C., & Stumme, G. (2006). *Information retrieval in folksonomies: Search and ranking*. Springer Berlin Heidelberg.

Irwin, A. (1995). *Citizen science: A study of people, expertise and sustainable development*. Psychology Press.

Irwin, A., & Horst, M. (2015). Engaging in a decentred world. *Remaking Participation: Science, Environment and Emergent Publics*, 64.

Jakus, G., Milutinović, V., Omerović, S., & Tomažič, S. (2013). *Concepts, Ontologies, and Knowledge Representation*. New York: Springer. doi:10.1007/978-1-4614-7822-5

Jørgensen, M. S., Hall, I., Hall, D., Gnaiger, A., Schroffenegger, G., Brodersen, S., et al. (2004). *Democratic Governance Through Interaction between NGOs, Universities and Science Shops: Experiences, Expectations, Recommendations*. Final Report of INTERACTS. Retrieved from http://www.wilawien.ac.at/interacts/interacts_report_final2.pdf

Kullenberg, C., & Kasperowski, D. (2016). What Is Citizen Science?–A Scientometric Meta-Analysis. *PLoS ONE*, *11*(1), e0147152. doi:10.1371/journal.pone.0147152 PMID:26766577

Lanthaler, M., Sporny, M., & Kellogg, G. (2014). *JSON-LD 1.0*. W3C Recommendation. Retrieved from https://www.w3.org/TR/json-ld/

Lewenstein, B. V. (2004). *What does citizen science accomplish?* Cornell University. Retrieved from https://ecommons.cornell.edu/bitstream/handle/1813/37362/Lewenstein.2004.What%20does%20citizen%20science%20accomplish.pdf

Lukyanenko, R., Parsons, J., & Wiersma, Y. F. (2016). Emerging problems of data quality in citizen science. *Conservation Biology*, *30*(3), 447–449. doi:10.1111/cobi.12706 PMID:26892841

Lupia, A. (2015). *Uninformed: Why People Seem to Know So Little about Politics and What We Can Do about It*. Oxford University Press.

OpenScientist. (2011). *Finalizing a Definition of "Citizen Science" and "Citizen Scientists*. Retrieved from http://www.openscientist.org/2011/09/finalizing-definition-of-citizen.html

Ottinger, G. (2010). Buckets of resistance: Standards and the effectiveness of citizen science. *Science, Technology & Human Values*, *35*(2), 244–270. doi:10.1177/0162243909337121

Oxford Dictionaries. (2014). Oxford University Press. Retrieved from http://www.oxforddictionaries.com/definition/american_english/citizen-science

Oxford English Dictionary. (2014). Oxford University Press.

Park, P. (2014). Realizing the potential of citizen science. In *UNEP Year Book 2014* (pp. 36–41). UNEP.

Phillips, T., Ferguson, M., Minarchek, M., Porticella, N., & Bonney, R. (2014). *User's Guide for Evaluating Learning Outcomes in Citizen Science*. Ithaca, NY: Cornell Lab of Ornithology.

Pimm, S. L., Alibhai, S., Bergl, R., Dehgan, A., Giri, C., Jewell, Z., & Loarie, S. et al. (2015). Emerging technologies to conserve biodiversity. *Trends in Ecology & Evolution*, *30*(11), 685–696. doi:10.1016/j.tree.2015.08.008 PMID:26437636

Prestopnik, N. R., & Crowston, K. (2012). Citizen science system assemblages: understanding the technologies that support crowdsourced science. In *Proceedings of the 2012 iConference* (pp. 168-176). ACM. doi:10.1145/2132176.2132198

Reed, S. L., & Lenat, D. B. (2002). Mapping ontologies into Cyc. In *AAAI 2002 Conference Workshop on Ontologies For The Semantic Web* (pp. 1-6).

Scassa, T., & Chung, H. (2015). *Typology of citizen science projects from an intellectual property perspective*. Washington, DC: Woodrow Wilson International Center for Scholars —Commons Lab.

SciStarter. (2016). Retrieved from http://scistarter.com/page/Citizen%20Science.html

Serrano Sanz, F., Holocher-Ertl, T., Kieslinger, B., Sanz García, F., & Silva, C. G. (2014). *White paper on citizen science for Europe*. European Commission. Retrieved from http://ec.europa.eu/newsroom/dae/document.cfm?doc_id=6913

Shirk, J. L., Ballard, H. L., Wilderman, C. C., Phillips, T., Wiggins, A., Jordan, R., & Bonney, R. et al. (2012). Public participation in scientific research: A framework for deliberate design. *Ecology and Society, 17*(2), 29. doi:10.5751/ES-04705-170229

Silvertown, J. (2009). A new dawn for citizen science. *Trends in Ecology & Evolution, 24*(9), 467–471. doi:10.1016/j.tree.2009.03.017 PMID:19586682

Simpson, R. (2013). *Explainer: what is citizen science?* The Conversation. Retrieved from http://theconversation.com/explainer-what-is-citizen-science-16487

Sowa, J. F. (1999). *Knowledge representation: logical, philosophical, and computational foundations.* Brooks/Cole.

Terhune, K. (2013). Retired Justice Sandra Day O'Connor, in Boise, laments 'alarming degree of public ignorance'. *Idaho Statesman.* Retrieved from http://www.mcclatchydc.com/news/nation-world/national/article24755302.html

Van Rijsoort, J., & Jinfeng, Z. (2005). Participatory resource monitoring as a means for promoting social change in Yunnan, China. *Biodiversity and Conservation, 14*(11), 2543–2573. doi:10.1007/s10531-005-8377-y

Vyver, J. (2014). Maratus harrisi: the tiny peacock spider discovered by Canberra man Stuart Harris in Namadgi National Park. *ABC News, Canberra, Australia.* Retrieved from http://www.abc.net.au/news/2014-08-14/discovering-maratus-harrisi/5670424

Waxman, O. B. (2016, April 4). Pigeons That Track Pollution. *Time,* 23.

Wernand, M. R., Ceccaroni, L., Piera, J., & Zielinski, O. (2012). Crowdsourcing technologies for the monitoring of the colour, transparency and fluorescence of the sea. *Proceedings of Ocean Optics XXI.*

Wiggins, A., & Crowston, K. (2011). From conservation to crowdsourcing: A typology of citizen science. In *System Sciences (HICSS), 2011 44th Hawaii international conference on* (pp. 1-10). IEEE. doi:10.1109/HICSS.2011.207

KEY TERMS AND DEFINITIONS

Citizen Science: Citizen science is work undertaken by civic educators together with citizen communities to advance science, foster a broad scientific mentality, and/or encourage democratic engagement, which allows society to deal rationally with complex modern problems.

Civic Educators: Civic educators are people who believe that providing information to others, and/or creating opportunities for others to learn, are paths to greater civic competence and a better future.

Interoperability: Interoperability is a property of a system, whose interfaces are completely understood, to work with other systems, present or future, without any restricted access or implementation.

Metadata: Metadata are data that provide information about other data. Metadata are used to describe data using metadata standards specific to a particular discipline. Describing the contents and context of data or data files increases their usefulness.

Ontology: A formal and explicit specification of a shared conceptualization, which is readable by a computer. Practically, ontologies are used to represent information in a standardized way so that it can be organized, processed by machines or humans, and used as a shared reference.

Standardization: Standardization is the process of developing and implementing technical standards. Standardization can help to maximize compatibility, interoperability, repeatability, or quality. It includes the case of "spontaneous standardization processes", to produce *de facto* standards.

Web Ontology Language: The W3C Web Ontology Language (OWL) is a Web language designed to represent rich and complex knowledge about things, groups of things, and relations between things. OWL is a computational-logic--based language such that knowledge expressed in OWL can be exploited by computer programs, e.g., to verify the consistency of that knowledge or to make implicit knowledge explicit. OWL documents, known as ontologies, can be published in the World Wide Web and may refer to or be referred from other OWL ontologies.

Chapter 2
More Than Just Networking for Citizen Science:
Examining Core Roles of Practitioner Organizations

Claudia Göbel
Museum für Naturkunde Berlin, Germany

Gregory J. Newman
Colorado State University, USA

Jessica L. Cappadonna
Queensland University of Technology, Australia

Jian Zhang
East China Normal University, China

Katrin Vohland
Museum für Naturkunde Berlin, Germany

ABSTRACT

Citizen science activity is growing rapidly around the world and diversifies into new disciplines with recent advances in technology. This expansion is accompanied by the formation of associations and networks dedicated to citizen science practitioners, which aim at supporting citizen science as a research approach. This chapter examines how four such organizations in the United States, Europe, Australia, and China have begun to take shape, and are working with citizen science communities and stakeholders in respective regions and globally. Challenges and future plans of these groups are also discussed. This chapter identifies three core roles of citizen science practitioner organization: 1) establishing communities of practitioners, 2) building expertise through sharing of existing and developing new knowledge, and 3) representing community interests. By focusing on this hitherto neglected phenomenon, the authors aim to stimulate further research, discussion and critical reflection on these central agents in the emerging citizen science landscape.

INTRODUCTION

Citizen science projects are scientific research projects that rely on public participation (Bonney, Ballard, et al., 2009). Citizens have a long history of participation in diverse scientific investigation activities stretching back to the foundation of learned societies, natural history museums, and universities (e.g.,

DOI: 10.4018/978-1-5225-0962-2.ch002

Dickinson, Zuckerberg, & Bonter, 2010; Mahr, 2014). Today, such projects exist all over the world, and advances in computing and mobile communication technologies have allowed projects to expand in geographic scale and diversity (Sullivan et al., 2014; Dickinson et al., 2012). Bias and sampling errors that once plagued citizen science data can now be avoided by implementing rigorous design strategies (e.g., Bonney, Cooper, et al., 2009; Tinati et al., 2015), and by analyzing data with improved statistical models (e.g., Bird et al., 2013). Projects vary greatly in focus, activities performed, geographic scope and other factors (Kullenberg & Kasperovski, 2016; Shirk et al., 2012). Despite this heterogeneity, some trends in terms of types of activities and key actors have been identified in recent studies.

Haklay (2015), for instance, distinguishes the following levels of engagement and types of activity: Passive Sensing, Volunteer Computing, Volunteer Thinking, Environmental and Ecological Observations, Participatory Sensing and Civic/Community Science. This typology is coherent with a classification developed by the Socientize consortium in the White Paper on Citizen Science for Europe (Serrano Sanz, Holocher-Ertl, Kieslinger, Sanz Garcia, & Silva, 2014), which includes Data Collection, Analysis Tasks, Serious Games, Participatory Experiments, Grassroots Activities, Collective Intelligence and Pooling of Resources as prototypical citizen science activities. Other participatory approaches that overlap with citizen science in terms of methodologies and normative claims are the Do-It-Yourself (DIY) movement (Nascimento, Guimarães Pereira, & Ghezzi, 2014) and the maker scene (Walter-Herrmann & Büching, 2013).

With regard to prominent topic areas of citizen science projects, Kullenberg and Kasperovski (2016) categorized citizen science into three main clusters in a recent bibliometric study. The biggest cluster is in the natural sciences covering research on biology and often deals with environmental issues, such as nature conservation (e.g. flora and fauna monitoring projects) or urban living quality (e.g. water monitoring), and curiosity of natural phenomena (e.g. identifying astronomical anomalies or ways in which proteins fold). The second cluster is geographic information research and comprises approaches such as geographic information systems that include public participation (e.g., Sieber & Haklay, 2015). The third cluster includes research in social sciences and epidemiology, where a range of methods that involve citizen contributions to research is found, for instance participatory health research (e.g., Wright, Gardner, Roche, Unger, & Ainlay, 2010), participatory action research (Nielsen & Nielsen, 2006), and transdisciplinary research (Jahn, Bergmann, & Keil, 2012). Discussions of these approaches, however, appear to be rather limited to the respective social science subdomains and are not well linked to more general debates in citizen science (Crain, Cooper, & Dickinson, 2014). Digital humanities are another popular field for citizen science projects (Kullenberg & Kasperovski, 2016), which includes research in genealogy, history (e.g. Zooniverse project Ancient Lives, Williams et al., 2014), and linguistics (Newman, 2014). One might thus argue that citizen science constitutes a widespread phenomenon, which finds application in a number of topic areas and scientific disciplines, while it appears as a rather fragmented field of research practices with various subdomains developing distinctive yet overlapping methodologies and discussions in the respective research communities.

Although citizen science has gained substantial momentum regarding diversity, reliability, and recognition, several challenges remain. For example, the European Union identified funding, education and training, evaluation, and technology access, as well as data policy, dissemination, and support as key challenges that must be carefully taken into account when working towards the improvement of citizen science throughout the region (Serrano Sanz et al., 2014). Additional challenges include mechanisms for assuring policy impact for relevant citizen science studies, as well as data management, data sharing, data visualization, and professional development (Haklay, 2015). These and similar obstacles are not unique to Europe or the United States, but instead are common and universally experienced across associations around the world.

Simultaneously, citizen science facilitators, managers, and volunteers have begun to converge to evaluate topics considered important for the improvement of scientific rigor, inclusiveness, impact, and reputation of citizen science in various parts of the globe. Topics addressed by these practitioners included promoting best practices, understanding common challenges, developing communications, sharing resources, and synergies that could result from collaborations. Numerous independent practitioner associations and networks have emerged, which seek to advance citizen science. Four such groups include the Citizen Science Association (CSA; based in the United States), the European Citizen Science Association (ECSA), the Australian Citizen Science Association (ACSA), the Chinese Citizen Science Network (CCSN) (Table 1) as well as groups that are forming in other regions of the world.

This chapter is dedicated to exploring the strategic role of citizen science practitioner organizations in addressing the above mentioned challenges and advancing in the field of citizen science. The four organizations in the U.S., Europe, Australia, and China are compared as case studies to review how practitioner organizations have developed and are working to strengthen citizen science regionally and globally. The first section reviews the development history and central characteristics of each organization, such as aims, activities, and structure. The second section describes current and planned activities of the organizations regarding:

1. Environmental monitoring,
2. Publication, communication, and data infrastructures,
3. Best practice and capacity building, and
4. Linking citizen science to policy making and cooperative activities.

The last section analyzes the roles that practitioner organizations play to strengthen the citizen science field, discusses challenges that associations and networks are facing, and concludes by outlining future steps.

Such reflections on the development of citizen science associations and networks help address the often neglected aspects of networking, professionalization, and institutionalization of rapidly growing fields such as citizen science. Therefore, this chapter provides valuable insights for better understanding the recent global boom of citizen science, while remaining sensitive to regional specificities and contexts. In addition, the developments covered in this chapter also lay the groundwork for actors in other regions of the world to strengthen cooperation and further establishing citizen science.

OVERVIEW OF CITIZEN SCIENCE ASSOCIATIONS AND NETWORK

This section gives an overview of the four citizen science organizations in the U.S., Europe, Australia, and China, which are currently active in collaborative dialogue across the world.

Citizen Science Association

The Citizen Science Association (CSA) is a non-profit association supported by volunteers, membership, and a voluntary member-elected board of directors based in the United States with the aim of supporting all forms of citizen science globally (Table 1). Initial efforts began in 2007 when the Cornell Lab of Ornithology (CLO) hosted an invitational workshop. This event brought 50 practitioners together to discuss best practices and launch the website entitled CitizenScience.org (Table 1) to host tools and guidelines for project design. Evaluation of the workshop and website revealed that project leaders seek not only resources on best practices, but also the opportunity for dynamic engagement with these materials and peers, suggesting a need for continued and improved online resources. Field building continued in 2011 with a second invitational workshop, coorganized by the CLO and the American Museum of Natural History, entitled "Engaging and Learning for Conservation". Participants expressed strong desires to learn about what other practitioners were doing and share insights, suggesting that organizers create ways to encourage continued engagement (Heimlich, 2012). A third conference in Portland, Oregon in 2012, sponsored by the Bechtel Foundation and led by the CLO and the Schoodic Institute, was the first citizen science conference open to general attendance by anyone; it brought 300 practitioners together. Conference evaluation revealed that many practitioners feel disconnected from peers, need access to resources and best practices, and believe that their own insights and innovations are unrecognized and underappreciated (Benz et al., 2013; Heimlich, 2012). Discussions identified next steps, including developing online tools; providing more opportunities for professional development; starting a new open-access journal; and connecting people with tools for data management and data visualization (Benz et al., 2013). Finally, a milestone occurred at this third conference when participants endorsed the establishment of a global citizen science association. The CSA was formed by a steering committee in February 2014 and launched more formally with the election of its twelve-member board of directors in February 2015.

European Citizen Science Association

The European Citizen Science Association (ECSA) also started out as a loose network of stakeholders, mainly museums and research institutes, with experience in carrying out citizen science activities in the field of environmental monitoring. The idea to consolidate the informal and sporadic exchange as well as to involve like-minded individuals affiliated with institutions and projects from all over Europe was conceived in the United Kingdom (UK) and carried forward by the Open Air Laboratories network (Stack & Donkin, 2013). In 2012, preparatory meetings for forming a pan-European network were held in London and Copenhagen involving citizen science project managers and other stakeholders from the UK, Italy and Germany. In 2014, a non-profit association was registered under German law. The ECSA aims to strengthen and advance the citizen science movement through communication, exchange and cooperation, capacity building and research (ECSA, 2015e). According to its strategy, ECSA has adopted a comprehensive definition of the term "citizen science", covering a wide spectrum of participatory research formats and all scientific disciplines (ECSA, 2015e). The organization is active on the regional (European) level with its main target group being organizations and individuals who conceptualize and implement citizen science initiatives, which are referred to as practitioners or practitioner organizations.

Australian Citizen Science Association

In 2013, Earthwatch Institute Australia released a discussion paper detailing the benefits of establishing a national association for the citizen scientist community following in the footsteps of the recently developed associations in the United States and Europe (Earthwatch Institute Australia, 2013a; 2013b). This paper led to the development of an initial meeting and workshop in May 2014 hosted by the Queensland

Museum and supported by CSIRO, Inspiring Australia, University of Tasmania, University of Technology, Sydney, New South Wales Government, and University of South Australia. Over 90 individuals attended the meeting and voted to establish a national citizen science association for Australia. Volunteers immediately began developing a framework for the group now known as the Australian Citizen Science Association (ACSA). Since then, volunteers have continued advocating for citizen science and developing the association. For example, members of ACSA developed communications, determined a host institution, developed a governance structure, and drafted a 3-year Strategic Plan (Table 1). Additionally, ACSA members organized the first Australian citizen science conference in July 2015. The 198 delegates that attended the conference in Canberra were from across Australia as well as from Europe, the U.S., and Southeast Asia. The event started with a welcoming address from Australia's Chief Scientist demonstrating national support for citizen science, which included the announcement of an Occasional Paper (Pecl, Gillies, Sbrocchi, & Roetman, 2015). Delegates were also invited to attend the first Annual General Meeting of ACSA, where a new seven-member Management Committee was elected. Since the conference, the Management Committee has continued furthering development of ACSA, though actions such as establishing Working Groups to advance goals of the Strategic Plan and enabling citizen science networking. The goals outlined in the ACSA Strategic Plan aim to encourage participation, build partnerships, and facilitate a community that is guided by best practices in citizen science, as well as to ensure impacts of citizen science are realized, and to establish ACSA as a trusted and reputable hub.

Chinese Citizen Science Network

The Chinese Citizen Science Network (CCSN) was set up informally in November 2013 by six ecologists with different backgrounds (Table 1). Since the history of citizen science is relatively short in China, it is important to consider development of citizen science in other countries. Such a review reveals the great potential and benefits citizen science may have if implemented more widely throughout China, whether promoted by science, non-governmental organization (NGO), and government stakeholders (Zhang et al., 2013). In China, there are currently approximately 8,000 environmental NGOs. However, collaborations among NGOs, citizens and professionals are rarely seen, leading to poor quality and inadequate representation on collected data by citizens. Currently, the six CCSN members are planning on initiating several small citizen science projects related to biodiversity monitoring by collaborating with scientists and citizens, hoping that these projects could become successful examples for the development of CCSN.

MAJOR ACTIVITIES OF THE ASSOCIATIONS

This section illustrates major areas of work of CSA, ECSA, ACSA and CCSN:

1. Activities in the context of environmental monitoring as one of the most significant areas of citizen science practice around the world,
2. Infrastructure and communication services provided by the associations as means of support and further development of the respective communities of practice,
3. Examples of best practice collections and standards representing milestones in the work of the associations,
4. Policy related work of the associations, and
5. Cooperation between associations.

Table 1. A structure and status overview of the associations and network as of May 2016

	CSA	ECSA	ACSA	CCSN
Website	citizenscienceassociation.org and citizenscience.org	ecsa.citizen-science.net	citizenscience.org.au	citizenscience.cn
Registered office	Incorporated as a non-profit in Connecticut, USA; hosted by Schoodic Institute, Winter Harbor, Maine, United States of America	Incorporated as a non-profit association in Germany; secretariat hosted by Museum für Naturkunde, Berlin, Germany	The incorporation process is in progress; hosted by Australian Museum, Sydney, Australia	Unincorporated; hosted by Institute of Botany, Chinese Academy of Sciences, Beijing, China
Geographic scale	Global with an emphasis on North America	Europe (core area), members from abroad welcome	Australia	Mainland China
Membership & Outreach Numbers	3500+ individual members (mostly US); no membership fees to date; social media currently include over 1,740 twitter followers.	Nearly 200 members across 28 countries (mostly EU); organizations and individuals, which pay membership fees; larger network of supporters, newsletter (about 1000 subscriptions), Facebook (about 200), Instagram (2333).	No formal membership structure to date; ACSA and Australian citizen science updates are currently sent through an email list (over 500 subscribers) and social media, including Facebook (1,500), Twitter (1,700), as well as LinkedIn (160) and Google+ (80).	CCSN is still on the stage of infancy consisting of eight core members.
Mission	To advance citizen science through communication, coordination, and education.	Connecting citizens and science through fostering active participation.	To advance citizen science through sharing of knowledge, collaboration, capacity building and advocacy for citizen science.	No mission formally approved as yet.
Vision	A world where people understand, value, and participate in science.	In 2020, citizens in Europe are valued and empowered as key actors in advancing knowledge and innovation and thus supporting a sustainable development of our world.	A community that supports, informs, and develops citizen science.	No vision formally approved as yet.
Objectives	1. Establish a global community of practice for citizen science. 2. Advance the field of citizen science through innovation and collaboration. 3. Promote the value and impact of citizen science. 4. Provide access to tools and resources that further best practice. 5. Support communication and professional development services. 6. Foster diversity and inclusion within the field.	ECSA developed a strategy with three key areas of work each comprising specific actions: 1. Promoting sustainability through citizen science. 2. Building a Think Tank for citizen science. 3. Developing participatory methods for cooperation, empowerment and impact.	1. Encourage broad and meaningful **participation** in citizen science 2. through facilitating inclusive and collaborative **partnerships** 3. and a community of best **practice**, knowledge and tools, 4. to ensure the value and **impact** of citizen science and its outputs are realized 5. enabled by ACSA as an effective, trusted and well recognized organization and hub for citizen science in Australia.	No objectives formally approved as yet.

Networking and Exchange for Environmental Monitoring

A large proportion of citizen science projects around the world focus on investigating aspects of natural world, with many projects being underway for decades (e.g., see reviews for the UK: Silvertown, 2009 and for the U.S.: Miller-Rushing, Primack, & Bonney, 2012; Mahr, 2014). For example, among the oldest continuous datasets are phenological records kept by farmers and agricultural organizations documenting the timing of events, such as sowing, harvests, and pest outbreaks (Hopkins 1918). Early programs such as the North American Breeding Bird Survey, the U.S. National Weather Service's Co-operative Observer Program, North American Bird Phenology Program, and lilac monitoring programs have generated large-scale datasets of biological and physical data that could not have been collected otherwise (Miller-Rushing et al., 2012). Other projects that have arisen at a regional, state, and even local level, including programs that monitor water quality, plant, and/or wildlife (Miller-Rushing et al., 2012), have also generated data not possible to collect without the support of volunteer citizen scientists. Many states, for instance, have long relied on volunteers to monitor water quality, fish populations, or other recreational uses of rivers and lakes (Nerbonne & Nelson, 2008) and programs such as Save Our Streams that tackle local problems have been in existence for a long time (Firehock & West, 1995). Similarly, in Europe, environmental monitoring has a longstanding history of public participation and is of great contemporary importance to traditional science (Science Communication Unit, 2013). Given this deep historical expertise in environmental monitoring, citizen science associations and the Chinese network are working together to facilitating broad-scale and community-based networking for project managers and citizen scientists in this domain. These new connections and collaborations are now enabling the citizen science community to exchange information, methods, and research outcomes, improve projects, and develop new initiatives following best practices.

For example, several initiatives are underway in the United States (U.S.) to improve networking and information exchange. The CSA is working with the Federal Community of Practice on Crowdsourcing and Citizen Science – a group formed within the U.S. government - to develop data-sharing protocols and standards and link federal directories of projects with other directories such as SciStarter (www.scistarter.com). The CSA has also formed a Professional Development (PD) working group that is working with the U.S. National Park Service (NPS) to plan professional development workshops at these centers across the US. These initiatives mark a new direction for citizen science in the U.S. with a focus on PD, data sharing, and project interoperability.

Similarly, a majority of ECSA's founding members have expertise in environmental monitoring so there is a strong focus within the association for related citizen science activities. Related to the objective of promoting sustainability through citizen science (ECSA, 2015a), ECSA aspires to implement joint sustainability-focused citizen science projects between organizations across Europe. The ECSA also advocates for the relevance of citizen science as an approach for environmental policy with national governments and agencies as well as at the European level. For example, ECSA organizes presentations and workshops, publishes policy papers (ECSA, 2015b, 2015c, 2015d) and has been invited to collaborate with the citizen science task force of the European Environmental Agencies (EPAs) in order to develop better environmental governance.

Much like other regions, many citizen science projects in Australia engage participants with nature monitoring. As with all efforts, ACSA is working to promote citizen science that contributes to evaluating environmental health and patterns. In Australia, such projects may engage citizen scientists to share observations of animals, plants, and habitat with researchers, make identifications from images and acoustic data, or supply specimens from nature. These projects are often supported by museums, universities, governmental agencies, organizations, community groups, and individuals. Citizen science initiatives occur at national, state, local, and community levels as well. Members of ACSA are currently working to identify all projects being done across the nation, and as projects are found, information is shared with the community primarily through ACSA social media (Table 1) and shortly to an ACSA-specific project finder. The ACSA members are also looking to better understand where and how data are collected, stored, aggregated, and used in a citizen science context. This includes considering various information management platforms, standards, and applications that enable citizen scientist to engage, communicate, and contribute around the data they are collecting and sharing. One national biodiversity data repository, which is commonly used by researchers, government, natural resource managers, and citizen science managers alike, is the Atlas of Living Australia (ALA, Belbin & Williams, 2016). Data that are entered into the ALA are then incorporated to the Global Biodiversity Information Facility (GBIF), a global open data infrastructure (Samy et al., 2013). The ALA supports citizen science, as well as other research through providing web services and tools to support the mobilization, discoverability, and validation of citizen science data into the national biodiversity dataset. In addition, the ALA also released a citizen science project finder and biodiversity data collection tool BioCollect that supports citizen science projects undertaking structured surveys (e.g. flora and fauna surveys), as wells as activity-based interventions (e.g. revegetation, site restoration, and seed collection). The ACSA and ALA associates exchange information regularly.

The Australian Citizen Science Association also respond to many inquiries regarding development of projects, organizing workshops, improving citizen science outcomes, and connecting with other initiatives. As ACSA has been under development, members have been facilitating global collaborations. Additionally, ACSA members have driven the development of The Australian Guide to Running a BioBlitz, recruiting guide co-authors who are involved in citizen science and biodiversity data collection across the nation (Hepburn et al., 2015). A bioblitz is an event, usually within a day, designed for scientists and members of the public to intensively collect information on all forms of biodiversity (e.g. plants, animals, fungi, microbes, etc.) found within a pre-determined area (Robinson, Tweddle, Postles, West, & Sewell, 2013). These events are effective for engaging communities with the environmental sciences (Roger & Klistorner, 2016). As ACSA matures, the association aims to connect those groups and individuals currently conducting environmental monitoring to share ideas on a broad scale, and invite those new to citizen science to help make environmental discoveries.

Several citizen science projects in China also monitor the environment. One ongoing project is Chinese Field Herbarium (CFH, http://www.cfh.ac.cn/default-en.html), which was initialed in 2008 and is serving as a web platform for biological field observations and data management. One core CCSN member, Dr. Bin Chen, plays a key role in the CFH project. Currently, the CFH has deposited more than 6,569,000 photos from over 12,000 citizen scientists, with over 1,450,000 of the photos being geo-referenced. The number of citizen scientists and photos continues to increase daily. Another ongoing effort of CCSN is to develop a mobile application for plant phenology monitoring, which is a collaborative effort with researchers from Germany and United States. Meanwhile, CCSN promotes citizen science by organizing sessions in national conferences on conservation biology and ecoinformatics.

Publication, Communication, and Data Infrastructures

Beyond the field of environmental monitoring, the broader citizen science community wants best practices resources, greater project diversity, and inclusiveness in the field, as well as recognition for contributions, and want to help build the field (Heimlich, 2012; Crall, 2013). The associations aim to respond to these needs by providing platforms that enable knowledge and tool sharing, as well as providing opportunities for networking and exchange of experiences. Major plans and achievements are described below:

- **Open-Access Journal:** The CSA established a globally focused, peer-reviewed, open-access journal entitled "Citizen Science: Theory and Practice" (http://theoryandpractice.citizenscience-association.org) published by Ubiquity Press to meet the needs the citizen science community. Previously, high quality research related to citizen science may never be published because an appropriate journal did not exist. Papers that did get published are dispersed across restricted-access publications in diverse fields, making them difficult to find. The citizen science journal offers a consolidated home for peer-reviewed research papers, case studies, opinions, book reviews, and other manuscripts on topics aimed at improving the theory, methods, and practice of citizen science. The current editorial board includes members from mainly CSA, as well as ECSA, ASCA, and other affiliates from around the world.
- **Data and Metadata Standardization:** To meet the information needs of the citizen science community and ensure that websites, the journal, and workshops are as transformative and far reaching as possible, an innovative cyberinfrastructure (CI) will be developed by CSA in cooperation with other associations. This CI, along with accompanying work to engage the community in its use, has the aim to streamline information discovery, accessibility, and reuse, which will improve efficiency, reduce redundancy, and actively engage dispersed expertise. Currently, practitioners are burdened with updating project details and other information in multiple locations. The CI will make it easy for practitioners to keep information they may have posted on many websites up to date in a single place and simultaneously share and synchronize information across websites. The cyberinfrastructure will be powered by a citizen science-specific data exchange protocol. This protocol will seamlessly share, exchange, and synchronize core information (as metadata) applicable to citizen science projects globally. The CI will consist of integrated databases describing people, projects, best practices, publications, workshops, and outcomes along with information, visualizations, and metadata - all within an open-source framework that can be extended by developers and members of the citizen science community. The associations will develop the CI as a service-based architecture with related Application Programming Interfaces (APIs). The CSA has already developed and piloted a prototype of the data exchange protocol to connect existing project databases at CitizenScience.org, SciStarter, CitSci.org (http://citsci.org/), ALA and BioCollect, which provides a proof-of-concept for this approach. ECSA and ACSA are engaged in the respective CSA working group to contribute relevant knowledge on developments from other regions and give inputs on the worldwide applicability of standards. In addition to the exchange of citizen science project metadata, CSA, ECSA, ACSA, and other organizations around the globe are working towards the development of standards for citizen science metadata that would enable the exchange of observation and measurement data and results of analyses (Bowser et al., 2016; Joint Research Center, 2016).

- **Communication:** CSA, ECSA and ACSA each have independent email lists, where subscribers may post project updates, job post, events, surveys, or discussion topics. Both CSA and ECSA also share one-way association updates with members via electronic newsletters. All associations offer webpages, albeit with different levels of service and interactivity (Table 1). In the US, the web platform of the CSA, CitizenScience.org, is recognized as the go-to location for finding information and guidance on citizen science project design and management (Thompson, 2010). A need exists for more dynamic content and improved opportunities for users to share expertise and to interact around pressing questions and innovative ideas (Crall, 2013). The CSA aims to invigorate CitizenScience.org with fresh content that invites multiple contributors, and fulfills the needs of practitioners from diverse disciplines and project types. This website includes high-quality references and resources, such as the Citizen Science Toolkit, data management guide, and evaluation guide. Important goals are to create more dynamic content through writing blog posts and inviting guest bloggers; interviewing practitioners to develop case stories of projects and profiles of people in the field; and scheduling forum discussions and coordinating with guest moderators. Content will also include book reviews, summaries of journal articles, materials for media, newsletters, conference reports, job postings, and a calendar of events. The focus will be on content that synthesizes, assesses, or reviews resources and recommends best practices with a new emphasis on bringing attention to ideas, projects, and people who have something new to offer citizen science practitioners – as well as those on the margins of the community such as data managers, resource managers, public health professionals, and media personnel.

The ECSA website, which at present mainly presents the work of the association, is in a progressive update process with the aim of building a state of the art interactive platform for practitioners and stakeholders that shall contain a database of publications, best practice resources, and tools, as well as offer collaborative online working space and training opportunities. It is also planned to make a repository of citizen science initiatives in Europe available to the public with concise project descriptions in English linking to the individual projects in their native languages, as well as to national aggregator sites in order to provide an easily accessible information hub at the European level. The association also joined social media (Instagram and Twitter) in 2015 (Table 1).

Communication methods for the ACSA community currently consist of a basic website, an email list, and social media (Table 1). Future plans also include developing a citizen science project finder, similar in concept to the ALA biodiversity project finder and the global SciStarter project finder, but focusing on all citizen science affiliated with Australia. Discovery of projects fosters project participation, resource sharing, and collaboration, which can then promote project innovations and sustainability, as well as reduce redundancy in future project development.

The CCSN members are eager to explore the potential of using mobile applications for environmental monitoring in China (Zhang & Huang, 2015), along with training workshops and courses for both citizens and scientists, as done by the U.S.-based National Ecological Observation Network (http://citizenscienceacademy.org/online-courses) providing online professional development resources.

Best Practice and Capacity Building

The CSA, ECSA, and ACSA associations have agreed to foster the exchange of ideas and promote best practices. Similarly, these associations aspire to develop principles, standards, key indicators, and evaluation tools to reflect the value of citizen science to both engage the public and inform the scientific community.

For example, the CSA aims to promote the value and impact of citizen science by developing metrics of success and impact for citizen science programs and developing a data exchange standard for sharing such metrics. Specifically, the CSA has been developing a rubric of impacts and outcomes that include publications, white papers, use of data for decision making, presentations, and use of data in policy development, to name a few. The ECSA plans to collaboratively develop methods and standards for research and engagement to improve cooperation between professional researchers and citizens across different scientific disciplines (ECSA, 2015e). The ECSA working group on "Sharing Best Practice and Building Capacity" has recently published a list of ten principles that account for strong citizen science projects (ECSA, 2015a). Those guidelines have been elaborated through consultation of citizen science projects across Europe and validated by the community in several rounds of discussion (e.g. Robinson, 2014). Additionally, ECSA aspires to work with groups related to the citizen science field, such as science shops, Do-it-Yourself researchers, and grassroots movements, to learn from their respective methodologies, such as Transdisciplinary Research or Action Research, and channel good practice back into the citizen science community. The goals of ACSA are very closely aligned with the best practice and impact plans of the CSA. The Australian Guide to Running a BioBlitz, is an example of how ACSA members are working to promote citizen science (Hepburn et al., 2015).

Linking Citizen Science to Policy

Citizen science is increasingly recognized by decision makers and included in policies in regions such as the United States, Europe, and Australia. This section provides a few examples of how initiatives in these respective regions currently address citizen science.

Across Europe, the European Union (EU) primarily recognizes the importance of citizen science through both environmental and research policy fields. The potential of citizen science for environmental policy making is mainly linked to data collection for monitoring and stewardship purposes. Benefits related to environmental education, participatory governance, and environmental justice are also noted but considered more difficult to assess (Science Communication Unit, 2013). Regarding research policy, the European Commission, which proposes and implements EU legislation, recognizes citizen science as an important trend that may be leveraged for two current EU policy and funding initiatives. The first initiative is the Open Science agenda (European Commission, 2016), which addresses the transformation of research through advances in information and communication technologies (ICTs) Citizen science is promoted here as a participatory element of research along with Open Access, e-infrastructure development and other measures (European Commission, 2013a). The second initiative is the Responsible Research and Innovation agenda (European Commission, n.d.), which is intended to support public engagement in research and innovation in order to better align outcomes of research and innovation processes with societal expectations and needs, such as grand societal challenges in health, environment, and energy, etc. (European Commission, 2012). The heightened attention from decision makers at EU level manifests in the sponsoring of projects for research, technology development, and public engagement. Prominent examples include the five Citizens' Observatory projects that develop earth observation technology for citizen participation in environmental stewardship (Citizens' Observatory, n.d.),as well as the Socientize project that conducted several citizen science activities (Socientize, n.d.), and coordinated the creation of the White Paper on Citizen Science for Europe (Serrano Sanz et al., 2014).

Several European countries have implemented national legislation with reference to citizen science (Haklay, 2015). The Scottish Environmental Protection Agency, for instance, commissioned a decision framework for choosing a citizen science approach in biodiversity and environmental monitoring (Pocock, Chapman, Sheppard, & Roy, 2014). Similarly, the German Federal Ministry of Education and Research (BMBF) supported citizen science by sponsoring the development of a German citizen science web platform (www.buergerschaffenwissen.de) as well as a set of capacity building measures. Both initiatives are designed to build a national network of citizen science initiatives via a project finder, stakeholder roundtables, and the development of a citizen science strategy for Germany, as well as guidelines to implement citizen science projects for practitioners. Related initiatives, such as scoping projects, landscape studies, platform development and public communication campaigns, are also underway in several additional countries.

As a stakeholder in European research and environment policy, ECSA regularly advocates for citizen science at European and country-specific events, such as the EU Green Weeks in Brussels, at EU consultations, and conferences relating to research, environmental policy, and/or citizen science. The ECSA also develops policy papers showcasing how citizen science can improve the link between science and society. The aim is to raise awareness for citizen science and provide expertise to decision makers in order to contribute to the proliferation of citizen science methodologies and to better link all levels of European governance.

In the U.S., the federal government has established several cross-agency working groups, including the Federal Community of Practice on Crowdsourcing and Citizen Science and the U.S. Geological Survey sponsored Community for Data Integration (CDI) citizen science working group. The Federal Community of Practice on Crowdsourcing and Citizen Science recently released a Toolkit (https://crowdsourcing-toolkit.sites.usa.gov/). In addition, the Wilson Center Commons Lab is working to address associated legal and regulatory considerations for citizen science (Gellman, 2015), as well as assist in the creation of a federal directory of citizen science projects in collaboration with the U.S. General Services Administration (https://www.citizenscience.gov/) (Wilson Center, n.d.). In addition, legislation is being proposed to more tightly couple citizen science with policy. For example, the White House Assistant to the President for Science and Technology, and Director of the Office of Science and Technology Policy, recently released a memo encouraging the use of citizen science and crowdsourcing across all federal agencies (Holdren, 2013).

Members of ACSA are beginning to explore how citizen science is supported through current policies, and where potential exists to influence future policy development, including through the Australian Government's National Science and Innovation Agenda. On a national level, the Australian Government also listed citizen science as a key objective to engage citizens with science, technology, engineering and mathematics in 2013 (Office of the Chief Scientist, 2013), and endorsed citizen science while welcoming conference attendees at the 2015 Australian citizen science conference (Pecl, Gillies, Sbrocchi, & Roetman, 2015). The Atlas of Living Australia (ALA) supports a wide array of biodiversity research, including citizen science, and this national data repository that went live in 2010 is funded through the National Collaborative Research Infrastructure Strategy (NCRIS). The Great Barrier Reef Marine Park Authority is a federal agency which implements community reef monitoring through Eye on the Reef and Reef Guardian programs, to gain large scale measures of marine health (Great Barrier Reef Marine Park Authority, n.d.). Agency support for citizen science on a State or Territory level varies. NSW agencies support several initiatives, e.g. water quality or sport fish tagging (NSW Department of Primary Industries, n.d.), with the New South Wales Office of Environment and Heritage (NSW OEH) recently releasing a citizen science position statement (NSW OEH, 2015a) and strategy (NSW OEH, 2015b). The ACSA members aim to highlight the success of citizen science projects promoted under such policies, to further advocate for continued and expanded support of citizen science.

Collaboration between the Associations and Network

As demonstrated above, the similarity of objectives and tasks addressed by the associations and network have given rise to numerous collaborative activities between the respective boards, committees, and working groups. In addition to sharing knowledge on how to build successful organizations and how to position in the respective local contexts, the cooperation of the three associations has been formalized in a Memorandum of Understanding (European Citizen Science Association, Citizen Science Association, & Australian Citizen Science Association, 2015). This landmark document maps out three key areas in which the associations will work together to strengthen citizen science on a global level:

1. Promote scholarship of citizen science via the journal Citizen Science: Theory and Practice,
2. Organize joint conferences to directly link practitioners with each other, and
3. Collaborate on building digital infrastructure, as well as share online resources, such as tools and best practices.

Cooperation between the associations has a range of aims that were discussed extensively in a recent article for the above-mentioned journal (Storksdieck et al., 2016). In summary, objectives include fostering global collaborations between citizen science practitioners, facilitators, and volunteers from different regions and disciplines, as well as scaling regional activities up to the global level and addressing global challenges. This will eventually contribute to development of new citizen science projects and perhaps improvement of existing ones, as well as support mutual learning and understanding across the world. A fundamentally important aspect to this work is to preserve concerns for local regional specificities. Sharing knowledge and pooling resources can increase the efficiency and impacts of each organization's work, which permits leveraging each other's efforts for building capacity in citizen science. Additionally, such collaborations can increase credibility with the scientific community and the general public, as well as help to establish associations as interlocutors in global fora, such as the United Nations, in which citizen science can make important contributions.[1]

CONCLUSION

Analyzing the associations in United States, Europe, and Australia, as well as the network in China, one finds four organizations with similar goals and comparable activities underway, though to varying degrees and targeting different geographic scopes. It is obvious from the histories of these organizations that each one is in a different stage in establishing as an organization and it would be misleading to assume a singular trajectory of association development. This section presents initial observations and interpretations as a basis for further study.

Structurally, CSA, ECSA, and ACSA show considerable similarities with formal arrangements in place to govern membership and decision-making within the organization, which allow them to operate in the long run and recruit interest and future members within respective regions and beyond. In contrast, the CCSN is organized more loosely as a network without formal structures and only limited reach. With regards to the scientific disciplines that the associations and network are targeting, one can distinguish between disciplinary-focused and generalist organizations. CCSN mainly aims to work with practitioners from biodiversity and environmental sciences, while CSA, ECSA, and ACSA actively recruit and engage with citizen science groups from other disciplines, such as the social sciences and the humanities. On the geographical scale of membership, CCSN, CSA and ACSA have a national scope, while ECSA operates at the European level. A feature shared by all organizations is having a research institute as hosting organization – the Schoodic Institute for CSA, the Museum für Naturkunde Berlin for ECSA, the Australian Museum for ACSA, and the Institute of Botany of the Chinese Academy of Science for CCSN. One might hypothesize that such a linkage to an established academic institution constitutes an important source of reputation, which helps to establish credibility for citizen science. The fact that all host institutions are situated in the natural sciences and more specifically in the field of biodiversity and environmental sciences underscores both to the long tradition of public participation in research and to the current relevance of citizen science in those fields. Evaluating the specific nature of each of those host institutions – natural history museums in Europe and Australia, the partnership with the national parks network in the US, and the Academy of Science in China – promises insights on how the global trend of citizen science manifests in different research systems.

Regarding the content of activity programs, similar needs spurred the creation of each association and continue to shape the portfolio of activities each group undertakes. CCSN shows analogous motivations for further association building. The authors have identified the following key activities as common concerns across groups: 1) networking and exchange for environmental monitoring, 2) providing publication, communication, and data infrastructures, 3) developing of best practice resources and implementing capacity building activities, 4) establishing links to policy making, and 5) strengthening joint activities among them. Through these measures, each of the associations and the network reviewed herein are strategically positioned to address the challenges mentioned in the introduction as facilitators of citizen science, whether they are working directly with citizen scientists, and/or practitioners and their organizations. In what follows, the authors will show that the examples of association activities provided above are not only an important element of the contemporary expansion of citizen science, but can be interpreted as a contribution to the professionalization of citizen science and reflect the challenges these organizations are facing.

Roles of Citizen Science Practitioner Organizations

Recently, it has been argued that citizen science associations contribute to the professionalization of the field (Haklay, 2015; Storksdieck et al., 2016). This argument is mainly supported by referring to potential benefits that such organizations can and aim to realize to support the field. Storksdieck et al. (2016) identify various such aspired benefits that can be summarized as:

- Linking practitioners, promoting mutual learning and creating synergies between citizen science projects in order to avoid duplication of efforts and increase the quality of citizen science;
- Fostering inter- and transdisciplinary collaborations to stimulate innovation and increase the impact of research along with contributing to a greater responsiveness to societal demands;
- Addressing common challenges of practitioners and offering generic services, such as outreach, education and professional development, to the practitioner community;

- Working with practitioners to develop common norms and standards for the conduct, evaluation, validation and reflection of citizen science;
- Engaging with stakeholders in research, policy making, civil society, and business, to promote collaboration, make citizen science known and increase its credibility as approach for research and innovation.

Further research is needed to judge the claim of an increasing professionalization in citizen science by systematically exploring manifestations and mechanisms of professionalization along with critical reflections on the consequences of such developments. As a starting point for such endeavors, this chapter has attempted to mobilize material to underpin the assumed benefits of citizen science practitioner organizations by reviewing mayor current and planned activities of CSA, ECSA, ACSA, and CCSN. As a result of this analysis, the authors have identified three core roles of citizen science practitioner organizations that can serve as a conceptual framework for further exploring citizen science associations and professionalization:

1. Establishing communities of citizen science practitioners;
2. Building expertise through sharing of existing and developing new knowledge on the practice of citizen science; and
3. Representing community interests.
    - **Establishing Citizen Science Communities:** This first role has an internal focus on members of the community and refers to the provision of infrastructure and services to enable networking activities. In many regions of the world, practitioners, advocates, and participants of citizen science are geographically distant from one another, and without established channels for communication that allow for exchange of ideas, resources, and knowledge. There is often a similar lack of communication across disciplines. The associations and network work to offer infrastructures and services for organizations and individuals involved in citizen science to share information and foster collaborations. General elements like websites, email lists, newsletters, and social media activities are means to support the exchange of information, provide guidance on citizen science project management, and facilitate discussion around pressing questions and innovative ideas. Community specific resources like cyberinfrastructure and metadata, including citizen science-specific data exchange protocols and apps, are meant to streamline information discovery, accessibility, and reuse and thus to improve efficiency and reduce redundancy. The creation of a scientific journal dedicated to citizen science has the goal of improving the theory, methods, and practice in the field and thus to establish a proper community of peers, as well as to anchor citizen science more profoundly in academic research. In addition, the institutionalization of loose networks into formalized associations allows for the development of strategies, provides accountability, and enables the persistence of the knowledge generated over time.
    - **Building Expertise through Sharing of Existing and Developing New Knowledge on the Practice of Citizen Science:** This second role addresses the shared purpose holding the community together, which is related to knowledge. Associations are hubs for the citizen science community to collect and promote best practice examples from projects to stimulate learning and refine methodologies. This is being achieved through professional development (e.g. conferences, workshops, and training events), online tools, and resources (e.g. publica-

tions and guidelines), which are anticipated to improve quality and overall impact of citizen science. Examples include CSA's development of metrics of success and impact for citizen science programs, as well as methods and standards for research, engagement, and communication like ECSA's "Ten Principles of Citizen Science" (ECSA, 2015a) and ACSA's best practice "BioBlitz guide" (Hepburn et al., 2015).

  - **Representing Community Interests:** This third role of practitioner organizations covers externally-focused tasks that link the community of practitioners to other stakeholders. Associations work with relevant groups outside citizen science including decision makers, research organizations, funders, industry, other civil society groups, media, and the general public to advocate for citizen science. They provide systematic information on developments in the field, represent the community in relevant discussions and negotiation processes, and offer a contact point helping to heighten awareness of citizen science and facilitating collaboration. The examples from the policy related work of the associations show that this advocacy work may take many forms, such as ACSA's role promoting success of citizen science projects run by local, state, and federal agencies, as well as non-profit, university, industry, and other organizations like the Atlas of Living Australia. Similarly, ECSA aspires to become a think tank for citizen science in European research and environment policy to provide expertise on citizen science to decision makers, and to improve links between European and national levels of governance.

The conceptual framework of three core roles of practitioner organizations also serve to illustrate how they address the challenges for citizen science that have been discussed above. Citizen science practitioner organizations build knowledge-based networks of peers facilitating access to expertise and resources to those involved in citizen science and represent community interests with stakeholders, such as policy makers and civil society. The discussion of activities of the associations in the contexts of environmental monitoring served to give an example of how these analytical roles manifest in a specific case, and how they are intertwined with each other. Finally, the discussion of the joint initiatives of the associations has shown how those activities can be scaled up to the global level and leveraged to reinforce each other.

Challenges for Citizen Science Practitioner Organizations

All the four associations and the network have gained visibility since beginning, and contribute to the acceptance of citizen science beyond the citizen science community by providing networking opportunities, information about existing projects, best practice guidelines, and other support. This increasing activity is linked to rising expectations of what the associations may deliver with regards to specific programs, resources, capacity building, and convenience power. Important challenges that have to be addressed in order to realize the potential of the citizen science associations and network include securing funding for the core business of organizations and for programmatic activities, as well as fostering inclusiveness in organization structures and a plurality of methodologies.

- **Funding:** A key challenge for all associations is fundraising to support communication activities, secretariat staff, the implementation of specific capacity building programs, the development (and maintenance) of infrastructure, such as dynamic websites and data exchange protocols, and research to better understand who is doing what, where, and why in citizen science regionally and globally. At present, all associations and the network discussed cover a significant share of their operational costs through in-kind contributions of their membership, whether through the administrational support of their headquarters, hosting of infrastructure, or voluntary contributions in working groups or at conferences.

The three typical sources of funding for similar organizations include membership fees, grants, and donations. Membership fees represent the most stable and independent way of funding. However in the early period of establishing an organization there is a trade-off between extending the organization's reach and raising fees. Of the organizations discussed in this chapter, the ECSA is the only association that has had a paid membership structure in place from its inception. Both CSA and ACSA are considering membership fees at a later stage of development.

A second important source of funding for citizen science is (research) grants, which could be particularly suited to pay for programmatic activities, i.e. joint thematic citizen science activities, or other finite projects. Grant funding has been acquired, for example, by ACSA from the Australian Government program Inspiring Australia as funding to support the setup of the association. The first CSA conference was initiated when a grant was awarded, and several sponsors then also helped cover remaining costs, making the event possible. The ECSA has received a grant for the first international conference and to date forms part of two Coordination and Support Actions under Horizon 2020 funding. One general concern in this regard is that citizen science has yet to become an established approach to research that is considered valid to the point of being fully integrated into research funding schemes in its own right (e.g. Germany: Pettibone, Ziegler, Bonn, & Vohland, 2015). Currently, citizen science is usually funded through accompanying streams, such as public engagement activities in the EU or science education programs in the US, which represent only a fraction of research budgets.[2] In the case of ECSA, one central motivation behind establishing the organization was to be able to bid for EU funding and thus to function as an accelerator for small citizen science projects to find partners through the ECSA network (in order to take part in a usual EU project, one needs consortium partners from at least three EU countries). Research project funding, however, can only cover the costs of maintaining an organizational infrastructure, including administrative personnel to a certain extend and does not ensure sustainability. This concern is likely to be augmented for regional associations, as cross border or international NGO funding appears to be even more scarce. In addition, managing such funding opportunities can also represent a challenge for the young organizations in terms of governance and the development of procedures for how to deal with competing interests within the associations.

Finally, a third source of funding is through donations and sponsorships, which could also provide a way to ensure continuity of basic organizational structure through the fluctuation of project funding. All of the associations have received some initial funding through such means. Given that each funding strategy has benefits and drawbacks, associations would likely benefit from having a diverse funding portfolio, not relying on any one particular source of income.

- **Inclusiveness and Plurality:** Citizen science associations and networks are developing to represent and advocate for the citizen science community as a whole. Citizen science combines research and civic engagement, and the community is interdisciplinary, also including stakeholders from a wide variety of backgrounds, cultures, positions, organizations, and experiences. Associations and networks must overcome a number of challenges to ensure inclusiveness. Already the use of the term "citizen science" can be challenging, and deserves consideration by associations and networks. While public participation in research has been occurring for centuries in many parts of the world, the term and current definition "citizen science" is relatively new, having been formed at the Cornell Lab of Ornithology (Bonney, Ballard, et al. 2009) in the United States (Pettibone, 2015). The contested nature of the term is evident at least at two levels, the linguistic one as well as the semantic one. The tension on the linguistic level is most apparent in non-English speaking countries where local terms and activities exist and the introduction of the English term is often debated and felt as unnecessary anglicization (Pettibone, 2015). On the semantic level, we find different interpretations of which approaches and activities count as citizen science and which ones don't. Great Britain offers the example of an English speaking country, in which "citizen science" has not been universally adopted for a full spectrum of participatory research activities. The term can be deemed unnecessary for long-standing activities involving volunteer domain experts (e.g. Pocock, Roy, Preston, & Roy, 2015), but is more frequently applied to engaging research activities open to the general public and so including people with low levels of expertise (M. Pocock & L. Robinson, personal communication, April 29, 2015). Apprehensions about adoption of the North American term can be encountered in Australia, though acceptance is growing rapidly. Across non-English speaking countries where "citizen science" has been translated into local languages, there are many different interpretations of what the term means[3]. More systematic research is needed to explore variations on the use of the term across the world, as well as to identify common denominators. For citizen science associations and networks the task is to promote multilingualism and to reach out to groups using other terminologies when investigating what citizen science is underway, so that a broad spectrum of participatory research activities can be equally represented and valued.

Apart from terminology, another inclusiveness challenge associations and networks grapple with is deciding on how to determine which approaches are included under the umbrella of citizen science, such as when evaluating membership applications. Factors such as participant engagement, activities, methods, academic disciplines, and outcomes may be considered when evaluating citizen science (for typologies see: Shirk et al., 2012; Haklay, 2015). While each organization has its own history in this respect, all of them have common roots in ecology, which has the potential to influence priories and activities of the associations. For instance, ECSA has a strong bias towards organizational members (usually research institutes) with environmental or biodiversity focus. Only recently, have groups practicing other approaches, such as cyberscience or participatory health research, been reached. Also the CCSN is currently focused on the advancement of citizen science in environmental protection and biodiversity monitoring. In the United States and Australia there are many projects focused on environmental monitoring. To address this imbalance, ECSA, CSA, and ACSA have adopted policies to actively reach out and offer services to a broad scope of projects, including all scientific disciplines and all approaches to citizen science design and implementation.

Another fundamental challenge of associations and networks is to bring citizen science to the mainstream culture, while not streamlining it in such a way that excludes other approaches to participatory knowledge generation that may be less common today. It is also necessary to critically reflect on the Western model of science and innovation, and reach out to those operating under different models to ensure all groups are represented rather than marginalized and excluded. Organizations advocating for citizen science are encouraged to develop inclusive conceptual frameworks and dialogue activities across different approaches to participatory knowledge generation, including for instance traditional knowledge, participatory action research, and transdisciplinary research. Inclusive organizations also require inclusive governance structures that encourage members to take an active part in the decisions and future of the associations and networks, as well as mechanisms that ensure transparency and accountability.

The Road Ahead

The discussion of the aims of the organizations presented in this chapter along with their activities in progress have shed light on their role as facilitators and their future potential. Associations and networks can function to professionalize citizen science and address gaps of knowledge to facilitate citizen science becoming an established approach for research and civic engagement. Substantial challenges remain to be addressed in order to be successful in supporting and advancing citizen science around the world.

Essential steps for the coming years will be to implement plans to ensure that ambitious visions of nurturing prosperous and diverse communities of practice are put into action. The associations and network will continue progressing with actions such as organizational and professional development activities. Members will continue synthesizing distributed knowledge into guidelines and standards, and developing cyberinfrastructure and associated tools, and implementing joint citizen science projects supported. The organizations will continue working to become trusted facilitators for the citizen science community by carefully engaging with a variety of different stakeholders whom hold different interests in citizen science as one of the most exciting movements in our globalizing techno-scientific civilizations. Extending and deepening the cooperation between the existing organizations, and engaging with newly developed and emerging groups, is a key ingredient to advocate for the global citizen science communities. Practitioner organizations must use resources with efficiency while working to advocate for citizen science as a reliable field of research and to maximize impact of citizen science and articulate the voices of citizens on global matters of concern. Citizen science organizations must remain sensitive to differences across stakeholder groups, respecting customs, languages, and differences of opinion, whether considering a local, state, national, or broader scale.

There is a growing body of literature on the results, methodologies and epistemologies of citizen science approaches, though few papers are published regarding citizen science networking. This chapter is one of the first overviews of citizen science associations and networks, focusing on the formation, roles, and challenges of such organizations. The aim of this chapter was to stimulate further research and discussion about citizen science communities and their networking efforts around the globe, by providing reflections on the composition and roles of the CSA, ECSA, ACSA and CCSN. Professionalization, regional specificities, and diversity have emerged as topics deserving further study in order to gain a more nuanced understanding of central agents in the emerging field and to contribute to their success.

REFERENCES

Belbin, L., & Williams, K. J. (2016). Towards a national bio-environmental data facility: Experiences from the Atlas of Living Australia. *International Journal of Geographical Information Science*, *30*(1), 108–125. doi:10.1080/13658816.2015.1077962

Benz, S., Miller-Rushing, A., Domroese, M., Ballard, H. L., Bonney, R., DeFalco, T., & Young, A. et al. (2013). Workshop 1: Conference on public participation in scientific research 2012; An international, interdisciplinary conference. *Bulletin of the Ecological Society of America*, *194*(1), 112–117.

Bird, T. J., Bates, A. E., Lefcheck, J. S., Hill, N. A., Thomson, R. J., Edgar, G. J., & Frusher, S. et al. (2013). Statistical solutions for error and bias in global citizen science datasets. *Biological Conservation*, *173*, 144–154. doi:10.1016/j.biocon.2013.07.037

Bonney, R., Ballard, H., Jordan, R., McCallie, E., Phillips, T., Shirk, J., & Wilderman, C. (2009). Public participation in scientific research: Defining the field and assessing its potential for informal science education. Washington, DC: Center for Advancement of Informal Science Education (CAISE).

Bonney, R., Cooper, C. B., Dickinson, J., Kelling, S., Phillips, T., Rosenberg, K. V., & Shirk, J. (2009). Citizen science: A developing tool for expanding science knowledge and scientific literacy. *Bioscience*, *59*(11), 977–984. doi:10.1525/bio.2009.59.11.9

Bowser, A., Schade, S., Newman, G., Göbel, C., Brenton, P., & Cavalier, D. (2016, April 14). *Reflections on the Federal Catalog of Crowdsourcing and Citizen Science Projects* [Blog Post]. Retrieved May 16, 2016, from https://wilsoncommonslab.org/2016/ 04/14/ reflections-on-the-federal-catalog-of-citizen-science-and-crowdsourcing-projects/

Citizens' Observatory. (n.d.). *About the Projects*. Retrieved March 6, 2016, from http://www.citizen-obs.eu/About.aspx

Crain, R., Cooper, C., & Dickinson, J. L. (2014). Citizen science: A tool for integrating studies of human and natural systems. *Annual Review of Environment and Resources*, *39*(1), 641–665. doi:10.1146/annurev-environ-030713-154609

Crall, A.W. (2013). *CitizenScience.org Beta Test: Findings and Recommendations*. Report prepared for the Cornell Lab of Ornithology.

Dickinson, J. L., Shirk, J., Bonter, D., Bonney, R., Crain, R. L., Martin, J., & Purcell, K. et al. (2012). The current state of citizen science as a tool for ecological research and public engagement. *Frontiers in Ecology and the Environment*, *10*(6), 291–297. doi:10.1890/110236

Dickinson, J. L., Zuckerberg, B., & Bonter, D. N. (2010). Citizen science as an ecological research tool: Challenges and benefits. *Annual Review of Ecology Evolution and Systematics*, *41*(1), 149–172. doi:10.1146/annurev-ecolsys-102209-144636

Earthwatch Institute Australia. (2013a, October 14). Call for the establishment of a national research and coordination network to advance public participation in Australian scientific studies 'Citizen Science Network (Australia)'. *Public Discussion Paper*. [Blog]. Retrieved May 30, 2016, from http://inspiringaustralia.net.au/citizen-science-network-australia/

Earthwatch Institute Australia. (2013b). *Summary of responses to Earthwatch's public discussion paper on the establishment of a National Citizen Science Network for Australia*. Discussion Paper.

European Citizen Science Association. (2015a). *10 Principles of Citizen Science*. Retrieved January 10, 2016, from http://ecsa.citizen-science.net

European Citizen Science Association. (2015b). *ECSA draft policy paper: Citizen science as part of EU policy delivery.* Retrieved March 6, 2016, from http://ecsa.citizen-science.net

European Citizen Science Association. (2015c). *ECSA draft policy paper: Endorsing the white paper on citizen science for Europe.* Retrieved March 6, 2016, from http://ecsa.citizen-science.net

European Citizen Science Association. (2015d). *ECSA policy paper: Endorsing the German green paper on citizen science.* Retrieved January 10, 2016, from http://ecsa.citizen-science.net

European Citizen Science Association. (2015e). *ECSA strategy*. Retrieved January 10, 2016, from http://ecsa.citizen-science.net

European Citizen Science Association. Citizen Science Association, & Australian Citizen Science Association. (2015). *Memorandum of Understanding between CSA, ECSA, CSNA*. Retrieved March 6, 2016, from http://ecsa.citizen-science.net

European Commission. (2012). *Responsible research and innovation: Europe's ability to respond to societal challenges*. Retrieved January 10, 2016, from https://ec.europa.eu/research/swafs/pdf/pub_public_engagement/responsible-research-and-innovation-leaflet_en.pdf

European Commission. (2013a). *Digital science in Horizon 2020.* Retrieved January 10, 2016, from http://ec.europa.eu/information_society/newsroom/cf/dae/document.cfm?doc_id=2124

European Commission. (2013b). *Factsheet: Horizon 2020 budget.* Retrieved January 10, 2016, from http://ec.europa.eu/research/horizon2020/pdf/press/fact_sheet_on_horizon2020_budget.pdf

European Commission. (2016). *Open Science*. Retrieved March 6, 2016, from https://ec.europa.eu/digital-agenda/en/open-science

European Commission. (n.d.). *Responsible research & innovation*. Retrieved March 6, 2016, from http://ec.europa.eu/programmes/horizon2020/en/h2020-section/responsible-research-innovation

Eye on Earth. (2015). *Summary of outcomes Eye on Earth summit 2015*. Retrieved January 10, 2016, from http://www.eoesummit.org/news/eye-on-earth-summary-of-outcomes/

Firehock, K., & West, J. (1995). A brief history of volunteer biological water monitoring using macroinvertebrates. *Journal of the North American Benthological Society*, *14*(1), 197–202. doi:10.2307/1467734

Gellman, R. (2015). *Crowdsourcing, Citizen Science, and the Law: Legal Issues Affecting Federal Agencies*. Washington, DC: Woodrow Wilson International Center for Scholars.

Great Barrier Reef Marine Park Authority. (n.d.). *Eye on the Reef program*. Retrieved March 6, 2016, from http://www.gbrmpa.gov.au/managing-the-reef/how-the-reefs-managed/eye-on-the-reef

Haklay, M. (2015). *Citizen Science and Policy: A European Perspective*. Washington, DC: Woodrow Wilson International Center for Scholars.

Heimlich, J. E. (2012). *Public participation in scientific research: 2012 conference evaluation*. Columbus, OH: Ohio State University Extension.

Hepburn, L., Tegart, P., Roetman, P., von Gavel, S., Niedra, S., & Roger, E. … Lambkin, C. (2015). *The Australian Guide to Running a BioBlitz*. Retrieved January 10, 2016, from http://csna.gaiaresources.com.au/wordpress/wp-content/uploads/2015/07/BIOBLITZ_Guidelines_WEB+-ACSA-logo.pdf

Holdren, J. P. (2013). *Memorandum for the heads of executive departments and agencies: Increasing access to the results of federally funded scientific research*. Washington, DC: Office of Science and Technology Policy.

Hopkins, A. D. (1918). Periodical events and natural law as guides to agricultural research and practice. *U.S. Department of Agriculture, Weather Bureau, Monthly weather review, Suppl. 9*.

Jahn, T., Bergmann, M., & Keil, F. (2012). Transdisciplinarity: Between mainstreaming and marginalization. *Ecological Economics, 79*, 1–10. doi:10.1016/j.ecolecon.2012.04.017

Joint Research Center. (2016). *Citizen Science: data and service infrastructure meeting: JRC-Ispra, 26th-27th January 2016* [Meeting Report]. Retrieved May 16, 2016, from https://ec.europa.eu/jrc/sites/default/files/Citizen_science_27022016_next_steps.pdf

Kullenberg, C., & Kasperowski, D. (2016). What is citizen science? A scientometric meta-analysis. *PLoS ONE, 11*(1), e0147152. PubMeddoi:10.1371/journal.pone.0147152

Mahr, D. (2014). *Citizen Science: Partizipative Wissenschaft im späten 19. und frühen 20. Jahrhundert*. Baden-Baden, Germany: Nomos; doi:10.5771/9783845253732

Miller-Rushing, A., Primack, R., & Bonney, R. (2012). The history of public participation in ecological research. *Frontiers in Ecology and the Environment, 10*(6), 285–290. doi:10.1890/110278

Nascimento, S., Guimarães Pereira, Â., & Ghezzi, A. (2014). *From citizen science to do it yourself science: An annotated account of an on-going movement (Science and Policy Report)*. Joint Research Centre of the European Commission.

Nerbonne, J. F., & Nelson, K. C. (2008). Volunteer macroinvertebrate monitoring: Tensions among group goals, data quality, and outcomes. *Environmental Management, 42*(3), 470–479. PubMeddoi:10.1007/s00267-008-9103-9

New South Wales Department of Primary Industries. (n.d.). *NSW DPI Game Fish Tagging Program*. Retrieved March 6, 2016, from http://www.dpi.nsw.gov.au/fisheries/recreational/saltwater/gamefish-tagging

New South Wales Office of Environment and Heritage. (2015a). *NSW Office of Environment and Heritage (OEH) Citizen Science Position Statement*. Retrieved May 18, 2016, from http://www.environment.nsw.gov.au/research/citizenscience.htm

New South Wales Office of Environment and Heritage. (2015b). *OEH Citizen Science Strategy 2016–18. Driving a new era of public participation in science to support OEH decision-making*. Retrieved May 18, 2016, from http://www.environment.nsw.gov.au/research/citizenscience.htm

Newman, G. (2014). Citizen CyberScience: New directions and opportunities for human computation[Editorial]. *Human Computation, 1*(2), 103–109. doi:10.15346/hc.v1i2.2

Nielsen, K. A., & Nielsen, B. S. (2006). Methodologies in action research: Action research and critical theory. In K. A. Nielsen & L. Svensson (Eds.), *Action research and interactive research: Beyond practice and theory* (pp. 63–87). Maastricht, The Netherlands: Shaker Publishing.

Office of the Chief Scientist. (2013). *Science, technology, engineering and mathematics in the national interest: A strategic approach*. Canberra, Australia: Australian Government.

Pecl, G. T., Gillies, C., Sbrocchi, C., & Roetman, P. (2015). Building Australia through citizen science. *Australian Government Office of the Chief Scientist: Occasional Paper Series, 11*.

Pettibone, L. (2015). *Constructing "citizens" in the citizen science literature: Who are they and what do they do?* Paper presented at Interpretive Policy Analysis Conference, Lille, France.

Pettibone, L., Ziegler, D., Bonn, A., & Vohland, K. (Eds.). (2015). *GEWISS Dialogforum: Forschungsförderung für Citizen Science*. GEWISS Bericht Nr. 7. Deutsches Zentrum für Integrative Biodiversitätsforschung (iDiv) Halle-Jena-Leipzig, Helmholtz-Zentrum für Umweltforschung – UFZ, Leipzig; Berlin-Brandenburgisches Institut für Biodiversitätsforschung (BBIB), Museum für Naturkunde, Leibniz-Institut für Evolutions- und Biodiversitätsforschung – MfN, Berlin in Kooperation mit der Leopoldina – Nationale Akademie der Wissenschaften. Retrieved January 10, 2016, from http://www.buergerschaffenwissen.de/sites/default/files/assets/dokumente/gewiss_7_foerderbericht.pdf

Pocock, M. J. O., Chapman, D. S., Sheppard, L. J., & Roy, H. E. (2014). *Choosing and using citizen science: A guide to when and how to use citizen science to monitor biodiversity and the environment*. Centre for Ecology & Hydrology. Retrieved January 10, 2016, from http://www.ceh.ac.uk/citizen-science-best-practice-guide

Pocock, M. J. O., Roy, H. E., Preston, C. D., & Roy, D. B. (2015). The Biological Records Centre: A pioneer of citizen science. *Biological Journal of the Linnean Society. Linnean Society of London, 115*(3), 475–493. doi:10.1111/bij.12548

Robinson, L. (2014): *Presentation at ECSA General Assembly April 2014*. Retrieved January 10, 2016, from http://ecsa.biodiv.naturkundemuseum-berlin.de/sites/ecsa.biodiv.naturkundemuseum-berlin.de/files/ECSA-GA-2014-04-14-Robinson.pdf

Robinson, L. D., Tweddle, J., Postles, M. C., West, S. E., & Sewell, J. (2013). *Guide to Running a BioBlitz 2.0*. Retrieved May 30, 2016, from http://www.nhm.ac.uk/content/dam/nhmwww/take-part/Citizenscience/bioblitz-guide.pdf

Roger, E., & Klistorner, S. (2016). BioBlitzes help science communicators engage local communities in environmental research. *Journal of Science Communication, 15*(3), 1–18.

Samy, G., Chavan, V., Ariño, A. H., Otegui, J. O., Hobern, D., Sood, R., & Robles, E. (2013). Content assessment of the primary biodiversity data published through GBIF network: Status, challenges and potentials. *Biodiversity Informatics, 8*(2), 1546–9735. doi:10.17161/bi.v8i2.4124

Science Communication Unit, University of the West of England, Bristol. (2013). *Science for Environment Policy In-depth Report: Environmental Citizen Science*. Report produced for the European Commission DG Environment. Retrieved January 10, 2016, from http://ec.europa.eu/environment/integration/research/newsalert/pdf/IR9_en.pdf

Serrano Sanz, F., Holocher-Ertl, T., Kieslinger, B., Sanz Garcia, F., & Silva, C. (2014). *White Paper on Citizen Science for Europe*. Socientize Consortium. Retrieved January 10, 2016, from http://www.socientize.eu/?q=eu/content/download-socientize-white-paper

Shirk, J. L., Ballard, H. L., Wilderman, C. C., Phillips, T., Wiggins, A., Jordan, R., & Bonney, R. et al. (2012). Public participation in scientific research: A framework for deliberate design. *Ecology and Society*, *17*(2), 29. doi:10.5751/ES-04705-170229

Sieber, R., & Haklay, M. (2015). The epistemology(s) of volunteered geographic information: A critique. *Geography and Environment*, 2(2), 122–136. doi:10.1002/geo2.10

Silvertown, J. (2009). A new dawn for citizen science. *Trends in Ecology & Evolution, 24*(9), 467–471. PubMeddoi:10.1016/j.tree.2009.03.017

Socientize. (n.d.). *Socientize*. Retrieved March 6, 2016, from http://www.socientize.eu/?q=eu/content/socientize-0

Stack, B., & Donkin, M. (2013). Teaching botanical identification to adults: Experiences of the UK participatory science project Open Air Laboratories. *Journal of Biological Education*, *47*(2), 104–110. doi:10.1080/00219266.2013.764341

Storksdieck, M., Shirk, J. L., Cappadonna, J. L., Domroese, M., Göbel, C., Haklay, M., & Vohland, K. (2016). Associations for citizen science: Regional knowledge; global collaboration. *Citizen Science: Theory and Practice*, *2*(1), 1.

Sullivan, B. L., Aycrigg, J. L., Barry, J. H., Bonney, R. E., Bruns, N., Cooper, C. B., & Kelling, S. et al. (2014). The eBird enterprise: An integrated approach to development and application of citizen science. *Biological Conservation*, *169*, 31–40. doi:10.1016/j.biocon.2013.11.003

Thompson, S. (2010, September 24). *Citizen Science Toolkit Project Evaluation*. Retrieved March 6, 2016, from http://www.informalscience.org/citizen-science-toolkit-project

Tinati, R., Van Kleek, M., Simperl, E., Luczak-Rösch, M., Simpson, R., & Shadbolt, N. (2015). *Designing for citizen data analysis: a cross-sectional case study of a multi-domain citizen science platform*. Paper presented at the Association for Computing Machinery's Human Computer Interaction Conference on Human Factors in Computing Systems, Seoul, South Korea. doi:10.1145/2702123.2702420

Walter-Herrmann, J., & Büching, C. (Eds.). (2013). *FabLab: Of machines, makers and inventors*. Bielefeld, Germany: Transcript; doi:10.14361/transcript.9783839423820

Williams, A. C., Wallin, J. F., Yu, H., Perale, M., Carroll, H. D., Lamblin, A., & Brusuelas, J. H. (2014). A Computational Pipeline for Crowdsourced Transcriptions of Ancient Greek Papyrus Fragments. In *IEEE International Conference on Big Data*. doi:10.1109/BigData.2014.7004460

Wilson Center. (n.d.). *The Commons Lab Inventory*. Retrieved March 6, 2016, from http://wilsoncenter.org/the-commons-lab-inventory

Wright, M. T., Gardner, B., Roche, B., von Unger, H., & Ainlay, C. (2010). Building an international collaboration on participatory health research. *Progress in Community Health Partnerships – Research. Education, and Action*, *4*(1), 31–36.

Zhang, J., & Huang, X. (2015). Environment: China needs more monitoring apps. *Nature, 520*(7548), 436. PubMeddoi:10.1038/520436d

Zhang, J., Chen, S., Chen, B., Du, Y., Huang, X., Pan, X., & Zhang, Q. (2013). Citizen science: Integrating scientific research, ecological conservation and public participation. *Biodiversity Science, 21*(6), 738–749.

KEY TERMS AND DEFINITIONS

Citizen Science Association: An incorporated organization with the aim of supporting citizen science in any given geography. The term is usually used to refer to an umbrella organization that is not limited to a specific methodology, e.g. participatory computing, or discipline, such as environmental sciences, but promoting the full spectrum of citizen science activities.

Citizen Science: Citizen science describes the involvement of ordinary citizens in scientific research processes. Although currently the majority of participants contribute to data collection in the environmental area, approaches to integrate citizens in the whole research process from developing research questions and methodologies up to interpreting and communicating results improves. In addition, research projects initiated by citizens and performed with or without institutional backing are also described as citizen science.

Community Monitoring: Community monitoring describes projects mainly in the field of environmental research, which are initiated by local communities and normally have also a political intention, such as reducing environmental pollution and/or conserving biodiversity.

Crowdsourcing: Crowdsourcing means that the "crowd", i.e. a high number of persons that can be anonymous, contribute data, objects, pattern recognition capacities or anything else to a task or project, such as solving a scientific question. For crowdsourcing approaches data quality is mainly assured via statistics of high numbers.

Do-it-Yourself (DIY) Science: This terminology is rooted in the hacker and maker community and has a bias to hands-on applied research, such as building sensors or developing new means of transport. A specific case is do-it-yourself biology, also called biohacking or garage biology, which more specifically addresses (molecular) biologists performing experiments outside institutional laboratories.

Participatory Action Research: One form of participatory research following the idea that researchers and stakeholders collaboratively develop and carry out a research project from the start to the end, for example in public health. Regarding the research design this approach shows similarities to transdisciplinary research or co-design, but the scientific communities which carry out such forms of participatory research differ: participatory action research is a term developed in social sciences, while co-design or citizen science represent terminologies used in the natural sciences. In addition, methodological and epistemological characteristics of these approaches differ.

Participatory Research: This term may be seen as meta-category for research conducted with the participation of members of the public. It is used mainly in the scientific meta-discourse about the impacts and preconditions of citizen science. The concept of participation increasingly becomes a topic of research, from reflections about its rhetoric potential up to the implicit power relations.

Practitioner: A person that is mainly involved in the organization of citizen science projects, usually as project and/or volunteer manager. As opposed to volunteers or participants, the term practitioner is used to highlight a coordinating role in designing project methodologies, managing as well as representing citizen science initiatives.

ENDNOTES

1 A recent example is the Eye on Earth Summit 2015 in Abu Dhabi, which recognized the importance of citizen science data as supporting information for environmental decision making in general and for reporting against the Sustainable Development Goals in particular (Eye on Earth, 2015).

2 For the European Union, for example, citizen science is seen as a science communication or public engagement measure both of which are covered by the „Science with and for Society" Programme in Horizon 2020, the EU Framework Programme for Research and Innovation, which receives 0,6% of the total budget (European Commission, 2013b).

3 In Germany, for example, there are debates on the hegemony of the expression citizen science. Some projects, such as the butterfly monitoring of the Helmholtz Center for Environmental Research (UFZ), or the bird monitoring by the umbrella organization of German ornithologists (DDA e.V.), consider themselves as citizen science projects. Others, such as the loss of night project of the Leibniz Institute of Freshwater Ecology and Inland Fisheries (IGB) or the mosquito atlas of Leibniz Center for Agricultural Landscape Research (ZALF), are also seen as crowd sourcing. Still other groups that usually operate independently from research institutions, such as the DIYbio community, hackers or makers, would rather not assign themselves to being part of the citizen science community. Moreover, in other disciplinary contexts different terminologies for comparable approaches may have already been established. For instance in the social sciences, participatory action research, transdisciplinary research, and so-called Mode 2 research or co-production and co-design are other expressions for participatory research formats.

Chapter 3
SciStarter 2.0:
A Digital Platform to Foster and Study Sustained Engagement in Citizen Science

Catherine Hoffman
SciStarter, USA

Eric B. Kennedy
Arizona State University, USA

Caren B. Cooper
North Carolina Museum of Natural Sciences, USA & North Carolina State University, USA

Mahmud Farooque
Arizona State University, USA

Darlene Cavalier
Arizona State University, USA

ABSTRACT

In this chapter, the authors focus on how SciStarter has developed a new digital infrastructure to support sustained engagement in citizen science, and research into the behaviors and motivations of participants. The new digital infrastructure of SciStarter includes integrated registration and contribution tracking tools to make it easier to participate in multiple projects, enhanced GIS information to promote locally relevant projects, an online personal dashboard to keep track of contributions, and the use of these tools (integrated registration, GIS, dashboard) by project owners and researchers to better understand and respond to the needs and interests of citizen-science participants. In this chapter, the authors explore how these new tools build pathways to participatory policymaking, expand access to informal STEM experiences, and lower barriers to citizen science. The chapter concludes with a design for a citizen-science future with increased access to tools, trackable participation, and integrated competencies.

INTRODUCTION

Individuals come to citizen science from different perspectives and preferences, and engage in a wide range of projects from data collection to public policy (Irwin, 1995; Bonney et al. 2009). Many participate exclusively online in crowdsourcing projects such as Galaxy Zoo, Fold-It, and Eyewire. Hundreds

DOI: 10.4018/978-1-5225-0962-2.ch003

of thousands gather data on-the-ground for academic and community research projects led by scientists at universities, government agencies/NGOs, and nonprofits. Untold others initiate their own projects in response to local environmental and health issues. All types of citizen science endeavors face challenges around data access and management, around access to tools, around relationships with scientists and project managers, and around making their scientific accomplishments visible as they accumulate over time. Among the most universal and ongoing challenges for all types of citizen science according to Crowston & Prestopnik (2013), however, "is attracting and retaining enough participants to make achievement of project goals possible" (p. 3). One known reason for this issue is the reality that individual projects and project types exist in "silos" –as self-segregated groups with particular shared interests (Dickinson, Zuckerberg, & Botner, 2010). SciStarter, an online citizen science "hotspot" and a research entity of Arizona State University (ASU), is building an innovative new platform to address these prolific issues in the field of citizen science.

As a research entity of ASU, SciStarter is helping researchers at the university's Center for Engagement & Training in Science & Society (CENTSS) address fundamental questions about citizen scientists and their motivations. The Center is structured to develop and implement new modes of engaging audiences in conversations about how science and society interact, while maintaining a robust research platform about these interactions.

SciStarter features more than 1,600 searchable citizen science projects and events from across the globe, added by researchers and project owners, serving an engaged community of more than 50,000 citizen scientists. SciStarter selects projects from its database to promote on its site, and to share with Discover Magazine, Astronomy Magazine, Philadelphia Inquirer, NPR, PBS, the United Nations, the National Science Teachers Association and others through open APIs that allow the database of projects to be shared with other sites. SciStarter also brings these projects to life through its syndicated blog network on the Public Library of Science and DiscoverMagazine.com, as well as in the pages of Discover magazine each month.

While SciStarter has demonstrated success in promoting projects and attracting potential participants for those projects, its development team can attest to the challenge of attracting and retaining participants over time. The next iteration of the site, known as SciStarter 2.0, supports the theory and practice of citizen science by:

1. Helping expand, deepen, and sustain public engagement in science and
2. Serving as an aggregating platform to enable research on the motivations and behaviors of participants across the enormous range of variation in citizen science experiences.

BACKGROUND

The design of SciStarter 2.0 is based on the premise that ownership of information and resources and the ability to organize and display contributions, combined with greater access to projects and opportunities to connect socially with fellow citizen scientists and professional researchers, can lead to the deep and sustained engagement in citizen science. Research will be done to understand which platform features motivate increased engagement, either deeper involvement in individual activities or broader participation in many activities, and what the differential impact of those features are across demographic categories, such as age, education, income, location, and project styles (contributory, community-based, etc.). SciStarter 2.0 will also enable researchers to explore and evaluate how varying types of scientific contextual content, such as humanizing profiles of scientists, historical narratives on research topics, and direct interaction with scientists impact interest, engagement, motivation, and behavior in terms of participation in science activities. Of the digital citizen science platforms in the field, SciStarter is the most suited to answer these crucial questions because of its large and growing community. The backbone of SciStarter is the community of 50,000+ citizen scientists that are ready to find new ways to engage in citizen science.

In 2013, the Sloan Foundation supported SciStarter's exploration of cloud-based mobile and web platforms for citizen science, which resulted in a widely used report (Cavalier et al. 2014) and in the development of wireframes for user profiles and dashboards. Also with support from Sloan, in February 2014, SciStarter organized a workshop in London at the Citizen CyberScience Summit where a design process was prototyped and participants were categorized and storylined into eight "personas" based on varying engagement level and motivation, from "dabblers" to "educators" to the most dedicated "super-users" involved in many projects. These personas are key to understanding how different people participate in citizen science projects and how different digital tools enhance their participation across projects and platforms. Additionally, the Knight Foundation supported a prototype tool that connected people to projects based on their location. This proof-of-concept tool provided the basis for the geographic matching tools implemented in SciStarter 2.0.

To further inform the design of SciStarter 2.0, a survey of more than 200 participants in the SciStarter community provided information to inform a baseline understanding of their engagement in citizen science. The results of the survey broadly supported the theory that participant engagement will be enhanced by integrated registration and a dashboard with associated portfolios, data access, project suggestion, and social engagement:

- 50% of respondents were active in more than one citizen science project
- 74% of those who participate in multiple projects are involved in projects on different topics
- 87% of respondents said they would use a dashboard on the site.
- 64% of respondents store their data independently as well as share it with a citizen science projects
- 57% of respondents don't know how to access their data once they submit it to a citizen science
- Project, and 84% of respondents would like to know how

These reports and surveys informed the design of SciStarter 2.0. The new platform includes:

1. An integrated registration and data transfer system for participants to more easily engage in one or multiple citizen science projects, across platforms
2. GIS implementation for project owners to define the geographic boundaries of projects so people can find them more easily
3. An online, personal dashboard for participants to track their projects, participation, and contributions to science, record interests in projects, create profiles, and find people and projects of interest to them
4. Use of these tools (integrated registration, GIS, dashboard) by project owners and researchers to better understand and respond to the needs and interests of citizen science participants.

SciStarter 2.0, with its contribution tracking and related programming, will enable the unprecedented ability to implement and study online practices that support and retain diverse types of citizen science participation. It is a smart collection of web components, including a dashboard and contribution tracking, designed to extend, enhance, and enrich participant experiences while at the same time supporting STEM research and enabling research on motivations and learning outcomes of participants. SciStarter currently includes blogs, newsletters, emails, project searching, project sharing, project information and promotional partnerships. Added to the existing system will be components that support participation in, and management of, multiple projects; continuity in, and sharing of, a broad and accessible citizen science community; and social components. Community and stakeholder research will ensure the development of a smoothly functioning system that will benefit thousands of citizen science projects, and hundreds of thousands of participants. In turn, this will allow better definition of the participant community to learn who is doing what and where, as well as what data, experiences, and interests they have. Previously, this has not been studied efficiently across research disciplines in a manner that enables researchers and practitioners to understand, and effectively respond to, the needs of their communities.

There is currently a broad base of work on what motivates a person to contribute to a single citizen science project, but little work on how a rewarding experience with one project motivates involvement in more citizen science projects and science experiences. By bringing the 1600+ citizen science projects on SciStarter together in one place, participants' efforts will no longer be siloed among the different projects, but can be examined together. SciStarter is the best place to answer questions about how participation in one project motivates participation in more or different projects.

Building Pathways to Participatory Policymaking and Expanding Access to Informal STEM Experiences

In its strongest form, the citizen science movement (alongside related efforts, like the open movement and other forms of public engagement) has the potential to radically transform how society makes decisions. From the particulars of how a municipality manages a park to high level decisions about where federal investments should be made, experiences in citizen science make clear that the public has a much more powerful role to play. This influence can occur in a variety of forms, including providing new and richer data, speaking on behalf of affected people and communities, and challenging traditional norms about who ought to be invited to the decision making table (Kennedy, 2016).

This optimistic vision, however – of citizens actively engaged in making decisions about science, technology, and society – requires a citizenry that has access to data, tools, peers, and expertise. It isn't enough to simply offer a pathway by which to access decision makers. Formal education has a role to play, but learning must be larger than the classroom. Informal science experiences are fundamental to the current state of science literacy in the U.S. (Lundh, Stanford, & Shear, 2014). A growing body of evidence demonstrates that Americans spend less than 5% of their lives in classrooms and learn most science outside of school, especially on the Internet (Falk & Dierking, 2010). In the words of Falk & Dierking (2010):

But is better schooling really the solution? ...the most important sources of scientific knowledge are not schools; [and] the informal infrastructure of museums, aquariums, broadcast programming and other sources of science exposure... is a far more potent source of public understanding of science than has been previously acknowledged. (p. 486)

To advance citizen science, therefore, requires developing a robust set of pathways through which citizens can think constructively and reflectively about the impact of science on their life and the lives of those around them; and feel empowered to contribute to these conversations and debates in productive ways. With respect to the first, SciStarter 2.0 can link interested members of the public with opportunities to learn and contribute more through direct engagement in citizen science. It also affords a venue wherein project participants can work through both practical challenges (e.g., ownership of information and access to resources, among others) and these more hefty reflections (including evaluating their impact, weighing the aims and objectives of varying projects, and contrasting their experiences during and after). More pragmatically, SciStarter 2.0 also expands access to STEM-related citizen science projects in the areas of biology, environmental science, astronomy, meteorology, ecology, and microbiology, making it easier to link these existing citizen science projects with informal learning opportunities.

Broadening Participation in STEM through Citizen Science

SciStarter begins to identify the elements required to engage citizen scientists in new or multiple projects, and to feel empowered in the process of citizen science. SciStarter 2.0 sets the stage for greater inclusion of previously marginalized groups in citizen science activities and will extend to all forms of public engagement in science in future iterations. To date, research suggests that participation in citizen science, while inspiring participant learning (Jordan, Gray, Howe, Brooks & Ehrenfeld, 2011; Price & Lee, 2013) and value to researchers (Cooper, Shirk, & Zuckerberg, 2014), is limited both relative to participant diversity (Pandya, 2012) and intensity and duration of engagement (Chu, Leonard, & Stevenson, 2012). The SciStarter 2.0 system addresses barriers to participation identified by Pandya (2012), Chu et al. (2012), and Rotman et al. (2014), including issues of local relevance (by adding much-needed GIS data), community bonds, and diverse channels of communication. Motivations of participants change over time, initially based strongly on personal interest and, for sustained participation, based on external factors such as attribution, acknowledgement, and relationships (Rotman et al. 2012; 2014). Pandya (2012) suggested that effective programs extend participant engagement beyond data collection to other aspects of the scientific process. SciStarter 2.0 will have the tools and standards to make it possible to begin to explore the relationships between inclusion, participation, and outcomes. A small, but growing body of research addresses participant motivation in relation to the sustainability of online citizen science projects (data analysis rather than data acquisition and sharing). SciStarter 2.0 will explore the importance of collective motives, norm-oriented motives, reputation, and intrinsic motives (Nov, Arazy, Lotts, & Naberhaus, 2013; Nov, Arazy, & Anderson, 2014), and will examine their efficacy when applied to a set of on-the-ground citizen science projects, unified in the technology-mediated SciStarter community. Studies of motivations of participation in on-the-ground citizen science projects tend to be focused on single projects, with a few exceptions, and they also find a variety of motivations, including motives of altruism, achievement, social, and esteem-building (Jacobson, Carlton, & Monroe, 2012; Nov, Arazy, & Anderson, 2011; Ryan, Kaplan, & Grese, 2001).

Building Pathways between Science, Citizenship and Decision-Making

Citizen science can create alternative co-evolutionary pathways from civics to science and from science to civics. For the civics to science pathway, SciStarter 2.0 can provide an on-ramp for lay citizens who participate in citizen panels, public forums and similar deliberative processes to move from participatory problem identification to participatory problem solving. Most specifically, SciStarter can take the output from participatory Technology Assessment (pTA), a process of engagement that seeks to improve the outcomes of science and technology decision-making through informed dialog with lay citizens (Sclove, 2010). The process specifically targets non-expert representative of the general population who—unlike political, academic, industry and organized interests—are routinely underrepresented in technology related policymaking. It has three steps:

1. Issue selection and problem framing,
2. Peer to peer deliberation,
3. Reporting and dissemination.

These processes identify data gaps (in addition to policy and participation gaps), which can be used by SciStarter to design, develop and seed, use-inspired citizen science connected with community priorities and concerns in a variety of areas from environmental hazards to public health concerns to emerging technologies. The pathway need not end there. In fact, this special class of community priority inspired citizen science can actually help frame the problems and issues taken up in a pTA exercise. Once set in motion, the iterative pTA to SciStarter and SciStarter to pTA process (Figure 1) could generate a broad range of benefits for the participants and their community. It would increase and deepen scientific and civic literacy, broaden citizen science participation in disadvantaged and marginalized communities and forge beneficial partnerships between citizen and the public, private and nonprofit entities designed to serve them.

Figure 1. SciStarter and possible pathway from civics to science and science to civics

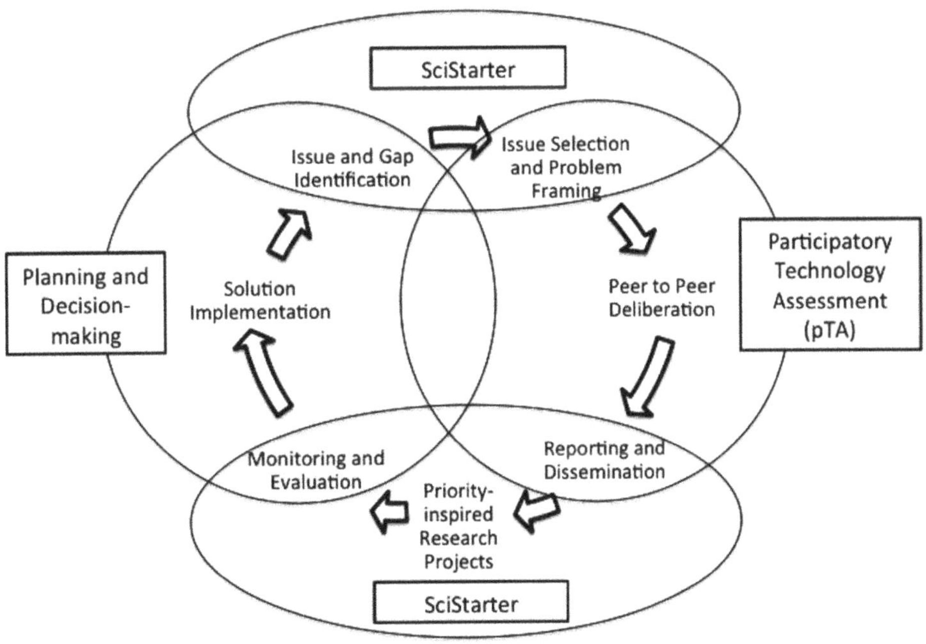

The pathways between science and civics are possible with SciStarter as it serves as a place for these communities to work together. By identifying the 1600+ projects and events in the database by location, people can find projects with local relevance that they are invested in to see through to the end goals. In their personal dashboard, citizen scientists can track their contributions to projects and to other participatory decision-making, building a full profile of experiences. As discussed later, the breadth of experience tracked in the dashboard can soon be used to develop contribution rewards.

Flipping the World of Citizen Science

The current structure of science has highly centralized control that makes it an elite, closed system of knowledge production in which scientists are authorities in making new knowledge which is produced in academia, industry, and government, and it is influenced by commercial and academic forces. A reformed structure has decentralized control that makes it an egalitarian, inclusive system that involves ordinary people in the process of knowledge production governed in ways that serve the public good via influence of public interests. In short, reforming science would mean the adoption of three pillars of public science by the scientific enterprise as a whole: open science practices, dialogue-based science communication, and participatory methods of citizen science.

SciStarter 2.0 develops a system for volunteer contributors that empowers them to participate more broadly, more deeply, and more meaningfully in science activities, while simultaneously enabling project owners to better study and manage (recruit, retain, and enhance experiences of) their contributors. Tools that give contributors control and power over their participation are transformational not only to the field of citizen science, but to public engagement in science overall (Overdevest & Mayer, 2008; Ottinger, 2010; Cooper, 2012). Technologies that can be used to address ethical and privacy challenges related to the sharing of location-based information, confidentiality, data ownership, and data submissions that cross legal jurisdictions are also needed for citizen science (Scassa & Sattler, 2011; Quigley & Roy, 2012; Bowser, Wiggins, Shanley, Preece, & Henderson, 2014). Tools that increase the capacity of project owners is predicated on the observation that projects are often coordinated by teams that include people with expertise in multiple fields from informal science education to informatics, whether from contributory-style projects or community-based efforts. It is difficult for any single project to have full capacity in all the disciplines that provide support for practice of citizen science. SciStarter 2.0 can help meet the communication and engagement needs of projects, thereby freeing resources for projects, big and small, to build capacity and sustainability in other areas. SciStarter 2.0 will provide technical components that research projects require for their "system assemblage," as described by Prestopnik and Crowston (2012), such as registration, sign-in, and collecting participant information. Researchers have only begun to explore the possibilities of citizen science as a tool for STEM engagement and learning. The nature of SciStarter 2.0 is such that, once the standards are developed, tested, and refined with partner projects, it will be possible to quickly expand a final product to a very wide range of citizen science participants and project owners and create a large and vibrant informal science learning community.

As a whole, these advances allow the pursuit of research questions about citizen scientists, questions about the subgroup of practitioners who are project owners and administrators, and questions about the role of this digital infrastructure as an intervention in the relationships among participants and project owners and how these components decentralize scientific knowledge.

Designing a Future with Low-Barriers, Trackable Participation, and Integrated Competencies

SciStarter 2.0 provides the tools, services, and research platform to enhance the citizen science community of participants, project owners, and researchers. The project vision includes the following goals:

1. Allow project owners to use the participant tools (sign-on, dashboard, GIS, and, soon the ability to loan or sell instruments through SciStarter) for the purposes of learning about if and how their current contributors move between projects, interact with citizen scientists in other projects, use the new online tools, access/use instruments, and strategically reaching out to potential contributors among the broad SciStarter community with related interests.
2. Improve researcher access to citizen science data by disseminating data from participants to researchers with related topics of interest.
3. Encourage growth of citizen science participation through software driven outreach (recommending projects with similar goals, in local areas, recommending appropriate instruments, etc).
4. Enhance citizen scientists' self-identification as stakeholders in the research process by making their research data available and shareable.
5. Evaluate how recognition and feedback, such as "information about where, how and to what extent the data [was] used," as recommended by Rotman et al. (2014), influence participants' behavior.
6. Provide contributors with (optional) information about others who have similar interests and/or are geographically nearby, thereby providing ongoing opportunities for the informal science community and the research community to reach out to, engage with, involve, and teach contributors.
7. Offer teachers turnkey tools for involving students in appropriate citizen science topics (and thus provide informal educators with aggregate info about students in the citizen science world).
8. Enable collaborative relationships among and between project owners and citizen science portals and providers.

Lowering Barriers to Participation

There are still significant logistical barriers to participating in citizen science projects including quick access to the required tools. SciStarter has the capacity to build lending libraries of tools by re-purposing the database as a searchable inventory of tools. Since many project owners, organizations, and agencies cannot recommend or sell products, SciStarter is in the unique role to provide this service to the broader community. The forthcoming "Build, Borrow, and Buy" system will make it easy for participants to get involved in projects right away with the necessary materials at their fingertips in the SciStarter system. Links directly on project pages will show which materials are needed and where to buy them, including from the SciStarter store.

The focus on the "Build" aspect integrates citizen science more deeply into the maker, hacker, and DIY science community, opening access to not only data collection, but tool development. SciStarter 2.0 has the capacity to host product ratings and reviews from participants that can be used as the Maker/Hacker/DIY community builds and refines tools. To further this effort, SciStarter is helping to co-organize the Makers Meet Citizen Science symposium at Arizona State University. This event and related projects will both advance the experience for citizen scientists and provide a necessary service, through ratings and reviews, to the "Build" community.

The "Borrow" aspect will be piloted with a lending library at the NC Museum of Natural Sciences, including with partner projects eMammal and Sparrow Swap. SciStarter is looking at software to provide the inventory tracking service. This service will be piloted through equipment loans for NC Candid Critters (a project of eMammal), Cat Tracker, Sparrow Swap, and project launching in fall with National Parks Service to record soundscapes, and will ideally be scaled up to a national level over time.

SciStarter is investigating partners to manage the retail sales of citizen science equipment so new participants can "outfit a project" in just a few clicks. These experimental activities, while incremental and timely, represent an important step forward to build sustainable revenue streams that will move SciStarter away from a dependency on grants.

By making citizen tools easier to build, borrow, and buy, one of the major barriers to participating in citizen science is significantly lowered. Furthermore, the contribution tracking system will allow us to measure the effect of accessible tools on citizen science participation across the broad spectrum of SciStarter projects.

Rewarding Contributions with SciStarter's Digital Infrastructure

As the community of citizen scientist builds and the skills that they develop grows, it is important to recognize their accomplishments. In a SciStarter online poll, 60% of the community says they completed college in a STEM major while 30% completed college but in a non-STEM major. Therefore, building acknowledgement for the skills and competencies developed in citizen science project must be beneficial for these two groups. First, participants that already have content knowledge in STEM can gain recognition for the new information and skills they acquire, possibly leading to advances in their own career. Second, participants without STEM backgrounds can build a reputation for their diverse knowledge and experiences outside of their current career. For example, competencies for citizen science projects could lead to credit in continuing education courses at community college, badges on career profiles like LinkedIn, or simply extrinsic motivations like free rewards from local businesses. SciStarter will work with experts to identify and map competencies to experiences and ultimately to projects and dashboards on SciStarter.

CONCLUSION

The SciStarter digital infrastructure radically changes the way people participate in citizen science and how researchers study those contributions. The platform allows participants to deepen and enhance their experience in citizen science while allowing researchers to track entry into citizen, movement among citizen science projects, and the spillover effects into participatory policy-making and additional STEM experiences. As the community of 50,000+ citizen scientists on SciStarter continues to grow in tandem with the 1600+ projects and events on the site, the opportunities to study public participation in scientific research will expand.

REFERENCES

Bonney, R., Ballard, H., Jordan, R., McCallie, E., Phillips, T., Shirk, J., & Wilderman, C. C. (2009). Public participation in scientific research: Defining the field and assessing its potential for informal science education. A CAISE Inquiry Group Report. Washington, DC: Center for Advancement of Informal Science Education (CAISE).

Bowser, A., Wiggins, A., Shanley, L., Preece, J., & Henderson, S. (2014). Sharing data while protecting privacy in citizen science. *Interaction, 21*(1), 70–73. doi:10.1145/2540032

Cavalier, D., Cheetham, R., Emanuele, R., Frankl, J., Johnson, M., Manik-Perlman, R.,... Redeschi, M. (2014). *Citizen Science Data Factory*. Retrieved from http://scistarter.com/research

Chu, M., Leonard, P., & Stevenson, F. (2012). Growing the base for citizen science: Recruiting and Engaging Participants. In J. L. Dickinson & R. Bonney (Eds.), *Citizen Science: Public Participation in Environmental Research* (pp. 69–81). Ithaca, NY: Cornell University Press. doi:10.7591/cornell/9780801449116.003.0005

Cooper, C. B. (2012). Links and Distinctions Among Citizenship, Science, and Citizen Science. A Response to "The Future of Citizen Science". *Democracy & Education, 20*(2), 13.

Cooper, C. B., Shirk, J., & Zuckerberg, B. (2014). The Invisible Prevalence of Citizen Science in Global Research: Migratory Birds and Climate Change. *PLoS ONE, 9*(9), e106508. doi:10.1371/journal.pone.0106508 PMID:25184755

Crowston, K., & Prestopnik, N. R. (2013). Motivation and data quality in a citizen science game: A design science evaluation. In *Forty-sixth Hawai'i International Conference on System Sciences (HICSS-46)*. doi:10.1109/HICSS.2013.413

Dickinson, J. L., Zuckerberg, B., & Bonter, D. N. (2010). Citizen Science as an Ecological Research Tool: Challenges and Benefits. *Annual Review of Ecology Evolution and Systematics, 41*(1), 149–172. doi:10.1146/annurev-ecolsys-102209-144636

Falk, J. H., & Dierking, L. D. (2010). The 95 Percent Solution School is not where most Americans learn most of their science. *American Scientist, 98*(6), 486–493. doi:10.1511/2010.87.486

Irwin, A. (1995). *Citizen science: a study of people, expertise and sustainable development*. London: Routledge.

Jacobson, S. K., Carlton, J. S., & Monroe, M. C. (2012). Motivation and satisfaction of volunteers at a Florida natural resource agency. *Journal of Park and Recreation Administration, 30*(1).

Jordan, R. C., Gray, S. A., Howe, D. V., Brooks, W. R., & Ehrenfeld, J. G. (2011). Knowledge gain and behavioral change in citizen-science programs. *Conservation Biology, 25*(6), 1148–1154. doi:10.1111/j.1523-1739.2011.01745.x PMID:21967292

Kennedy, E. B. (2016). When Citizen Science Meets Science Policy. In D. Cavalier & E. B. Kennedy (Eds.), *The Rightful Place of Science: Citizen Science*. Tempe, AZ: Arizona State University.

Lundh, P., Stanford, T., & Shear, L. (2014). *Nano and Society: Case Study of a Research-to Practice Partnership between University Scientists and Museum Professionals*. Menlo Park, CA: SRI International.

Nov, O., Arazy, O., & Anderson, D. (2011). Dusting for science: motivation and participation of digital citizen science volunteers. *Proceedings of the 2011 iConference*. New York: ACM doi:10.1145/1940761.1940771

Nov, O., Arazy, O., & Anderson, D. (2014). Scientists@ Home: What drives the quantity and quality of online citizen science participation? *PLoS ONE*, *9*(4), e90375. doi:10.1371/journal.pone.0090375 PMID:24690612

Nov, O., Arazy, O., Lotts, K., & Naberhaus, T. (2013). Motivation-targeted personalized UI design: a novel approach to enhancing citizen science participation. *ECSCW 2013: Proceedings of the 13th European Conference on Computer Supported Cooperative Work*. London: Springer. doi:10.1007/978-1-4471-5346-7_15

Ottinger, G. (2010). Buckets of Resistance: Standards and the Effectiveness of Citizen Science. *Science, Technology & Human Values*, *35*(2), 244–270. doi:10.1177/0162243909337121

Overdevest, C., & Mayer, B. (2008). Harnessing the power of information through community monitoring: Insights from social science. *Texas Law Review*, *86*, 1493–1526.

Pandya, R. E. (2012). A framework for engaging diverse communities in citizen science in the US. *Frontiers in Ecology and the Environment*, *10*(6), 314–317. doi:10.1890/120007

Prestopnik, N. R., & Crowston, K. (2012). Citizen science system assemblages: understanding the technologies that support crowdsourced science. In *Proceedings of the 2012 iConference*. New York: ACM. doi:10.1145/2132176.2132198

Price, C. A., & Lee, H. S. (2013). Changes in participants scientific attitudes and epistemological beliefs during an astronomical citizen science project. *Journal of Research in Science Teaching*, *50*(7), 773–801. doi:10.1002/tea.21090

Quigley, K., & Roy, J. (2012). Cyber-Security and Risk Management in an Interoperable World An Examination of Governmental Action in North America. *Social Science Computer Review*, *30*(1), 83–94. doi:10.1177/0894439310392197

Rotman, D., Hammock, J., Preece, J., Hansen, D., Boston, C., Bowser, A., & He, Y. (2014). Motivations affecting initial and long-term participation in citizen science projects in three countries. In iConference 2014 Proceedings. London: ACM.

Rotman, D., Preece, J., Hammock, J., Procita, K., Hansen, D., Parr, C., & Jacobs, D. et al. (2012). Dynamic changes in motivation in collaborative citizen-science projects. In *Proceedings of the ACM 2012 conference on Computer Supported Cooperative Work*. London: ACM doi:10.1145/2145204.2145238

Ryan, R. L., Kaplan, R., & Grese, R. E. (2001). Predicting Volunteer Commitment in Environmental Stewardship Programmes. *Journal of Environmental Planning and Management*, *44*(5), 629–648. doi:10.1080/09640560120079948

Scassa, T., & Sattler, A. (2011). Location-based services and privacy. *Canadian Journal of Law and Technology*, *9*(1 & 2).

Sclove, R. (2010). *Reinventing technology assessment: A 21st Century model*. Washington, DC: Woodrow Wilson International Center for Scholars.

KEY TERMS AND DEFINITIONS

Competency: Skills that may be acquired through participating in citizen science projects.

Contribution Tracking: A digital system to record contributor activity in citizen science projects including data collection, data input, analysis of images/photos, etc.

Dashboard: A digital place to visualize a contributor's interests, skills, location, past participation, and contributions to themselves as well as the rest of the citizen science community.

Digital Infrastructure: The network, data, devices, and software that provides a service or product to an online community.

Integrated Registration: A digital system that makes it easier for contributors to login to different citizen science projects. This may include registering with oAuth providers like Facebook or Google or it may send information like name, email, and zip code to projects a contributor is joining.

Participatory Policymaking: A way of informing government policy by engaging citizens in the complete process of developing new policies.

Chapter 4
What Drives Citizens to Engage in ICT-Enabled Citizen Science?
Case Study of Online Amateur Weather Networks

Mohammad Gharesifard
UNESCO-IHE, The Netherlands

Uta Wehn
UNESCO-IHE, The Netherlands

ABSTRACT

In order for citizen science initiatives to pan out well, various actors need to be willing to engage in citizen science activities. The particular interest in this chapter lies with the citizens and their motivations to participate in ICT-enabled citizen science since, arguably, without citizen participation, there is no citizen science activity. The authors examine in detail what determines citizens' interest to share their weather-related data collected with Personal Weather Stations via online amateur networks and how these citizen activities could be up-scaled to address prevalent hydro-meteorological data gaps. A decision making theory is used to guide empirical research in three European countries. The results indicate no regional differences between the main drivers and incentives and raise the question whether weather observation is still a male-dominated activity in the digital age which would have implications for upscaling this citizen science initiative.

INTRODUCTION

Citizen science is being heralded as the means for overcoming many challenges: data scarcity (Muller et al., 2015), science education (Harjanne, Ervasti, Karhu, & Tuomenvirta, 2015) and citizen participation in science (Franzoni & Sauermann, 2014), in decision making and planning (Wehn, Rusca, Evers, & Lanfranchi, 2015), policy making (Haklay, 2015) and in monitoring and forecasting (Lanfranchi, Wrigley, Ireson, Ciravegna, & Wehn, 2014). Nevertheless, in order for citizen science initiatives to pan

DOI: 10.4018/978-1-5225-0962-2.ch004

out well, various actors need to be willing to engage in citizen science activities which often start out as pilot projects before they are established and institutionalized. Depending on the particular set up, distinct actors are involved, such as spatial planners and other decision makers from various local or national authorities, policy makers, scientists in academic, educational and applied professional environments, and of course citizens, often stemming from distinct interest or contributor groups. These actors are subject to (distinct) incentives and drivers. The particular interest of this research lies with the latter – the citizens – and their motivations to participate in ICT-enabled citizen science since, arguably, without citizen participation, there is no citizen science activity. Moreover, their involvement in citizen science is typically required and desired not once, but on a continuous basis.

In this chapter, the authors examine in detail a particular case: citizens' willingness to collect weather-related data using Personal Weather Stations and to share them via online amateur weather networks. The increasing availability of user-friendly and affordable weather stations (Bell, Cornford, & Bastin, 2013) as well as online weather networks for sharing the collected weather observations appears to have given new impetus to the long-established practice of amateur weather observation.

Citizen observations of the weather are particularly relevant in view of the gradual but steady decrease of ground-based hydro-meteorological observations by national water resources government agencies since the 1980s, as observed by the World Bank (García, Rodríguez, Wijnen, & Pakulski, 2016), owing to budget constraints and related lack of maintenance as well as political turmoil that leads to the destruction of equipment, prevents readings or terminates funding. The resulting gaps in real-time and long term data records cannot be filled by satellite observations alone (García et al., 2016). At the same time, long term data records are urgently needed for policy and planning purposes and real-time data for monitoring and forecasting: for two consecutive years (2015 and 2016), the World Economic Forum has ranked water crises and the failure to address Climate Change-related mitigation and adaption as among the top three threats facing the world's population (WEF, 2015, 2016).

To better understand what determines citizens' interest to participate in online amateur weather networks and how their activities could be up-scaled to address prevalent hydro-meteorological data gaps, the lens of a decision making theory is used to guide empirical research in three European countries (United Kingdom, The Netherlands, and Italy). The findings show that there are no regional differences between the main drivers and incentives for citizens to share their PWS data; they also raise the question whether weather observation is still a male-dominated activity (Endfield & Morris, 2012; Manley, 1952; Subkowski, 2006) in the digital age which has implications for upscaling this citizen science initiative.

The chapter is structured as follows. In the second section, the conceptual framework for this research is introduced, followed by the third section in which the methods for selecting relevant locations and respondents for the empirical research are presented. In the fourth section, the results of the empirical research are used to analyze what influences citizens' willingness to share personally-collected weather data and how this is manifested. In the fifth section, the findings are discussed regarding the most/least frequently mentioned drivers; regional differences and similarities; and gender. The final section concludes the chapter with recommendations for citizen science initiatives.

CONCEPTUAL FRAMEWORK

The basic principle behind citizen science initiatives is not only the observation of specific phenomena (e.g. birds, the weather, flora, fauna, etc.) but the act of sharing such observations with others. Following a brief review of the literature, Müller, Thoring, and Oostinga (2010) suggested general motives for users to participate in citizen science activities, namely money, altruism, usefulness and fun. A more thorough review and subsequent survey on participation in water quality monitoring by Minkman (2015) found that altruism and fun were strong drivers while money (financial compensation) was a weak driver and lack of time a major obstacle.

For this research, and building on the authors' earlier review of decision making theories that could be utilized to conceptualize and understand the data sharing behavior of weather amateurs (Gharesifard & Wehn, 2016), the model of data sharing (developed by Wehn de Montalvo (2003a, 2003b) was selected which is based on the Theory of Planned Behavior (Ajzen, 1985) as the framework for this study. This enabled a systematic investigation and explanation of the conditions (i.e. both drivers and obstacles) under which citizens are willing and able to share weather-related data that they collected so as to gauge whether and how this citizen science activity can be scaled up. This resulted in the following definitions of the components (Box 1).

A combination of beliefs (behavioral, normative and control beliefs) stemming from these components forms the intention or willingness to share data. In general, a combination of more positive and favorable attitudes, stronger positive social pressure and greater perceived behavioral control will lead to stronger motivations and intentions to share data. Perceived control over data sharing is stipulated to also have a direct influence on actual data sharing behavior, as illustrated in Figure 1 below, since circumstantial factors (such as a functioning Internet connection) might limit actual data sharing, even though the willingness to share may be very high. In this study, the beliefs underlying the intention to share PWS data via online platforms are investigated based on qualitative empirical research, as explained in the following section.

Box 1. Conceptual definitions

Behavior
The behavior to be examined by the Theory of Planned Behavior (TPB) can be defined by taking into account four elements: action, target, context and time (Ajzen, 1985, 1991). The data sharing behavior at the core of this research is defined as follows:
- Action: sharing PWS data via online platforms
- Target: hydro-meteorological data collected with Personal Weather Stations
- Context: online amateur weather networks
- Time frame: present (the period of undertaking this empirical research (November 2014 - January 2015).

Attitude
Expectations about the positive and negative outcomes of resulting from sharing PWS data via online networks (behavioral beliefs).

Social pressure
Comprised of the normative beliefs of others and their (dis)approval of data sharing via online amateur weather networks (normative beliefs).

Perceived control
Perceptions about the absence or presence of specific factors that impede or facilitate data sharing (Wehn de Montalvo, 2003b) (control beliefs).

Figure 1. Basic model of spatial data sharing
Wehn de Montalvo, 2003b

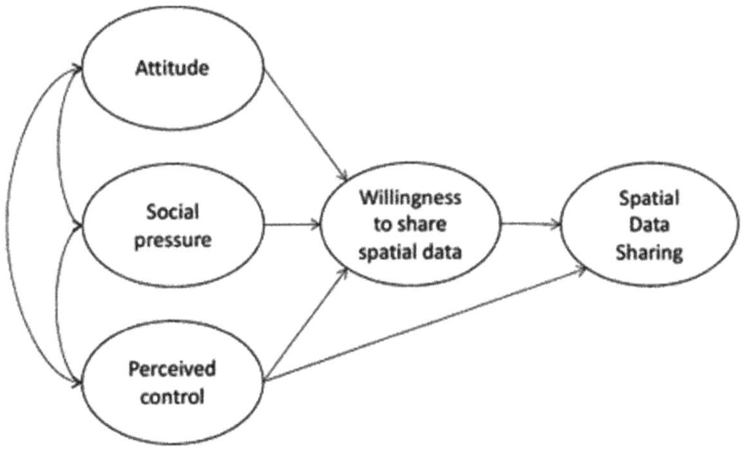

METHODS

Research Design

In order to acquire an in-depth understanding of citizens' beliefs about sharing Personal Weather Station (PWS) data via online platforms, a case study approach and qualitative research methods were chosen, following Plengsaeng's et al. (2014) and Ngo Thu and Wehn's application of the TPB to data sharing (Ngo Thu & Wehn, 2016). The empirical research for this study was undertaken in three countries in which the case studies of the EU-funded WeSenseIt project (Citizen Observatories of Water; funded under FP7, (2012-2016)) were located: Delfland in the Netherlands, Doncaster in the UK and the cities of Padua and Vicenza (Alto Adriatico) in Italy.

Two major groups of citizens were targeted for the data collection:

1. PWS data-sharers (station owners who were already engaged in sharing their station data on one or more amateur weather network) and
2. The general public or citizens who either did not have a PWS or had the equipment but did not share the data (the 'non-shares').

The research instruments included interview protocols (for the general public) and an online survey (to collect data from the PWS data-sharer group) that was prepared in English and translated to Italian (for the Italian case study). Both instruments contained open questions about:

1. Advantages and disadvantages of citizens sharing their PWS data via online networks.
2. People or organisations who push citizens or hold them back from doing so.
3. Opportunities or constraints that render it easy or difficult for citizen to share their PWS data via online networks.

Other sources of primary data that fed into the analysis include project report and observations during the interviews. The results of the data collection from primary sources were complemented with findings of previous studies in the areas of citizen science and knowledge sharing in online communities. Finally, all of the above mentioned data sources were analyzed to develop a model of behavioral determinants for sharing PWS data via online weather networks.

Selection of the Interview Locations in Each Case Study

A number of data collection locations were selected for each of the three case studies. One of the main selection criteria for conducting the interviews and the online surveys was to include areas with different densities of Personal Weather Stations. The intention was that this would capture diverse views about the reasons why PWS data sharing is being practiced (or not). Therefore a density map of PWS stations was developed for each country. The stations that were used to generate these maps were primarily selected from the Citizen Weather Observer Program (CWOP) and Weather Observations Website (WOW) networks; but many of these stations also contributed data to other networks such as Weather Underground (WU), Weather Observations Website (WOW), European Weather Networks (EWN), Personal Weather Stations Network (PWSweather) and Davis WeatherLink network. The final maps (Figures 2, 3, and 4) were generated by overlaying municipal administrative areas of the three cases with the coordinates of

Figure 2. PWS frequency map of the Netherlands
Created by authors, January 2015

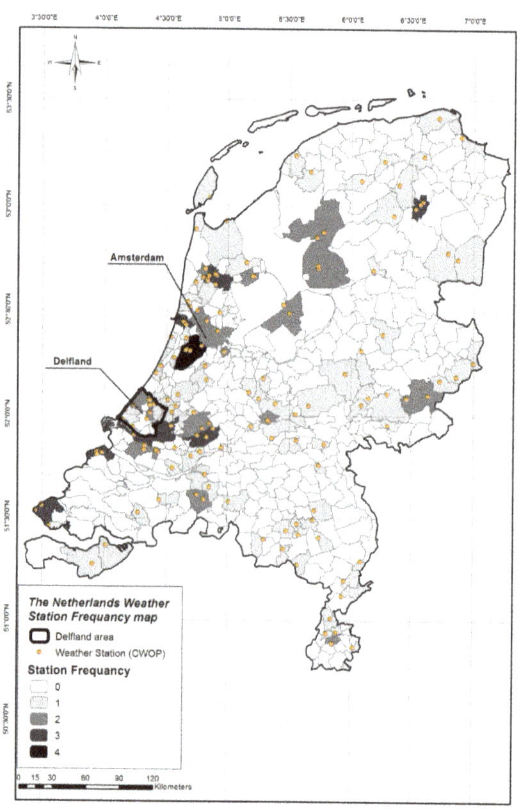

Figure 3. PWS frequency map of the UK
Created by authors, January 2015

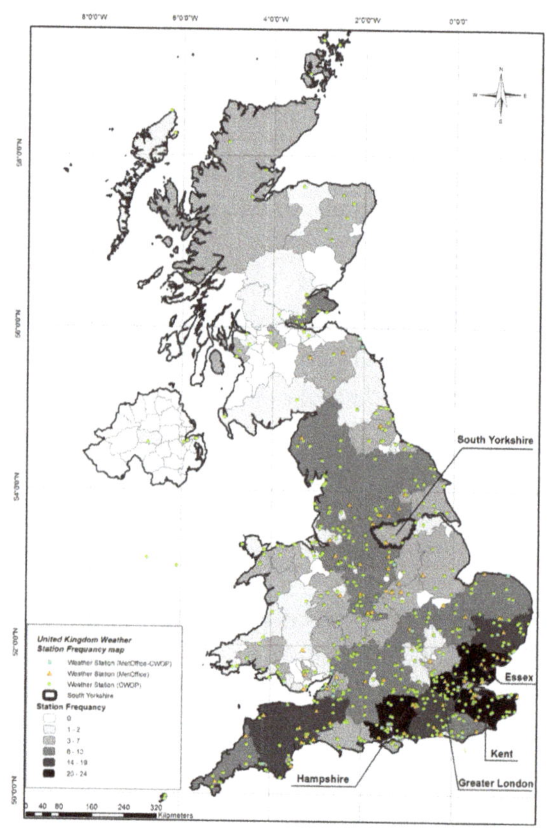

Figure 4. PWS frequency map of Italy
Created by authors, January 2015

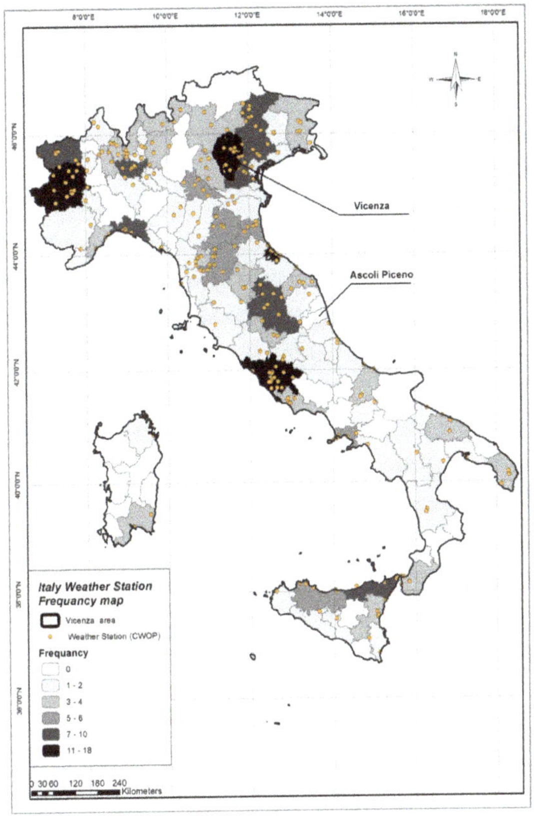

the above mentioned stations in ArcMap. In total, more than one thousand stations (149 in the Netherlands, 626 in the UK, and 292 in Italy) were included to generate these maps.

Based on the PWS frequency maps, in addition to the case study areas of the WeSenseIt project, an additional location with a contrasting density of stations was selected per country.

In the Netherlands case, given the low to medium density of stations in the Delfland area, the only municipality with highest number of stations was selected as the alternative location for conducting face to face interviews in the Netherlands. This municipality is Haarlemmermeer which is located in the province of North Holland with four CWOP stations (see Figure 2). This region was reclaimed from water in the 19th century and includes Schiphol Airport and Hoofddorp which is the main town of this municipality.

In the UK case, South Yorkshire (where Doncaster is located) with a total number of five stations was initially categorized as an administrative area with medium frequency of stations. However, none of these five stations were actually located in Doncaster. This meant that the second location for conducting interviews needed to be chosen from locations with high concentration of stations. According to Figure 3, there are four regions that have very high frequency of stations. These regions are; the Greater London and three counties in South and South-East UK (Hampshire, Essex and Kent). Based on the importance of the Greater London area in terms of its population and also encompassing the capital city of the UK, this area was chosen as the second location for conducting face to face interviews in the UK.

The WeSenseIt project locations in Italy are Padua and Vicenza (Alto Adriatico); these two cities are located in the Province of Vicenza. Based on the PWS frequency map (Figure 4), Vicenza was one of the four provinces with very high frequency of stations and therefore the second location for conducting the interviews was ideally a province with no stations. The Province of Ascoli Piceno in the Marche region was thus selected as the alternative location for conducting the interviews in this case.

Selection of the Participants

In all three case studies, general considerations to include respondents from different age and gender groups were taken into account. Moreover, interviews were conducted in different locations such as shopping centers, parks, train stations, restaurants, etc. at various times of the working days and also during the weekends in order to create reasonable chances for different members of the general public to be approached for the interviews.

The potential respondents for the online surveys were selected from the pool of more than 1000 stations available in the Netherlands, UK and Italy. Per case, 100 invitation emails were sent and the main criteria that were considered for selecting these potential respondents were:

1. Inclusion of at least some stations from the six previously mentioned interview locations,
2. A balanced inclusion of possible respondents from regions with different station frequency categories,
3. Spatial coverage of the rest of the stations across the country, and
4. Availability of contact information of the station owner.

Participants in the Netherlands Case

For the Netherlands case, 11 face to face interviews with the general public were conducted at the two empirical research locations (6 interviews in the Delfland area and 5 in Haarlemmermeer). The bar chart presented in Figure 5 illustrates the gender frequency of different participant age groups for the face to face interviews in the Netherlands case.

In total, 13 valid responses were received from the PWS data-sharer group in the Netherlands. Figure 5 summarizes the online survey results, in this case based on the gender and age group of the participants. As the table shows, all of the participants are male and older than 35 years.

Participants in the UK Case

In the UK case, face to face interviews were conducted with 10 respondents from the general public group (5 in Doncaster and 5 in London). Figure 6 summarizes the frequency of these respondents, based on different gender and age groups.

The total number of valid responses received in the online survey in the UK case was 14. The bar chart presented in Figure 6 illustrates the gender frequency of different participant age groups for the UK case. In this case, similar to the Dutch case, all of the valid responses came from PWS data-sharers older than 35 years and only one of these respondents was female.

Figure 5. Gender frequency and age groups (interviewees and online survey participants) in the Netherlands case

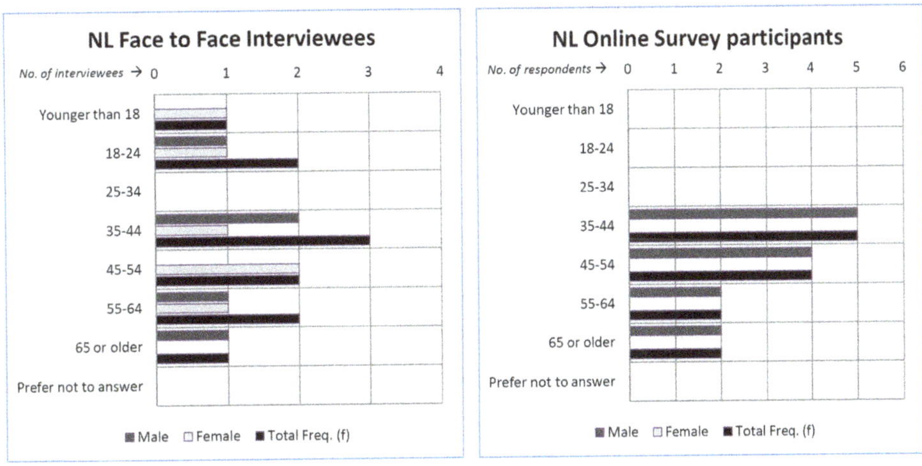

Figure 6. Gender frequency and age groups (interviewees and online survey participants) in the UK case

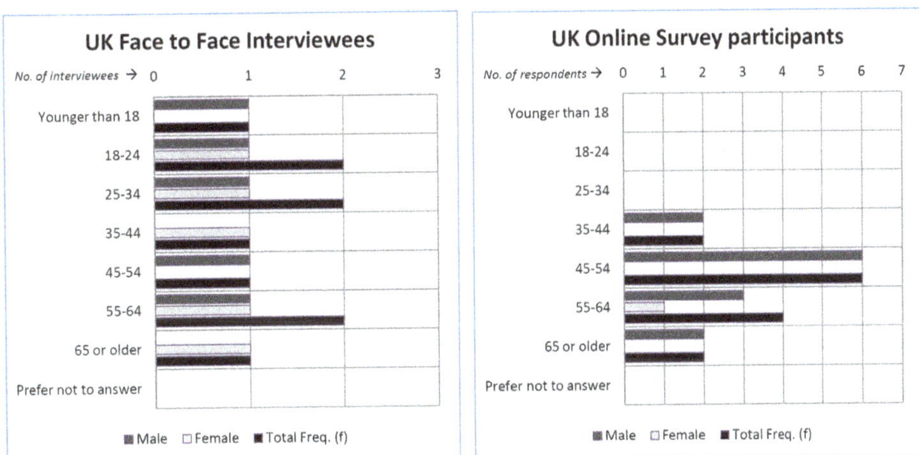

Participants in the Italy Case

In total, 9 interviews were conducted via phone/Skype in the two empirical research locations in the Italian case (4 in the province of Vicenza and 5 in the province of Ascoli Piceno). Figure 7 summarizes all the phone/Skype interviewees, based on their gender and age groups.

The total number of valid responses received in the Italian case was 16. The bar chart presented in Figure 7 illustrates the gender frequency of different participant age groups for the UK case. As the corresponding bar chart shows, in this case all of the valid responses came from male PWS data-sharers, but unlike the Netherlands and UK cases, three of the respondents were younger than 35 years old and none older than 65.

Figure 7. Gender frequency and age groups (interviewees and online survey participants) in the Italy case

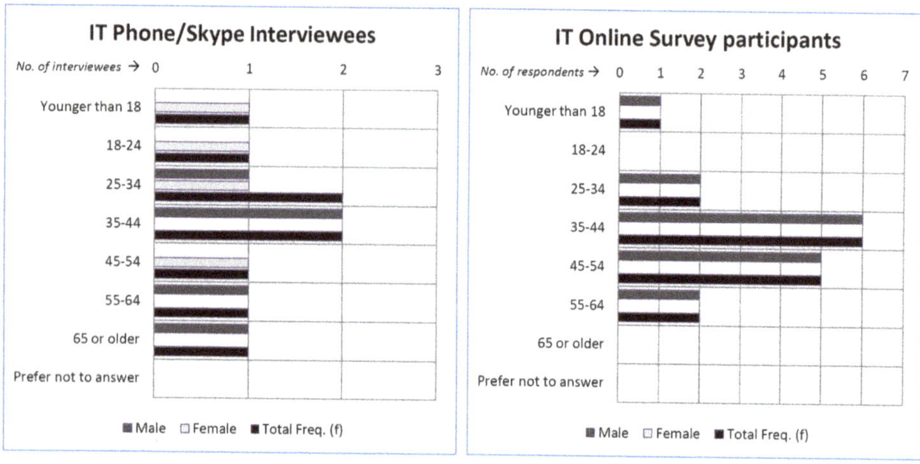

Citizen's Willingness to Share Personally-Collected Weather Data: What Influences it and How?

Citizens' willingness towards sharing PWS data via amateur weather networks is a function of their 'attitude' towards this behavior (beliefs about gains and losses), the 'social pressure' that they perceive from members of society and also their 'perceived control over the behavior' as a result of the presence or absence of influential factors. The following sections will describe different domains that were identified by respondents from the general public and the PWS data-sharers in the three case studies.

Attitude-Related Factors

Four relevant domains of beliefs were elicited about gains and losses or the expectations of citizens about the outcome of sharing PWS data via amateur weather networks, namely:

1. Tangible personal outcomes,
2. Intangible personal outcomes,
3. Societal outcomes, and
4. Interpersonal trust.

Tangible Personal Outcomes

'Tangible personal outcomes' is identified as the first domain of the attitude and refers to the actual or approximate gains and/or losses that a person perceives as the result of sharing PWS data via amateur weather networks.

Several other studies in the areas of public participation, citizen science activities and online communities, have identified personal outcomes as a significant influential factor on the intentions for participation. Some authors refer to this as 'personal outcome expectations' (Chiu, Hsu, & Wang, 2006; M.-H. Hsu, Ju, Yen, & Chang, 2007); others as 'personal gains' (Hew & Hara, 2007) or 'perceived relative advantage' (Chen & Hung, 2010; M.-J. J. Lin, Hung, & Chen, 2009). McLure Wasko & Faraj distinguish between two fundamentally different types of personal outcomes: tangible and intangible returns. They argue that the personal outcomes of an action can be actual and extrinsic (tangible) or in contrast intrinsic and in the form of self-actualization or satisfaction (intangible) (McLure Wasko & Faraj, 2000). The former falls within this attitude domain, while the latter is discussed in the second domain of the attitude component.

One of the first questions that might cross one's mind when asked to spend time and money to collect weather data and share it with others via web-platforms would be: what's in it for me? In this regard enjoying the collected data for different personal purposes such as leisure, outdoor activities, sports, weather-related businesses (e.g. farming, railways, and construction industry), possible financial gains for the data-sharers, information for car drivers, travelers and tourists were mentioned as the positive examples of beliefs. On the other hand, especially for those whose daily activities are not very dependent on the weather, and also those who have no interest in the subject, the availability of as they say 'enough official data' seemed to create a sense of reluctance about the necessity of collecting and sharing such data. This is the negative perception about the tangible usefulness of the collected data and as expected was found only in the respondents from the general public.

The second category of behavioral beliefs focuses on the privacy and security issues. One of the main concerns of both PWS owners and the general public was the fear of theft. The instruments needed for collecting and sharing data must be installed outdoor in the backyard, garden, roof, etc and therefore not easy to protect at all times. These devices may cost from a couple of hundreds to more than a thousand Euros. Due to the fact that the location of any stations is easily retrievable using the web-platforms and Google Earth, the issue of security is certainly a tangible outcome that may hinder the participation. This argument is also true for the privacy related issues and the possibility of being located by any other unwelcome visitors, for example marketers, vendors, researchers, etc. Another relevant issue that was emphasized by the respondents is often referred to as cyber security. Access to the web almost always involves increased vulnerability to cyberattacks, especially since one need to open more ports and run software 24/7 which might have security leaks. Examples that reflected concerns about cyber security were only mentioned in the Netherlands and UK cases.

Intangible Personal Outcomes

The second Domain of the Attitude component as explained in the previous section is 'intangible personal outcomes' that refers to intrinsic gains in form of self-actualization or inner-satisfaction.

A sense of 'belonging to a community of friends with shared interests/visions' was elicited as one of the intangible outcomes of sharing PWS data via amateur weather networks. As a result of the effort that the PWS owner puts into collecting and sharing the data, he or she is welcomed and included in a virtual community of citizens who share an interest or have a similar vision and this generates a sense of self-actualization that may be a good source of motivation for participation. Previous studies have also recognize having a shared vision and interest as "a bonding mechanism" (Tsai & Ghoshal, 1998) and mention that "virtual communities are groups of people brought together by common interests and goals" (Chiu et al., 2006). During the empirical research phase, a number of PWS-data-sharers and respondents from the general public identified and mentioned this belief as a source of motivation for participation.

'Learning from each other' is the second cluster of behavioral beliefs in this domain. The researchers have categorized this behavioral belief as an intangible personal outcome, because it is mainly about the sense of enjoyment from sharing knowledge with others and learning from them. A reciprocal sense of satisfaction is normally generated for both sharer and recipient of the information. This process of learning can happen through one to one communications between the virtual community members or via group communications in online forums, Facebook pages, etc. Respondents from the general public and PWS-data-sharers seemed to perceive value in learning from other society members and considered this as an incentive for participation.

The third category of behavioral beliefs is labeled as 'recognition by others'. This form of belief that was mentioned only by the station owners and in a negative form refers to the fact that PWS data-sharers find themselves worthy of receiving some sort of commendation and acknowledgement from other members of the society and especially those who enjoy this service; an expectation that is not fulfilled in most cases and therefore translates into a sense of disappointment and thus considered as a negative outcome. This category of beliefs was identified during the online surveys where PWS-owners stated that other sites may use their data without permission or acknowledgement or national weather service organizations use these data for free without any sort of gain for the data-sharer. Some respondents explicitly mentioned that valuation of data does not have to be monetary per say and intangible values are just as important.

The last category of this domain is 'interest in the weather' and refers to the sense of enjoyment, entertainment and satisfaction that one gains from observing the weather and sharing the data on web-platforms. Not surprisingly, this was mentioned in the positive form by the PWS data-sharers and in the negative from by the general public. The first group mentioned the fun factor as a driving force while the second group highlighted their lack of interest in the weather observation as a preventing factor and mentioned that they simply do not enjoy this activity. An example that was mentioned by one of the PWS data-sharers in Italy case was; "I find it very interesting to monitor and evaluate the small variations that exist in weather attributes between different areas; even if they are very close to each other, these variations still exist".

Societal Outcomes

The third domain of the attitude component is the 'societal outcomes'. This belief is closely related to the definition of morality or the evaluations or implications of the behavior on the society at large. Relevant secondary literature about participation in citizen science activities and online communities had also elicited this domain as a proxy for attitude towards behavior. Different terminology is used in these literatures such as 'community-related outcome expectations' (Chiu et al., 2006; M.-H. Hsu et al., 2007); 'community interest' (McLure Wasko & Faraj, 2000); or 'Altruism' that can be considered as a subset of the 'societal outcome' domain (Cho, Chen, & Chung, 2010; Hew & Hara, 2007).

The first behavioral belief in this domain is related with the perceived applicability of PWS data for 'reduction/mitigation of environmental risks'. Several respondents from the general public and PWS data-sharer groups identified relevant examples of such applications, here are a few example: One of the interviewees from the general public in the Netherlands case mentioned that the Dutch are mostly living under the sea level and linked this to the potentials of such data for flood risk reduction in the Netherlands. In the UK case, one of the respondents from the general public stated; "I believe that the overgrowing problem of global warming should enable citizens to share their data so we can get a better understanding" of this phenomena. As the third example, an online survey participant in Italy mentioned that; "to improve the quality of life in the territory, climate change and pollution are two factors that should be monitored closely".

Benefiting for society at large through 'creating knowledge about the weather' is the second category of behavioral belief in this domain. The examples that were mentioned during the interviews and online survey are; contribution to citizen values and well-being, creating a collective knowledge about the weather and climate (that is not individually possible), creating a complementary source of data to the official observations both in terms of spatial and temporal distribution, economically efficient weather data for the government and the whole society, and creating an alternative source of data for research purposes.

Interpersonal Trust

'Interpersonal trust' is the last domain of the attitude component. Several literatures in areas of knowledge sharing in online communities have recognized the importance of trust as a determinant of intention to participate in these communities (Chow & Chan, 2008; C.-L. Hsu & Lin, 2008; H. F. Lin, 2008). Due to the fact that in this study the main actors are the citizens, a specific type of trust called 'interpersonal trust' was identified as one of the domains that influences the perception of the citizens about sharing PWS data. Chen and Hung described the interpersonal trust in the context of knowledge sharing in online communities as; "a degree of belief in good intentions, benevolence, competence, and reliability of members who share knowledge" (Chen & Hung, 2010). The issue of trust and its relation with attitude can be articulated as the expectation of the trustor that his/her act will not have any harmful outcomes for him/her (Barber, 1983; Pavlou & Fygenson, 2006). Optimistically speaking, it may also refer to the assumption that the interests of trustor will be protected by the trustee (Hosmer, 1995; Pavlou & Fygenson, 2006). In the case of this research, interpersonal trust implies the extent to which society members believe in good intentions, competence and reliability of citizens as non-professionals to engage in collecting and sharing weather related data. The stronger this trust is; the more inclined people are expected to be towards engagement in this activity. Two categories of behavioral beliefs related to interpersonal trust were elicited during the face to face interviews and online surveys:

The first one is labeled as 'competence and reliability' and is related to the competence of the citizens as non-professionals in collecting and sharing weather related data and as a result; the reliability of this type of data. If interpersonal trust does not exist society members may not perceive any advantage for their engagement and this may affect their intention for participation. The issue of trust in the competence of non-professionals and the quality of the data that they produce were mentioned several times during the interviews and the online survey. Contradictory opinions seemed to exist among the respondents; some respondents trusted the official data more than the personal observations and mentioned that data contributors may intentionally or unintentionally falsify the data. Others had a sense of mistrust in the official data and therefore perceived the personally collected data as a more reliable source of information about the weather. There were also a third group of respondents who trusted in both source of data and considered the personally collected data as a complementary data stream for the official observations.

The second category of beliefs in this domain relates to the 'intentions of the data sharing promoters'. Some respondents believed in the good intentions of the data sharing promoters, while others, believed that data sharing promoters have their own reasons (agenda) for supporting PWS data sharing such as promoting specific businesses, conveying certain messages or selling their own products.

Social Pressure

'Social Pressure' is the next component of the conceptual framework of this research. In the context of this research, social pressure refers to the beliefs regarding the perception of other individuals or group of individuals (referents) about sharing PWS data and whether they will approve or disapprove participating in it. These beliefs are also referred to as normative beliefs (Ajzen, 1991). The accessible normative beliefs that were elicited by respondents in this research are categorized into five different domains:

1. Public/private organizations,
2. Scientific community,
3. Weather enthusiast community,
4. Other society members, and
5. Moral norms and altruism.

Likewise, each Domain is formed from a set of normative beliefs that will be introduced in the following sections. The normative beliefs reflect the perception of both PWS owners and the general public about the referents view on sharing PWS data via amateur weather networks and can be positive or negative in nature. The 'social pressure' domain is highly dependent on the behavior in question and less extendible from relevant secondary literature (in comparison with the 'attitude' domain); therefore this section is mostly based on the findings from this empirical research.

Private/Public Organizations

The first domain of the 'social pressure' relates to the perceived pressure from public/private organizations and whether they will be in favor of, or against sharing personally-collected weather data via amateur weather networks. Respondents from the PWS data-sharer group and the general public elicited several examples of public and private organizations that they believed may approve or disapprove engagement in sharing such data. They mainly based their judgments on whether these organizations or companies may gain or lose authority, power or income because of this behavior. In some cases, different respondents had opposing beliefs about the same organization or company; perceiving it in favor of or against sharing PWS data via online amateur weather networks. Four different groups were elicited by the respondents from the general public and PWS data-sharers;

1. 'New weather-related commercial actors'; such as manufacturers of the personal weather stations and application developers. This group was identified as supporter of PWS data sharing, because of the direct benefits that they have in increased engagement of the citizens;
2. 'Traditional weather-related commercial actors'; this group includes long-established institutions and organizations such as news agencies/channels and private weather forecast organizations. Respondents from both group of participants in all three cases mentioned that these commercial actors may approve this behavior because they can benefit from using the data (e.g. in forecasts) and/or may disapprove it because it may affect their business by questioning its necessity;
3. 'Weather-related (inter)governmental organizations'; Similar to the second category, these organizations may also approve or disapprove sharing personally-collected weather data. Several examples of these organizations were mentioned by both groups of respondents in the three cases.

Here are a number of examples that were mentioned in the three case studies; KNMI (the Dutch Meteorological Institute), Waterschappen (the Dutch Water Boards), Milieudefensie (the Dutch Environmental Defence Organization), NASA (the National Aeronautics and Space Administration of the United States), ESA (the European Space Agency), Met Office (the UK Meteorological Institute), the United Nations, Italian Meteorological Service, ARPA (the Regional Environmental Protection Agency), WWF (the World Wide Fund for Nature) and Italian Military Air Force;

4. 'Other industrial sectors'; these are the industrial sectors that may somehow be affected by the weather and therefore may approve and/or disapprove the behavior of sharing personally-collected weather data via web-platforms (e.g. agriculture, energy, tourism, construction, transport and insurance industries and also other industries that might harm the environment).

Scientific Community

'Scientific community pressure' was elicited as the second domain of the 'social pressure' component. According to the respondents from the general public and PWS data-sharers this group may also approve or disapprove the behavior of PWS data sharing via online weather networks. A number of respondents from both respondent groups believed that the scientists and researchers will enjoy this source of freely available data and will perceive it as a complementary data stream to the available official observations and therefore will welcome and support this behavior. This also includes schools and universities that might use the data for educational purposes. On the other hand some interviewees from the general public believed that scientific community may be against involvement of general public in collection and sharing weather related data because of mistrust in the capability of general public in doing so and therefore may disapprove it. In summary, the social pressure in this domain was perceived from two main group; scientists and educational institutes.

Weather Enthusiast Community

The third domain of social pressure relates to the individuals or groups of individuals who are interested in weather data for different reasons. Respondents from both groups in all three cases elicited 'weather enthusiast individuals' as independent members of the society who may motivate each other and the general public to further engage with this behavior. According to the PWS data-sharers in all three cases, actual and virtual 'weather networks' and 'weather-related hobby clubs' (such as ham radio clubs, aviation clubs, sailing clubs, etc.) are also two other groups that will approve collecting and sharing weather related data by the citizens.

Other Society Members

The forth domain that may influence the citizens to engage in sharing PWS data via online amateur weather networks is 'other society members'. By other society members refer to individual citizens who may not gain or lose directly like private/public and also do not belong to the scientific community and/or weather enthusiast community but still may approve or disapprove this behavior for different reasons. This group can be divided to three categories;

1. **Citizen Science/Big Data Critics:** This category that was identified by PWS data-sharers and the respondents from the general public in all three cases refers to the existing negative pressure from

citizen science/ big data critics who have some sort of mistrust in the competence of the citizens to collect and share data with an acceptable quality and also are concerned about the privacy and security issues resulted from that;

2. **(Anti) Environmentalist Community:** This group contains both positive and negative beliefs from rather self-explanatory groups within societies who perceive certain environmental-related benefits or losses for this type of data;
3. 'Family and Peers' is the last group of this Domain.

This includes family members, neighbors and friends of the one who shares data that may support this activity or stand against it based on their personal opinion or circumstances.

Moral Norms and Altruism

The last domain with regards to the 'social pressure' component is labeled as 'moral norms and altruism'. Morality can also be considered as moral obligations to perform or not perform a behavior (Sabini, 1995) and therefore may be categorized as a 'Social Pressure' antecedent. These types of beliefs, especially when related with the risks, are closely linked with altruism and may be considered as a source of inner approval to perform the behavior. This is also true when the behavior performer provides a useful for society at large. Based on the results of the interviews with the general public and online survey with the PWS data-sharers in all three cases, 'risk prevention' and 'Benefit for society at large' were elicited as the relevant sets of beliefs for this category.

Perceived Control

The final main component of the conceptual model of this study is 'Perceived Behavioral Control' (PBC), and is also argued to be a function of beliefs. These beliefs are formed based on the perception of an individual about the absence or presence of certain factors that may impede or facilitate performing the behavior and are referred to as control beliefs. The PBC component is essential while studying behaviors that are not under full 'volitional control' (Ajzen, 1991). In this study, the behavior of sharing PWS data via online amateur weather networks is being investigated and a number of circumstances and factors may interfere with the performer's control over this behavior. These factors or circumstances can be further divided into two groups based on their relation to the individual who performs the behavior; internal factors or external ones (Ajzen & Madden, 1986; Wehn de Montalvo, 2003b). Some examples of internal factors are personal abilities, knowledge and skills while opportunity, time and dependence on other's cooperation may be categorized as external factors (Ajzen & Madden, 1986). In this study, four different domains were identified under the 'Perceived Behavioral Control' component, namely;

1. Technical skills,
2. Knowledge self-efficacy,
3. Resource control and
4. Opportunities.

Based on the above explanations, the first two categories are internal factors while the last two are external to the individuals.

Technical Skills

The first Domain of the 'Perceived Behavioral Control' relates to the control beliefs about the presence or absence of technical skills of the individual who want to participate in sharing PWS data via online amateur weather networks. Many literatures in the areas of citizen science and knowledge sharing in online communities have highlighted the importance of technical skills (Hew & Hara, 2007; Kaufmann, Schulze, & Veit, 2011; McLure Wasko & Faraj, 2000). This is very much relevant in the case of using PWSs to collect and share weather related data in the sense that a range of technical skills exists, that their presence may facilitate citizen's participation and at the same time their absence is likely to impede their engagement. Based on the empirical research results in all three cases, two different categories of technical skills were identified; technical skills about 'setting up and maintenance' of the PWSs and 'IT skills' in general that includes basic computer skills (hardware and software), using the Internet to send and receive data and in some cases managing a personal webpage.

Knowledge Self-Efficacy

The second elicited domain of the PBC is 'knowledge self-efficacy' that has been identified as one of the main subordinates of Perceived Behavioral Control (Ajzen, 2002; Armitage & Conner, 1999; Manstead & Van Eekelen, 1998). According to Bandura, Perceived self-efficacy is defined as "people's beliefs about their capabilities to exercise control over their own level of functioning and over events that affect their lives" (Bandura, 1991). It is important to mention that 'self-efficacy' is different from 'controllability' that is another subordinate of Perceived Behavioral Control and Ajzen makes this distinction clear by linking the first one to the "ease or difficulty of performing a behavior" and the second one to "beliefs about the extent to which performing the behavior is up to the actor" (Ajzen, 2002). In the case of sharing personally-collected weather data via web-platforms a specific type of self-efficacy was identified that relates to the perception of the citizens about their knowledge of the methods of data collection and weather observation in general and its effect on the ease or difficulty of their participation. This is closely linked with citizens' "confidence in an ability to provide knowledge" (Chen & Hung, 2010) related to the weather. Some members of the society tend to believe that normal citizens do not have the required knowledge about observing the weather and collecting the relevant data and thus do not feel confident to participate in these activities. On the other hand others may have this knowledge self-efficacy and this may act as an enabling control factor for their participation.

Generally speaking many members of the society seem to believe that they know enough about meteorological phenomena and weather related events when they watch the news on TV or check the forecasts on websites or their mobile Apps. It even forms an integral part of some people's daily conversations, however when they imagine themselves in the position of the data provider, some of them may not be as confident as before. This perception about lack of meteorological knowledge self-efficacy was mentioned more specifically by some participants. These are two examples of the responses received in this regard; "there is a need for educating people about weather phenomena and this can even start from primary schools"; "I think 'a high level of knowledge is required to do this".

Familiarity with 'data collection methods' and the technical knowledge that may be required for that is another control belief of this domain. This is closely related to the perceived knowledge and confidence about the perception of an individual about the difficulty of these methods. An example answer that was received during the interviews with the general public in the Netherlands case was; "citizens should be well instructed how to gain data, otherwise they will be incorrect and therefore not useful".

Resource Control

The fourth domain of the PBC component is resource control. In the case of sharing PWS data via online amateur weather networks, control over resources refers to the perception of the individuals about the extent to which performing this behavior depends on their access to external resources such 'equipment', 'Internet connection', 'finance', 'time', 'usability of web-platforms and apps' and 'PWS installation location'.

The first category of resources is the required equipment to observe the weather and share the data on web-platforms. These are the hardware required for citizens' engagement. Several participants from both respondent groups mentioned this category of beliefs when asked about the factor that may facilitate or enable them to participate.

The second group of resources is the Internet connection that is essential for sharing the collected data. The majority of respondents in the Netherlands and UK cases believed that the Internet connection is not an issue and some even mentioned this in their answers, however in both cases there were participants who still believed that this may positively or negatively affect the citizens engagement and mentioned for example that: there are "there is a lack of good Internet access in some areas". This issue seemed a bit more highlighted in the Italy case as a larger proportion of the respondents from the general public and PWS data-sharer groups mentioned this as a factor that hinders their engagement. An example that was mentioned by one of the PWS owners is; "the inefficiencies in our infrastructure system (Internet for all) make it difficult to share a big amount of data".

Finance is the next cluster of resources and was elicited as an enabling (if present) or disabling (if absent) factor. This category of resources includes both initial capital investment and the ongoing operation and maintenance costs of having a PWS. This is also evident while visiting some of the personal WebPages where station owners ask for donations from their visitors; an anonymous example is provided in Figure 8.

Time is the forth category of resources that were identified based on the collected responses during the interviews with the general public. The responses suggest that availability of time to collect and share the data and maintain a PWS is factor that citizens will consider before engaging in the activity.

Both PWS data-sharers and respondents from the general public in all three cases mentioned easy to use web-platforms and applications may facilitate citizens' participation. On the other hand they also identified complex and hard to understand web-platforms and applications as a barrier for their engagement.

Figure 8. An example of a donation request note from an amateur website

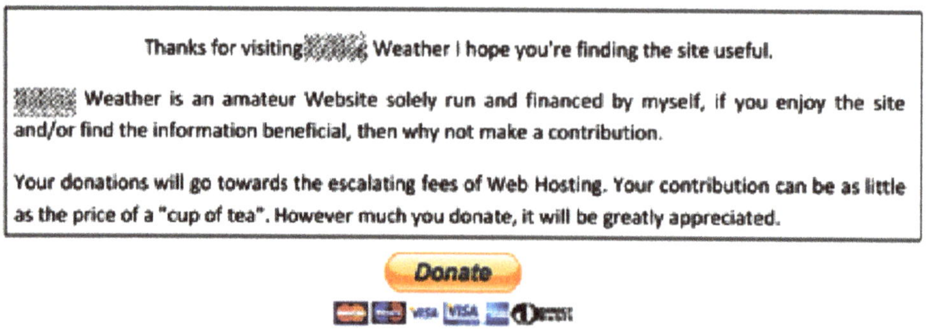

The sixth control factor of the resource control Domain is the suitable location for installing the weather stations. This is an important issue for the ones who live in apartments or perhaps live in sheltered locations where high rise buildings and trees cause limitations for standard weather observations. As an example, one of the PWS owners in the UK case, stated that; "unfortunately my garden has high trees on the West and South elevations so my wind speed and direction is not always accurate. I do make this clear on my personal weather site".

Opportunities

The second category of external control factors and the last domain of Perceived Behavioral Control is labeled as opportunities. Absence of these series of 'circumstantial factors' is not expected to affect the behavior (Wehn de Montalvo, 2003b) but their existence may facilitate sharing personally-collected weather data. This belief seemed to exist only among PWS owners as it was not elicited by any member of the general public in neither of the cases. PWS data-sharers identified two categories of opportunities;

1. 'Incentives provided by web-platforms'; the examples that were mentioned for the this cluster are: receiving feedbacks from the web-platform operators about the data shared by their station (in terms of quality, possible errors, etc.), a certificate that indicates they have provided this data for a certain period (e.g. after one year), an excursion to official weather station in their locality, and a small annual retainer fee to those who have joined a network.
2. 'Opportunities to gain and exchanging knowledge'; for example, one of the PWS owners in the UK case stated; "I have found that sharing weather data and learning has opened up many new avenues of useful teaching experience for my students".

The Model of Sharing PWS Data Via Online Amateur Weather Networks

The model presented in Figure 9 summarizes the discussions about citizens' willingness to share personally-collected weather data, based on the findings and the theoretical framework of this research. Initially and prior to conducting the empirical research, two possible approaches were envisioned for developing this model. The first one was to develop individual models for each case (or any combination of cases) and the second one was to develop a common model for the three cases; and the choice was totally dependent on the level of differences in the findings across the three case studies. As it is further discussed in the cross case comparison section, the results did not show significant difference across the Netherlands, UK and Italy case studies, and thus a common model (Figure 9) was developed for the three cases.

DISCUSSIONS

Most/Least Frequently Mentioned Influential Factors

In this section, the most and least frequently mentioned influential factors for sharing PWS data via online amateur weather networks are presented. The discussions are based on a compilation of interview and online survey results in the three case studies. In total, 43 PWS data-sharers and 30 participants from the

Figure 9. The model of sharing PWS data via online amateur weather networks (in the Netherlands, UK and Italy)

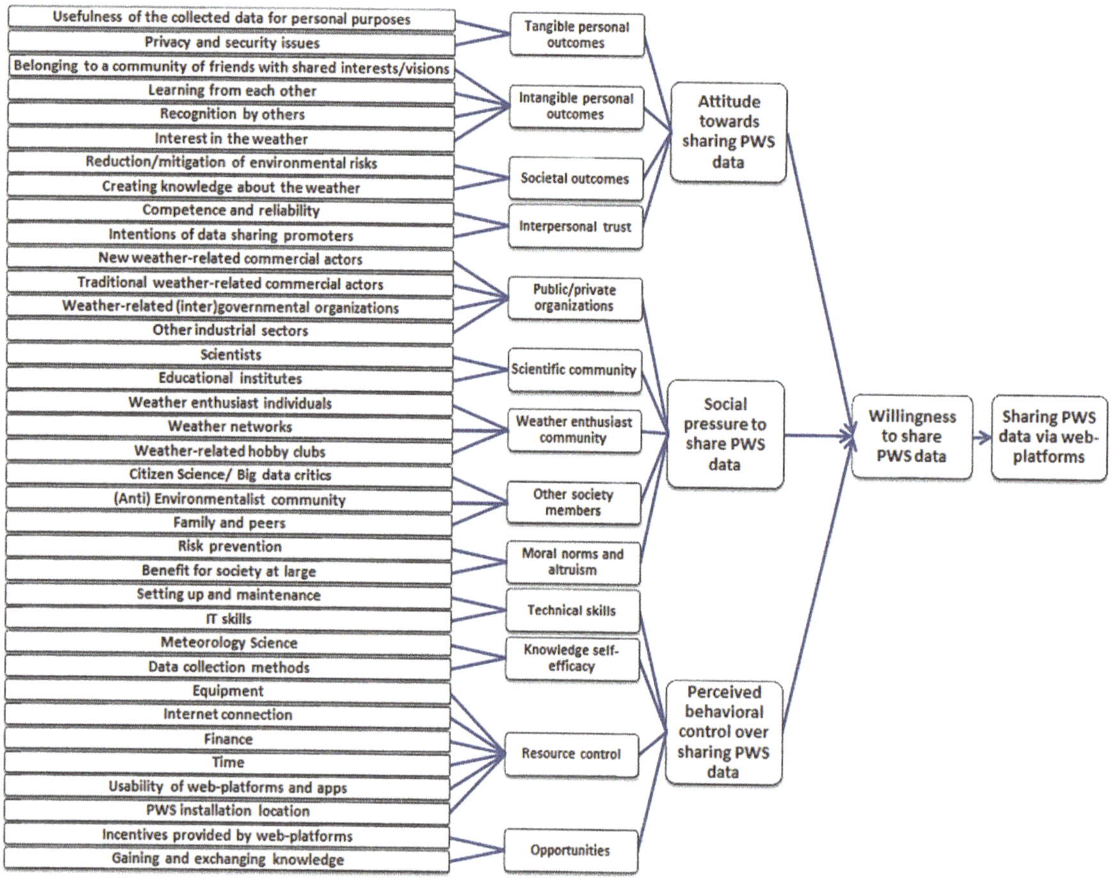

general public participated in this study. A tabular format is utilized to illustrate the most and the least frequently mentioned behavioral, normative and control beliefs that may positively or negatively affect the citizen's willingness to share PWS data via online amateur weather networks.

Among the identified positive beliefs about the outcomes of sharing PWS data, benefits of for the society at large through creating knowledge about the weather was the most frequently mentioned belief (see Table 1). This implies that more than 50% of the respondents from both groups (PWS data-sharers and the general public) believed that sharing PWS data will help enhance the current spatial and temporal weather databases. Furthermore, one of the most contested beliefs was the competence and reliability of the citizens to participate in this activity. There were 37 positive and 23 negative responses about this influential factor and this clearly shows the diversity of views about this issue among the respondents.

Interestingly, the most frequently mentioned referent that encouraged respondents to participate in sharing PWS data appeared to be their inner-self (Table 2). Benefiting society at large which is linked to the moral norms and altruism domain was mentioned by more than 50% of the total respondents in all three cases. On the other hand, in general, for all three cases, Citizen Science/ Big data critics seemed to impose the most negative social pressure on the citizens, discouraging them from participation.

Table 1. Elicited outcomes (attitude)

Component	Perceived outcomes	Positive/negative beliefs about the outcomes	No. of total responses	
			positive	Negative
Attitude	Tangible personal outcomes	Usefulness of the collected data for personal purposes	16	7
		Privacy and security issues	0	11
	Intangible personal outcomes	Belonging to a community of friends with shared interests/ visions	11	0
		Learning from each other	5	0
		Recognition by others	0	6
		Interest in the weather	10	7
	Societal outcomes	Reduction/mitigation of environmental risks	13	0
		Creating knowledge about the weather	40	0
	Interpersonal trust	competence and reliability	37	23
		Intentions of data sharing promoters	13	4

Table 2. Elicited sources of social pressure

Component	Social pressure by key referents	Perceived pressure (not) to share	No. of total responses	
			To share	Not to share
Social Pressure	Public/private organizations	New weather-related commercial actors	4	0
		Traditional weather-related commercial actors	8	8
		Weather-related (inter)governmental organizations	17	5
		Other industrial sectors	14	8
	Scientific community	Scientists	8	5
		Educational institutes	7	0
	Weather enthusiast community	Weather enthusiast individuals	10	0
		Weather networks	10	0
		Weather-related hobby clubs	5	0
	Other society members	Citizen Science/ Big data critics	0	17
		(Anti) environmentalist community	1	2
		Family and peers	2	1
	Moral norms and altruism	Risk prevention	13	0
		Benefit for society at large	40	0

Generally speaking, the presence of each control factor is perceived to make sharing personally-collected weather data easier and its absence make it more difficult. Resources such as equipment, finance, and usable web-platforms and mobile applications were among the most frequently mentioned factors, in that their absence is perceived as a barrier and their presence is perceived as a facilitator for sharing PWS data via online amateur weather networks (see Table 3). Furthermore, with regards to the

Table 3. Elicited perceived control factors

Component	Perceived control factors	Beliefs about (presence/absence of) control factors	No. of total responses	
			Easy/ Present	Difficult/ Absent
Perceived Behavioural Control	Technical skills	Setting up and maintenance	10	11
		IT skills	11	15
	Knowledge Self-efficacy	Meteorology science	0	3
		Data collection methods	0	20
	Resource control	Equipments	19	5
		Internet connection	10	12
		Finance	14	24
		Time	6	6
		Usability of web-platforms and Apps	17	12
		PWS installation location	3	3
	Opportunities	Incentives provided by web-platforms	10	0
		Gaining and exchanging knowledge	4	0

knowledge self-efficiency domain, unfamiliarity with data collection methods was frequently mentioned as an influential control factor for not participating in this activity.

This study did not aim for a quantitative validation and ranking of the revealed beliefs. Future research should validate the model presented here, using a survey approach to allow respondents to rank each belief. Based on a large sample of respondents, a quantitative analysis of their responses will result in generalizable insights of the most influential beliefs and of the extent to which specific behavioral beliefs have a direct influence on sharing PWS data via online amateur networks.

Cross Case Comparison

The purpose of this section is to compare the findings of the research, across the three case studies, with the aim of identifying similarities and differences in the responses received from the participants. For this purpose, the results from the three cases are first compared and contrasted at the domain level to identify, the major differences (if any); and then zoomed in at the belief level to discuss the major similarities and differences in positive and negative perception about sharing PWS data via online amateur weather networks, across all three cases.

The review of the responses at the domain level clearly demonstrates that the groups of beliefs (domains) about sharing personally-collected weather data via web-platforms that were identified in the Netherlands, UK and Italy cases are basically the same. To summarize, for each case, four Attitude domains were elicited, namely: 'tangible personal outcomes', 'intangible personal outcomes', 'societal outcomes' and 'interpersonal trust'. Five different domains represented the Social Pressure component; 'public/private organizations', 'scientific community', 'weather enthusiast community', 'other society members' and 'moral norms and altruism'. Finally the Perceived Behavioral Control has four domains, namely; 'technical skills', 'knowledge self-efficacy, 'resource control' and 'Opportunities'.

The result of the analysis showed a great deal of similarities in the range of elicited beliefs by respondents from the general public and PWS data-sharers in all three cases. The behavioral and control beliefs fully matched for the Netherlands, UK and Italy cases. However, the normative beliefs in the three cases, showed two marginal differences. Firstly, in the Italy and UK cases, some respondents mentioned that they perceive social pressure from environmentalist and individuals or groups and also from those who may harm the environment for their personal benefit (anti-environmentalists). This was not mentioned during the interviews with the general public and online surveys with PWS owners in the Dutch case. The second difference was identified in the Italy and Netherlands cases when the respondents mentioned the issue of approval or disapproval of sharing personally-collected weather data via web-platforms by family members and peers as these may consider it as a waste of time and money or, on the other hand, support it for personal reasons. This was not elicited from any of the respondents in the UK cases.

A Predominantly Male Activity?

The bar charts presented in Figures 5, 6, and 7 for the online survey respondents reveal that only one female respondent (in the UK case) in all three case studies participated in the online surveys for this research. Given the total number of valid responses (i.e. 43 for all three cases), this represents less than 3% of the total valid responses that were collected from the PWS data-sharer community.

The qualitative nature of this study puts inherent limits on the sample size which is therefore not representative of the entire population (in each of the three countries or all of Europe). This study therefore cannot draw firm conclusions about the gender (im)balance regarding the participation in ICT-enabled amateur weather observations. Nevertheless, given the random selection procedure of the participants, this finding raises a question about the gender dimension of the larger online amateur weather observers' community in these countries.

Recent research carried out within one of the UK's amateur weather observation communities (i.e. UK-Climatological Observer Link) indicates that both, amateur and professional weather observations, are a predominantly male preserve (Endfield & Morris, 2012). Several reasons for have been suggested for this in the literature. Gordon Manley (controversially) argued long ago that "prolonged maintenance of daily observations demand an odd and uncommon type of enthusiasm" (Manley, 1952, p. 300) that is best found within the male community. Diverse factors come into such as the invisibility of female efforts in maintenance, recording and sharing the data (the so-called 'invisible technicians' (Endfield & Morris, 2012)), and the tendency of men to enjoy their "closed off universe" (Subkowski, 2006, p. 386) and to get involved in long-term and continuous data collection, analysis and storage efforts (Endfield & Morris, 2012).

With advancements in ICTs and the availability of automatic amateur weather stations as well as dedicated apps, further research is required to generate sound insights into the current situation of gender balance regarding participation in ICT-enabled weather observations, so as to inform the individuals and organizations involved in setting up and – especially - in scaling-up citizen science projects.

CONCLUSION

In this chapter, a decision making theory was used to serve as a framework for a systemic investigation of a specific type of citizen science activity, namely collecting weather observations with Personal

Weather Stations and sharing these via online amateur weather networks. Based on empirical research in distinct locations of three European countries with a carefully selected sample of participants, the following conclusions can be drawn.

First, there appear to be no substantial regional differences between the main drivers and incentives for citizens to share their PWS data. This suggests that the reasons and obstacles for sharing PWS data via online platforms are fairly homogeneous in the studied countries. Nevertheless, while some of the drivers and obstacles for sharing weather observations pertain across citizen science activities, others differ. The mistrust in the competence of the citizens to collect and share data of an acceptable quality is a commonly discussed concern (Bonney et al., 2014; Crall et al., 2011; Nature, 2015)which is manifested as negative social pressure, holding back participation. However, in the case of sharing weather observations, altruism and societal benefits appear to play a significant role whereas in other citizen science initiatives such as biodiversity monitoring, citizens seem to be driven more by personal returns and show greater reluctance to share the data (Ganzevoort & van den Born, 2016). It remains to be seen whether the Big Data-related anxieties regarding privacy and security uncovered here also pertain across different citizen science initiatives. This leads to the second conclusion that the reasons for participating in citizen science activities are thematically bound and therefore need to be carefully considered in context when aiming to scale up specific citizen science initiatives.

Finally, if in the digital age observing the weather is still a male-dominated activity, so is the participation in online amateur weather observation networks and communities; such gender bias would have implications (and inherent limitations) for upscaling this citizen science initiative. Similarly, despite ever advanced apps and ever easier to use weather stations, the technical knowhow and capabilities required for collecting and sharing hydro-meteorological data do matter and can constitute a tangible barrier for continued participation. Affordable equipment, training for particular citizen segments and use-friendly web-platforms and mobile applications still seem a must if citizen science is to close the data gaps.

REFERENCES

Ajzen, I. (1985). From Intentions to Actions: A Theory of Planned Behavior. In J. Kuhl & J. Beckmann (Eds.), *Action Control* (pp. 11–39). Springer Berlin Heidelberg. doi:10.1007/978-3-642-69746-3_2

Ajzen, I. (1991). The theory of planned behavior. *Organizational Behavior and Human Decision Processes*, *50*(2), 179–211. doi:10.1016/0749-5978(91)90020-T

Ajzen, I. (2002). Perceived Behavioral Control, Self-Efficacy, Locus of Control, and the Theory of Planned Behavior. *Journal of Applied Social Psychology*, *32*(4), 665–683. doi:10.1111/j.1559-1816.2002.tb00236.x

Ajzen, I., & Madden, T. J. (1986). Prediction of goal-directed behavior: Attitudes, intentions, and perceived behavioral control. *Journal of Experimental Social Psychology*, *22*(5), 453–474. doi:10.1016/0022-1031(86)90045-4

Armitage, C. J., & Conner, M. (1999). Distinguishing Perceptions of Control From Self-Efficacy: Predicting Consumption of a Low-Fat Diet Using the Theory of Planned Behavior. *Journal of Applied Social Psychology*, *29*(1), 72–90. doi:10.1111/j.1559-1816.1999.tb01375.x

Bandura, A. (1991). Social cognitive theory of self-regulation. *Organizational Behavior and Human Decision Processes Organizational Behavior and Human Decision Processes*, *50*(2), 248–287. doi:10.1016/0749-5978(91)90022-L

Barber, B. (1983). *The logic and limits of trust* (Vol. 96). Rutgers University Press New Brunswick.

Bell, S., Cornford, D., & Bastin, L. (2013). The state of automated amateur weather observations. *Weather*, *68*(2), 36–41. doi:10.1002/wea.1980

Bonney, R., Shirk, J. L., Phillips, T. B., Wiggins, A., Ballard, H. L., Miller-Rushing, A. J., & Parrish, J. K. (2014). Next steps for citizen science. *Science*, *343*(6178), 1436–1437. doi:10.1126/science.1251554 PMID:24675940

Chen, C.-J., & Hung, S.-W. (2010). To give or to receive? Factors influencing members knowledge sharing and community promotion in professional virtual communities. *Information & Management*, *47*(4), 226–236. doi:10.1016/j.im.2010.03.001

Chiu, C.-M., Hsu, M.-H., & Wang, E. T. G. (2006). Understanding knowledge sharing in virtual communities: An integration of social capital and social cognitive theories. *Decision Support Systems*, *42*(3), 1872–1888. doi:10.1016/j.dss.2006.04.001

Cho, H., Chen, M., & Chung, S. (2010). Testing an integrative theoretical model of knowledge-sharing behavior in the context of Wikipedia. *ASI Journal of the American Society for Information Science and Technology*, *61*(6), 1198–1212.

Chow, W. S., & Chan, L. S. (2008). Social network, social trust and shared goals in organizational knowledge sharing. *Information & Management*, *45*(7), 458–465. doi:10.1016/j.im.2008.06.007

Crall, A. W., Newman, G. J., Stohlgren, T. J., Holfelder, K. A., Graham, J., & Waller, D. M. (2011). Assessing citizen science data quality: An invasive species case study. *Conservation Letters*, *4*(6), 433–442. doi:10.1111/j.1755-263X.2011.00196.x

Endfield, G. H., & Morris, C. (2012). Well weather is not a girl thing is it? Contemporary amateur meteorology, gender relations and the shaping of domestic masculinity. *Social & Cultural Geography*, *13*(3), 233–253. doi:10.1080/14649365.2012.677470

Franzoni, C., & Sauermann, H. (2014). Crowd science: The organization of scientific research in open collaborative projects. *RESPOL Research Policy*, *43*(1), 1–20. doi:10.1016/j.respol.2013.07.005

Ganzevoort, W., & van den Born, R. J. G. (2016). Beyond 'data drones': Citizen scientists' concerns and motivations. Paper presented at the Citizen Observatories for Water Management Conference, Venice, Italy.

García, L., Rodríguez, D., Wijnen, M., & Pakulski, I. (2016). *Earth Observation for Water Resources Management: Current Use and Future Opportunities for the Water Sector*. World Bank Publications. doi:10.1596/978-1-4648-0475-5

Gharesifard, M., & Wehn, U. (2016). To share or not to share: Drivers and barriers for sharing data via online amateur weather networks. *Journal of Hydrology (Amsterdam)*, *535*, 181–190. doi:10.1016/j.jhydrol.2016.01.036

Haklay, M. (2015). Citizen Science and Policy: A European Perspective. The Wodrow Wilson Center, Commons Lab.

Harjanne, A., Ervasti, T., Karhu, J. A., & Tuomenvirta, H. (2015). Combining science education with citizen science - Experiences from a research institute led science education project. *LUMAT*, *3*(7), 948–959.

Hew, K. F., & Hara, N. (2007). Knowledge sharing in online environments: A qualitative case study. *ASI Journal of the American Society for Information Science and Technology*, *58*(14), 2310–2324. doi:10.1002/asi.20698

Hosmer, L. T. (1995). Trust: The connecting link between organizational theory and philosophical ethics. *Academy of Management Review*, *20*(2), 379–403.

Hsu, C.-L., & Lin, J. C.-C. (2008). Acceptance of blog usage: The roles of technology acceptance, social influence and knowledge sharing motivation. *INFMAN Information & Management*, *45*(1), 65–74.

Hsu, M.-H., Ju, T. L., Yen, C.-H., & Chang, C.-M. (2007). Knowledge sharing behavior in virtual communities: The relationship between trust, self-efficacy, and outcome expectations. *International Journal of Human-Computer Studies*, *65*(2), 153–169. doi:10.1016/j.ijhcs.2006.09.003

Kaufmann, N., Schulze, T., & Veit, D. (2011, August 4 - 7). *More than fun and money. Worker Motivation in Crowdsourcing-A Study on Mechanical Turk.* Paper presented at the AMCIS, Detroit, MI.

Lanfranchi, V., Wrigley, S., Ireson, N., Ciravegna, F., & Wehn, U. (2014). *Citizens' Observatories for Situation Awareness in Flooding.* Paper presented at the 11th International ISCRAM Conference (Information Systems for Crisis and Response Management), May 2014, University Park, PA.

Lin, H. F. (2008). Determinants of successful virtual communities: Contributions from system characteristics and social factors. *Information & Management*, *45*(8), 522–527. doi:10.1016/j.im.2008.08.002

Lin, M.-J. J., Hung, S.-W., & Chen, C.-J. (2009). Fostering the determinants of knowledge sharing in professional virtual communities. *Computers in Human Behavior*, *25*(4), 929–939. doi:10.1016/j.chb.2009.03.008

Manley, G. (1952). The Weather and Diseases: Some Eighteenth-Century Contributions to Observational Meteorology. *Notes and Records of the Royal Society of London*, *9*(2), 300–307. doi:10.1098/rsnr.1952.0017

Manstead, A. S. R., & Van Eekelen, S. A. M. (1998). Distinguishing Between Perceived Behavioral Control and Self-Efficacy in the Domain of Academic Achievement Intentions and Behaviors. *Journal of Applied Social Psychology*, *28*(15), 1375–1392. doi:10.1111/j.1559-1816.1998.tb01682.x

McLure Wasko, M., & Faraj, S. (2000). It is what one does: Why people participate and help others in electronic communities of practice. *The Journal of Strategic Information Systems*, *9*(2-3), 155–173. doi:10.1016/S0963-8687(00)00045-7

Minkman, E. (2015). Citizen Science in Water Quality Monitoring: Developing Guidelines for Dutch Water Authorities for Contributory Mobile Crowd Sensing. TU Delft, Delft University of Technology.

Muller, C. L., Chapman, L., Johnston, S., Kidd, C., Illingworth, S., Foody, G., & Leigh, R. R. et al. (2015). Crowdsourcing for climate and atmospheric sciences: Current status and future potential. *JOC International Journal of Climatology*, *35*(11), 3185–3203. doi:10.1002/joc.4210

Müller, R. M., Thoring, K., & Oostinga, R. (2010). Crowdsourcing with semantic differentials: A game to investigate the meaning of form. *AMCIS Proceedings*.

Nature. (2015). Rise of the citizen scientist. *Nature, 524*(7565), 265-265. doi:10.1038/524265a

Ngo Thu, H., & Wehn, U. (2016). Data sharing in international transboundary contexts: The Vietnamese perspective on data sharing in the Lower Mekong Basin. *Journal of Hydrology Journal of Hydrology*, *536*(2), 351–364.

Pavlou, P. A., & Fygenson, M. (2006). Understanding and predicting electronic commerce adoption: An extension of the theory of planned behavior. *Management Information Systems Quarterly*, 115–143.

Plengsaeng, B., Wehn, U., & van der Zaag, P. (2014). Data-sharing bottlenecks in transboundary integrated water resources management: A case study of the Mekong River Commissions procedures for data sharing in the Thai context. *Water International*, *39*(7), 933–951. doi:10.1080/02508060.2015.981783

Sabini, J. (1995). *Social Psychology* (2nd ed.). New York: W. W. Norton & Company, Inc.

Subkowski, P. (2006). On the psychodynamics of collecting. *The International Journal of Psycho-Analysis*, *87*(2), 383–401. doi:10.1516/4UMF-YF9G-FVFR-JM09 PMID:16581582

Tsai, W., & Ghoshal, S. (1998). Social capital and value creation: The role of intrafirm networks. *Academy of Management Journal*, *41*(4), 464–476. doi:10.2307/257085

WEF. (2015). *Global Risks 2015, Insight Report* (10th ed.). World Economic Forum.

WEF. (2016). *Global Risks 2016, Insight Report* (11th ed.). World Economic Forum.

Wehn, U., Rusca, M., Evers, J., & Lanfranchi, V. (2015). Participation in flood risk management and the potential of citizen observatories: A governance analysis. *Environmental Science & Policy*, *48*(0), 225–236. doi:10.1016/j.envsci.2014.12.017

Wehn de Montalvo, U. (2003a). In search of rigorous models for policy-oriented research: A behavioural approach to spatial data sharing. *Journal of the Urban and Regional Information Systems Association*, *15*(1), 19–28.

Wehn de Montalvo, U. (2003b). *Mapping the determinants of spatial data sharing*. Ashgate Pub Ltd.

KEY TERMS AND DEFINITIONS

Amateur Weather Data: Weather data that are collected and shared by members of the general public (can also include experts in meteorology science) using Personal Weather Stations (as compared to official weather observations).

ICT-Enabled Citizen Science: Citizen Science activities that are employing Information Communication Technologies (ICTs) such as low-cost and innovative sensor devices, smart phones and social media to facilitate data collection and sharing by the members of the general public.

Knowledge Self-Efficiency: The level of confidence that someone has in her/his own knowledge about a certain topic (in this case, about Meteorology Science and weather data collection methods).

Online Amateur Weather Network: A virtual network of amateur weather observers that hosts, aggregates and visualizes amateur weather data on an online platform.

Personal Weather Station (PWS): A set of sensors and instruments that enables the measurement of different weather attributes (often in an automated way) and which is normally installed at the user's home or work place.

Societal Outcomes: The evaluations or implications of a certain behavior on society at large, as compared to its implications for the performer of the behavior.

Weather Enthusiast Community: Individuals or groups of individuals who are interested in weather data for different and often personal reasons.

Chapter 5
The Social Function of Citizen Science:
Developing Researchers, Developing Citizens

Luis Arnoldo Ordóñez Vela
Fundación InterConectados, Venezuela

Giovanna Lombardi
Universidad Central de Venezuela, Venezuela

Enrico Bocciolesi
eCampus University, Italy

Robin M. Urquhart
Independent Researcher, USA

ABSTRACT

This chapter focuses on the risk that, when citizen science is introduced in social environments different from those in the Global North where it originated, it may be subject to the error of providing the right answer to the wrong question. To avoid this type of errors, it is necessary to train those who participate in citizen-science studies: citizens as well as researchers. Otherwise, we may encounter new forms of scientific dependence that benefit knowledge accumulation and policy decision-making in the Global North, without contributing to the quality of life of those who carry out the studies. This chapter analyzes the relationship between civic development, citizen science and ways of implementing research conclusions through public policies, given the characteristics of political and citizen participation in the Global South. Here, the introduction of citizen science is seen as an opportunity to construct a more inclusive and participatory society, and to reduce the risk of returning to paternalistic, passivity-inducing and purely instrumental approaches to development.

INTRODUCTION

It is ironic indeed that the UK Government is credited with providing 'good quality, accessible and accurate health information on Ebola' for its own UK population, while making no mention of what is much more important: the chaotic situation of information and education in countries affected by Ebola. (Neil Pakenham-Walsh, 2016, HIFA, personal communication)

DOI: 10.4018/978-1-5225-0962-2.ch005

The Global South in general, and Latin America in particular, are scientifically interesting but poverty-ridden regions (España, 2004). It is not surprising that Latin America suffers from endemic forms of a series of social problems. The information needed for defining appropriate public policies to address issues ranging from poverty, authoritarianism and lack of education, to educating the population in citizenship, may be generated by research. But complex problems can only be studied and analyzed in complex ways. In this respect, Citizen Science has much to contribute: it can contribute information that is useful for defining social policies while increasing social capital and engagement in the communities where it is applied.

Any intervention in a "developing region" must take "development discourse" into consideration: this encompasses the

...set of techniques and power-knowledge relationships (that) has been operating in different ways in the Third World ever since Development was defined as "a response to the problematization of poverty that took place in the years following World War II, and not a natural process of knowledge accumulation that gradually uncovers problems and deals with them; as such, development must be seen as a historical construct that provides a space in which poor countries are known, specified and intervened in" (Escobar, 1995:44-45in Aguilar, 2010).

With this in mind, the problem that will be addressed in this chapter is how to develop Citizen Science in the Global South in such a way as to help develop citizenship. This implies two things: developing collaborative scientists on the one hand and collaborative citizens on the other. These are difficult tasks since the authoritarian past in the Global South has led to scientists and university teachers being perceived as members of the ruling elite, so they have an uneasy relationship with the communities where the "citizens" are to be recruited for Citizen Science.

BACKGROUND

The intention of this book, Citizen Science in Modern Research, is to generate discussion of this new discipline, Citizen Science, in order to formalize it in its early stages. This would lead to better cooperation among citizen-science initiatives, allow participants to address topics which are seldom explored, and lead to an examination of the ways in which Citizen Science relates to other sciences. It is expected that the book will provide inspiration to researchers who have designed tools for specific uses to seek new uses for them and to *realize their**potential for social innovation*** (bold type is the authors').

Many definitions of Citizen Science emphasize its potential for social innovation, for example, projects in which volunteers partner with scientists to answer real-world questions (Cornell Lab of Ornithology); through Citizen Science, people share and contribute to data monitoring and collection programs. Usually this participation is done as an unpaid volunteer (National Geographic): it can contribute to collaborative local environmental management, build community capacity, develop scientific literacy and increase citizen stewardship (Kerry Riddell); it can create a nexus between science and education ... and ...expand ...public engagement.(Newman et al. 2012). The truth is that neither citizen participation in volunteer activities nor the consequences of such participation for community development should be taken for granted (Hyatt, 2001). On the one hand,

It is necessary to improve our understanding of what motivates citizen scientists, since the amount and quality of participation vary from one project to another, and "factors that enhance participation fre-

quency may not necessarily lead to enhanced contribution quality". Careful investigation of the effects of motivation is required. (Nov, Arazy & Anderson, 2014).

On the other, it is necessary to examine whether a particular Citizen Science project is conceived of as a panacea –

- a panacea refers to a blueprint for a single type of governance system (e.g., government ownership, privatization, community property) that is applied to all environmental problems(...). Practitioners and scholars who fall into panacea traps falsely assume that all problems of resource governance can be represented by a small set of simple models, because they falsely perceive that the preferences and perceptions of most resource users are the same. (Ostrom, Janssen & Anderies, 2007).

Therefore it is necessary to ask whether Citizen Science as proposed in many studies is a panacea in Ostrom's sense (Ostrom, 2007; Ostrom et al, 2007) and what steps should be taken in order to obtain the desired benefits for both scientists and citizens.

INCURRING IN TYPE III ERRORS

Type III errors occur when researchers provide the right answer to the wrong problem. (Kimball, cited by Dunn, 2015)

A critical issue of problem structuring is how well... formal problems actually correspond to the original problem situation. If most problem situations in fact contain whole systems of problems (messes), then a central requirement of policy analysis is the formulation of... formal problems that adequately represent that complexity (Dunn, W. N., 2015).

With this in mind, to develop Citizen Science in the Global South and Latin America in particular, it is necessary consider when and how to introduce Citizen Science into communities. In Latin America, there are two separate worlds that coexist culturally in nations, that of the "people" and that of the "masters" (the rich, the educated, the powerful, including university teachers and politicians). With such social inequality, there are two problems to be faced if Citizen Science is to fulfill its potential as a builder of community capacity and citizen stewardship: how to develop collaborative scientists and how to develop collaborative citizens when introducing Citizen Science into a given context, as depicted in Figure 1. Citizen science and development. Factors to be considered

Figure 1. Citizen science and development: Factors to be considered

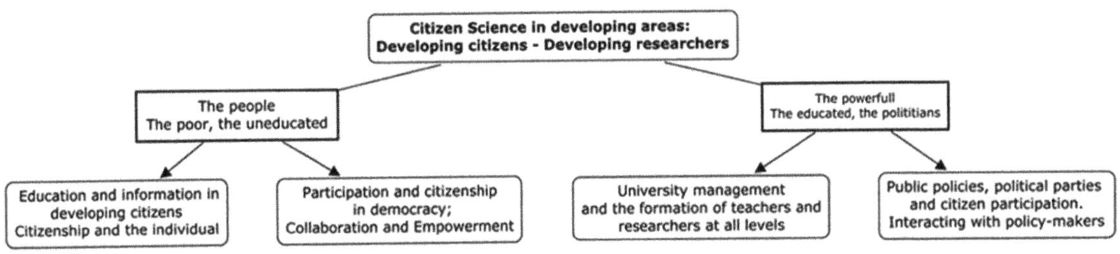

CITIZEN SCIENCE AND "DIFFERENT APPROACHES TO RESEARCH"

Various aspects of scientific research - mainly data collection and analysis - are labor-intensive, time-consuming, and consequently costly. Online Citizen Science reduces the costs of scientific research, increases the resources available to research teams, fosters a partnership between citizens and scientists, and enhances public understanding of science. (Nov, Arazy, & Anderson, 2014)

Though Citizen Science would seem to offer many advantages, it is not the only possible approach to research in communities, as can be seen from Figure 2, based on an interview with John Gaventa from AURA (African University Research Approaches).

From this figure, the authors conclude that Citizen Science is only one of many possible kinds of research "with" and "for" the people. The degree of citizen participation can vary widely, and one approach, derived from studies of habitats, has suggested that research can be classified according to a typology of monitoring categories defined by their degree of local participation, ranging from studies with no local involvement where professional researchers undertake the monitoring, to entirely local efforts where the monitoring is undertaken by local people.

The type of monitoring that will be required is an important aspect to be considered when designing Citizen Science projects. According to Danielson et al. (2009),

Locally based monitoring is particularly relevant in developing countries, where it can lead to rapid decisions to solve the key threats affecting natural resources, empower local communities to better manage their resources, and refine sustainable-use strategies to improve local livelihoods. Nevertheless, we recognize that the accuracy and precision of the monitoring undertaken by local communities in different situations needs further study and field protocols need to be further developed to get the best from the unrealized potential of this approach.

Figure 2. Approaches to research from the point of view of the community

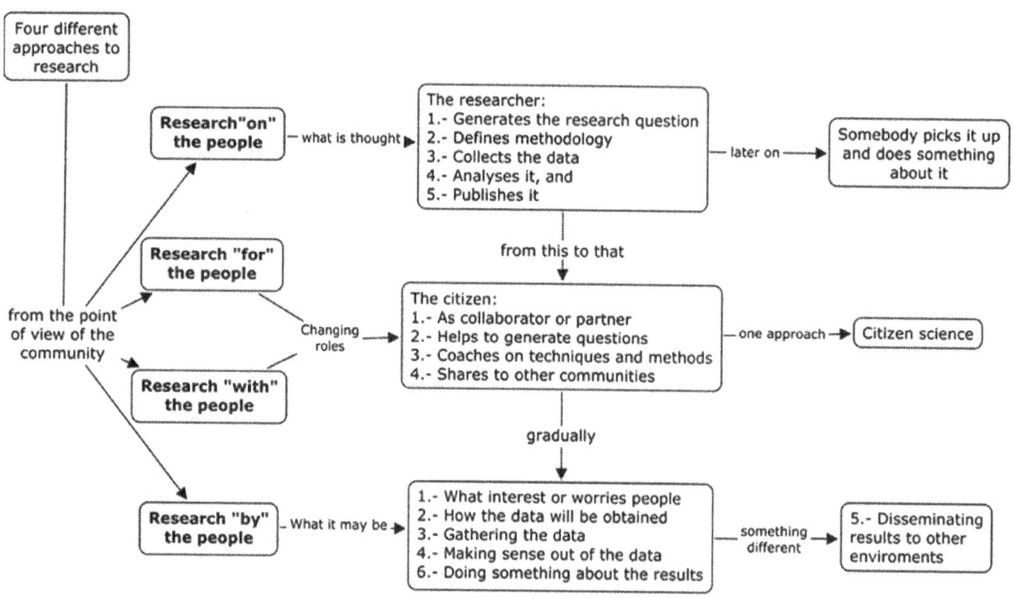

Another aspect that requires careful consideration before introducing Citizen Science into a community is the kind of "community development" that is desired. Every person is immersed in the concepts and categories of their own particular culture, and words like "poor", "emergent" (used by the International Monetary Fund and the United Nations Development Programme), "undeveloped" (used by the World Bank), "middle income", "sustained partner of US" (used by the United States Agency for International Development), "lower middle class", "poverty line", "modernization", and "vulnerable" mean different things in different places. Aguilar (2010) points out that this presents a particular problem for researchers working in countries or even communities other than their own. If one of the aims of Citizen Science is to achieve some kind of community development in poor and underdeveloped regions, researchers must be aware of these categories that inhibit appropriate "problem structuring".

A similar conclusion can also be drawn from the development discourse that appears in academic papers or public agencies, as shown in Figure 3.

These authors reach the conclusion that "social development" (whatever that is) is not necessarily going to derive from, or be stimulated by, Citizen Science or any other kind of research undertaken in communities. As can be seen from Figure 3, social capital is not an automatic result of generating knowledge through research in a given community.

Thus, one of the first challenges encountered when designing a particular Citizen Science project, is to determine what type of development is desired and how to achieve it. There are many possibilities and the magnitude of the hurdle to be surmounted is suggested by the following list of concepts:

- **Communities of Practice (CoP):** A community of practice is a group of people who share a craft and/or a profession. The concept was first proposed by cognitive anthropologists Jean Lave and Etienne Wenger in 1991 (Wikipedia). A CoP can evolve naturally because of the members' common interest in a particular domain or area, or it can be created deliberately with the goal of

Figure 3. Achieving development through Citizen Science

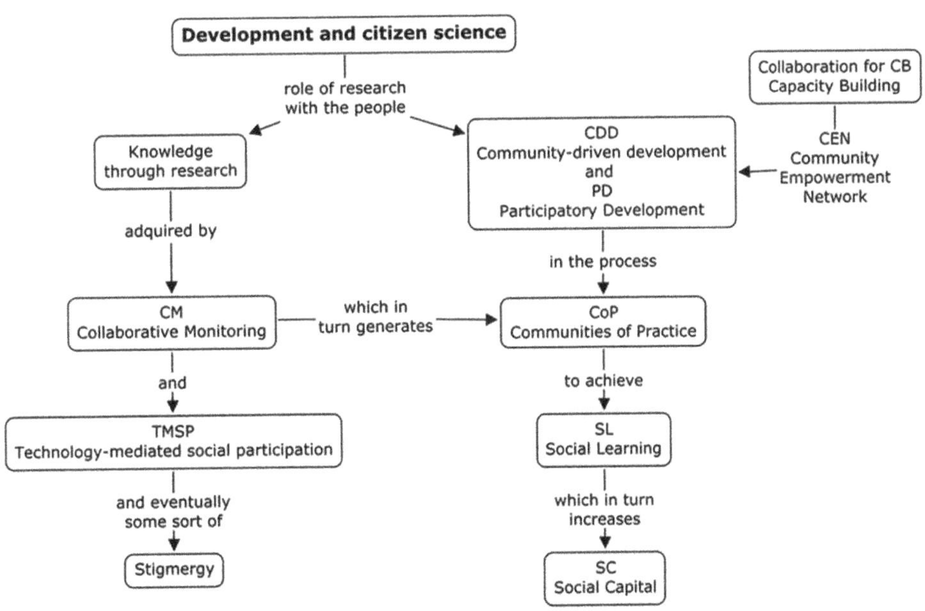

gaining knowledge related to a specific field. It is through the process of sharing information and experiences with the group that the members learn from each other, and have an opportunity to develop themselves personally and professionally (Wenger, 1998).

- **Capacity Building (CB):** Capacity building refers to efforts to strengthen economies, governments, institutions and individuals through education, training, mentoring, and the infusion of resources. Capacity building aims at developing secure, stable, and sustainable structures, systems and organizations, with a particular emphasis on using motivation and inspiration for people to improve their lives (Jones, 2014).
- **Community-Driven Development (CDD):** CDD is used to describe projects that encourage local communities to control the direction of their own development, from project conception through implementation and evaluation: a kind of community of practice (Caldwell Johnson, 2005); a sub-form of social capital that can have a positive effect on citizen empowerment and development (Esman, 2003).
- **Community Empowerment Network (CEN):** This network has the intention of building the capacity of communities and their development partners to design and deliver more effective CDD projects that respond to the needs of the poor. This goal includes efforts to: increase understanding of what makes CDD projects succeed or fail; enhance access to relevant capacity development materials; and enhance coordination among development partners participating in capacity development activities (Caldwell Johnson, 2005).
- **Collaborative or Participatory Monitoring (CM):** This is a type of oversight that involves multiple individuals or organizations with different interests and forms of expertise in the design and implementation of monitoring (Fernandez-Gimenez et al., 2008). It provides participants with the opportunity to develop themselves personally and professionally (Wenger, 1998).
- **Development:** It can be conceptualized as an arena, meaning a socially constructed and bounded field, or space, within which actors negotiate, and struggle for, the wide spectrum of resources ranging from material ones (e.g., salaries of development workers or infrastructural outputs of a project) to political and symbolic ones (e.g., prestige and authority) (adapted from Cima, 2015).
- **Participatory Development (PD):** PD seeks to engage local populations in development projects.
- **Social Capital (SC):** Refers to social organization that enables collective action and thereby citizen empowerment (Esman, 2003).
- **Social Learning (SL):** To be considered social learning, a process must: (1) demonstrate that a change in understanding has taken place in the individuals involved; (2) demonstrate that this change goes beyond the individual and becomes situated within wider social units or communities of practice; and (3) occur through social interactions and processes between actors within a social network (Reed et al., 2010).
- **Stigmergy:** This is the phenomenon of indirect communication mediated by modifications of the environment (Marsh & Onof, 2008).
- **Technology-Mediated Social Participation (TMSP)**: Recent years have seen a substantial growth in the scale and scope of TMSP projects, such as Wikipedia and Linux, which rely on volunteers who contribute their time, energy and skills for the creation of a public good.

GOING BEYOND PANACEAS IN CITIZEN SCIENCE

Regarding the development problems of the Global South, it is not easy to define the role, if any, that might be assigned to citizen-driven research in producing, or at least catalyzing, "cultural community transformations". As Ostrom (2007) indicates, "Moving beyond panaceas to develop cumulative capacities to diagnose the problems and potentialities of linked social–ecological systems (SESs) requires serious study of complex, multivariable, nonlinear, cross-scale, and changing systems."

For example, in on-line Citizen Science, there is a need to study how people use the internet and the web. Online Citizen Science is based on two pillars:

1. A technological pillar, which involves developing computer systems for managing large amounts of distributed resources, and
2. A motivational pillar, which involves attracting and retaining volunteers who would contribute their skills, time, and effort to a scientific cause.

As Nov, Arazy, and Anderson (2014) point out,

While the technological dimension has been widely studied, the motivational dimension of Citizen Science received little attention to date [...] As the sustainability of online Citizen Science projects depends on volunteers [...] Building on the social movement participation model, findings from a longitudinal empirical study in three different Citizen Science projects reveal that quantity of contribution is determined by collective motives, norm-oriented motives, reputation, and intrinsic motives. Contribution quality, on the other hand, is positively affected only by collective motives and reputation.

Obviously, this kind of study may only be performed in certain areas, more akin to the Global North, and the conclusions drawn may not be extrapolated to other communities.

The motivational dimension of Citizen Science has cultural roots. The case of Wikipedia is an almost perfect example of stigmergic achievement and, given its widespread utilization, could be considered a good indicator of cultural differences in collaborative participatory effort, as may be appreciated in Table 1.

These data show that the ratio of speakers to articles is very different for different languages/cultures. Swedish has the largest relative production of Wikipedia articles: 1 page is produced for every 3.1 speakers. Chinese has the smallest relative production of Wikipedia articles: for every article produced there are 1371.6 speakers. The largest relative production of articles is for languages confined mainly to highly developed and prosperous European countries that had small or no empires and therefore did not export their languages to other heavily populated places (Polish, Italian, French, Dutch and Swedish). The relative production of articles for languages exported to many parts of the world by countries with large empires (Spain, Britain and Portugal) is much smaller. This difference may be due to the fact that many ex-colonies are still developing and their populations tend to be less educated and have less access to technology for economic reasons. (e.g. Argentina, Brazil, India).

The relationship between development and the ratio of speakers per article seems to be confirmed when comparing English and Spanish. The total number of English speakers in the world is comparable to the number of speakers of Spanish (505 million vs 470 million). However, the numbers of speakers per article are 98,5 and 375, respectively. Many of the English speaking countries formerly in the British empire are developed countries (e.g. USA, Canada, Australia) while very few countries from

Table 1. Wikipedia, number of speakers per article in a number of selected countries

Wikipedia 2013			
Language	**Articles**	**Speakers**	**Speakers/articles**
Swedish	1.300.000	12.000.000	10
Italian	1.000.000	60.000.000	60
English	4.000.000	300.000.000	80
Spanish	1.000.000	400.000.000	400
Wikipedia 2016			
Language	**Articles**	**Speakers**	**Speakers/articles**
Swedish	3.000.000	9.000.000	3
Italian	1.300.000	60.000.000	50
English	5.000.000	500.000.000	100
Spanish	1.300.000	500.000.000	400

Comparative data 2013 – 2016.

the former Spanish and Portuguese empires are developed countries. Thus the number of speakers per article would seem to be related to the level of development of the countries where those languages are spoken. Since development implies higher levels of education and greater access to technology, it can be concluded that educational level and technological access are the cultural factors with the greatest effect on people's motivation to participate in collaborative projects such as Wikipedia. Other factors could be considered, such as the cultural management of information, cultural or economic difficulties for the transfer of technology, or the greater or lesser degree of transparency at the level of Government, but at this point they are merely speculation. All in all, more research is required on the cultural factors affecting people's motivation before introducing Citizen Science into a given culture.

CITIZENSHIP, PARTICIPATION, AND EMPOWERMENT

The interplay of different elements in promoting a favorable environment for development is shown in Figure 4. Education and information in developing citizens.

In general, citizenship, participation and democracy provide a favorable environment for development. According to the definition given by Marshall (1949) on the UNESCO website (2016), citizenship is a complex and difficult concept that includes rights, knowledge and culture. It can be defined as

...the status of having the right to participate in and to be represented in politics. (Baylis & Smith, 2001). It is a collection of rights and obligations that give individuals a formal juridical identity. Marshall, whose work has long dominated the debates about social citizenship, considered citizenship as: a status bestowed on those who are full members of a community. All who possess the status are equal with respect to the rights and duties with which the status is endowed. (UNESCO, 2016).

However, as Lloyd, Lipu, & Kennan (2010) point out, "A prerequisite for participation, inclusion, and informed citizenship is the ability to develop knowledge from information about the social, economic,

Figure 4. Education and information in developing citizens

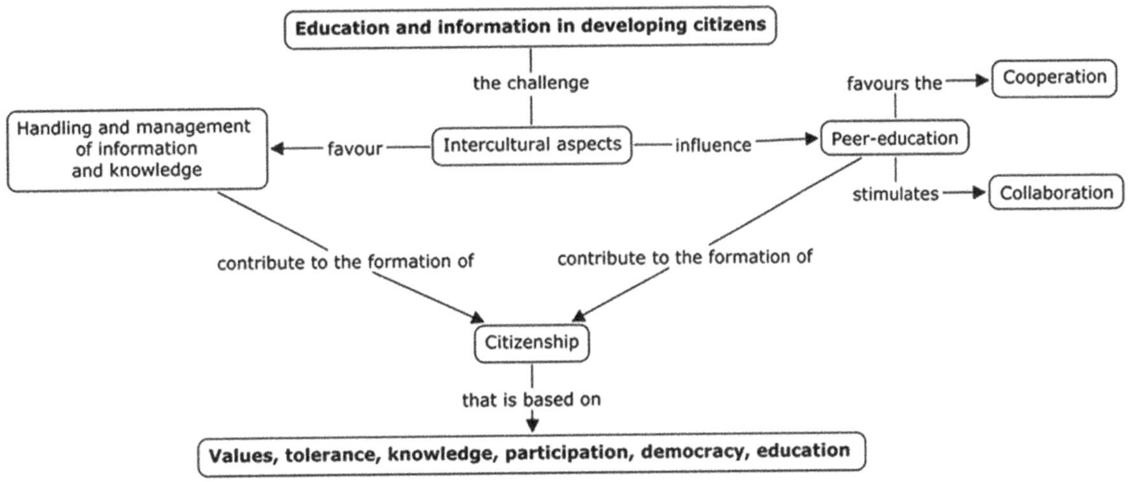

and community dimensions through which modern... society is constituted." The acquisition of such knowledge and the values underlying citizenship require the collaboration of many educational actors. To be a citizen means to be aware of one's rights and duties, but teaching these rights and duties is not a simple undertaking. Families, schools and the media all need to make their contributions. And when Citizen Science is introduced into a given community, it has the potential to act in synergy with these educational forces.

The book *Citizenship and Education in Twenty-eight Countries: Civic Knowledge and Engagement at Age Fourteen* (Torney-Purta et al, 2001), offers interesting perspectives on the relationship between citizenship and education. They asked, *What does democracy mean to young people in different parts of the world? What is their implicit theory regarding what a democracy is and what is likely to strengthen or to weaken it? They are exhorted to be good citizens, but what does that concept imply? If young people read that 'the government' should (or should not) be expected to take certain responsibilities, what do they think that means?* To answer these questions, the authors analyzed the replies of young people to a questionnaire about what is good and what is bad for democracy. They concluded that *Educational practices play an important role in preparing students for citizenship. Schools that operate in a participatory democratic way foster an open climate for discussion within the classroom and invite students to take part in shaping school life and are effective in promoting both civic knowledge and engagement. Many students, however, do not perceive this participatory climate in their classrooms or these opportunities in their schools.* But it is necessary to ask whether the results of such a study would be the same in the present-day global and internet-ridden scenario, or whether and in what way they might differ.

Citizenship training is a cornerstone of adult education as it is practiced today. For example, in his book Pedagogy of the Oppressed, Freire (1968) describes his approach to adult education, which links the identification of issues to positive action for change and development. His ideas invite researchers to rethink the subtle interplay of communication between society and the individual, the relationship of formal and informal education at all levels, and the way information and knowledge intermingle in a given culture. Freire's approach is also valid for Citizen Science. Although Citizen Science is not an educational initiative in itself, it offers an opportunity for learning about citizenship, since active participants in the joint improvement of their community's living space learn to link knowledge to action.

To foster conscious citizenship, Citizen Science could also benefit from incorporating new practices that have been found effective in mainstream education. In 1984, William Damon wrote *Two different forms of peer learning, "peer tutoring" and "peer collaboration," are distinguished. Each has its potential use: peer tutoring for transmitting information and drilling special skills; peer collaboration for facilitating intellectual discovery and the acquisition of basic knowledge.[...] The recent interest in peer-based learning has arisen from a number of converging trends within psychology and education, trends that are quite distinct and often opposing* (Damon, 1984, p.331). Peer education might be a useful element that could be incorporated into Citizen Science in order to foster conscious citizenship.

Being a citizen is, first of all, an act of awareness, to be aware of the possibilities offered by your own particular environment, and of ways to contribute to, and help others to structure, a common living space. Only if this awareness is developed inside a community can Citizen Science achieve its full potential, that is, helping science and helping the communities where it is inserted.

And within a changing world, new educational and social perspectives have come to the fore, such as integration and inter-culturalism, as has the awareness that citizen participation must be valued and continuously and consistently supported. This is important for Citizen Science because of its intercultural nature, in that that researchers and citizens from different cultures interact within projects. What's more, according to the *Conceptual and Operational Framework UNESCO* of 2013 (pp.4-5) "intercultural interactions have become a constant feature of modern life, even in the most traditional societies, [...] Hence the growing awareness among policy-makers and civil society that intercultural competences may constitute a very relevant resource to help individuals [...]. Such intercultural competences can be defined as abilities to adeptly navigate complex environments marked by a growing diversity of peoples, cultures and lifestyles, [...]" (Fantini & Tirmizi, 2006). Thus it behooves researchers involved in Citizen Science to become aware of, assess and build intercultural competences that must become intercultural values not only in the communities that are the object of Citizen Science, but in their research teams as well.

Nowadays, knowledge and information management is a particularly important intercultural competence for achieving a favorable social environment and citizenship development. Every citizen needs to be able to access tools and resources and for this they need to acquire skills in information retrieval and processing. In the 21st century, people speak of the Information Society, of complexity. In this type of environment, technology influences many things, and offers alternatives and possibilities. As Harrison and Stephen, (1996, p.205) point out, "Computer networks are having a major impact on enhancing and transforming teaching and learning relationships, opportunities, and outcomes. Traditional educational structures are being dramatically altered by new communication and information technologies."

An interesting example of this may throw light on the effects of different and innovative variables, conditions, modes of work and information management. King & McGrath (2002) point out that

In some countries, women with as little as four years of education are more likely to choose to have smaller, healthier families and decide to send their own children to school. Education contributes significantly to the improvement of health status by enhancing the capacity of men and women to care for their own health and that of their families, and to make more effective use of preventive and curative services. A good primary education can help to foster agricultural innovation and improve the capacity of the poor to make use of their environment in a sustainable way. (DFID, 2000c, p.2)

However, in many places, awareness and management of knowledge still does not seem sufficiently developed (Ugalde et al., 2004), even though, as McGrath and King (1994) point out, education and content knowledge must be understood as necessary tools for the improvement of citizenship, the community and its social and cultural environment. This is so even though there is still some confusion about the definitions and use of related terms, such as education, knowledge and information (Bocciolesi & González, 2015).

A similar confusion seems to occur with terms such as social capital and empowerment, and their impact on community development. From Alguezaui and Filieri (2010) it can be concluded that while social capital may have several positive impacts as far as information access is concerned, there is also a number of less beneficial impacts... social capital might not be beneficial as it makes it difficult to obtain novel information; it might not consider certain new information or information from external sources (as various stories of technology blindness remind us);and lead to norms of control that turn into resistance to innovation (as in the case of the famous "not invented here" syndrome). Such factors can also affect Citizen Science. In sum, citizenship-derived participation evolves in complex ways to generate the synergy capable of improving local living conditions along with local organizational capacity, as shown in Figure 5. Participation and citizenship empowerment

It is a well- established principle in political sociology that empowerment of the poor and the weak is usually the product of a struggle. The ability to struggle, to overcome resistance, to achieve some measure of control over their lives, to secure a fairer share of public services and government resources depends on the formation of social capital, usually in the form of organization. Organization enables ordinary men and women to mobilize their collective energies in pursuit of common goals. Thus the principal expression of social capital is through autonomous social organization. Numerous examples can be cited of its contribution to empowerment. (Esman, 2003) These examples strongly support the proposition associated with the theory of mass society that the existence of voluntary associations increases the democratic potential of a society. Democracy depends upon citizen participation, and it is clear that organizational membership is directly related to such participation (Almond & Verba, 2003).

Figure 5. Participation and citizenship empowerment

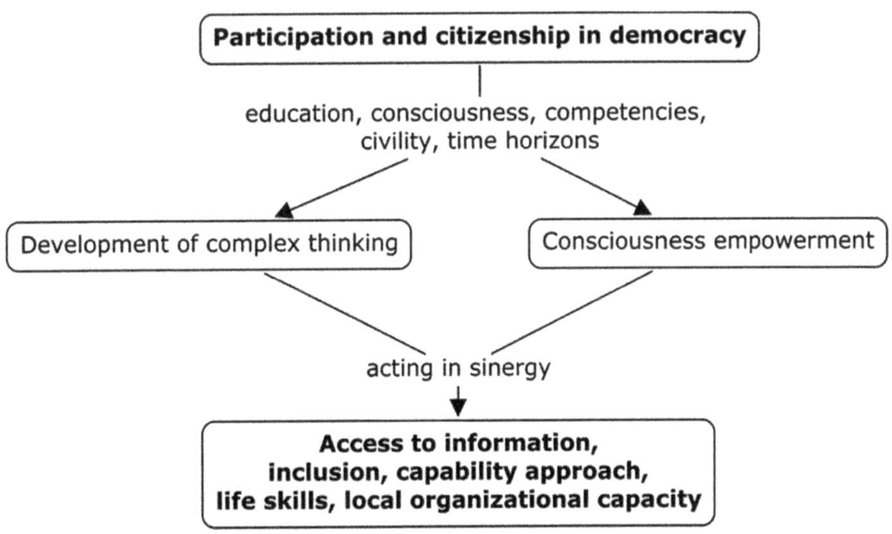

Thus, voluntary associations where members communicate and exchange ideas are extraordinary mechanisms for building citizenship. As Dewey says in his essay Democracy and Education, "Society not only continues to exist by transmission, by communication, but it may fairly be said to exist in transmission, in communication" (1916, p.8).

How and why do people participate in public life? How do they get involved in community and political affairs? How do the associations and networks that connect people to one another and structure their social and political interactions determine their civic values? An interesting approach is put forward in the book Connected (Christakis & Fowler, 2009) where the possibilities and consequences of being connected are explored. These authors indicate that topics such as the relationship between e-government and participatory culture come to be natural areas for exploration, and research in these areas could shed new light on the connectivity generated through civil society and social capital, their impact on the way democracy functions, and on the stability and change of political regimes. A project created by the World Bank promises to be illuminating in this respect. Now that municipal governments are beginning to create websites that enable citizens to interact with governments, the World Bank is interconnecting the municipal websites of 10 capital cities in Latin America, in order to encourage the exchange of experiences as well as public engagement. (Narayan-Parker, D., 2002) It will be interesting to study the impact of this kind of development on civic life. Only when these issues become better understood will researchers in the Global South be able to undertake studies such as the one described by Nov, Arazy and Anderson, (2014) as follows:

The present study contributes to theory in two ways. First, it increases our understanding of what motivates citizen scientists. The second contribution goes beyond the unique Citizen Science context to the study of TMSP in general, and concerns the differences between contribution quantity and quality in their motivational underpinnings. Our findings from one Citizen Science project provide preliminary evidence that factors that enhance participation frequency may not necessarily lead to enhanced contribution quality, and in some cases actually detract from quality.

DEVELOPING TEACHERS AND RESEARCHERS: THE SOCIALLY RESPONSIBLE UNIVERSITY

Citizens are made, not born, as Tertullian said of Christians. To be and feel oneself to be a citizen is not a product of nature – it is the result of a cultural process in the personal history of each individual and the collective history of society. (François-Xavier Guerra, 1999)

In their work on Communities of Practice, Hart and Wolff (2006) state, "To illustrate what they believe these 'new forms of engagement' to be producing, Gibbons and colleagues (1994) set out a distinction between Mode 1 knowledge and Mode 2. Mode 1 knowledge is pure, disciplinary, homogenous, expert-led, supply driven, hierarchical, peer-reviewed and almost exclusively university-based… Mode 2 is applied, problem-centered, transdisciplinary, heterogeneous, hybrid, demand-driven, entrepreneurial, network-embedded and often (increasingly) handled outside of higher education institutions." Mode 1 seems to prevail in the Global South, and is perhaps responsible for the general perception that universities are not involved in bringing about the changes needed in society; this is in contrast with past perceptions of universities as protagonists of social change. In part because of the insularity of the traditional activities of universities (generating, transferring, and applying knowledge), and in part for socio-historical reasons, universities are now perceived as part of the problem as well as part of the solution of the social inequities in Latin America (Vallaeys, 2007). This perception interferes with efforts to introduce Citizen Science into communities. The concept of Knowledge Mobilization (Shaxson, 2012) "a two way process that makes use of the existing stock of knowledge and co-creates new knowledge to help foster change" could act as *a leitmotif* to help translate/transfer university-based knowledge to help citizen groups. Thus, discovering how to achieve Knowledge Mobilization should be a top priority for Citizen Science researchers.

This does not mean that generating Mode 1 knowledge should be censured. All it means is that there is a need for universities and communities to establish collaborative projects as part of the constant quest to strengthen teaching and research, and as a way to comply with the university's social responsibility. But in order to do that, universities in the Global South must break the vicious circle of teachers formed in an authoritarian culture who in turn form elite authoritarian citizens incapable of becoming the type of researchers who will attempt to carry out the research the world demands, as indicated in "Towards the European Higher Education Area: responding to challenges in a globalized world". This document explicitly establishes that "Higher education should play a strong role in fostering social cohesion, reducing inequalities and raising the level of knowledge, skills and competences in society. Policy should therefore aim to maximize the potential of individuals in terms of their personal development and their contribution to a sustainable and democratic knowledge-based society." (Bologna Process, Declarations London Communiqué, 2007, 2.18, p.5)

It may be asked why, despite this diagnosis and clear vision of the challenges to be tackled, there have been no institutional changes made that could ensure the desired results. One hypothesis is that the present university paradigm, where teaching, research and extension are managed in sealed compartments, leads to tackling the problem with ad hoc overlapping subjects or mandatory courses. This situation is much more evident in Latin America with its tendency to bureaucratic "insularity" (Waissbluth, 2003). Such an approach accentuates a dichotomy between researchers and citizenship, science and ethics. *This sense of separation and opposition between morality and efficiency will always be fatal for the consolidation of the skills and attitudes of students, since their basic training urges them to be "efficient and effective professionals". The theoretical and practical challenge is to show students that what is ethical is also effective, and that immoral strategies always end up being very inefficient and harmful* (Vallaeys, 2003, p.4). This problem is as old as society's acknowledgement of the importance of education for transmitting the values of established systems. And according to Napoleon, *Of all political questions, this is perhaps the most important. There will be no stability in the state until there is a body of teachers with fixed principles. Till children are taught whether they ought to be Republicans or Monarchists, Catholic or Unbelievers, and so on, there may indeed be a state, but it cannot be a nation* (Note outlining an Order of Teachers, in Napoleon Bonaparte: His Rise and Fall, by J.M. Thompson (Oxford: Blackwell, 1953), p. 206 quoted by Derek Heater in Citizenship: The civic ideal in world history, politics and education, 3rd edition, 2004)

One possible approach to understanding what is being asked of universities and how they could achieve it may be appreciated in Fig. 6: University Social Responsibility: What is required from the institution? (After Vallaeys, 2007)

The problem to solve at this point is how to obtain university teachers and researchers who practice this kind of social responsibility and therefore may be capable of teaching students the values and practices of citizenship as well as basic professional knowledge. Only teachers of this sort are in a position to undertake research within the values of University Social Responsibility (USR) as is the case of Canada

Figure 6. University social responsibility: What is required from the Institution?

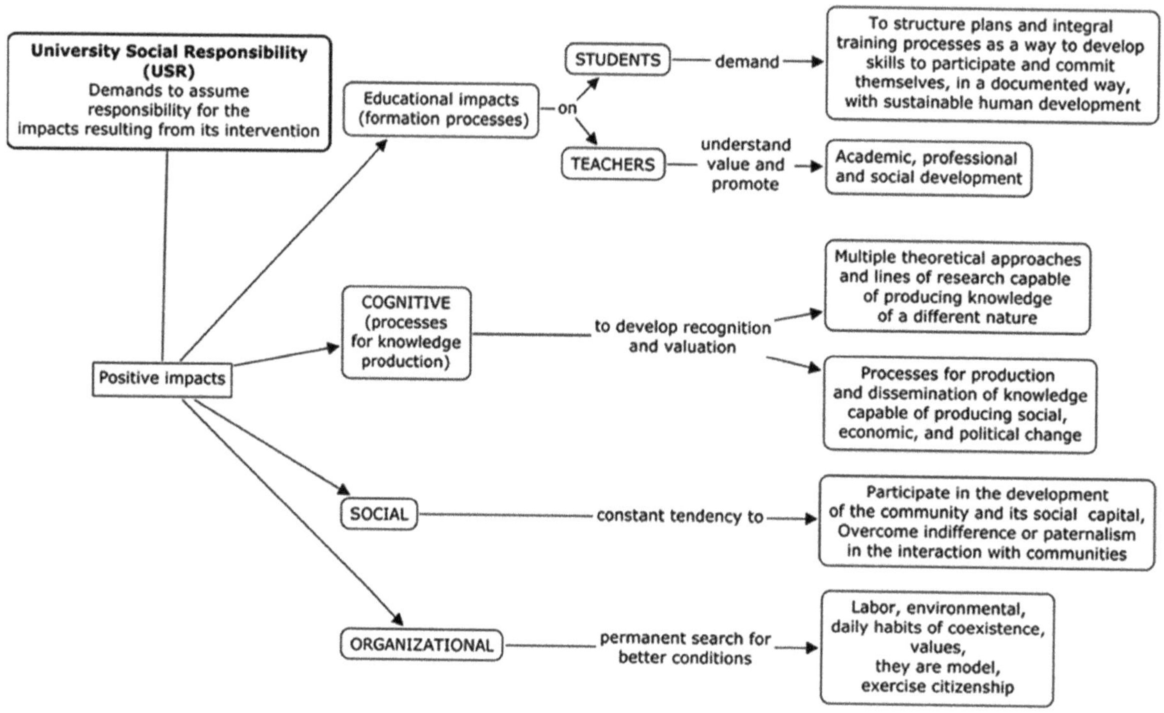

and England (Lall, 2011) where academics are committed to having students learn about social problems such as poverty, education, security, health, among others, problems that affect society as a whole but particularly marginalized populations.

The preparation of teachers and researchers with these characteristics requires comprehensive training, that is, training first *as a person, second, as a professional in a specific area of work and, third, as a citizen with what could be called civic skills* (Villa & Villa, 2007, p.19). Thus instructional design should offer learning experiences in which integration of all three areas can occur. In this respect, Villa and Villa (2007) propose that curricular design should cover three main areas and their intersections. The main areas are Academic, Social and Professional, with the intersection between the social and academic areas generating Personal Development, the intersection between the social and professional areas generating socio-professional responsibilities, and the intersection between the professional and academic areas generating applied competences.

When maximum priority is assigned to academic and professional training in teacher education, it is prevented from incorporating the social component that enables the exercise of citizenship, and therefore, this component is usually assumed as an isolated learning objective. This approach, where there is a lack of balance among the three areas, translates into training plans for teachers that focus on the development of professional competencies and lead to hyper specialization. What is more, even with an appropriate curricular design, there is a disconnection between such a curriculum and the traditional teaching model. In the traditional teaching model, education is understood as a process of transfer and acquisition of accumulated knowledge and the learner is understood as a receiver of information. The consequences of this model lead to the conceptualization of the teacher as a specialist in a field or discipline, which goes against the generation of Mode 2 knowledge as proposed by Gibbons et al. (1994)

Perhaps what is needed is some kind of capacity building that can empower university students in training and their teachers. *The contemporary view of capacity-building goes beyond the conventional perception of training. The central concerns of environmental management and community building – to manage change, to resolve conflict, to manage institutional pluralism, to enhance coordination, to foster communication, and to ensure that data and information are shared – require a broad and holistic view of capacity development. This definition covers both institutional and community-based capacity-building (Capacity Building in* Learning for Sustainability). Therefore for universities to make a greater contribution to the region's development, they must help develop the capacities and empower students and researchers.

An example of a way to do this stems from the introduction of a new law in Venezuela in 2005. This established community service as a requirement for graduation from Venezuelan universities (Gaceta Oficial, 2005). The application of this law has generated a curricular space that seeks to promote participation in society, and to develop social capital understood as: *a tangible public, but not visible, good, which generates symbolic or material benefits to the community and is, based on social relationships in order to generate links and collective capital* (Klisberg, 2001, pp. 14-17). Community Service must be a social project linked to vocational training using a methodology that links learning and service. It must be carried out in a community, have a minimum duration of 120 hours, and be completed in not less than 3 months. The student must be supervised both by a community and by an academic tutor.

These community service projects are concrete manifestations of learning and knowledge exchange between the university and the community. Their intention is for students to acquire systematized knowledge and also become citizens (Saltmarsh, 2005), so that the number of people who are "socio-economically excluded" is not surpassed by those who are "excluded cognitively", that is, those who have not developed skills for appropriating information (Froes, 2002). In these projects, learning occurs from reflection and synthesis that interrelate theory and practice. This conscious interrelationship should produce new knowledge (Correia & Bleicher, 2008) which must be returned to communities (Mitchell, 2008).

Theoretically, the projects in which students participate could be real challenges and require research and understanding of the causes of social problems and actions that could contribute to change (Mitchell, 2008). However, the production of knowledge is closely related to the ability to systematize practices, so it is important to consider how Community Service should be organized so as to produce knowledge. This question could be explored in an original Citizen Science project or be incorporated into projects that are already underway in other latitudes. If this approach is adopted, the student could act as a "coach" to the community, and the university teacher would be the manager of a team of coaches distributed among many communities.

As may be appreciated in Figure 7. Community Service projects in the Science Faculty of the Central University of Venezuela, UCV. (Michinel, 2011) project management in the UCV has been both deductive and inductive. In deductive management:

1. An overarching project was defined;
2. Lines of action were defined;
3. Macro-projects were proposed; and
4. Micro-projects in the communities were developed and carried out.

Figure 7. Community service projects in the science faculty, UCV

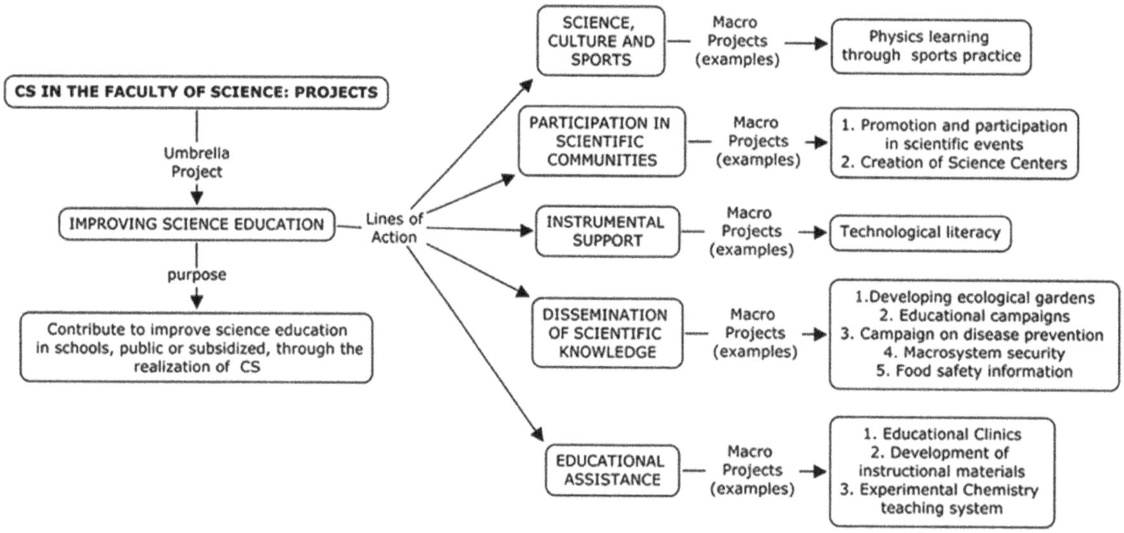

In projects managed inductively, micro-projects begun by students in the initial period of the Law (September 2007), were organized and synthesized. This may be appreciated in Figure 8. Community Service projects in the Economics and Social Science Faculty, UCV. (Rodriguez, 2012)

But such a process may only be tackled by motivated and well prepared teachers. For this reason, a group of UCV researchers has proposed introducing Collaborative Research to graduate students. To do this they have formed an interdisciplinary team of researchers from several Latin American and European countries in order to create a Mass Open Online Course (MOOC) on "Interdisciplinary Collaborative Research in social sciences in the 21st century", aimed at graduate students and to be offered through a network of graduate programs. This type of research seeks to: a) to develop the capacity to generate efficient collaborative teams, such as those composed of the students and teachers who will work in

Figure 8. Community science projects in the economics and social sciences faculty, UCV

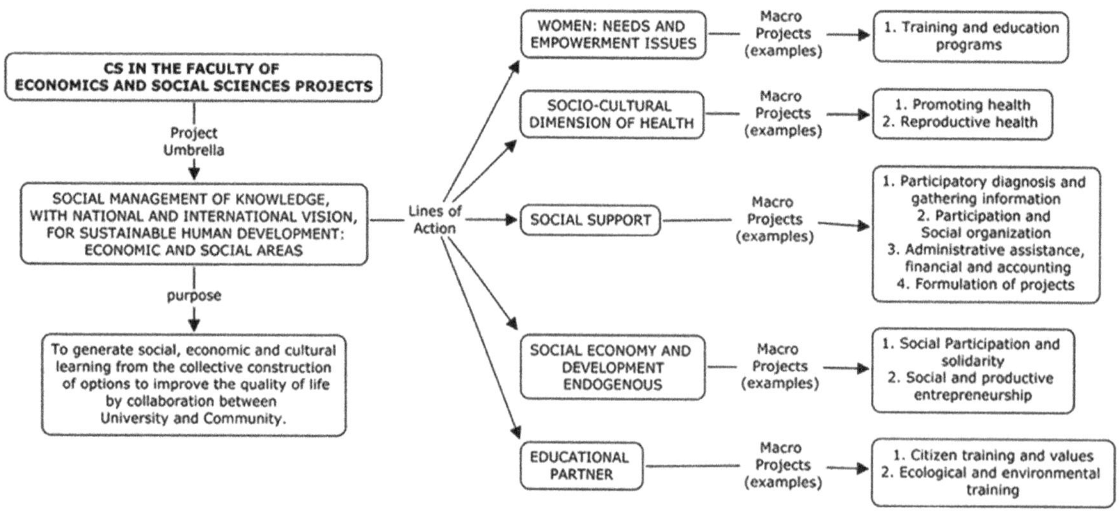

a particular community, and b) to develop the capacity to identify and study "real world" problems in communities, by working together with the people that inhabit them (Ordóñez et al, in press).

The aim is to provide "civic skills" both to university teachers and researchers and to university graduates, in order to begin the process of creating a civil society capable of interacting efficiently with the institutions that generate public policies, and in which communities acquire knowledge about their problems and generate ways to influence the decisions and public policies defined by those institutions. This aim is represented in Figure 9. University Social Responsibility and the generation and mobilization of Knowledge with communities

The first step should be to generate "capacity building" among the university teams. As may be appreciated from the preceding paragraphs, a possible way to introduce this in traditional universities is by interrelating science education and citizenship education, as suggested by Davies (2004), in Figure 10. The overlapping of Science and Citizenship Education

Interrelating science education and citizenship education is a very demanding task which involves the whole of society. For this reason, the next section will deal with the political systems and politicians in Latin America.

INTERACTING WITH POLITICIANS AND POLITICAL INSTITUTIONS

...the discourse of Evidence-Based Policy (EBP) offers poor guidance to those who seek to ensure that social policy making is informed by the findings of social science. EBP discourse relies on a technocratic, linear understanding of the policy making process and on a naïve empiricist understanding of the role of evidence. This renders it unable to engage with the role of the underlying discursive frameworks and paradigms that render evidence meaningful and invest it with consequence. EBP discourse does not help us understand either how policy changes, or what is at stake in dialogue across the 'research-policy divide'. (Andries du Toit, 2012)

Figure 9. University social responsibility and the generation and mobilization of knowledge with communities

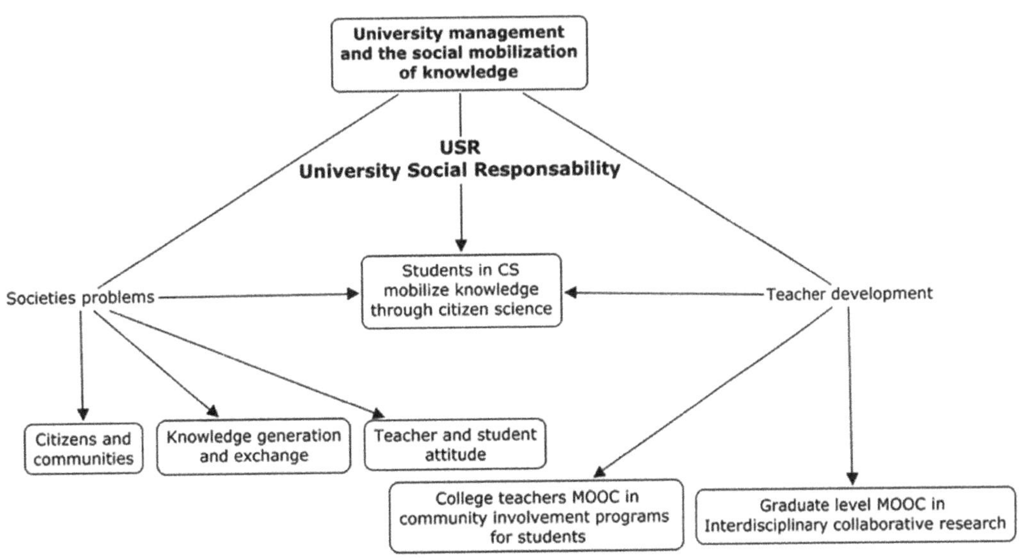

Figure 10. The overlapping of science and citizenship education

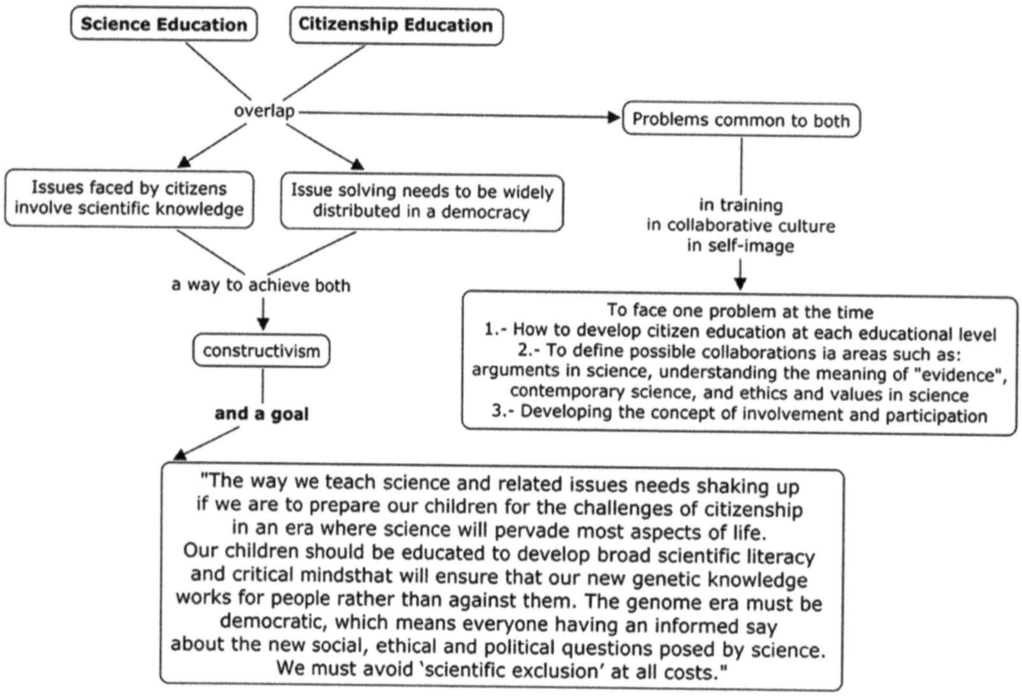

In order to get community support, researchers must be able to interact with and be part of the political system as represented by political parties and Parliament, who do not decide by themselves alone, but are dependent on the community. However, the authoritarian legacies of Latin America make such interaction difficult, as is indicated by Cesarini and Hite (Bermeo, 2004) in their book "Authoritarian legacies and democracy in Latin America and Southern Europe". They conclude it by saying,

The range of legacies is quite vast, including the discursive frames of the postcolonial period that continue to affect today's political dynamics in Latin America's democracies; the severed bond of trust between the state and its citizens; the traumatic past of human rights violations and its associated impunity; the economic policies, institutions and networks of interest representation left by dictators; the style of policing inherited from authoritarianism; and the military's patterns of authoritarian domination (bold type is the authors').

It is within this complexity of political processes in Latin America and their impact on civil society that communities perceive any attempt to introduce Citizen Science into their environment. Understanding "the severed bond of trust between the state and its citizens" is critical for convincing people that the data obtained through Citizen Science will contribute to the wellbeing of their community and serve as leverage for achieving a better society through better decision making.

When citizens and politicians distrust one another, policy can be misinterpreted. Scientists are not exempt from this "...there is a gap between policy on the one hand and how scientists perceive it on the other. Policy documents put forward a broad notion of impact, but scientists perceive them as focusing too narrowly on commercial impacts. Scientists are further puzzled by how societal impact is evaluated and organized, and their perceptions frame their behavior" (de Jong, Smit & van Drooge, 2015). Distrust of the possibility of changing public policies can hamper the incorporation of Citizen Science, especially since historical practice has often led to choosing public policies aimed at solving "problematic sensations" based on the previous experiences, prejudices and stereotypes of decision makers, rather than on solid evidence-based analysis, as depicted in Figure 11, Dunn's phases of policy making: the "plunging in" short circuit

The situation is very similar to that of lack of trust in the police, since politicians, like the police, are somehow not part of the community, but represent "power". As Goldsmith, (2005) points out,

Police reform is widely undertaken in developing and post-authoritarian countries. The starting point for analysis of this phenomenon, it is suggested, is the absence of public trust in police that characterizes police-community relations in these countries. Without public trust in police, 'policing by consent' is difficult or impossible and public safety suffers... In addition to building trust, ways of institutionalizing distrust are needed.

According to Vigoda (2002)... *despite the fact that citizens are the formal "owners" of the state, ownership will remain a symbolic banner for the governance and public administration–citizen relationship in a representative democracy. The alternative interaction of movement between responsiveness and collaboration is more realistic for the years ahead.*

The most important thing for research teams interested in introducing Citizen Science in a given community is the need to be conscious of their role as gatekeepers between members of the research team itself (students, field workers and so on), communities, and local, regional or national authorities involved in achieving the project's objectives. Furthermore, if Citizen Science is going to have a positive impact on decision makers, it must not neglect media intervention, and therefore, it is advisable to incorporate journalistic know-how into the team. Research teams may be well advised to think of themselves as negotiating teams and to implement a shared strategy when negotiating with all the actors

Figure 11. Dunn's phases of policy making: the "plunging in" short circuit
(after Russo, Schomaker & Russo, 1990)

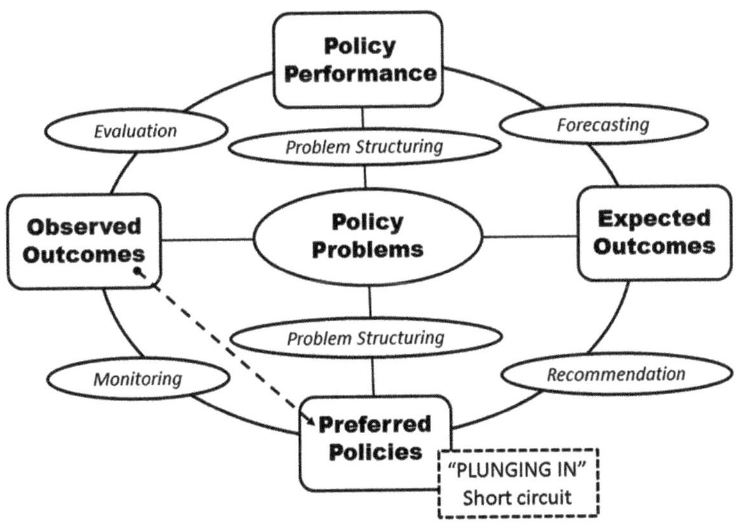

Figure 12. Topics to be considered in public policies, political parties and citizen participation

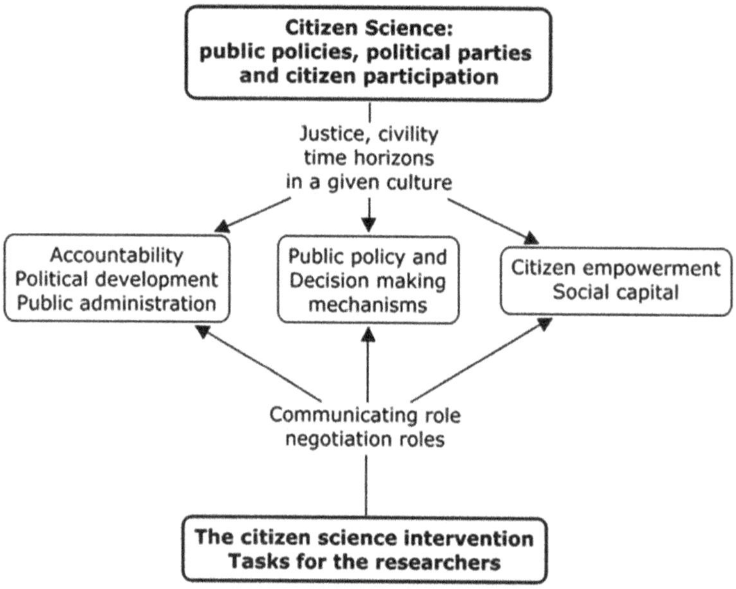

involved. All these elements are considered in Figure 12. Topics to be considered in public policies, political parties and citizen participation

What is the desirable outcome of Citizen Science? As can be seen in Figure 2, there are many possible ways of disseminating research results once the data have been obtained and processed. And the results should not be published exclusively in research journals, since this does not help build citizenship, or motivate further collaboration from the citizens involved. What is needed is to find ways to use the results, "doing something about the results" and "disseminating the results to other environments", through a combination of publication, sharing with other communities and media involvement, so that citizens and decision makers become aware of the problem, the findings, and the possibilities opened up by the knowledge generated. An informed public is needed to approve the public policies generated to "fix" the problematic situation being studied. In the Global South at the present time, the best approach might be for researchers to seek the collaboration of the media in order to introduce citizen research simultaneously to the population and to the political community. However, Habermas (2006) indicates two "critical conditions for this: mediated political communication in the public sphere can facilitate deliberative legitimation processes in complex societies only if a self-regulating media system gains independence from its social environments and if anonymous audiences grant a feedback between an informed elite and a responsive civil society." It cannot be guaranteed that both of these conditions are present in today's Global South or specifically in Latin America. But any Citizen Science project needs to address them. In this, scientific journalism initiatives may be good allies.

SOLUTIONS AND RECOMMENDATIONS

In their Guide for trans-boundary research partnerships (Stöcli et al, 2012) the Swiss Academy of Sciences mentions 11 principles that may help guide researchers considering or planning to engage in fair and equal partnerships:

1. Set the Agenda together
2. Interact with stakeholders
3. Clarify responsibilities
4. Account to beneficiaries
5. Promote mutual learning
6. Enhance capacities
7. Share data and networks
8. Disseminate results
9. Pool profits and merits
10. Apply results
11. Secure outcomes

Within this context, the Swiss Academy also mentions other actors who must be involved, or with whom links and bridges must be established: scientific communities active in the same field of research; agencies commissioning the research and/or who will use the results; users and beneficiaries of the research outcomes; and the general public interested in the field of the research.

It is obvious that, from the point of view of the community where a Citizen Science project is going to be implemented, the responsibilities are even greater than those addressed by the Swiss Academy, since ethics require that the project contribute to solving the community's social problems. But Citizen Science is not a panacea. If social problems are defined as *the breach... between the legal order (all citizens are equal under law) and the inequalities of existence (poverty, unemployment, hunger)* (Aguilar, 2010), it must be concluded that native research teams involved in Citizen Science in Latin America, in addition to acquiring the data needed for their particular research purpose, must ensure their project's contribution to answering a social problem. This has to be considered before the research project is implemented and ways of evaluating its effects should be included in the research design. Once the project is over, its effects should be evaluated *ex post*.

If the original research team is based in a developed country, it should take many of the aspects referred to in this chapter into consideration. However, the main responsibility for this falls on the research-academic systems of the developing region. Furthermore, the developing region's research systems must direct their analysis and efforts to the generation, management and transfer of knowledge between the university and society. Those working in authoritarian and traditionalist academic and political environments find it difficult to admit their lack of knowledge about a problem unless they deem it an "unimportant issue". This represents an important obstacle to the process of recognizing problems, finding information about them, advancing explanatory hypotheses that allow exploration of solutions, or accepting any errors detected in monitoring and evaluation in order to try to correct them in subsequent actions; more so when the problems are social ones.

Training researchers, no matter their specific discipline, to acknowledge cultural factors when interacting with communities is basic....*virtually all human learning occurs in a culturally influenced, if not culturally created environment ...Skill in cultural analysis, which can be developed in teacher training (both institutional and in-service), makes possible the identification of significant cultural influences on the conditions of learning. Such identification should, when coupled with other kinds of knowledge available through psychology and sociology, enhance the predictions of consequences...* (Spindler, 1959). Until the completion of their doctoral research, scientists' training occurs frequently in protected (laboratory) environments. This experience hampers successful interactions when they attempt to study "real problems" that present

...challenges associated with transferring innovations to community systems, changing program delivery from an experimental context controlled by researchers to program delivery controlled by community organizations, and sustaining long-term effects of interventions. It is suggested that researchers who develop and implement community interventions [...] need to confront several issues:

1. Fostering effective long-term relationships between researchers and the communities they study and in which they intervene and

2. Designing and implementing interventions that are useful to community systems after the formal phase of research ends (Altman, 1995).

In this regard, researchers are recommended to consult publications from the health care system, which has had to deal with similar problems in the past, for example, *Culturally competent healthcare systems: a systematic review.*(Anderson et al., 2003)

This chapter has raised some questions and an attempt has been made to provide answers, but some questions may not have an answer, and the few answers provided are not the only possible ones. The main question may be: why Citizen Science? Though there are many ways to 'do science', Citizen Science is attractive because it offers the possibility of democratizing knowledge and empowering citizens, by allowing participants to learn, make decisions, and act on the problems of community life in a documented and rational manner.

However, given its special cultural conditions, it would not be appropriate for the Global South, particularly Latin America, to copy the models of Citizen Science implemented in the developed world, which could end up giving "the right answer to the wrong problem", because they do not take into account the cultural specifics of the region, the predominant model of knowledge production (mode 1), and the characteristics of its institutions of higher education that favor curriculum designs concentrating almost exclusively on professional and academic formation. Though Latin American universities may pay lip-service to comprehensive education and the production of socially relevant knowledge, their awareness of the importance of these issues is very limited.

Institutions of higher education have multiple opportunities for transforming their practices and educational processes (for students as well as teachers). By incorporating Citizen Science into their day-to-day activities, they can increase inter-disciplinary collaboration and contribute to the development of citizenship. However, Citizen Science is not the only mechanism for teaching citizenship, but one option among many.

CONCLUSION

The introduction of Citizen Science into communities in developing countries offers opportunities to build more inclusive and participatory societies. However, to fulfill the potential of Citizen Science, researchers must take two things into account: the specific cultural characteristics of the communities where their projects are to be carried out, and the political institutions where decision makers must factor in the evidence accumulated through research. In order to prepare local scientists with the managerial skills and appropriate attitudes for Citizen Science, universities must assume their social responsibility as knowledge mobilizers in the societies in which they are embedded.

REFERENCES

Aguilar, P. L. (2010). *Development Discourse and Social Policies: Challenges to Theory and Contemporary Debates*. Retrieved April 27, 2016 from: https://www.scribd.com/doc/283054476/Aguilera-2010-Development-Discourse-and-Social-Policies-Challenges-to-Theory-and-Contemporary-Debates

Alguezaui, S., & Filieri, R. (2010). Investigating the role of social capital in innovation: Sparse versus dense network. *Journal of Knowledge Management*, *14*(6), 891–909. doi:10.1108/13673271011084925

Almond, G. A., & Verba, S. (1963). Organizational membership and civic competence, from The Civic Culture In *The Civic Society Reader*. Academic Press.

Altman, D. G. (1995). Sustaining interventions in community systems: On the relationship between researchers and communities. *Health Psychology*, *14*(6), 526–536. doi:10.1037/0278-6133.14.6.526 PMID:8565927

Anderson, L. M., Scrimshaw, S. C., Fullilove, M. T., Fielding, J. E., & Normand, J. (2003). Task Force on Community Preventive Services. Culturally competent healthcare systems: A systematic review. *American Journal of Preventive Medicine*, *24*(3), 68–79. doi:10.1016/S0749-3797(02)00657-8 PMID:12668199

Baylis, J., & Smith, S. (2001). *The Globalization of World Politics: An introduction to international relations*. Oxford University Press.

Bermeo, N. G. (2004). Authoritarian legacies and democracy in Latin America and Southern Europe. *University of Notre Dame Press*.

Bocciolesi, E., & González, S. J. G. (2015). Educar a leer entre realidad y complejidad. Las variables de la contemporaneidad entre México, España e Italia. *Memorias del Encuentro Internacional de Educación a Distancia*.

Bolivariana de Venezuela, C. R., & Oficial, G. (2005). *Law of Community Service, Gaceta Oficial de la República de Venezuela, Nº 38272, 14 de Septiembre, 2005*. Retrieved April 15, 2016 from: http://www.ciens.ula.ve/scciens/organizacion/ley.php

Bonaparte, N. (2004). His Rise and Fall (3rd ed.). Oxford, UK: Blackwell.

Caldwell Johnson, E. (2005). *Communities of Practice for Local Capacity in Central Asia: The Community Empowerment Network*. Retrieved April 16, 2016 from http://learningforsustainability.net/capacity-building-empowerment/

Christakis, N. A., & Fowler, J. H. (2009). *Connected: the surprising power of our social networks and how they shape our lives*. Little Brown.

Cima, O. (2015). Expanding spaces of participation: Insights from an infrastructural project in rural Nepal. *Geographica Helvetica*, *70*(4), 353–361. doi:10.5194/gh-70-353-2015

Citizen Science. (n.d.a). Retrieved April 16, 2016, from https://en.wikipedia.org/wiki/Citizen_science

Citizen Science. (n.d.b). In *National Geographic*. Retrieved April 16, 2016 from http://education.nationalgeographic.org/encyclopedia/citizen-science/

Citizen Scientist. (n.d.). In *Open Scientist*. Retrieved April 16, 2016 from http://www.openscientist.org/2011/09/finalizing-definition-of-citizen.html

Correia, M. G., & Bleicher, R. E. (2008). Making connections to teach reflection. *Michigan Journal of Community Service Learning*, *14*, 41–49.

Damon, W. (1984). Peer education: The untapped potential. *Journal of Applied Developmental Psychology*, *5*(4), 331–343. doi:10.1016/0193-3973(84)90006-6

Danielsen, F., Burgess, N., Balmford, A., Donald, P., Funder, M., Jones, J., & Yonten, D. et al. (2009). Local Participation in Natural Resource Monitoring:a Characterization of Approaches. *Conservation Biology*, *23*(1), 31–42. doi:10.1111/j.1523-1739.2008.01063.x PMID:18798859

Davies, I. (2004). Science and citizenship education. *International Journal of Science Education*, *26*(14), 1751–1763. doi:10.1080/0950069042000230785

de Jong, S. P., Smit, J., & van Drooge, L. (2015). Scientists' response to societal impact policies: A policy paradox. *Science & Public Policy*.

Defining Citizen Science. (n.d.). In *The Cornell Lab of Ornithology*. Retrieved April 16, 2016, from http://www.birds.cornell.edu/citscitoolkit/about/definition

Dewey, J. (1916). Democracy and Education. In The middle works of John Dewey (vol 9). Academic Press.

du Toit, A. (2012). *Making sense of "evidence": Notes on the discursive politics of research and pro-poor policy making', Working Paper 21*. PLAAS, UWC.

Dunn, W. N. (2015). *Public Policy Analysis*. Routledge.

Esman, M. J. (2003). *Social capital and empowerment*. Cornell University. Retrieved April 12, 2016 from: http://siteresources.worldbank.org/INTEMPOWERMENT/Resources/486312-1097679640919/think_pieces_esman.pdf

España, L. P. (2004). La cultura y las causas de la pobreza en Venezuela. In Luís Ugalde, Luis Pedro España y otros: Detrás de la Pobreza. Percepciones, Creencias, Apreciaciones. Editan: Asociación Civil para la Promoción de Estudios Sociales y la Universidad Católica Andrés Bello.

Fantini, A., & Tirmizi, A. (2006). *Exploring and assessing intercultural competence*. Academic Press.

Fernandez-Gimenez, M. E., Ballard, H. L., & Sturtevant, V. E. (2008). Adaptive management and social learning in collaborative and community-based monitoring: A study of five community-based forestry organizations in the western USA. *Ecology and Society*, *13*, 4.

Freire, P. (2005). *Pedagogía del oprimido*. Siglo XXI.

Fróes, B. T. (2002). Análise Contrastiva: memória da construção de uma metodologia para investigar a tradução de conhecimento científico em conhecimento público. *DataGramaZero - Revista de Ciência da Informação, 3*, 16. Retrieved April 20, 2016 from: http://repositorio.ufba.br/ri/handle/ri/2985

García Hamilton, J. I. (2002). *El autoritarismo y la improductividad en Hispanoamérica*. Buenos Aires.

Gaventa, J. (Interviewee). (2015). *Professor John Gaventa for AURA Research 1*. Retrieved May 10 from https://www.youtube.com/watch?v=HEy4YTnY4II

Goldsmith, A. (2005). Police reform and the problem of trust. *Theoretical Criminology*, *9*(4), 443–470. doi:10.1177/1362480605057727

Guerra, F. X. (1999). *El soberano y su reino. Reflexiones sobre la génesis del ciudadano en América Latina. In Ciudadanía política y formación de naciones: perspectivas históricas de América Latina.* México: Fondo de Cultura Económica.

Güss, C. D., Tuason, M. T., & Gerhard, C. (2010). Cross-National Comparisons of Complex Problem-Solving Strategies in Two Microworlds. *Cognitive Science*, *34*(3), 489–520. doi:10.1111/j.1551-6709.2009.01087.x PMID:21564221

Habermas, J. (2006). Political communication in media society: Does democracy still enjoy an epistemic dimension? the impact of normative theory on empirical research1. *Communication Theory*, *16*(4), 411–426. doi:10.1111/j.1468-2885.2006.00280.x

Harrison, T. M., & Stephen, T. (Eds.). (1996). *Computer networking and scholarly communication in the twenty-first-century university*. SUNY Press.

Hart, A., & Wolff, D. (2006). Developing local communities of practice through local community–university partnerships. *Planning Practice and Research*, *21*(1), 121–138. doi:10.1080/02697450600901616

Hyatt, S. (2001). From citizen to volunteer. In The new poverty studies: The ethnography of power, politics, and impoverished people in the United States. Academic Press.

Jones, R. C. (2014). Creating new technical universities in emerging countries. In *Interactive Collaborative Learning (ICL), 2014 International Conference on* (pp. 493-498). IEEE. doi:10.1109/ICL.2014.7017822

King, K., & McGrath, S. (2002). *Globalisation, enterprise and knowledge: Education, training and development in Africa.Symposium Books Ltd.*

Kliskberg, B. (2001). *El Capital Social*. Caracas: Editorial PANAPO.

Lall, N. (2011). Estructuras de investigación colaborativa comunidad universidad: aproximación a su posible impacto. *Rizoma-Freireaano, 9*. Retrieved April 22, 2016 from: http://www.rizoma-freireano.org/index.php/estructuras-de-investigacion-colaborativa-comunidad-universidad-aproximacion-a-su-posible-impacto--nirmala-lall

Lloyd, A., Lipu, S., & Kennan, M. A. (2010). On becoming citizens: Examining social inclusion from an information perspective. *Australian Academic & Research Libraries*, *41*(1), 42–53. doi:10.1080/00048623.2010.10721433

Marsh, L., & Onof, C. (2008). Stigmergic epistemology, stigmergic cognition. *Cognitive Systems Research*, *9*(1), 136–149. doi:10.1016/j.cogsys.2007.06.009

McGrath, S., King, K., Leach, F., & Carr-Hill, R. (1994). *Education and Training for the Informal Sector.* UK Department for International Development, Education Research Paper, 11.

Michinel, J. L. (2011). Un proyecto paraguas para la realización del servicio comunitario por estudiantes de la Facultad de Ciencias de la UCV: mejoramiento de la educación en escuelas públicas (y privadas subsidiadas) a través del SC. *Revist@serviciocomunitario, 1.* Retrieved March 22, 2016 from: http://saber.ucv.ve/ojs/index.php/rev_scfc/article/view/631/578

Mitchell, T. (2008). Traditional vs. Critical Service-Learning: Engaging the Literature to Differentiate Two Models. *Michigan Journal of Community Service Learning*, *2*, 50–65.

Narayan-Parker, D. (Ed.). (2002). *Empowerment and poverty reduction: A sourcebook.* World Bank Publications.

Newman, G., Wiggins, A., Crall, A., Graham, E., Newman, S., & Crowston, K. (2012). The future of Citizen Science: Emerging technologies and shifting paradigms. *Frontiers in Ecology and the Environment*, *10*(6), 298–304. doi:10.1890/110294

No somos suecos… ¡somos latinoamericanos!. (n.d.). In *Interconectados in Interconectados.* Retrieved April 18, 2016 from: http://interconectados.org/no-somos-suecos-somos-latinoamericanos/)

Nov, O., Arazy, O., & Anderson, D. (2014). Scientists@ Home: What drives the quantity and quality of online Citizen Science participation? *PLoS ONE*, *9*(4), e90375. doi:10.1371/journal.pone.0090375 PMID:24690612

Olson, J. S., & Olson, G. M. (2003). Culture surprises in remote software development teams. *Queue*, *1*(9), 52. doi:10.1145/966789.966804

Ordóñez Vela, L. A., Álvarez, G. C. L., Bocciolesi, E., Lombardi L, G., Martínez, A. K., & Said-Hung, E. (in press). Cultura, conectividad y colaboración: las tres "C" que marcan el futuro de la ciencia en el siglo XXI. In E. Said-Hung (Ed.), Educación, participación y escenarios digitales. Debates sobre la mediación digital en el siglo XXI. Barranquilla: Ediciones Uninorte.

Ostrom, E. (2007). A diagnostic approach for going beyond panaceas. *Proceedings of the National Academy of Sciences of the United States of America*, *104*(39), 15181–15187. doi:10.1073/pnas.0702288104 PMID:17881578

Ostrom, E., Janssen, M. A., & Anderies, J. M. (2007). Going beyond panaceas. *Proceedings of the National Academy of Sciences of the United States of America*, *104*(39), 15176–15178. doi:10.1073/pnas.0701886104 PMID:17881583

Participatory Development. (n.d.). In *Wikipedia.* Retrieved April 16, 2016 from https://en.wikipedia.org/wiki/Participatory_development

Process, B., & Communiqué, D. L. (2007). *Towards the European Higher Education Area: responding to challenges in a globalised world.* Retrieved from http://www.ehea.info/Uploads/Declarations/London_Communique18May2007.pdf

Reed, M., Evely, A. C., Cundill, G., Fazey, I. R. A., Glass, J., Laing, A., et al. (2010). What is social learning? *Ecology and Society*. Retrieved April 25, 2016 from https://research-repository.st-andrews.ac.uk/bitstream/handle/10023/1624/Fazey2010EcologySociety15WhatIsSocialLearning.pdf?sequence=1&isAllowed=y

Riddell, K. (n.d.). Retrieved April 16, 2016, from http://www.wcmt.org.uk/sites/default/files/report-documents/Riddell%20K%20Report%202014.pdf

Rodriguez, M. A. (2012). *Proyectos de Servicio Comunitario en Facultad de Ciencias Económicas y Sociales de la UCV*. Comunicación Personal.

Russo, J. E., Schoemaker, P. J., & Russo, E. J. (1990). *Decision traps: Ten barriers to brilliant decision-making and how to overcome them*. New York, NY: Simon & Schuster.

Saltmarsh, J. (2005). The Civic Promise of Service Learning. *Liberal Education, 91*(2), 50–55.

Shaxson, L., Bielak, A., Ahmed, I., Brien, D., Conant, B., Fisher, C., & Phipps, D. (2012). Expanding our understanding of K*(KT, KE, KTT, KMb, KB, KM, etc.). A concept paper emerging from the K* conference held in UNU-INWEH, Hamilton, ON.

Spindler, G. D. (1959). *Learning in Culture: Anthropological Perspective*. Educational.

Stöckli, B., Wiesmann, U., & Lys, J. A. (2012). A guide for transboundary research partnerships: 11 principles. Berne: Swiss Commission for Research Partnership with Developing Countries (KFPE).

Torney-Purta, J., Lehmann, R., Oswald, H., & Schulz, W. (2001). Citizenship and education in twenty-eight countries: Civic knowledge and engagement at age fourteen. *Herengracht, 487*, 1017.

Type III Error. (n.d.). In *Wikipedia*. Retrieved April 16, 2016, from https://en.wikipedia.org/wiki/Type_III_error

Ugalde, L., España, L.P., Lacruz, T., De Viana, M., González, L., Luengo, N., & Ponce, M. (2004). *Detrás de la pobreza*. Universidad Católica Andrés Bello.

UNESCO, Leeds-Hurwitz, W., & Stenou, K. (2013). *Intercultural Competences: Conceptual and Operational Framework*. UNESCO.

Vallaeys, F. (2003). Orientaciones para la enseñanza de la ética, el capital social y el desarrollo en las universidades latinoamericanas. *Revista venezolana de Gerencia, 21*, 38-73.

Vallaeys, F. (2007). *La Responsabilidad Social Universitaria: ¿Cómo entenderla para quererpracticarla?* Retrieved April 16, 2016 from: https://www.academia.edu/21125213/LA_RESPONSABILIDAD_SOCIAL_UNIVERSITARIA_C%C3%93MO_ENTENDERLA_PARA_QUERERLA_Y_PRACTICARLA

Vigoda, E. (2002). From responsiveness to collaboration: Governance, citizens, and the next generation of public administration. *Public Administration Review, 62*(5), 527–540. doi:10.1111/1540-6210.00235

Villa, S. A., & Villa, L. O. (2007). El aprendizaje basado en competencias y el desarrollo de la dimensión social en las universidades. *Educar, 40*, 15–48.

Waissbluth, M. (2003). La insularidad en la gestión pública latinoamericana. *Revista del CLAD Reforma y Democracia, 27*, 75–88.

Wenger, E. (1998). Communities of practice: Learning, meaning, and identity. New York: Cambridge. doi:10.1017/CBO9780511803932

KEY TERMS AND DEFINITIONS

Culture: A defining aspect of what it means to be human, encompassing the range of practices and accumulated knowledge and ideas that are transmitted through social learning in specific human societies.

Development: In Social Sciences, processes of change in societies and the initiatives taken through practices and academic disciplines to improve various aspects of local communities to improve the standard of living.

Empowerment: Measures designed to increase the degree of autonomy and self-determination in people and in communities.

Knowledge Mobilization: A two-way process that makes use of the existing stock of knowledge and co-creates new knowledge to help foster change.

Latin America: Countries in the Americas south of the United States where Spanish or Portuguese predominate.

Panacea: In Greek mythology, Panacea was a goddess of Universal remedy. In medicine, a substance meant to cure all diseases. In public policy analysis, tendency to assume simple solutions to complex problems.

Chapter 6
Geographical Information Systems in Modern Citizen Science

Laia Subirats
Eurecat, Spain

Joana Simoes
GeoCat, The Netherlands

Alexander Steblin
Eurecat, Spain

ABSTRACT

This chapter shows how citizen-science initiatives have been known to exist for a long time, but only recently they were further enhanced thanks to technological and societal developments, such as the availability of mobile devices, the widespread use of the internet and the low cost of location devices. These developments shaped the geographic information system (GIS) world as it is known today: a group of technologies that allows retrieving, storing, analyzing and sharing spatial information, by people who are not necessarily GIS professionals. This chapter starts with a general background about GIS, adding then more detail in topics of particular relevance in the context of citizen science. The rest of the chapter is focused on reviewing and classifying the use of GIS in citizen-science initiatives; and some use cases are described in order to provide practical examples of the use of these technologies for solving specific spatial problems. The chapter closes with a brief discussion of the future of GIS in citizen science, in the light of current technological trends.

INTRODUCTION

Citizen science has a great focus in environmental issues such as pollution, habitat and biodiversity (Castell et al., 2014; Uhrner et al., 2013), and thus it has an eminent geographic component. The phenomena that concern citizen scientists do occur in a certain place and at a certain scale, and they are often interrelated with other phenomena that are also linked to places and scales. GIS is the "glue" which

DOI: 10.4018/978-1-5225-0962-2.ch006

enables us to relate these different places, scales and phenomena in a system where the information can be effectively stored, managed and analyzed.

Scientists have been known to use spatial analysis for a very long time. In 1854, British epidemiologist John Snow (Goodchild, 2007), plotted the reported cholera deaths onto a street map (a technique similar to what is currently called a *mash-up*), in order to find out the source of the infection: the public water pump on Broad Street.

Although this early example already illustrates the application of GIS principles to solve a societal problem, it was not until recently that the use of GIS became generalized among the citizen science community, along with other communities. Undoubtedly, modern GIS developments such as interoperability, the growth of Free and Open Source Software (FOSS) and digital Volunteered Geographic Information, amongst others, have played an important role for this widespread technology adoption. These aspects are discussed within the "Background" section, which aims at familiarizing the reader with GIS technologies. Unfortunately it would be too ambitious to provide a complete introduction to GIS

Figure 1. Original map by John Snow, depicting the clusters of cholera cases in the London epidemic of 1854
Drawn and lithographed by Charles Cheffins (Snow, 1854).

in the scope of this chapter, so the authors focused instead in the most recent events, and in particular in those technologies, movements and ideas which are more relevant in the context of citizen science.

The rest of the chapter is entirely devoted to describing and understanding the use of GIS in citizen science projects. The next section, "GIS in the Context of Citizen Science", starts with an extensive list of citizen science GIS projects, and it aims to showcase the diversity of citizen science applications powered by GIS, as well as exposing common aspects across many of them. The following section elaborates a bit more about these commonalities across infrastructures, by proposing a Citizen Science GIS taxonomy. This is followed by a description of use cases, where the use of GIS is portrayed with more detail, within real-world, well-defined, citizen science problems. These use cases are examples, and they were chosen based on their representativity, as well as on the availability of information about those projects.

The last section is more practical, and it provides a set of recommendations on how-to setup a typical GIS infrastructure for a citizen science project. Finally the authors close the chapter with some general conclusions and an overview of GIS trends which could impact the development of citizen science in the upcoming years.

Although not all citizen science relates to GIS, there is an important overlap between the two areas and it could be stated that there is also a positive feedback, in the sense that citizen scientists are not only contributing to GIS, but also pushing for its use and contributing towards its development. Apart from demonstrating that GIS technologies can support the development and further enhancement of citizen science projects, the objectives of this chapter were to understand and describe which are these technologies, and "why" they are relevant, which in turn can help us to understand "why" they were adopted by citizen scientists.

BACKGROUND

Although GIS has been erroneously described as a *container of a map in a digital form*, it is actually much more than that (Longley et al., 2005). A more complete definition of GIS would be *an integration of data, hardware, and software designed for management, processing, analysis and visualization of georeferenced data* (Neteler & Mitasova, 2008).

In recent years, GIS has evolved from being a highly specialized niche area, restricted to a group of experts, to a technology that most people use to manage different aspects of their lives (Neteler & Mitasova, 2008). This may include trivial tasks such as searching for a restaurant nearby, navigating through a road network, or even locating a missing or stolen phone.

Nowadays most GIS users use software that runs either on their desktop, or on the web (Longley et al., 2005); in fact, these usage patterns can be used to derive the two main categories of GIS: Desktop GIS and Web GIS. Desktop GIS has been around since the mid 90s, and provides rich functionality to the end user, which includes viewing, querying, updating and analyzing spatial information and related data. The applications are normally large, and they may be referred as "fat" clients. QGIS (http://qgis.org/en/site/), gvSIG (http://www.gvsig.com/en) and GRASS (http://grass.osgeo.org/) are examples of desktop GIS that run in Windows, Linux or OS X. Spatial libraries such as GDAL\OGR (http://www.gdal.org/) are the backbone of desktop GIS.

Internet GIS, or web GIS, sometimes also referred as web mapping, or Geoweb, has developed significantly in the last two decades. The architecture of internet GIS is two tiered, comprising a server,

Figure 2. Screenshot of the user interface of a Desktop GIS (QGIS)

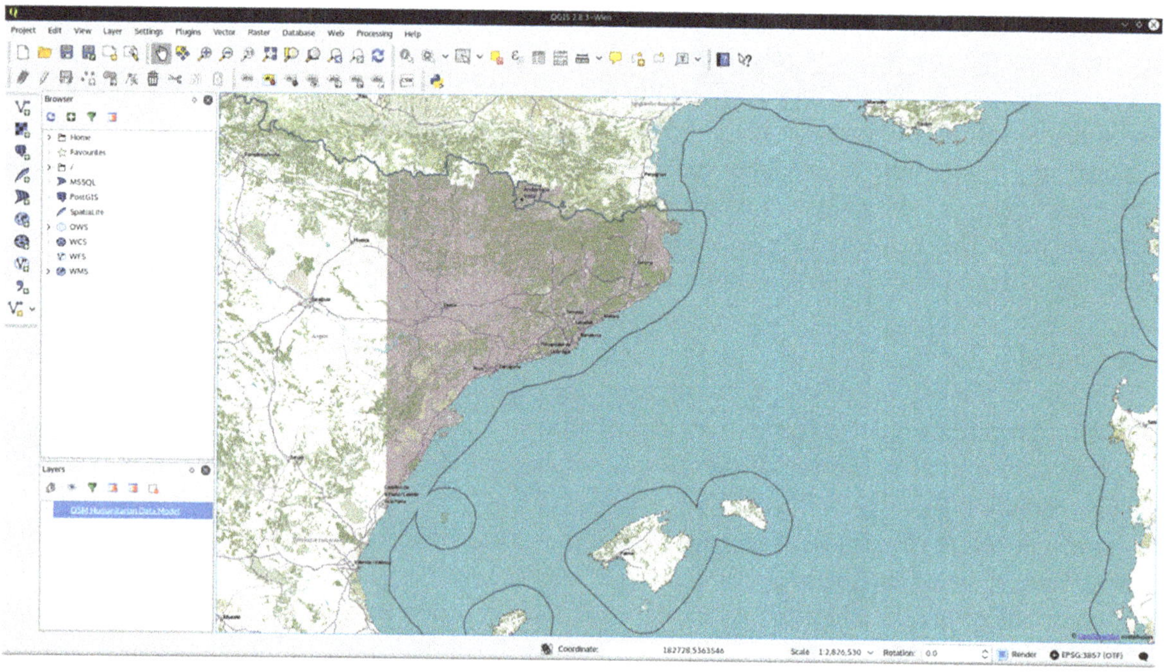

which runs on a computer server and can handle multiple requests, and a range of networked clients (Longley et al., 2005). These clients can be as simple as a web page, as the most complex functionality is built into the server, and thus they are referred as "thin" clients. This kind of architecture has the potential for a very large user base with a very low cost per user, and unsurprisingly it has an important role in supporting public participation.

The first generation of web mapping can be dated to the late 90s, and it was based on simple HTML protocols that delivered mostly static maps, and resulted in a poor user experience (Minghini, 2014). In the following ten years a second generation of web mapping arrived, based on DHTML, Java and ActiveX, which resulted in websites with increased interactivity (e.g.: MapQuest, Yahoo! Maps). During this period, many organizations (in particular from the public sector) entered the GeoWeb 1.0, and server side products were developed, including the Free and Open Source alternative for serving maps, MapServer (http://www.mapserver.org/). The delivery of spatial data over the web was increasingly more sophisticated, but a number of factors still limited its diffusion; these were basically the complexity in developing applications, the limited bandwidth and the cost of the base cartography (Minghini, 2014). This limited the number of people who developed applications, and delegated most users to a passive role; a top-down approach was a key characteristic of this generation of web mapping.

The Web 2.0, with the advent of technologies such as SOAP, AJAX and APIs, together with hardware improvements such as the low cost of GPS receivers, was the driving force behind the Geo Web 2.0. The third generation of web mapping is iconically represented by Google Maps (https://www.google.com/maps), which arrived in 2005. The rise of the mapping APIs, which ease the development of map websites and at the same time give access to base data, allowed a large community of users to create and share spatial information (Minghini, 2014). The birth of the APIs is associated to the Google map platform, which was reverse engineered by hackers, in order to create *mash-ups*, applications that integrated the

Figure 3. HousingMaps (Rademacher,2005) overlaid Craiglist records on a map; it is considered the first mashup

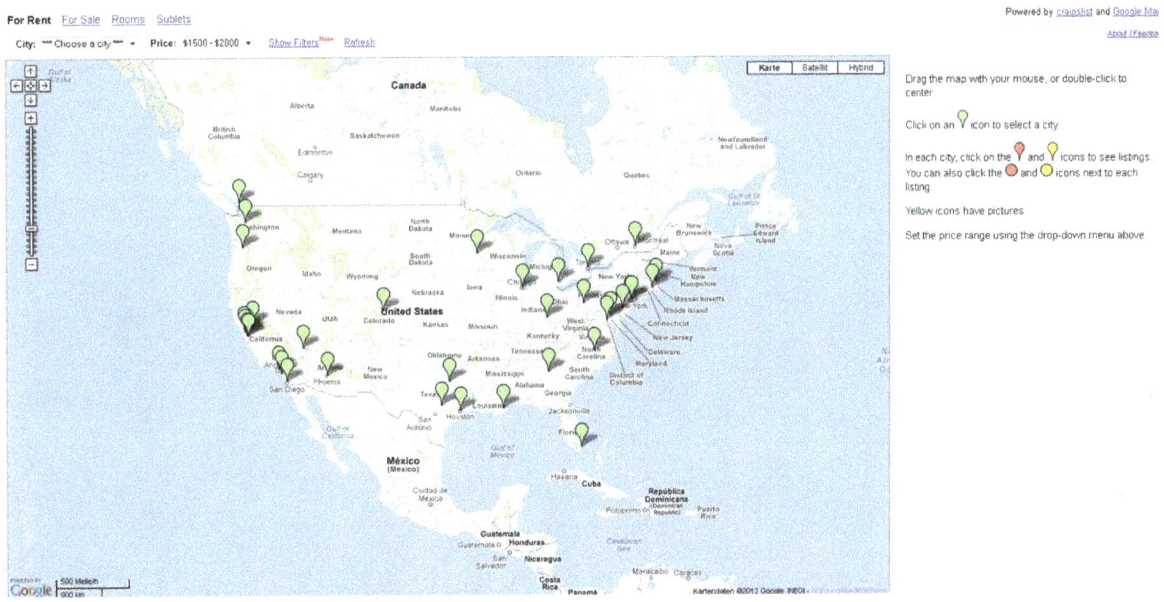

map layers from Google, together with information from other services and data. Shortly after, Google released a public API and it was quickly followed by other competitors (e.g.: MapQuest, Yahoo!).

The most recent web mapping clients are independent from the server-side applications, and can consume different services provided by servers. They support browsing maps (e.g.: zooming, panning), and layer interaction (e.g.: switching layers on/off) out-of-the-box. OpenLayers (http://openlayers.org/) is the most well-known and widely used web mapping client; it was developed with the idea of providing a Free and Open Source alternative to the Google Maps client, and it is entirely written in JavaScript, providing a rich experience in most modern browsers, without any server side dependencies. OpenLayers 3 includes the latest HTML5 and CSS3 features, and it supports 3D virtual globe capabilities through the use of WebGL.

As OpenLayers capabilities increase, it tends to become more difficult to use. Leaflet (http://leafletjs.com/) was released in 2010, with a main focus on developing mobile-friendly applications, and it has attracted a lot of attention since then, mostly due to its simplicity, performance and usability. Its objective is not to provide such advanced functionality as OpenLayers but to satisfy the basic needs of the majority of the users. This approach may explain its popularity, which is showcased by its adoption by major websites such as FourSquare, Pinterest, Flickr and OpenStreetMap.

Unlike the traditional approach to GIS, where professional cartographers would use complex desktop GIS and deal with advanced concepts and functionality, sometimes perceived as "boring", the GeoWeb 2.0 introduced techniques which enabled a more pleasurable and effective user experience, attracting a more casual and far wider usage of geographic information. This new approach is summarized in the term *neogeography*, coined by Eisnor (2006) to describe the non-professional practice of mashing up multiple sources of geographic information. Citizen science shares this aspect of the GeoWeb 2.0, of including non-professional scientists, voluntarily participating in a scientific project.

Figure 4. Screenshot of the FourSquare website, powered by Leaflet (Foursquare, n.d.)

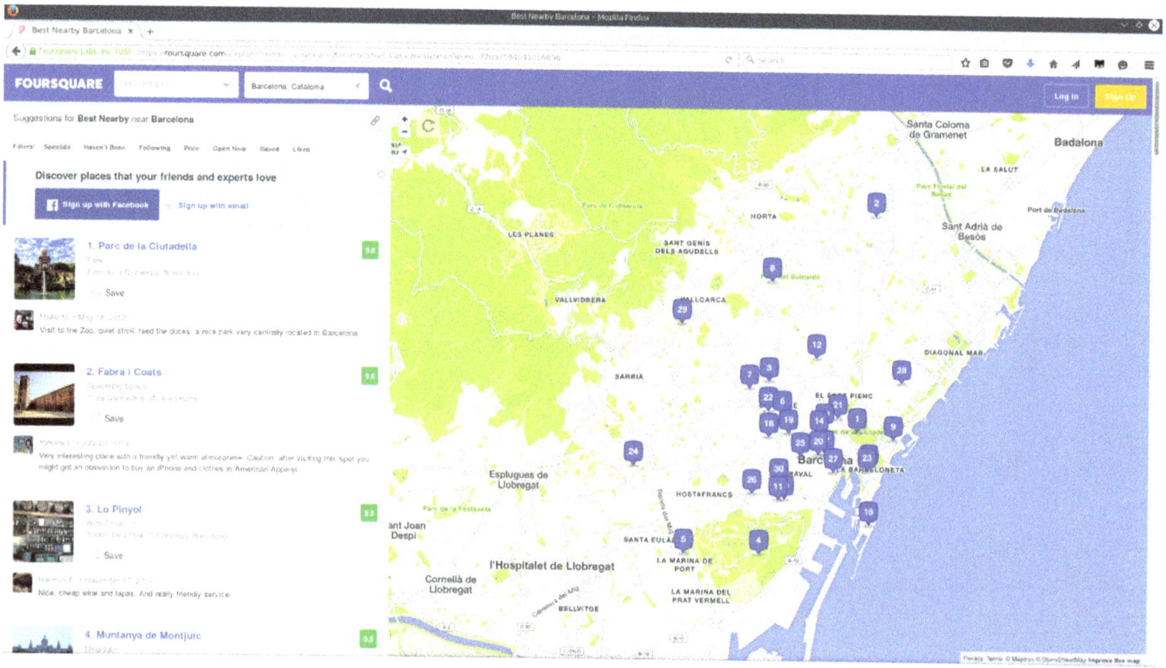

STANDARDS AND INTEROPERABILITY: WHY DO THEY MATTER?

Standards are technical documents, that define interfaces and encodings that can then be used by developers to implement applications that use these interfaces and encodings (Open Geospatial Consortium, n.d.a). They are very important, because they are the way to guarantee that data collected by different applications and different use cases is interchangeable, and that different services can talk to each other. This is particularly relevant in crowd-sourced initiatives where there are a lot of players involved, in different places, who may even use different devices. It is important to enforce standards, in order to ensure that these data will be discoverable, viewed, accessed and integrated, or else there is the risk to create data silos.

In the case of the geospatial community, two organizations have played a key role in setting a well-defined standardization framework: the Open Geospatial Consortium (OGC) and the International Organization for Standardization (ISO), and in particular its technical committee 211 (ISO/TC 211).

The ISO/TC 211 was first established in the mid 90s, and it was set to define a structured set of standards to describe georeferenced objects and phenomena. Among others, it created the ISO 19115 (International Organization for Standardization, 2003), which defines a schema for describing geographic data (metadata) and the 19119 (International Organization for Standardization, 2005), which identifies and describes the architecture patterns for services interfaces used for geographic information.

The OGC was also established in the mid 90s, mostly focusing on interoperability through standard development, but also featuring a compliance program with the goal of providing resources, tools and policies, for improving software implementation in compliance with these standards. The idea behind developing standard-based products, is that these products should work together effortlessly, without further debugging. As of 28th February 2016, there are 46 OGC standards, listed on their official website (Open Geospatial Consortium, n.d.a).

OGC Web Services (OWS) is a set of standards created for use in web applications; they include the Web Map Service (WMS), the Web Feature Service (WFS) and the Web Coverage Service (WCS). These standards enabled the development of vendor-agnostic server side products, which serve vector and raster data. MapServer and GeoServer are examples of such products. They also enabled the development of vendor-agnostic web mapping clients, which can combine different types of server-side services, as is the case of OpenLayers.

The Catalog Service for the Web (CSW) is another OGC standard, which is used to expose a catalog of geospatial records over the internet. These services support the publishing and searching of metadata about geospatial data, services and related information. They also support the discovery and retrieval of metadata (harvesting). The CSW profile must be consistent with the core metadata elements defined by ISO 19115 and its XML implementation given by ISO/TC 19139. The services metadata elements must be consistent with ISO 19119.

The Sensor Web Enablement (SWE) OGC standards (Open Geospatial Consortium, n.d.b), enable developers to make all types of sensors and sensor data, discoverable, accessible and usable through the web. These standards are very generic, and although they provide all the needed functionality for crowdsourcing campaigns, they also provide unnecessary functionality for this particular context. With the goal of simplifying the application of these standards to citizen science projects, the Citizen Science Observatory (http://www.citizen-obs.eu/), a network of five EU sponsored citizen observatory projects, has compiled a set of best practices and made them available on a GitHub repository. The "SWE4CitizenScience" (Citizen's Observatory, n.d., Higgins, 2016) is a remarkable effort for enforcing the interoperability in crowdsourcing campaigns, at the same time it eases the application of these standards in the context of citizen science.

Although OGC specifications are the golden standard for interoperability of geographic information, they are sometimes overlooked by the neogeography community (including citizen scientists), due to its complexity and relative obscurity (Minghini, 2014). This explains why the birth of the GeoWeb 2.0 relied mostly on APIs, rather than on OGC standards, which were available much earlier and could have enabled the same type of applications. Nevertheless, most Free and Open Source GIS relies heavily on OGC standards, and its level of popularity has been responsible for a wider adoption of standards, and thus an increasing interoperability among the GIS community.

The Role of Free and Open Source GIS

The term "Free Software" (use interchangeably with "Free and Open Source Software") was coined by Richard Stallman in 1984 (Stallman & Gay, 2009). It is based on the idea of guaranteeing to the users four basic freedoms, which are designed to promote software evolution (e.g.: bug fixing, capabilities improvement):

- The freedom to run the program, for any purpose.
- The freedom to modify the program to suit their needs (which requires access to the source code).
- The freedom to redistribute copies of the program, either free of charge, or subject to a fee.
- The freedom to distribute modified versions of the program, so that the overall community can benefit from their improvements.

These freedoms are enforced by a method called *copyleft* (Stallman & Gay, 2009), which is implemented through a family of software licenses. The first *copyleft* license was the GNU General Public License (GPL), written by Richard Stallman in 1984, and in its unmodified form it still dominates the FOS community.

The access to the source code is particularly relevant in GIS, because some algorithms are complex and it may be important to review them, in order to understand them (Neteler & Mitasova, 2008). Proprietary software falls short in terms of examinability of algorithms, but also in the distribution of newly implemented models, since the original software is required to run the model (Steiniger & Hay, 2009). Bug fixing and feature improvement are another consequence of the FOS model. Although the average user may not be able to solve bugs and improve features, there are a number of people who can do this, and the overall community can support them by testing and issuing bug reports, and then benefit from the improvements. FOSS development tends to be very dynamic, and sometimes it evolves quicker than proprietary software.

It is important to dissociate "free" software from "free of cost", as FOS can be distributed for a fee, and software that is *gratis*, may not guarantee the four basic freedoms to the user. Likewise, it is important to not mistake FOSS with "Open-source" software, as this definition does not specify if the source code can be modified and distributed.

In recent years the paradigm of FOSS development has taken root in the GIS community, resulting in the creation of several very sophisticated software projects that implement the different components of a complete GIS stack (Steiniger & Hay, 2009); ranging from map server applications (e.g.: MapServer, GeoServer), and spatial database management systems (e.g.: PostGIS), to desktop GIS (e.g.: QGIS, GRASS, SAGA) and web mapping libraries (e.g.: OpenLayers, Leaflet). These days FOS GIS have nothing to envy from proprietary systems, both in terms of functionality and ease of use. Unsurprisingly, all these features at zero cost and supported by a vibrant community have resulted in a rising uptake of FOS GIS among different user groups, including universities, public authorities, private companies, journalists and non-profit organizations.

According to Steiniger and Bocher (2009), the rise in popularity of free GIS tools can be measured using four indicators. The first one is the number of projects started, which is exposed on the number of entries added to the website FreeGIS.org (featuring 356 entries, on 27th February, 2016). The second indicator is the increasing financial support to FOS GIS projects, given by governmental organizations. GRASS, gvSIG and Jump are examples of projects which received such funding. The third indicator is the rate of downloads of FOS GIS software. Steiniger and Bocher (2009) mention as an example, SAGA GIS, which experienced an average increase of downloads in its documentation section, between 2005 and 2008, from 700 to 1300 per month. The fourth indicator is the increasing number of use cases of FOS GIS software (for instance those documented on Ramsey (2007)).

The Open Source Geospatial Foundation, OSGeo, (http://www.osgeo.org/) was created in 2006, driven by the need of a public representation of the FOS GIS community (Steiniger & Bocher, 2009). Since then it has led important initiatives such as publishing a journal, hosting a number of software projects, founding an education and curriculum committee, donating Sol Katz award to individuals who have demonstrated leadership in the FOS GIS community and organizing its yearly international conference, FOSS4G. The FOSS4G conference in Seoul, in 2015, has attracted 562 attendees from 48 different countries (Brovelli & Raghavan, 2015). The presentations include research talks, software development reports, and business case studies, and there is also space for user group meetings, developer workshops and software demonstrations.

Overall, the FOSS paradigm fits very well with the citizen science paradigm, since both are bottom-up initiatives; thus it does not come as a surprise that the citizen science community has been deeply involved in FOSS and vice-versa (Okolloh, 2009).

Volunteered Geographic Information

The term Volunteered Geographic Information (VGI) was created by Goodchild (2007), to describe the widespread engagement of large numbers of private citizens, often with little in the way of formal qualifications, in the creation of geographic information. It is important to note, that this approach of using citizens as sensors is opposed to the top-down approach taken for centuries, where this would be a role reserved for official agencies (e.g. mapping agencies).

Although some VGI themes such as land use, or soil classification may require sophisticated mapping skills, others activities such as collecting place names and geographic features are much less demanding, and have been widely adopted in citizen science initiatives. Technology improvements, which resulted from the advent of Web 2.0, and the low cost of GPS devices, contributed for enabling the collection of rigorous geographic information, without requiring an advanced knowledge in topics such as coordinate reference systems. Ushahidi (http://www.ushahidi.com/) is a good example of a platform that eases the process of *activist mapping*, a blend of activism, journalism and geographic information. It provides a user-friendly user interface, through which citizens can submit their local observations (using phones or computers) and a backend where it stores this spatio-temporal information. The aggregated results are then displayed in a web page. Ushahidi has been at the core of many citizen science initiatives (Meier & Brodock 2008; Liu et al., 2010).

The "democratization" of data collection, has also been associated with a "democratization" in the access to datasets, expressed by the Open Data philosophy (Auer et al., 2007). In line with Open Hardware, Open Software (see section "The Role of Free and Open Source GIS"), Open Content and Open Access, it expresses the idea that data should be freely available to everyone to use and republish as they wish, without any type of constraints. Although there is a very clear association between bottom-up datasets (for instance OpenStreetMap, n.d.), that is traditionally not the case with top-down, governmental datasets. Nevertheless, in recent years there has been an increasing pressure from citizens and communities, in order to promote the release of official data with open licenses. The governments, the United Nations and the European commission have responded to this effort with initiatives such as Data.gov (https://www.data.gov/), OpenDataBCN (http://inspire-geoportal.ec.europa.eu/), the FAO GeoNetwork (http://www.fao.org/geonetwork/srv/en/main.home) and the INSPIRE geoportal (http://inspire-geoportal.ec.europa.eu/).

The reactions to hurricanes Irene and Sandy, that recently affected the US, illustrate the use of *public data for public good*; *i.e.*: they show how the combination between VGI and Open Public Data can provide a critical, updated, information, which can potentially save lives. Hurricanes have always been a major concern, in particular around central and North America, and they often left a trace of destruction and death. For instance it is estimated that hurricane Galveston, which affected Texas in 1900 has provoked around 8000-1200 deaths; sixty three years later, hurricane Flora, which affected Florida has taken a death toll of 7,193 (Wikipedia, n.d.). The difference nowadays, is that more than 50% of American adults use social networks, 35% have smartphones and 78% are connected to the Internet (Loukides et al., 2013). For both citizens and local authorities, the internet turns into a two-fold pool for resources: in one hand for submitting information and on the other for receiving data about the hurricane.

The "Google Crisis map for Hurricane Sandy" (Google Crisis Response, 2012) is an interactive map which predicts the future of the "Frankestorm" in real-time; this impressive mashup is fed by public Open Data (e.g.: government weather data), as well as social data gathered by citizens (e.g.: location of volunteering opportunities, location of shelter and recovery centers). During the hurricane, this map was used by a large number of individuals for instructing other people about what to expect, and what to do.

Likewise to what happened during hurricane Irene, during hurricane Sandy there was a huge pressure from citizens towards local authorities regarding the release of official information about the hurricane. The New York City and the U.S. federal government decided to publish all their evacuation plans as open data, and apart from publishing their official evacuation maps, they also enabled the community to develop their own mashups and apps. By releasing open data, these organisms acted as a platform for public media, civic entrepreneurs and nonprofits to enable people to help themselves and one another at a crucial time (Loukides et al., 2011).

On the other hand, by acting as a network of connected sensors, citizens can share the effects of the hurricane in real time, providing local and regional authorities unprecedented insights into what is happening. The city of Maryland has taken a deliberate step towards promoting and organizing this VGI and invited their citizens to share and view hurricane data throughout the state (http://gopi.maryland.gov/). For this purpose they used a web tool and free mobile app, seeclickfix (http://en.seeclickfix.com/), that allows citizens to report neighborhood issues to local authorities. The app uses phone GPS in order to collect georeferenced information, which is later aggregated in maps.

Official data (for instance administrative maps) is still important and relevant for GIS, as it provides extremely accurate information, with a consistent degree of coverage. Nevertheless, this data is costly

Figure 5. Google Crisis Map for Hurricane Sandy, showing emergency alerts, FEMA disaster declared areas, the location of volunteer opportunities, shelters and recovery centers
(Google Crisis Response, 2012)

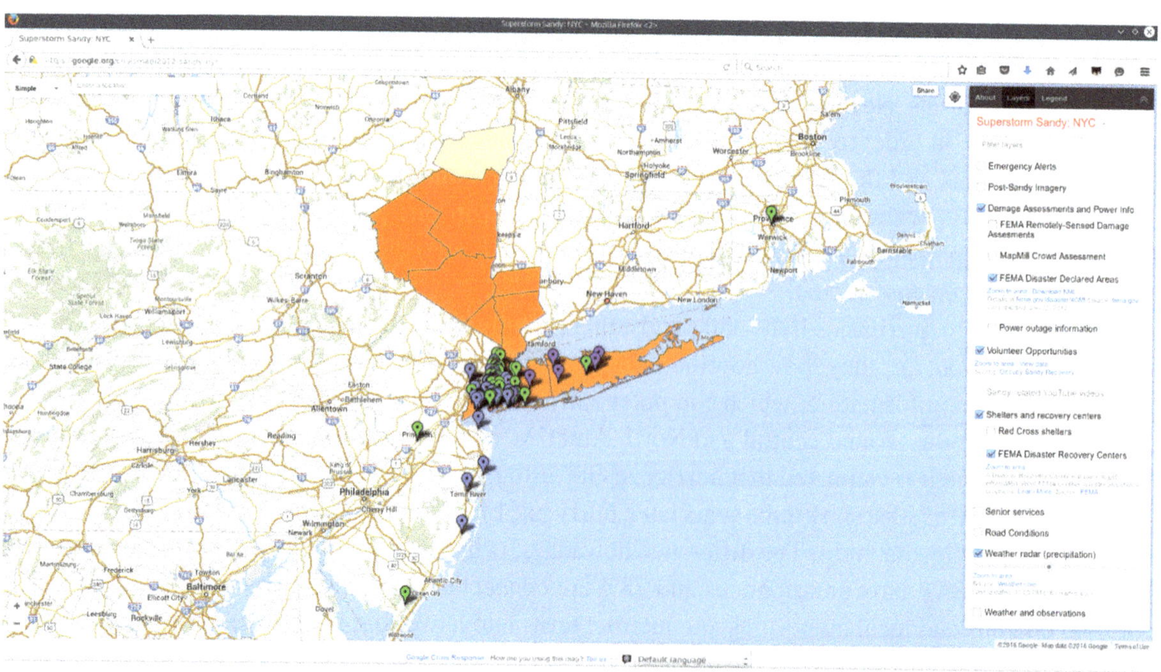

Figure 6. Evacuation zones map, produced by WNYC data news editor John Keefe
(The WNYC Data News Team, 2015)

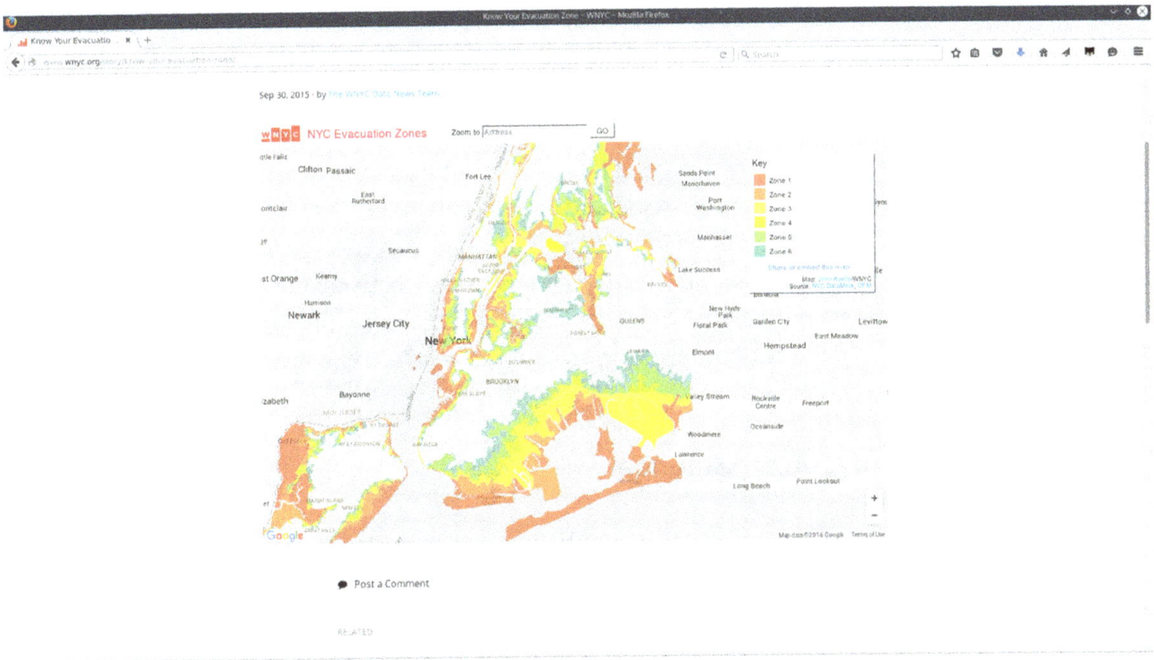

to produce, and often takes a long time to create it. Most applications demand data availability and up-dated information, and they are turning to *unofficial* sources, such as OSM, Twitter or other examples of VGI. This information may not be as accurate as the official versions, but it is extremely updated, it may cover some themes or geographical areas that are not covered, and can provide a great complement to official information. Public authorities are also understanding this and they are incorporating it, or using it to improve their own datasets.

In the early nineties, the Map Science Committee of the National Research Council issued a report describing the concept of National Spatial Data Infrastructure (NSDI); among other things it introduced the notion of *patchwork*, the idea that mapping agencies should no longer attempt to provide uniform coverage of the entire country, but instead they should provide standards and protocols, under which groups of individuals may emerge a global coverage, with varying detail, according to their needs (Goodchild, 2007). This directive provides the perfect ground for nurturing VGI under an official data infrastructure, and it exposes the growing interest of public authorities in this type of bottom-up data.

GEOGRAPHICAL CITIZEN SCIENCE

As mentioned in the “Introduction” section, a great deal of the information collected through citizen science projects has a geographic component. The intersection between citizen science and VGI is often called *Geographical Citizen Science*, a term coined by Haklay (2013) to describe citizen participation in scientific projects which involve the collection or analysis of geospatial information. Although geographic information can be, and still is, collected accurately using non digital technologies (e.g.: *paper* maps), *citizen cyberscience* is the branch of citizen science which takes advantage of GIS and of the Web 2.0, and thus it will be the focus of this chapter.

Geographical citizen science volunteers can be classified according to their role (Haklay, 2013). While active volunteers contribute consciously to the observation or analysis, passive volunteers normally behave as "observation platforms", without being actively engaged in the data collection process. An example of such behaviour would be an user that carries a phone with an app, which records location tracks.

Geographical Citizen Science can also be classified according to the explicit nature of geographic information (Haklay, 2013). While some activities are directly aimed at collecting geographic information (for instance the location of an observation of a species), in other citizen science activities this information can be collected unintendedly, as it may be the case of some geotagged pictures.

As citizen science projects often target geographically spread observations, and thus require a large, often sparse, network of volunteers, VGI results in an extremely attractive approach for supporting those projects.

GIS in the Context of Citizen Science

GIS has been at the core of many citizen science initiatives. As in many VGI projects, citizen scientists use the internet as their primary communication channel, and thus web GIS is their privileged tool to share spatial information. Unsurprisingly there are already some initiatives that provide general resources and information such as the citizen science association (CSA), the European citizen science association (ECSA) and the Australian citizen science association (ACSA). These global initiatives demonstrate that citizen science is building a momentum across the world.

In addition, there are guidelines for successful web mapping applications in citizen science, such as Newman et al. (2010). Moreover, an increasing number of initiatives use mobile-based applications to enhance public participation and engagement in citizen science (Brovelli et al., 2015).

This section lists different citizen science initiatives which make use of GIS, detailing a few use cases. It also analyses some aspects of the use of GIS in citizen science and provides generic guidelines, which the authors consider useful when starting a project.

Examples of GIS Enabled Citizen Science Projects

In order to illustrate better the current GIS trends within citizen science it has been collected a list of citizen science projects which have presence in diverse research areas. Some projects have been selected from the Barcelona city council's citizen science booklet (Barcelona City Council, 2016). These projects cover different topics such as zoology ("Atrapa el Tigre", iNaturalist, JellyWatch, NatureWatch, Sea-Watchers UrbanGene), air quality (Safecast), water quality (Citclops, EyeOnWater, Riunet, SecchiDisk), ornithology (eBird) and coral health monitoring (CoralWatch). There are different types of entities behind these initiatives, including universities (in most cases), local associations, companies, research centers, and public authorities. Most of these initiatives use web mapping and rely on OGC standards.

"Atrapa el Tigre" (http://atrapaeltigre.com), promoted by Centre for Advanced Studies of Blanes of the Spanish National Research Council (CEAB-CSIC), was created to survey and monitor the population of the Asian tiger mosquito (Kampen et al., 2015). The project includes a database where it stores all the data related to this species, a service to serve these data, and an interactive web client with a map, where it displays the information to the final user. Figure 7a) presents a screenshot of the web mapping client.

The map features colored markers displaying different information, depending on the probability of a mosquito being the target species. The samples are grouped by clusters, and when clicking on a cluster, it zooms to its bounds. Each cluster features the number of samples associated to it. Clicking on a sample triggers the display of a pop-up menu with the photo of the mosquito, along with the answers to some questions that help to identify the tiger mosquito, and whether this sample was validated by an expert.

Citclops (http://citclops.eu) is a research project which focuses on assessing the color, transparency and fluorescence of the sea (Ceccaroni et al., 2015). This project also includes a database where it stores all the information related to watercolor (measured in Forel-Ule (FU) units), a service to serve these data, and an interactive web client with a map where it displays the information. Figure 7b) shows a screenshot of the web mapping client, featuring a map with different color markers, which correspond to different FU colors; the number and color of the cluster indicates the average FU color for the samples in that cluster. More information about the Citclops objectives and how GIS technologies contribute to achieve these objectives are detailed in the "Use Cases" section.

Figure 8 shows screenshots of the web mapping clients of the CoralWatch and eBird projects.

CoralWatch (http://www.coralwatch.org) was developed by the University of Queensland, with the objective of quantifying coral health (Alabri, 2010). The CoralWatch website presents an interactive map

Figure 7. Web mapping clients of the a) Atrapa el Tigre and b) Citclops projects

Figure 8. Web mapping clients of a) CoralWatch and b) eBird

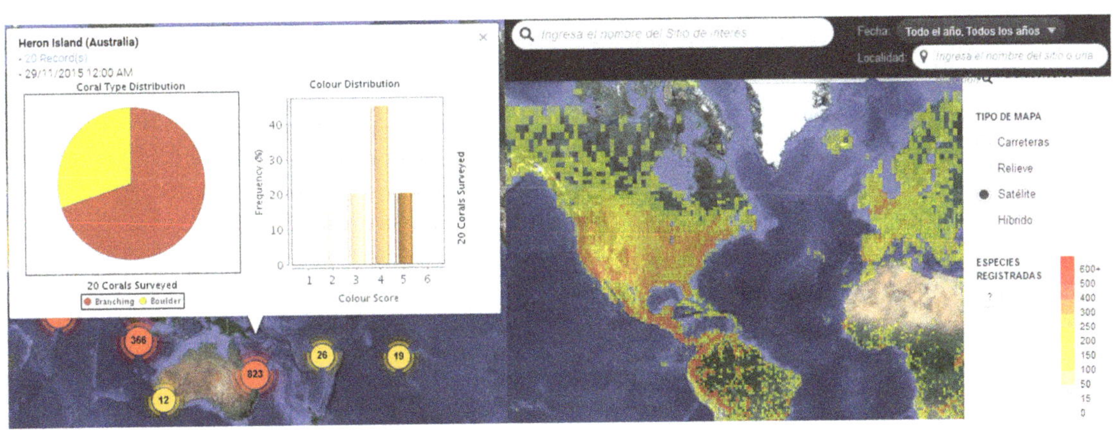

featuring calculated clusters of corals (Figure 8a). The numbers associated with the clusters indicate the number of samples, and the color scale depicts the number of samples. When clicking on a cluster, some additional information is displayed (the number of analyzed colors, a pie chart featuring the coral type distribution and an histogram with the color distribution). Clicking on the survey details will extend the information to each of the records in the survey. More information about the backend technologies of CoralWatch is given at the use cases section.

The eBird project (http://ebird.org) is an initiative of the Cornell Lab of Ornithology (National Audubon Society), with the objective of monitoring birds (Sullivan et al., 2009). The project, which started in 2002, presents a website with an interactive heatmap of the number of registered species. The web application offers different ways of plotting the distribution of the birds: by region, hotspots (Figure 8b), by species, bar charts (e.g.: explore the bird population distribution, according to the time of the year, or the location), line graphs (explore different metrics of species observations in a region or location) and a submission map, sighting submissions from people across the globe, are displayed in real time.

Figure 9 shows the web mapping application of EyeOnWater (http://eyeonwater.org) developed by Citclops / MARIS. The underlying motivation for this project was to assess the color and clarity of natural waters, and the physiological state of Ulva (Sea Lettuce). The website, launched in 2015, displays an interactive map with the different observations of FU color, the users who submitted them, and the most recent observations. The color of the samples indicates the Forel-Ule color of the observation. Lists of the most active users and most recent observations complement the information on the map. When clicking on a user, its observations are automatically displayed. One of the most interesting aspects of this project is the application of user engagement techniques. For instance, scoring citizens who submit observations, ranking them, and rewarding the most active data collectors with a special status, that can range from Jellyfish to Neptune.

Figure 9. Web mapping client of EyeOnWater

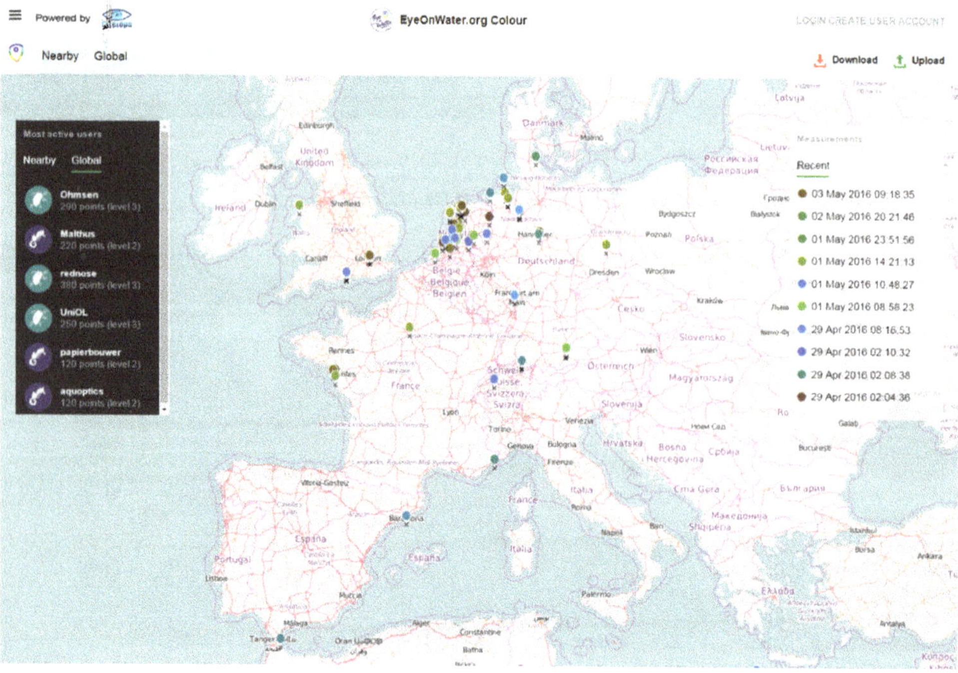

Figure 10. shows the web-mapping application of iNaturalist (http://www.inaturalist.org), a project started in 2008 by the California Academy of Sciences. The idea behind this initiative is to share observations of animals and plants. The map displays the observations of different species (in grey circles) and supports filtering by different parameters (e.g.: observation, species, identifiers and observers). The most recent samples are listed on the side of the map.

Naturewatch (http://naturewatch.org.nz) was developed by the New Zealand Bio-Diversity Recording Network Trust, with the purpose of monitoring nature (e.g.: Botany, Entomology, Ecology, Ornithology, Phenology and Zoology). The web mapping interface is similar to the one used in iNaturalist. As in iNaturalist, it supports filtering by observations, species, identifiers and observers. When clicking on an observation on the map, a photo of the observation is displayed, along with the observer and the respective research grade.

JellyWatch (http://www.jellywatch.org), developed by the Monterey Bay Aquarium Research Institute, is a project whose aim is to monitor jellyfish. The web mapping application shows the locations of jellyfish around the world, along with the information about other marine organisms. The different species of jellyfish are represented by different icons.

Riunet.net (www.riunet.net) was developed by the Freshwater Ecology and Management Research Group, of the Universitat de Barcelona, with the objective of evaluating the ecological quality of river ecosystems. The website features an interactive map plotting the distribution of the different ecological samples of a river. Figure 11 shows a screenshot of Riunet, featuring the markers that indicate the ecological status of the river. When clicking on a sample, a pop-up is automatically triggered, displaying additional information (a photo, the observer's name, and the description of the different parameters measured in the sample).

Safecast (http://safecast.org) was started in 2011, by the Momoko Ito Foundation; its main concern is to monitor radiation and air quality (Yasuhiko, 2014). The project's website shows the level of radiation and air quality collected by sensors. The amount of radiation in a zone, is depicted by a color scale. When clicking on one sensor, the evolution of the radiation for that sensor is displayed.

Figure 13 shows screenshots of the web mapping clients of Seawatchers and Secchi Disk projects. Sea watchers (http://www.observadorsdelmar.cat) was developed by the Institute of Marine Sciences of the Spanish National Research Council (ICM-CSIC) with the purpose of monitoring the current state of seas and oceans (Azzurro et al., 2013). The Secchi Disk study (http://www.secchidisk.org), started in

Figure 10. Web mapping client of iNaturalist

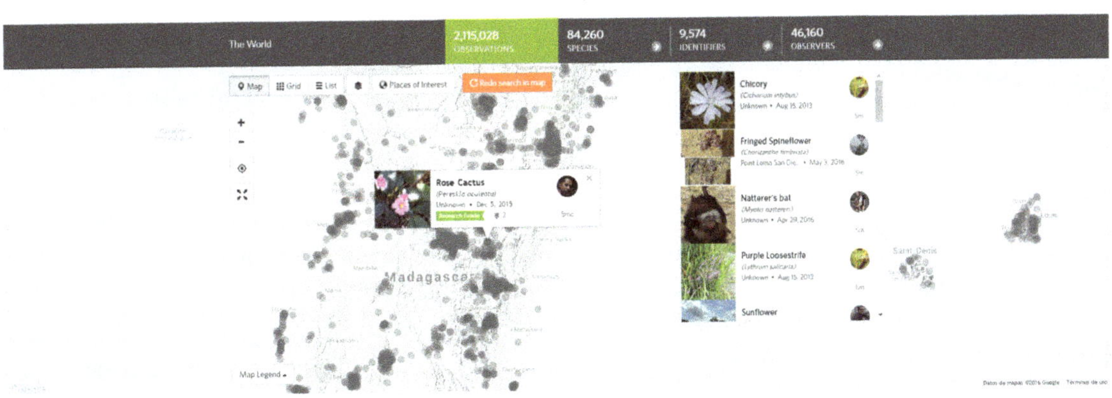

Figure 11. Web mapping client of Riunet.net

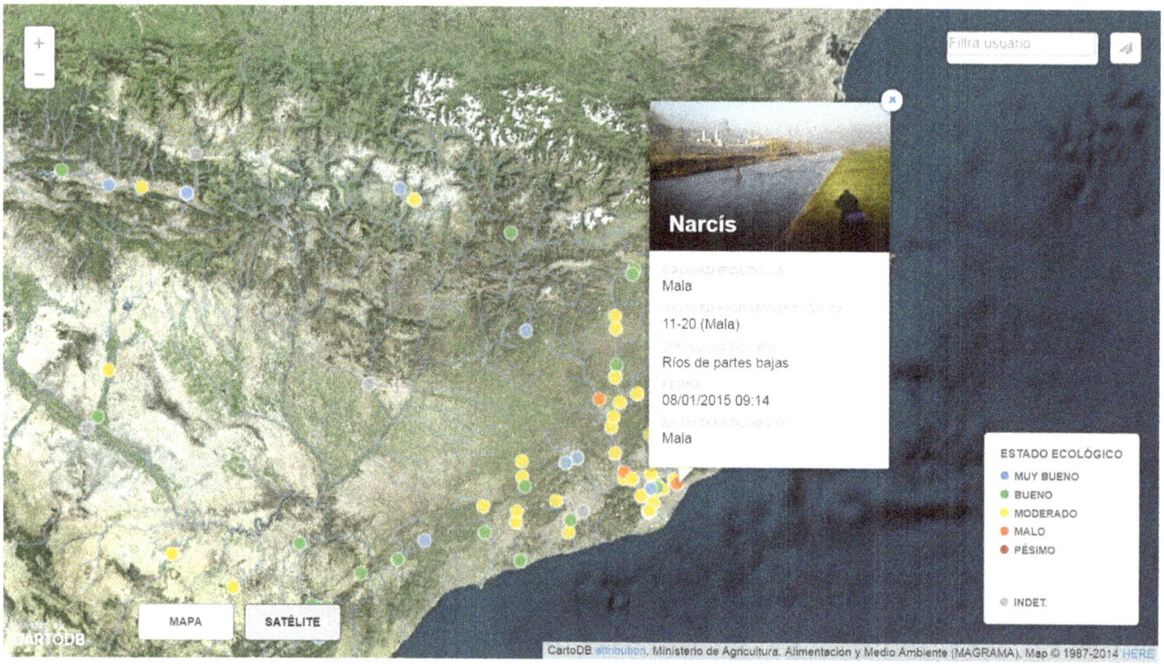

Figure 12. Web mapping client of Safecast

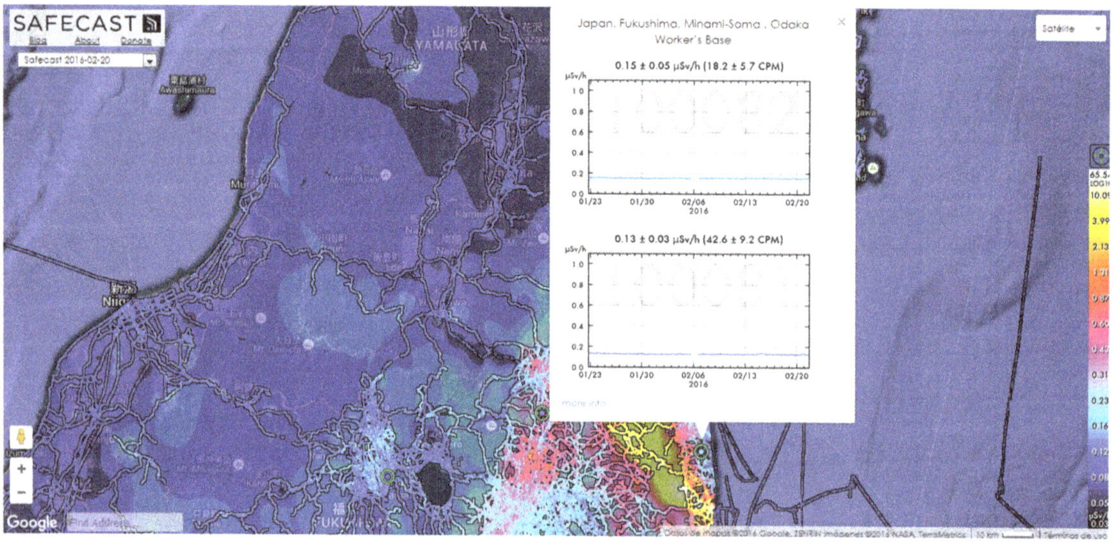

2013, focuses specifically in monitoring changes in the marine phytoplankton. In Seawatchers the different colors indicate different species, while in Secchi Disk indicate the different levels of transparency.

Urban Gene (http://urbangene.epfl.ch) was developed by the École Polytechnique Fédéral de Lausanne. The overall idea of this project was to assess the current state and dynamics of biodiversity, and to identify ecological, socioeconomic and socio-demographic factors that may influence dispersal, gene

Figure 13. Web mapping clients of a) Seawatchers and b) Secchi Disk project

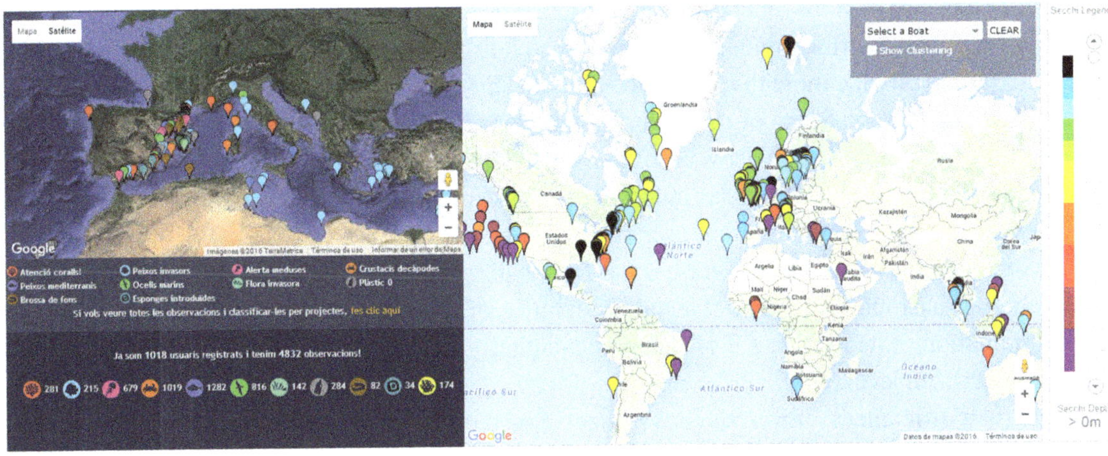

flow and adaptation of plant and animal species in urban areas (Joost et al., 2014). The developed web mapping application (in Figure 14), displays clusters of species, along with an indication of the number of samples on each cluster. The backend of UrbanGene is described in more detail in the use cases section.

Classification of Citizen Science Projects

Any type of citizen science project based on GIS demands a high level of interaction between the citizen user and the technological platform provided by the project. Thus there is a requirement to enable a kind of access door, or gateway, in order to exchange the generated knowledge between two sides, users and GIS platform. Through this entrance there is a dataflow in both directions. The direction from the user to the platform firstly generates all the necessary information required by the project such as the samples, registers, observations or sensor data and secondly receives user requests for access to this information. The direction from the platform to the user responds to requests for information by downloading, view-

Figure 14. Web mapping client of Urban Gene

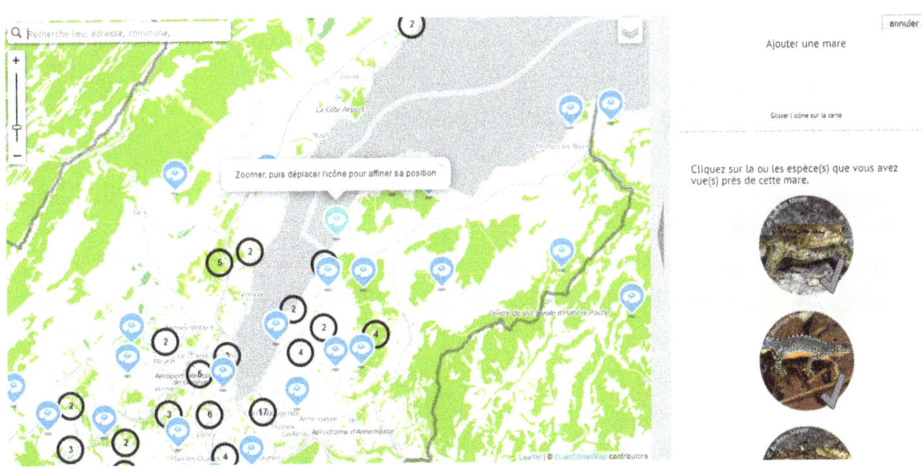

ing or analysis. Examples of information request could be downloading raw data, samples display on a web map or data analysis by generating reports or graphs. This "door" can be interpreted as a citizen science interface. It is the first things a user sees, when he/she joins a citizen science project. Moreover, this interface could provide more than one access door. This gateway could take the form of an app, web map, web form or sensors. Some of them can generate, display data or both.

As can be noted, all interaction between citizens and GIS platform is carried out through this interface. Therefore, this interface has to be fit, as much as possible, for an audience of non-experts. The usability and user experience are very important aspects to consider when designing this type of interface. The tools provided to citizens should be very easy to use, understandable and intuitive. In this way it can be assured a higher user engagement, universal acceptance of the initiative pursued by the project and a wide dissemination of its results. All of this will eventually lead to a more pronounced impact on both scientific and citizen level.

Despite their differences, citizen science projects based on GIS share something in common. On one hand they share the interface that has been mentioned above. And on the other hand they share the actions performed through this interface that are the interaction with data and its visualization. In Figure 15 a summary of the interrelation between these elements is displayed.

Figure 15 shows that first of all, the only interaction door between the user and the technology platform in question is the citizen science interface. Analyzing the structure below the interface in greater depth, one will notice that it is formed by two large blocks: data and visualization. The data block represents all actions, components and technologies that have some relation to it such as data input, data storage, data analysis, etc. The visualization block represents the actions, components and technologies related to the graphical presentation of the data or results after its analysis through web mapping, clustering and layer display techniques, etc.

Figure 15. Concept components used by citizen science projects

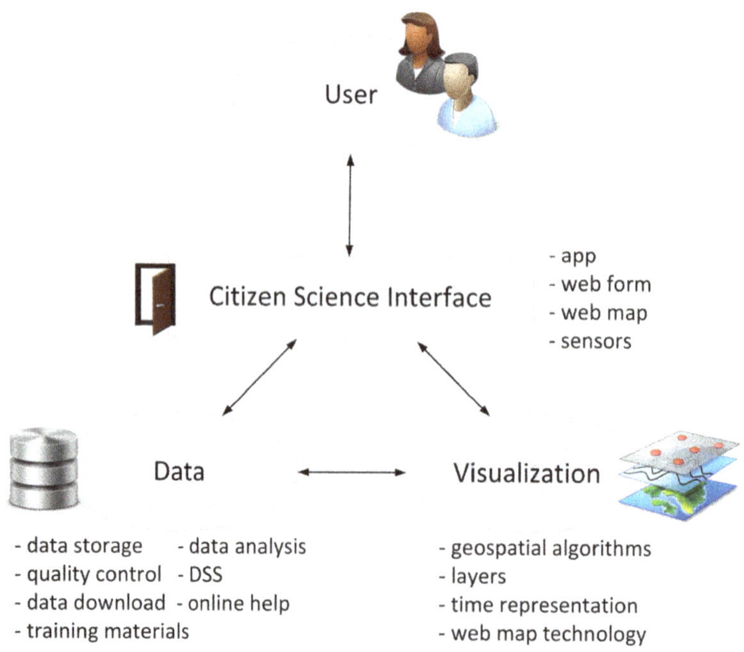

The next two sections contain a more detailed explanation of these two groups citing several examples of technologies and citizen science projects. Note that this classification does not attempt to make a technology separation, but concepts (in this case, data usage and data display concepts) to which different types of technologies are applied.

Visualization-Driven Classification

The analyzed projects are grouped according to the following classification topics: geospatial algorithms, layer display, representation of time and web technology.

Clustering data points according to its location is a common interpretation of collected data used by GIS technologies. In citizen science, there are several initiatives that cluster samples such as Atrapa el Tigre, Citclops and CoralWatch (Alabri, 2010). Other common geospatial algorithms include the calculus of point trajectories, for instance used in Citclops - Barcelona World Race 2014-15 initiative (Ceccaroni et al., 2015), heatmaps (e.g.: Safecast) and the generation of new samples by clicking in the map, which it is shown in Figure 14 (UrbanGene, Joost et al., 2014).

The organization of information in layers, which supports combining different thematic units in a single view, is a core feature of GIS. In citizen science initiatives, there are several projects which take advantage of this approach to expose important information, such as Atrapa el Tigre, Citclops, iNaturalist and Seawatchers.

Time is an important variable in citizen science, which unlike space does not have a standard native support in GIS, but can be represented with different strategies. Some web mapping applications represent the time dimension with widgets such as a calendar (Citclops) or a time slider; others support time filtering (Atrapa el Tigre, Citclops, iNaturalist and Sea watchers). Finally eBird is an example of a web mapping application which displays information in real-time.

Web mapping clients, can also be classified according to the map service they use. Many citizen science initiatives use Bing (Citclops), while others use OpenStreetMap (Atrapa el Tigre and UrbanGene), Google Maps (iNaturalist, NatureWatch, Sea watchers and Secchi Disk study) or CartoDB (Riunet).

Web Mapping Application Classification

Newman (2010) presents another classification, based on web map application features and the role they support. These features include: data entry, quality control, data download, analysis, decision support, training, and technical support.

Data entry can be achieved by entering data forms (Citclops), an app (like Secchi Disk study or Citclops), sensors (Citclops, Citi-Sense, Omniscientis, Wesenseit) or by interactively clicking on the map (UrbanGene).

Quality control can be performed in different ways, by specific algorithms (Citclops), by assigning a reputation to the user who collects the sample (EyeOnWater), or through expert validation (Atrapa el Tigre).

Data download is generally not supported in citizen science projects (although some of them such as EyeOnWater have this functionality). Most of them do not allow the download of the complete dataset, and not even of a subset.

Regarding data analysis there are several factors that can be displayed:

- Calculate the number of contributions per year by project. In iNaturalist and NatureWatch you can see the total number of contributions, species and observers.
- Create project statistics as charts. This is done in some projects such as CoralWatch.
- Report the most frequently reported species. Projects like NatureWatch an iNaturalist show the most recent found species.
- Display and reward the member with the most contributions. For example EyeOnWater display most active users in one side of the screen. Others like NatureWatch and iNaturalist enable to look for observers and rank them according to the number of observations.
- Display and reward the project with most contributions. Projects like NatureWatch and iNaturalist rank the species according to the contributions when clicking to the species tab.

Decision support is not supported in most projects. They lack functionality such as activating early warning email alerts, defining locations of interest for early alerts, defining species of interest for early alerts or creating predictive models. Citclops offers decision support capabilities in a module called 'Citclops Data Explorer'.

Regarding training materials, there are plenty of them in most initiatives, such as online videos or tutorials.

Although online help is implemented in most initiatives, telephone support is a more difficult functionality to find.

Use Cases

In order to demonstrate the diversity of GIS technologies and how they can be applied in the context of citizen science, the authors selected a few use cases, in which the underlying technologies were disclosed. The next paragraphs contain a detailed description of each use case.

Citclops

Citclops is a European project (2012-2015) that was developed under the umbrella of FP7. This project was led by Eurecat, with the collaboration of the following institutions: Spanish National Research Council (CSIC), Carl von Ossietzky Universität, Oldenburg, Royal Netherlands Institute for Sea Research, Kinetical Business SL, TriOS Mess, MARIS, Noveltis SAS, Coastwatch Europe and Deltares. The Citclops project aims to develop systems to retrieve and use data on seawater color, transparency and fluorescence, using low-cost sensors combined with people acting as data carriers, contextual information and a community-based Internet platform. GIS technologies have an important role in this context, because they allow the visualization, analysis and interpretation of collected data.

As a result, a decision support system named 'Citclops Data Explorer' has been developed, which provides several interfaces to visualize collected data: the citizen observations, the marine data analyzer and the marine data repository (Ceccaroni et al., 2015). For each interface different GIS technologies are used. The citizen observations' interface, allows citizens to visualize the observations collected through the "Citclops-Citizen water color monitoring" app. It enables the user to share his/her localization and see the samples nearby. The marine data analyzer is aimed at a more advanced user. In this case, it is shown more in-depth information regarding the state of seawaters, such as the historical state of bathing waters, seawater color obtained from the satellite data and historic weather data. The data in this interface can be graphically visualized in time series plots. Finally, the marine data repository allows a more extensive search and data downloading according to different parameters.

To develop these interfaces several GIS technologies have been used such as Bing Maps and Leaflet for the front-end and Geoserver, PostgreSQL and PostGIS for the back-end. GIS technologies were used mostly for visualization, storage and delivery of information while other technologies have been applied for developing different features (for instance Octave for quality control and Python language with Flask framework for data upload).

Figure 16 shows the general diagram of the internal structure of the Citclops server, with its components and dataflows. The arrowed lines indicate the flow of the observation data. Dotted lines depict the paths, when an observation is being inserted into the Citclops Server; solid lines depict the paths, when viewing the observation-data through the Citclops Data Explorer (Ceccaroni et al., 2015).

During the insertion action the citizen can provide his or her observation through the "Citclops-Citizen watercolor monitoring" app or the web form available through the Citclops' website. Both components are connected to the Upload service located on Citclops Server. This service has been developed in Python language using Flask framework. Once the sample reaches the server it is split in two parts; the graphical part (watercolor images) goes to a dedicated directory and the information part (geospatial and description metadata) goes to the PostgreSQL database where the PostGIS extension has been activated.

During the viewing process the citizen uses the web-based Citclops Data Explorer. The backend part of this component is based on GeoServer, which also uses the same database to read the metadata, in order to locate the observations' markers (with watercolor images taken from the dedicated directory) on the map. A part of the color samples GeoServer also serves the available satellite images with watercolor processed data. These samples have been collected through Medium Resolution Imaging Spectrometer

Figure 16. Application components and data-flows in the Citclops Server

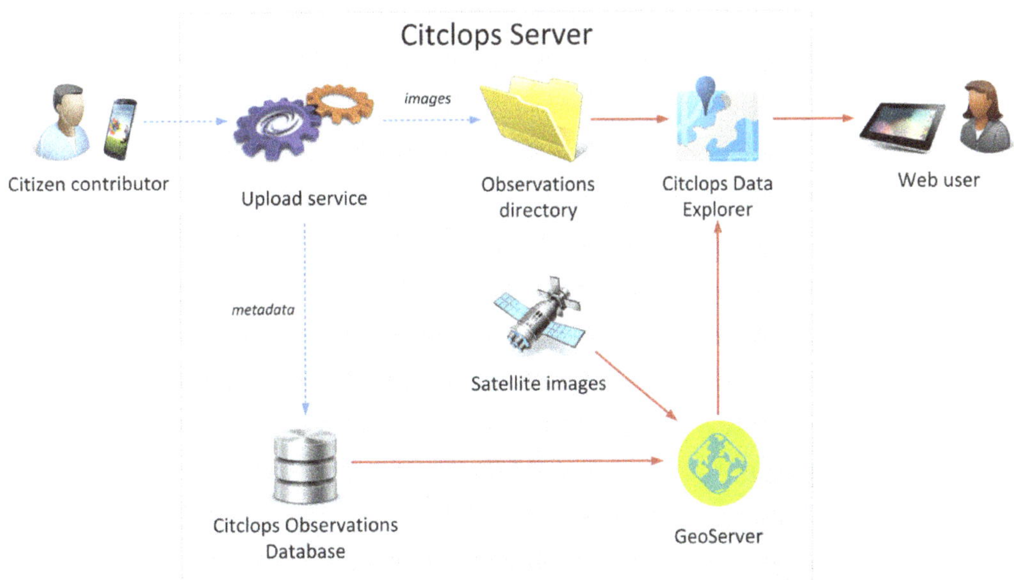

(MERIS) instrument, which has been launched by the European Space Agency (ESA) onboard its polar orbiting Envisat Earth Observation Satellite.

The frontend part of the Citclops Data Explorer is based on Bing maps, using the satellite view layer. The interactive map was developed using the javascript library Leaflet. In order to show the watercolor samples contributed by citizens, GeoServer returns the messages with information on their location, metadata and a link to the image. This information is then added to the web map by the Leaflet API. MERIS satellite images are also included as layers. Besides visualizing time series, the user can also query a color image by date, using a calendar.

EyeOnWater

This project was born during the execution of the Citclops project. It is an evolution of the results obtained in Citclops, counting with several improvements. These improvements include progress on the issues of usability, user experience and user engagement.

The idea and functionality remain more or less similar. The concept involves the collection by citizens of watercolor samples through the smartphone app and the visualization and analysis of these samples on the web map.

As for the technologies used in this project, a summary of components is shown in Figure 17. The backend includes MapServer where the use of WMS and WFS standards is made. The PHP language is used for functionality unrelated to the maps. The data storage part uses SQLServer database. The Frontend includes OpenStreetMap for web mapping and OpenLayers 3 stack for the API interaction. Javascript is also used for the GUI web elements.

In the case of the mobile application it has been developed in the project for various purposes, such as observing color or transparency of the water. The tools that have been used for its development are Apache Cordova framework with Crosswalk project and HTML5.

Figure 17. EyeOnWater components

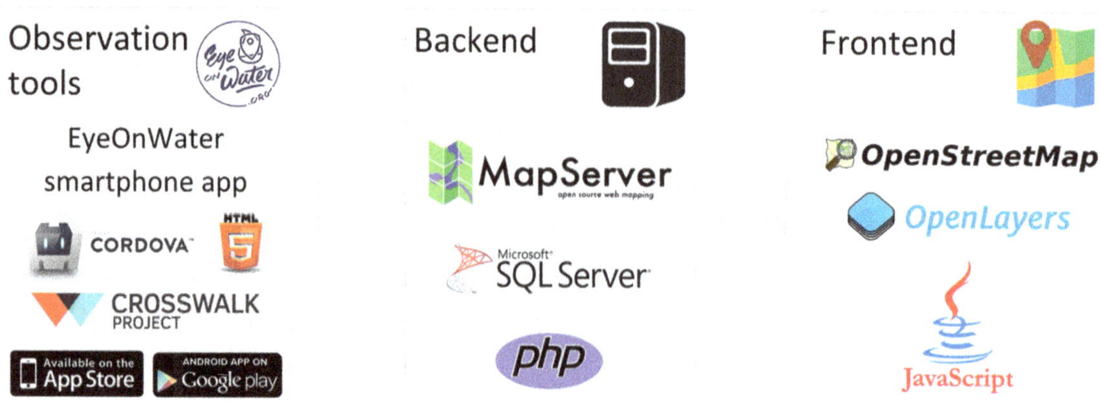

CoralWatch

In the case of CoralWatch the technologies used are slightly different, but the architecture remains similar. Figure 18 describes in detail, the architecture and components used in this project.

On the backend side, CoralWatch integrates Java based tools. On the web server the Apache FreeMarker template engine is used to generate HTML output based on templates and changing data. PostgreSQL with PostGIS, is the selected database engine. A different aspect in this case here is the use of PL/R Statistical processing, which performs statistical analysis of coral data to determine whether a bleaching event has occurred. In addition the CoralWatch server includes a component called "Integration tools", which consist of correlating the CoralWatch data with other data repositories using an ontology. Finally, there are one or more connectors to provide the ability to send samples using the smartphone.

On the frontend side there is an interesting combination of Google Maps API and TimeMap javascript library. On one side the Google Maps enables the geospatial representation, and on the other side, the TimeMap allows the integration of spatial and temporal objects into a map and a timeline.

As a final result, a simple but effective citizen science tool has been implemented, which gives the ability to report information regarding the status of corals and visualize the results in very intuitive way.

UrbanGene

In UrbanGene the authors found similar technologies and a similar structure of the solution. On the backend side two main elements have been detected; the web server, which is developed in PHP, and data storage based on the PostgreSQL/ PostGIS stack, which adds support for geographic objects allowing location queries to be run in SQL.

Figure 18. CoralWatch system architecture extracted from Alabri (2010)

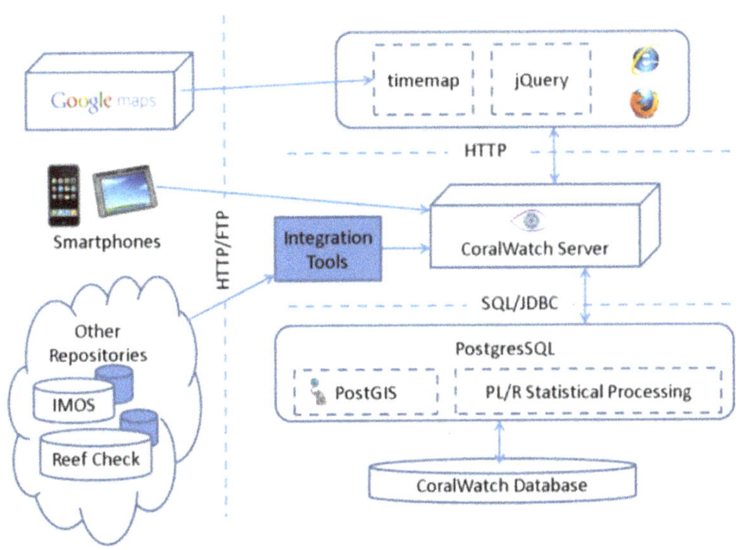

The part of the frontend uses a fairly standard combination consisting of OpenStreetMap to enable the geospatial representation, Leaflet API to enable the data web visualization and jQuery UI library to manage some user interactions like the drag & drop (Joost et al., 2014).

GIS INFRASTRUCTURE FOR CITIZEN SCIENCE

After reviewing all these solutions, one can ask the question: "which technology should I use for my citizen science project?" In this section the authors compare different GIS technologies in order to support a more informed decision regarding these options. To help understanding the comparison made in this section, the reader may refer to Figure 19, where a summary chart of all components is presented.

The first step could be to choose a map server, to provide the base maps. There are several popular options such as Google Maps (proprietary), Bing Maps (proprietary), MapQuest (proprietary), OpenStreetMap (ODbL), Here (proprietary) and Apple Maps (proprietary). In case you need to show high-resolution satellite maps, Bing Maps provides a resolution slightly higher than other solutions. Apart from this, one could also use an OGC standard (e.g.: WMS, WFS, WMTS) to access maps served by a non-proprietary server (or even choose to serve them).

The next step is to choose a web map API for data web visualization. As stated earlier in the "Background Section" the most well-known web map APIs are Leaflet (BSD-2-Clause license) and OpenLayers (FreeBSD license). As mentioned before, Leaflet has the advantages of being simple and easy to use, having a good look and feel, being light, offering mobile support and having a good performance. On

Figure 19. General GIS architecture within a citizen science project

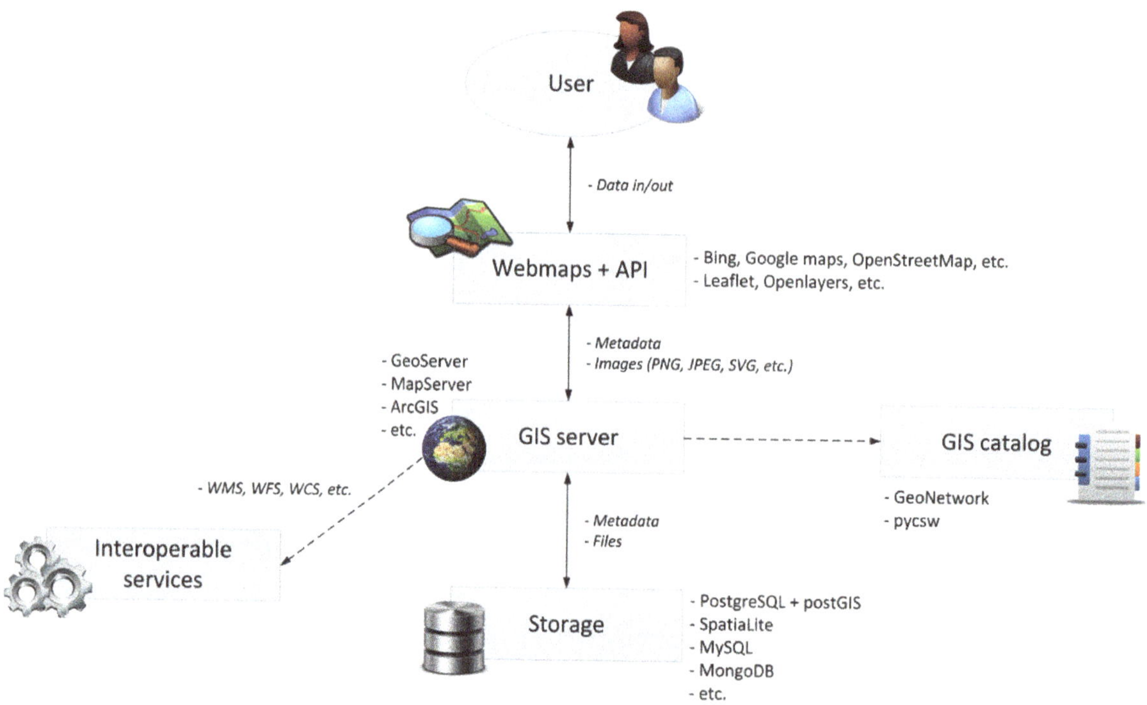

the other hand, OpenLayers supports all WebGIS protocols, has a better documentation and community support. It is also a more mature and tested project, it features more map controls, it has better integration of projections, and in most cases plugins are not needed.

The next step would be to decide in which database, the data and the images should be stored. Most citizen science initiatives use PostgreSQL (PostgreSQL license) because of the PostGIS extension (Citclops, CoralWatch and UrbanGene). PostGIS is a spatial database extender for PostgreSQL object-relational database. It adds support for geographic objects allowing location queries to be run in SQL. It should be noted that there are other databases with spatial extensions such as MySQL (GPL version 2), MongoDB (GNU and AGPL), Oracle Spatial (proprietary) or SQLite (public domain) with SpatiaLite extension, but they are not nearly as popular.

The next step would be to choose a server to publish the produced information to the web. Most initiatives use a GeoServer (GPL license) (Citclops, Brovelli et al., 2014, Castell et al., 2014) to share, process and edit geospatial data. Others use Mapserver (X/MIT license) that supports the creation of spatially enabled internet applications (Fritz et al., 2012). There are many other solutions, both free and open source (e.g.: MapGuide Open Source), as well as proprietary (e.g. ArcGIS server).

Finally, there is the option of publishing the created services in a Geospatial web catalog, which can be consulted and used in an interoperable way through the CSW standard (see "Standards and Operability: Why They do Matter"). In this manner it enables open access to geo-referenced databases and related metadata generated by the citizen science projects. The most widely used application to catalog these services and manage spatially referenced resources is GeoNetwork (GPL license) (http://geonetwork-opensource.org). It provides powerful metadata editing and search functions as well as an interactive web map viewer (Fritz et al., 2012). In addition there is another application that has similar functionality; pycsw is an OGC CSW server, which allows for the publishing and discovery of geospatial metadata, providing a standards-based metadata and catalogue component of spatial data infrastructures. Both solutions are FOSS and run on all major platforms.

Quality Control in GIS Enabled Citizen Science Projects

Quality control is another aspect which should be considered when building a GIS infrastructure for citizen science. The information collected within the framework of a citizen science project is usually freely and voluntarily provided by users and the tools used in this process are also distributed in an open and public manner. This can lead to the situation where users can send erroneous, incomplete or inappropriate information. To ensure the minimum quality of provided samples and prevent the collapse of citizen science platform is advisable to integrate a layer of quality control within the project's architecture.

There are different techniques to check and ensure the incoming data to the system. But first of all it should be decided which data will be controlled and why. For example, in case if the samples are images reported by a mobile application, the quality of what has to be analyzed? If the image actually contains the sample/subject to be analyzed or not? If metadata of the image is complete or not? If the coordinates where the image has been taken are correct or not? The possibilities are multiple and are directly related to the context of the project.

There is a brief mention of quality control in the "Web mapping application classification" section. For example, one of the techniques used in the Citclops project to filter the citizen observation data consists in a software module which analyzes the images collected via app and computes the water color (Novoa et al. 2014 and Novoa et al. 2015). In this way it becomes possible to compare the results that have been reported manually by the citizen on the one hand and calculated by the algorithm on the other hand (Ceccaroni et al., 2015). Another quality control process used in the Citclops project consists in collaborative techniques. In this case there is a functionality implemented in the user interface that allows to flag images that do not correspond to the water (Ceccaroni et al., 2015). The same type of technique has been detected in the web viewer of EyeOnWater project.

But one of the most complete designs of quality control module has been observed in the COBWEB project. The quality assurance (QA) is one of the key components of its system architecture and is provided on several levels through the QA workflow service which have been developed basing on Business Process Model and Notation (BPMN) and the OGC's Web Processing Service (WPS) standard. Very interesting approach used in this solution consists of seven pillars that mark the topology of quality assessment in citizen science (Higgins, 2016).

CONCLUSION AND FUTURE DIRECTIONS

In citizen science it is very important to promote user engagement and user participation. Modern GIS technologies enable citizens to visualize, collect and analyze geospatial data in a user-friendly way. Because of the high level of graphical feedback of GIS interfaces, they produce a high engagement with the end user of citizen science projects. As a consequence, GIS technologies used in citizens science projects create a unique user experience which empowers the end users to use them more, and provide more valuable information for this type of projects. In certain parts of the world, the majority of citizens are equipped with smartphones, which enables them at the very least, to be passive contributors to citizen science projects, and a great deal of them has a presence in social networks, an extremely powerful tool for sharing information; the citizen science community has taken advantage of these favorable conditions, by creating mobile apps that collect and display georeferenced information; these apps have the effect of increasing user engagement, encouraging them to contribute to these citizen science initiatives.

Interoperability is a key property of citizen science projects because information generated by these projects should be reusable and exchangeable. Moreover, the aggregation of new types of information and results is much easier. Therefore, with interoperability, citizen science initiatives can provide more holistic information to the user. With the objective to contribute to interoperability, some standards such as WMS, WFS or WCS were created and promoted by international institutions such as OGC. Standards such as the Sensor Observation Service (http://www.opengeospatial.org/standards/sos), or SWE are particularly relevant for those citizen science projects in which GIS information is collected through sensors, in raw data format. When combined with web mapping APIs, the backend - normally a spatial database - gives support to the visualization of sensor data. Thus, this type of information generated by sensors, is on one hand more understandable for citizens, and on the other hand more interoperable and easy to analyze by researchers.

As the citizen is the main actor in a citizen science project, the initiatives aim to support an open participation and easy access to the tools. Through these tools the delivery of new information to the project (such as new samples or observations) and its future visualization and analysis (such as reports) have become of a global and universal usage. Nowadays the access to this kind of technology is extended to most people, both in terms of cost and ease of use.

There are technological trends that could have a positive impact in GIS, supporting further expansion and improvement of citizen science; one of these trends is Big Data. This trend refers to a transition, supported by cloud, mobile and social technologies, which is leading us to a *data-enabled* world (Loukides et al., 2013). However, this growing availability of data does not come without challenges. The four V's of Big Data, attempt to summarize some of these challenges: volume, velocity, variety and veracity; a fifth V has also been mentioned: value (Xu & Yang, 2014). Citizen science projects often use data collected by sensors, sometimes with a low latency, which can be near real-time. The low latency, combined with the bottom-up nature of these projects (run by sometimes large groups of volunteers) means that these datasets have the potential to become huge. Thus velocity and volume may already be a challenge for citizen scientists, and will most certainly be one in the future, as more people join the initiatives and larger time series datasets accumulate. Scalability and distributed operations have been successfully implemented in the NoSQL world, but GIS still tends to rely on single computers and relational databases (Xu & Yang, 2014). Nonetheless there are already some attempts to give support to Big Spatial Data, for instance by extending the Hadoop framework (Aji et al., 2013), and it is expected an increaseing developing activity on this area, in the near future.

The variety and veracity aspects, on the other hand, are well-known challenges in citizen science projects, which do not mean they have been fully addressed. The variety of data is a constant in citizen science projects, as they often mix environmental information with other elements (for instance societal), in a range of different formats that are related to the way this information was collected. At least to some extent, this aspect has been successfully addressed by the use of interoperable standards, such as those from OGC or ISO. Finally, the last v ("veracity") is a common concern for VGI projects, where there are different degrees of trustworthy. One solution could be to enforce embedding the degree of quality in the metadata, and to display it on the map, but this type of approach is not yet generalized, and it still lacks standard support. Although the quality of data, and in particular OpenData, is already a concern there is still a lack of strategies to approach it, and this aspect is expected to get some attention in the near future. This attention should be translated first in sets of best practices, and eventually in standards related to quality control.

As we have seen through this chapter, citizen scientists accumulate the role of both, data creators and data consumers. Nowadays, empowered by modern technologies, they contribute to the generation of often large and complex datasets, which they may want to relate with existing large and complex datasets (e.g.: from social networks, sensors), that share common spatial attributes. Big Data technologies are key to unlock the consumption of this information, and extract value from it. Thus, as it happened before with other modern GIS developments, citizen scientists could be among those who would benefit from Big Spatial Data support.

As both GIS and citizen science are rapidly evolving fields, it is very difficult to predict exactly how they will be in the future, but looking at the current synergies, it is very likely that they will keep evolving together.

REFERENCES

Aji, A., Wang, F., Vo, H., Lee, R., Liu, Q., Zhang, X., & Saltz, J. (2013). Hadoop GIS: A high performance spatial data warehousing system over mapreduce. *Proceedings of the VLDB Endowment*, *6*(11), 1009–1020. doi:10.14778/2536222.2536227 PMID:24187650

Alabri, A. (2010). Enhancing the quality and trust of citizen science data. IEEE eScience 2010.

Auer, S., Bizer, C., Kobilarov, G., Lehmann, J., Cyganiak, R., & Ives, Z. (2007). *Dbpedia: A nucleus for a web of open data*. Springer Berlin Heidelberg.

Azzurro, E., Broglio, E., Maynou, F., & Bariche, M. (2013). Citizen science detects the undetected: The case of Abudefduf saxatilis from the Mediterranean Sea. *Management of Biological Invasions*, *4*(2), 167–170. doi:10.3391/mbi.2013.4.2.10

Barcelona City Council, Citizen Science. (2016). *20 projects to promote the city*. Retrieved from: https://issuu.com/bcnlabcienciaciudadana/docs/llibret_icub__v.eng_

Brovelli, M. A., Dotti, L., Minghini, M., Pancaldi, M., & Zamboni, G. (2014). Volunteered Geographic Information For Water Management: A Prototype Architecture.*International Conference on Hydroinformatics*.

Brovelli, M. A., Minghini, M., & Zamboni, G. (2015). Public participation in GIS via mobile applications. *ISPRS Journal of Photogrammetry and Remote Sensing*.

Brovelli, M. A., & Raghavan, V. (2015). Reflections on FOSS4G Seoul 2015. *Directions Magazine*. Retrieved from: http://www.directionsmag.com/entry/reflections-on-foss4g-seoul-2015/455554

Castell, N., Kobernus, M., Liu, H. Y., Schneider, P., Lahoz, W., Berre, A. J., & Noll, J. (2014). *Mobile technologies and services for environmental monitoring: The Citi-Sense-MOB approach*. Urban Climate.

Ceccaroni, L., Velickovski, F., Steblin, A., & Subirats, L. (2015) Citclops: data interpretation and knowledge-based systems integration.*Environmental Information Infrastructures and Platforms Workshop*.

Citizen's Observatory. (n.d.). *swe4citizenscience*. Retrieved from: http://github.com/opengeospatial/swe4citizenscience

Eisnor, D. (2006). *What is neogeography anyway?* Retrieved from: http://platial.typepad.com/news/2006/05/what_is_neogeog.html

Foursquare. (n.d.). Retrieved from: https://foursquare.com/

Fritz, S., McCallum, I., Schill, C., Perger, C., See, L., Schepaschenko, D., & Obersteiner, M. et al. (2012). Geo-Wiki: An online platform for improving global land cover. *Environmental Modelling & Software*, *31*, 110–123. doi:10.1016/j.envsoft.2011.11.015

Goodchild, M. F. (2007). Citizens as sensors: The world of volunteered geography. *GeoJournal*, *69*(4), 211–221. doi:10.1007/s10708-007-9111-y

Google Crisis Response. (2012). *Superstorm Sandy: NYC*. Retrieved from: http://google.org/crisismap/2012-sandy-nyc

Haklay, M. (2013). Citizen science and volunteered geographic information: Overview and typology of participation. In *Crowdsourcing geographic knowledge* (pp. 105–122). Springer Netherlands. doi:10.1007/978-94-007-4587-2_7

Higgins, C. I., Williams, J., Leibovici, D. G., Simonis, I., Davis, M. J., Muldoon, C., & O'Hare, G. et al. (Manuscript submitted for publication). Citizen OBservatory WEB (COBWEB): A Generic Infrastructure Platform to Facilitate the Collection of Citizen Science data for Environmental Monitoring. *International Journal of Spatial Data Infrastructures Research.*

International Organization for Standardization. (2003). *ISO 19115:2003 - Geographic information - Metadata.* Retrieved from: http://www.iso.org/iso/catalogue_detail.htm?csnumber=26020

Joost, S., Baumann, R., Ertz, O., Ingensand, J., Widmer, I., & Rappo, D. (2014). *A participatory WebGIS platform to support biodiversity inventory in the Geneva cross-border area.* Third Open Source Geospatial Research & Education Symposium (OGRS), Espoo, Finland.

Kampen, H., Medlock, J. M., Vaux, A. G. C., Koenraadt, C. J. M., van Vliet, A. J. H., Bartumeus, F., & Werner, D. et al. (2015). Approaches to passive mosquito surveillance in the EU. *Parasit Vectors, 8*(1), 9. doi:10.1186/s13071-014-0604-5 PMID:25567671

Liu, S. B., Iacucci, A. A., & Meier, P. (2010). Ushahidi Haiti and Chile: next generation crisis mapping. *ACSM Bulletin, 246.*

Longley, P. A., Goodchild, M. F., Maguire, D. J., & Rhind, D. W. (2005). Geographic Information Systems and Science. John Wiley & Sons.

Loukides, M., Warden, P., Watters, A., & Croll, A. (2013). *Big Data Now - Current Perspectives from O'Reilly Radar* (A. Noren, Ed.). O'Reilly Media.

Meier, P., & Brodock, K. (2008, October 23). Crisis mapping Kenya's election violence: Comparing mainstream news, citizen journalism and Ushahidi. *iRevolution Blog.*

Minghini, M. (2014). *Multi-dimensional geoweb platforms for citizen science and civic engagement applications.* PhD thesis.

Neteler, M., & Mitasova, H. (2008). Open Source GIS: A GRASS GIS Approach (3rd ed.). Springer. doi:10.1007/978-0-387-68574-8

Newman, G., Zimmerman, D., Crall, A., Laituri, M., Graham, J., & Stapel, L. (2010). User-friendly web mapping: Lessons from a citizen science website. *International Journal of Geographical Information Systems, 24*(12), 1851–1869. doi:10.1080/13658816.2010.490532

Okolloh, O. (2009). Ushahidi, or 'testimony': Web 2.0 tools for crowdsourcing crisis information. *Participatory Learning and Action, 59*(1), 65-70.

Open Geospatial Consortium. (n.d.a). *Open Geospatial Consortium.* Retrieved from: http://www.opengeospatial.org/

Open Geospatial Consortium. (n.d.b). *Sensor Web Enablement (SWE).* Retrieved from: http://www.opengeospatial.org/ogc/markets-technologies/swe

OpenStreetMap. (n.d.). *Copyright and License.* Retrieved from: http://www.openstreetmap.org/copyright

Pycsw. (n.d.). Retrieved from: http://pycsw.org/

Rademacher, P. (2005). *HousingMaps*. Retrieved from: http://www.housingmaps.com/

Ramsey, P. (2007). *PostGIS case studies*. Retrieved from: http://www.refractions.net/expertise/whitepapers/postgis-case-studies/

Simonis, I. (2015). *Citizen Observatories: A Standards Based Architecture*. EGU General Assembly 2015, Vienna, Austria.

Snow, J. (1854). *On the Mode of Communication of Cholera* (2nd ed.). Retrieved from http://matrix.msu.edu/~johnsnow/images/online_companion/chapter_images/fig12-5.jpg

Stallman, R. M., & Gay, J. (2009). *Free Software, Free Society: Selected Essays of Richard M. Stallman*. Paramount, CA: CreateSpace.

Steiniger, S., & Bocher, E. (2009). An overview on current free and open source desktop GIS developments. *International Journal of Geographical Information Science*, *23*(10), 1345–1370. doi:10.1080/13658810802634956

Steiniger, S., & Hay, G. J. (2009). Free and open source geographic information tools for landscape ecology. *Ecological Informatics*, *4*(4), 183–195. doi:10.1016/j.ecoinf.2009.07.004

Sullivan, B., Wood, C., Iliff, M., Bonney, R., Fink, D., & Kelling, S. (2009). eBird: A citizen-based bird observation network in the biological sciences. *Biological Conservation*, *142*(10), 2282–2292. doi:10.1016/j.biocon.2009.05.006

The WNYC Data News Team. (2015). *Know Your Evacuation Zone*. Retrieved from: http://www.wnyc.org/story/know-your-evacuation-zone/

Uhrner, U., Grosso, G., Romain, A. C., Hutsemekers, V., Delva, J., Kunz, W.,... Ledent, Ph. (2013). Development of an Environmental Information System for Odour using Citizen and Technology Innovative Sensors and Advanced Modelling. *CEUR Workshop Proceedings*.

Wikipedia. (n.d.). *List of deadliest Atlantic hurricanes*. Retrieved from: https://en.wikipedia.org/wiki/List_of_deadliest_Atlantic_hurricanes

Xu, C., & Yang, C. (2014). Introduction to big geospatial data research. *Annals of GIS*, *20*(4), 227–232. doi:10.1080/19475683.2014.938775

Yasuhiko, A. (2014). Safecast-Radiation Risks. *The Asia-Pacific Journal*, 11.

Chapter 7
Citizen Science and Its Role in Sustainable Development:
Status, Trends, Issues, and Opportunities

Hai-Ying Liu
Norwegian Institute for Air Research, Norway

Mike Kobernus
Norwegian Institute for Air Research, Norway

ABSTRACT

The chapter aims to analyse the role of citizen science in sustainable development, including case studies implementation, with specific focus on its suitability of citizen science in environmental sustainability. The authors structured this chapter in five sections: Background; Main focus; Solutions and recommendations for designing and executing citizen science initiatives; Future research directions with thoughts on the future role of citizen science; and Conclusion. In section of main focus, first, the authors reviewed the state of citizen science in sustainable development and explored the potential of citizen science for environmental research and governance. Second, authors identified and elaborated the core components that support the role of citizen science and demonstrated the practical approach to realize its objective. Third, using several citizens' observatories studies from various regions in Europe and within diverse environmental fields, authors highlighted the lessons learned, and reflected on major outcomes, challenges and opportunities.

INTRODUCTION

Participation of the public in scientific and action research, independently or in cooperation with scientists, is often referred to as 'citizen science' (Hand, 2010). Additional terms, such as 'crowd science', 'crowd-sourced science', 'civic science', or 'networked science' may be viewed as synonyms (Hand, 2010). Citizen science itself can be said to have existed since the very start of scientific practice, where it has developed in many different guises. As the internet continues to proliferate in the world, new tools like social media, mobile devices and mobile sensors are becoming the norm (Lanfranchi, Wrigley, Ire-

DOI: 10.4018/978-1-5225-0962-2.ch007

son, Ciravegna, & Wehn de Montalvo, 2013)). Lately, it has been witnessed a global increase of citizen science projects and citizens engaging in projects as amateur researchers, as sensors, as advocates and even watchdogs (Haklay, 2015). Further, crowdsourcing methods, data processing and visualization technologies are developing rapidly, leading to a wide range of new opportunities for public participation in a compelling range of topics (Buytaert et al., 2014). As an example, it can be seen that a tremendous increase of environmental observation activities in this area, i.e., various citizens' observatories that encompass different models of citizen science and span a diverse range of subjects (e.g., biodiversity, water, air, climate change, agriculture, disaster, etc.), empowering average people to monitor their environment, collectively generate scientific data and support environmental risk response (Liu, Kobernus, Broday, & Bartonova, 2014).

The objective of this chapter is to analyze the role of citizen science in sustainable development, including case studies in citizen science implementation, with specific focus on the suitability of citizen science in environmental sustainability. It comprises the following five sections: Background; Main focus of the chapter; Solutions and recommendations for designing and executing citizen science initiatives that help inform and create environmental action, based on evidence, sound science and citizens' needs; Future research directions with thoughts on the future role of citizen science; and The Conclusion. In the section, 'Main Focus', the authors have divided it into the following substructure:

1. First, the authors reviewed the state of citizen science in sustainable development and explored the potential of citizen science to complement more traditional ways of scientific data collection and knowledge generation for environmental research and governance;
2. Second, the authors identified and elaborated the core components that support the role of citizen science and demonstrated a practical approach to realize its objective of environmental sustainability;
3. Third, using several citizens' observatories case studies from various regions in Europe and within diverse environmental fields, the authors highlighted the lessons learned, and reflected on major outcomes, challenges and opportunities in the integration of environmental-oriented citizen science within environmental management, the role of scientific knowledge in the decision-making process, and the potential contestation to established community institutions posed by the co-generation of knowledge.

BACKGROUND

Citizen participation with environmental science has a long history, before it was termed 'citizen science' (Roy, 2012). In fact, until the late 19th century, there were no professional scientists as it is well known them today (Kight, 2012; Miller-Rushing, Primack, & Bonney, 2012). Research was typically undertaken by amateurs, often Gentlemen of Leisure or men of the cloth (Miller-Rushing et al., 2012). Charles Darwin, for example, was not a trained scientist (Kight, 2012; Dillon, 2014). Likewise, the Swedish botanist and physicist, Linnaeus, was not only a Lutheran minister for a period, but worked with a wide network of citizen volunteers who sent samples to him, a not uncommon practice (Scyphers et al., 2015). One could argue that he exemplifies the very nature of citizen science. However, these men were at the pinnacle of the scientific community, so it might be somewhat disingenuous to claim they were amateurs. Nonetheless, citizen science was an activity that many average citizens engaged in at the avocational, or hobby level. Long before social media made citizen science accessible to the wider population in the digital age, amateurs found a way, adopting the participatory science model to their circumstances.

The first 'modern' large scale citizen participation began in 1999, at the University of Berkeley, California, where data from radio telescopes was analyzed using computers that otherwise would have been idle (Korpela et al., 2011). These computers belonged to citizens who simply wanted to be part of something, and by contributing their computer's CPUs, they helped crunch data. And as budgets become increasingly tight within research communities, alternative solutions to data gathering were needed and the Internet and mobile technologies provided the answer. The development of communication technologies offered many new avenues for citizens to contribute in a number of ways. Personal perceptions, comments, and even using the phone itself as a mobile sensor, all have a place. Such is the popularity of participatory sensing now that it has become a relatively common component to many projects and research studies (Lanfranchi et al., 2013). You can find active citizen participation in a wide range of topics, where people provide data on biodiversity, land use, weather, personal health, perception of cleanliness of buildings, and so much more (Liu et al., 2014).

The need to have a comprehensive understanding of environmental integrity, including function and structure, is often confounded by a lack of, or inadequate and incomplete, data and monitoring initiatives by professional scientists and government agencies. To fill the void, non-professionals and citizen organizations have emerged the world over to track trends and to work towards effective and meaningful planning, management, and stewardship (Conrad & Hilchey, 2011). This is especially facilitated by the innovative tools and ICT. For example, with the appearance of new low-cost sensor technologies, monitoring the environment is now, in a literal way, in everyone's hands (Liu et al., 2014; Kobernus et al., 2015). Novel sensor technologies open the opportunity to monitor the environment at spatial resolutions not possible to achieve with traditional monitoring systems (Liu, Skjetne, & Kobernus, 2013). Citizens can now contribute to monitor the environment in a direct and meaningful manner.

Formally, the role of citizen science has been defined as research co-design/co-creation, the systematic collection and analysis of data, development of technology, testing of natural phenomena, and the dissemination of these activities together with scientists on a primarily avocational basis (Hand, 2010). In recent years, citizen science has been viewed as an increasingly essential tool to support and promote community-based environmental governance, as it attempts to provide citizens with a voice and accountability in the future of sustainability research (Liu et al., 2014). Obviously, citizen science enables a more active role for citizens concerning understanding the environment, since citizens are traditionally considered as 'consumers' of information services at the very end of the information chain and not as 'data providers'. However, it is no longer sufficient to develop and provide passive lists of environmental indices or reports to inform citizens about changes in their environment. There is a need to engage citizens to find out how they can inform the community, and to empower citizens to improve their own health and well-being through actively making informed choices via citizen science activities and outcomes. Involving citizens at the local level by developing knowledge pools can help create an atmosphere of active participation and generate a sustainable movement that can build over time. Citizens have expectations to interact and participate in the decision-making processes, and to be engaged in a dialogue about their communities, preferences and future (Liu et al., 2012a). According to sociological research, the recent increase in pro-active participants of social IT media, with a particular focus on environmental issues, results from a shift from materialism to post-materialism where more and more people are showing increasing interest in a renewable and sustainable life style, which is the key objective of environmental governance and lies at the heart of 'sustainable development' (Salonen & Åhlberg, 2013; Liu et al., 2014).

'Sustainable development' is development that meets the needs of the present without compromising the ability of future generations to meet their own needs (World Commission on Environment and Development, 1987). Citizen science is crucial in bridging the gap between environmental governance and the public, and can play a key role in sustainable development (Liu et al., 2014). However, there is no globally agreed and understood definition of 'environmental governance' and it can be interpreted in many different ways. In principle, environmental governance comprises the rules, practices, policies and institutions that shape how humans interact with their environment (UNEP, 2009). It can refer to the processes of decision-making involved in the control and management of the environment for the purpose of attaining environmentally-sustainable development. Good environmental governance takes into account the role of all actors that impact the environment, from governments to Non-Governmental Organizations (NGOs), the private sector and civil society, the individual and citizen groups. Cooperation between actors is critical to achieving effective governance that can help us move towards a more sustainable future. In this context, the key aspect of citizen science is the direct involvement of ordinary citizens, and not just that of scientists/professionals. This can be in data collection as well as harnessing the citizens' collective intelligence, i.e., the distributed information, experience and knowledge embodied within individuals and communities. This will help to bridge gaps that many areas of environmental management are still suffering from. Namely, citizen science should enable citizens' participation in environmental monitoring in order to contribute to environmental governance by providing relevant data and information that can help decision-makers make sound decisions. This can be advanced by providing citizens with a voice and supporting them with information on their environment, consequently raising their awareness and leading to higher levels of involvement. This reveals key components that underpin the core objectives of citizen science in sustainable development, i.e., raising citizens' awareness and consciousness, enabling dialogue and supporting data exchange/knowledge sharing amongst citizens, scientists, policy/decision makers, and other stakeholders (Liu et al., 2014). These three components of citizen science can explain the major links between citizen science and environmental governance, and in fact are the three pillars that comprise citizen science enabling it to support environmental governance (Kobernus et al., 2015). In the context of citizen science's contribution to environmental governance, it is important to recognize that citizens are not a monolithic group, all alike. There are stakeholders/user groups including individuals or groups of volunteers, scientists, government authorities, emergency services, etc. Hence, various actors in citizen science have different behaviors, intentions, interrelations, agendas, interests, as well as influence, resources and power on decision-making and political processes. Moreover, there are a number of questions about its potential as a democratizing force in environmental policy and management (Fore, Paulsen, & O'Laughlin, 2001; Liu et al., 2014). However, given the many contending conceptions within democratic theory (e.g., direct, representative, participatory, minimal, deliberative, aggregative, etc.), it should be noted that this aspect is a complex subject with no "one size fits all" solution. Nevertheless, citizen science has shown its potential to address issues of environmental equity and sustainability, and to improve social justice and responsibility, which is directly linked to the principles and treaties on sustainable development, including economic development, social development and environmental protection.

MAIN FOCUS OF THE CHAPTER

Status and Trends of Citizen Science in Sustainable Development

Due to the increasing significance, utility and function of citizen science for environmental research, governance and practice (Conrad & Hilchey, 2011; Roy et al., 2012), a review of existing citizen science projects ranging from small one-off local surveys to large-scale long-term programs by focusing on its status and trends including consensus, divergence and gaps is presented here. In order to comprehensively review the state of citizen science in academic journal articles, the authors reviewed citizens' observatories in varying environmental and social contexts. A review of the vast academic-and non-academic-based literature from around the globe indicates (Grossberndt & Liu, 2016):

1. For Professionals: Lack of funding and staffing for environmental monitoring; For governments: more environmental monitoring data and stakeholders needed for decision-making processes. These have led to an increase in the use of amateur scientists around the globe.
2. There are differences in the monitoring intent in different parts of the world. In parts of Europe (particularly in Britain), many of the citizen science initiatives focus on collecting data about species and habitats, whereas there is a predominance of monitoring efforts on ecosystem functions and environmental quality in North America.
3. The issue of monitoring interest often transcends government boundaries, and many NGOs responsible for cross-state, cross-province, or cross-country concerns have increased their use of citizen science.
4. Citizen science relationships with academic institutions and universities have increased, perhaps due to their capacity to provide training, lab facilities, free space, and funding.

This review is not intended to be inclusive of all types of citizen science initiatives, especially given the enormous proliferation of citizen science activities (e.g., there are more than 1,100 active and searchable global citizen science projects listed on SciStarter (Wikipedia, 2016)), but instead is representative of a variety of citizen science campaigns across the globe. Citizen science activities can differ in focus, approach or technique. The review indicates that many citizen science initiative focus on monitoring water quality, birds count, butterflies, mosquitoes, wildlife, air quality, algae, amphibians, plants, fish, worms, odors, and ice (Grossberndt & Liu, 2016).

In addition, citizen science initiatives have engaged both the commodity-based monitoring (e.g., resource fishery, resource forestry, etc.) and non-commodity-based monitoring (e.g., recreational fishery, monitoring water quality, monitoring air quality, etc.) (Grossberndt & Liu, 2016). Commodity-based monitoring deals with issues of economic as well as social and environmental importance. Non-commodity-based monitoring focuses on issues that may not seem to be directly economically important.

Citizen science initiatives also differ in the types of monitoring activities the organization undertakes (Grossberndt & Liu, 2016). Citizen science monitoring activities include many different types of assessment of the environment:

1. Status assessment (e.g., air quality status);
2. Impact assessment (e.g., effect of air pollution);
3. Adaptive management (e.g., managing based upon monitoring).

Citizen science monitoring activities also include different aspects of the environment monitored (Grossberndt & Liu, 2016):

1. Environmental composition (e.g., air, water, soil, etc.);
2. Structure (e.g., environmental analysis, key pollutants, etc.); and
3. Processes (e.g., linking human health with environmental problem). Process-based monitoring is suggested as being most desirable in many studies.

Furthermore, in a context of sustainable development, almost by definition (Bastiaensen et al., 2005), the poorer people will have difficulty to voice their concerns and even more to articulate them to (scientific) outsiders. So more often than otherwise, it therefore be expected that participative citizen science will address more main stream concerns of the local majority, or even of small dominant local elites (Buytaert et al., 2014). But in view of the post-research application of scientific findings, this usually has the advantage to facilitate rallying more and more relevant local support to implement recommendations through social action, while evidently risking to bypass critical concerns of disadvantaged groups and even to contribute to the deepening of their dispossession and exclusion. This constitutes a tricky, but often unavoidable dilemma in the implementation of citizen science, and requires a careful analysis of the social interface between the citizen science practices and the learning and struggle in local political arenas. By no means, however, does this dilemma reduce the potential contribution that science and citizen science can make to local bricolage toward better governance (Buytaert et al., 2014).

According to Buytaert et al. (2014), despite being an intrinsic part of the scientific discovery and knowledge generation process, the concept and potential of citizen science in itself only recently received increasing scientific attention. New technological developments are supporting novel and more efficient methods for data collection and processing, visualization and communication. These opportunities make reflecting upon the challenges and opportunities of citizen science, especially in a context of managing natural resources and leveraging them for human well-being, timely and relevant. This is particularly the case for water resources, which is often one of the most fundamental ecosystem services and a significant bottleneck for sustainable development and poverty alleviation. Their review of technologies reveals a large potential for increasing involvement of citizens in data collection because of the availability of inexpensive, robust and highly automated sensors, and the possibility to combine them with powerful environmental models to create rich and interactive visualization methods (Buytaert, Dewulf, De Bièvre, Clark, & Hannah, 2016). In the fields of hydrology and water resource management, Buytaert et al., (2014) highlighted that citizen science has the potential of being more inclusive methods to generate knowledge on improved uses of ecosystem services that supports sustainable development and improves human well-being. Their discussion focuses particularly on:

- The motivation and level of engagement by the non-scientists.
- The type of data monitored, and potential for participatory modelling and knowledge exchange.
- The legitimacy of the generated knowledge.
- The sustainability of the citizen science activity.
- Potential uptake in decision making/resource governance.

Core Components of Citizen Science and Its Practical Approach

In recent years, there has been a boom in citizen science projects, which allow members of the public to play an active part in monitoring and recording their environment across the globe. The increase in the prevalence of citizen science globally has been mirrored by a growth in the number of variables that are monitored, the number of monitoring locations and the different types of participating citizens (Liu et al., 2014; Kobernus et al., 2015; Grossberndt & Liu, 2016). This calls for a more integrated approach to handle the emerging complexities involved in this field (Liu et al., 2012b); but before this can be achieved, it is essential to establish a common foundation for citizen science and their usage. There are many aspects related to a citizen science project. One view is that its essence is a process that involves environmental monitoring, information gathering, data management and analysis, assessment and reporting systems. Hence, it requires the development of novel monitoring technologies and of advanced data management strategies to capture, analyze and survey the data, thus facilitating their exploitation for policy and society (Liu et al., 2014; Kobernus et al., 2015). Practically, there are many challenges in implementing the citizen science approach, such as ensuring effective citizens' participation, dealing with data privacy, accounting for ethical and security requirements, and taking into account data standards, quality and reliability. These concerns all need to be addressed in a concerted way to provide a stable, reliable and scalable citizen science program. On the other hand, the citizen science approach carries the promise of increasing public awareness for risks in their environment, which has a corollary economic value, and enhancing data acquisition at low or no cost (Liu et al., 2014; Kobernus et al., 2015).

Type of Citizen Science

Citizen science does not only serve scientific purposes, but also decision-making, since the collected data can contribute to informed policy-making. In addition, citizens can benefit by addressing the environmental issues that affect them directly in the context of participatory decision-making. In this way, the value of lay knowledge should not be underestimated (Science Communication Unit, University of the West of England, 2013). According to Haklay (2015), the following three policy dimensions can be distinguished:

1. Level of geography;
2. Policy domains; and
3. Level of engagement and type of citizen science activity.

Citizen science initiatives can influence policy decisions in a specific geographical area, i.e., local, regional, national and international. Usually, problems that affect people directly, lead to more engagement since people are both more aware and concerned (Haklay, 2015). This potential can be used to build upon observation activities through citizen science initiatives. Local citizen science is often linked to environmental activism and supports community management by working towards effective and meaningful management planning, management and stewardship (Conrad & Hilchey, 2011). Local citizen science can also apply the so-called community-based monitoring approach (CBM). CBM describes a process where concerned citizens, public authorities and further stakeholders collaborate to monitor, track and respond to issues that arise from common community concerns (Whitelaw, Vaughan, Craig, & Atkinson, 2003).

Environmental issues are very complex and would actually require additional resources. Thus, there is an increasing need for communities to fall back on citizen science approaches (or CBM respectively) and include different stakeholders with their diverse knowledge and experience into the decision-making processes (Conrad & Daoust, 2008). In addition to the decision-making bodies' saving time and money, the societal benefits of CBM will be creating environmental democracy, social capital, and increase in scientific literacy and inclusion in local issues (Conrad & Hilchey, 2011).

Policy areas can be manifold and partially overlapping. For example, city-scale infrastructure contains public transport, environmental quality, education, infrastructure and public health. Thus, cities can be a canvas for a potpourri of local monitoring activities, originating from different concerns but using the accumulated data to see the bigger picture (Grossberndt & Liu, 2016). Moving citizen science projects to regional, national or even international level is likely to meet even more challenges than there already are. Since bottom-up initiatives usually dispose of limited budgets only, it will be less likely to find community science approaches with an active involvement of citizens in all parts of the participation cycle, i.e., citizens will rather only be asked to share observations or viewpoints on certain issues. Nevertheless, national and even international initiatives including citizen science are possible and do exist. Projects funded by the European Commission (EC) and formation of international organizations like the European Citizen Science Association (ECSA) provide frameworks for national initiatives and NGOs to create synergies to promote citizen science on a larger scale and call on international institutions such as the European Environmental Agency (EEA) to promote citizen participation also at an international level (Haklay, 2015).

However, how can people actually be involved in citizen science? Haklay (2015) distinguished different levels of engagement within citizen science. As it can be seen, these categories resemble those participation levels presented earlier in this chapter. The first level is called passive sensing and describes a process where information is collected without any effort on the participants' part through their own devices. Volunteer computing is a method where participants allow scientists to use their unused computing resources on, e.g., their PC or smartphone for complex computer models while the device is not in use. The next level, volunteer thinking makes use of the citizens' cognitive abilities not used during passive leisure activities, e.g., people contribute by recognizing patterns while watching TV. The obtained data will then be further used in scientific projects. Environmental and ecological observations describe the traditional form of citizen science, where volunteers monitor and/or observe their personal environment, often based on protocols that are designed by scientists. In participatory sensing activities, the citizens are more actively involved in designing both data collection and analysis. The last activity type can be summarized as civic/community science. In this bottom-up approach, the participants can choose their level of engagement and can even be involved in the analysis and interpretation of the results and their publication/utilization. This level of engagement does not necessarily require the involvement of scientists (Grossberndt & Liu, 2016).

Citizen science initiatives should be tailored to match both interest and skills of the participants. This is crucial in order to develop successful initiatives. Although motivations vary widely, the authors find that enjoyment and enthusiasm for the goals of the initiative are two of the main drivers. As known from other research areas, the feeling of control over the scientific process can also lead to long-lived engagement (Roy et al., 2012).

Citizen Science Framework

Overall, citizen science is based on a general framework linking citizens and scientists together, allowing the implementation of a more comprehensive scientific approach. The framework can address a particular question, tests hypotheses with appropriate data collected from a well-designed protocol, allow the interpretation and understanding of new results and can help to derive and spread original educational outputs (Kobernus et al., 2015).

To establish a citizen science program and to make it useful for society, there is a need for researchers to collaborate with citizens, citizen groups and their representatives, and with the representatives of the local authorities, to identify interests and needs (Liu et al., 2014). In a citizen science program, all parties shall be encouraged to be engaged as active participants, to create knowledge of the environmental situation in a participatory manner and to contribute to dealing with the situation (Landranchi et al., 2013). One of the main aspects of a citizen science program is the need to effectively address citizens participation in data collection, data interpretation and information delivery. According to Liu et al. (2014), this can be expressed as the following set of five sequential aspects that underlie the citizen science skeleton and support effective citizen participation (Figure 1):

a. Citizens' participation in identifying what citizens want and what a citizen science program can offer to provide information and knowledge in response to public concerns. This is achieved mainly by a dialogue among the stakeholders;

Figure 1. Sequential aspects of a citizen science program

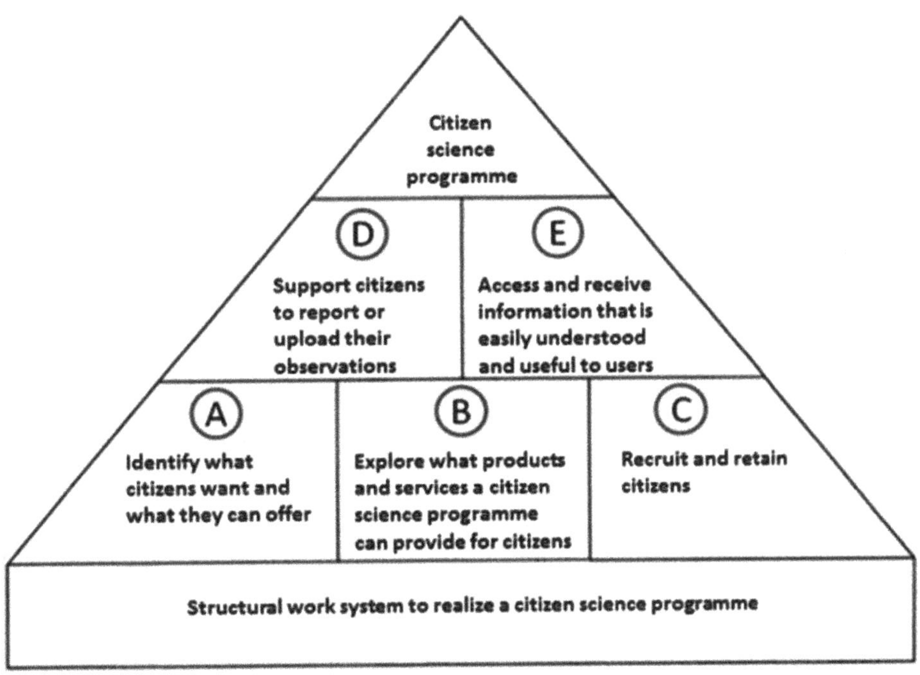

b. Citizens' participation in exploration of what products and services a citizen science program can provide for the citizens. This involves systematizing and structuring citizens-created content to make it appealing for use by citizens during their normal daily life;
c. Recruiting and retaining citizens to participate in and contribute to environmental governance: further clarify the purpose, scope and expected impact of the citizen science program, identify motivations that will promote citizens to contribute to and take part in the citizen science activities, and encourage public participation in data collection and interpretation;
d. Obtaining public participation in the relevant decision making and/or in changing their related personal priorities and behavior by gaining access to environmental data, knowledge and experience, and by using tools that can support citizens to report or upload their objective/subjective observations, inference and concerns;
e. Supplying tools to access/receive timely information on the relevant environmental issues in a manner that is both easily understood and useful to the users.

Among critical factors of success, the simplicity of each component of the framework, the structure of the scheme, the regular feedback of results to participants and a good communications strategy are crucial. Moreover, for long-term projects, the sustainability of the framework (often demanding sources of funding and people with permanent positions) should be carefully thought through (Devictor, Whittaker, & Beltrame, 2010).

Citizen Science Model in Practice

A key concept in the operation of citizen science program is the idea of a campaign that specifies stakeholders' concerns by defining the types of data that need to be collected and by describing the goal, expected feedback and analysis outcomes in terms of maps, statistical results, etc. The raison d'être of the observatory then is to "run and orchestrate" a campaign in order to assure that sufficient data is gathered, both qualitatively and quantitatively, and to enable non-expert stakeholders to define, monitor and analyze campaigns in a way understandable to them (D'Hondt, Zaman, Philips, Boix, & De Meuter, 2014; Zaman et al., 2014).

Figure 2. illustrates a common model of citizen science program: A set of concentric circles that are characterized by different types of information needs and of information gathered and shared (left part). The circles represent different types of stakeholders: environmental information and services providers and people actively involved in citizen science program (e.g., public, researchers, policy-makers, small and medium-sized enterprises (SMEs), etc.). In addition, the circles also represent different tools that different stakeholders may use to provide their observations and share their data and information, e.g., innovative low-cost physical sensors, citizens as sensors, and social media, etc. The aim of the citizen science approach is to support all stakeholders by designing various tools and applications that support co-participation (Landranchi et al., 2013).

In addition to the different stakeholders and the way to engage such stakeholders, Figure 2 also illustrates the four core components of the data flow model in various citizen science projects (right part):

1. **Citizens**: Different types of stakeholders, including the general public;
2. **Monitoring Tools and its Platforms**: Technologies for environmental monitoring (i.e., smartphone, low-cost monitoring tools and its platforms);

Figure 2. A common citizen-science model in practice

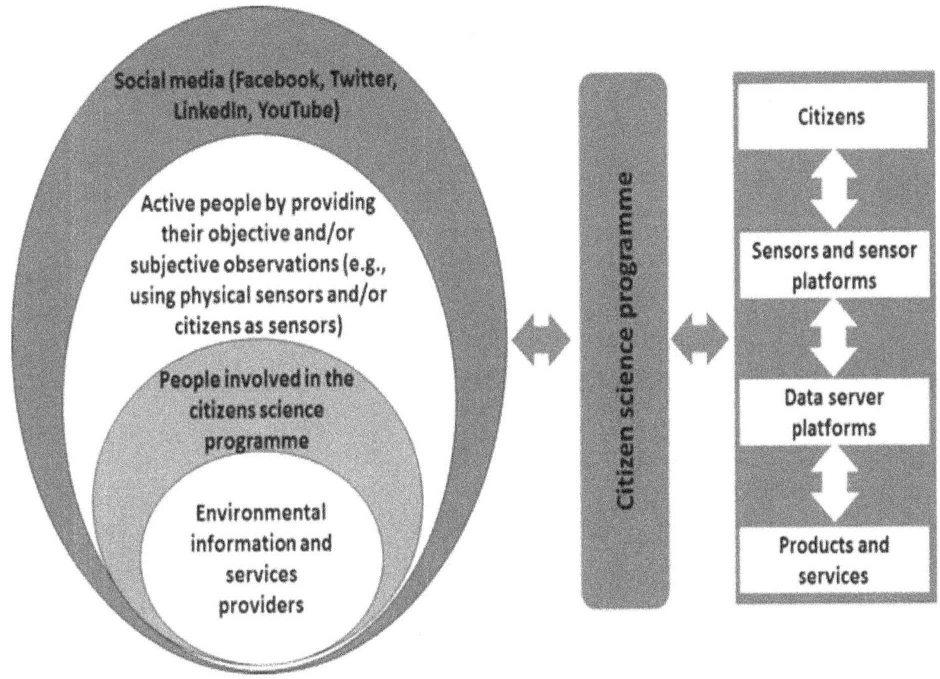

3. **Data Server Platforms**: Information and communication technologies (e.g., web-featured service);
4. **Products and Services**: Information products and services, i.e., mobile apps, web applications, social media networks, public environmental survey, etc.

The first three components (citizens, monitoring tools and its platforms, data server platforms) are the main elements of participatory sensing, reflecting the concept of communities or other groups of people contributing information in the form of data and knowledge (Dua, Bulusu, Feng, & Hu, 2009). The fourth component (products and services), linked to the usage of obtained data and knowledge, is designed and expected to be used by various stakeholders for different purposes. These components can be considered as the four pillars that support various citizen science programs in practice (Engelken-Jorge et al., 2014; Liu et al., 2014). They relate to the stakeholders' experience, expertise and expectations and are clearly important. Within these four pillars, various actors may use the tools or instruments to produce raw data that leads to a range of new products or services as well (Lanfranchi et al., 2013; Liu et al., 2014). Here, the authors present a common model of citizen science program which is being been translated into practice in several existing citizen science projects, i.e., Citclops, CITI-SENSE, Citi-Sense-MOB, Cobweb, EVERYAWARE, Omniscientis and WeSenseIt. All of these projects put an emphasis on delivering highly innovative technologies to support citizens, communities and authorities by getting 'near real-time' and/or 'up-to-date' data about environmental conditions (e.g., air quality, water quality, etc.) and situations (e.g., flooding), while respecting the different information needs and actions respond. On a technical level, these projects consists of a combination of crowdsourcing and custom applications designed to empower and foster participation with the objective of creating an enriched information and knowledge base to facilitate decision making while increasing opportunities for citizen engagement in their community (Ciravegna, Huwald, Lanfranchi, & Wehn de Montalvo, 2013; Lanfranchi et al., 2013; Liu et al., 2014). These projects base data capture on one or several of the following tools:

1. A set of innovative low cost sensors for monitoring environment that are designed and can be used directly by both professionals and citizens alike;
2. A set of public environmental perception surveys ("citizens as sensors") designed to be used by citizen participants;
3. Tools that enable the exploitation of the citizens' collective intelligence through crowdsourcing (e.g., Volunteered Geographic Information (VGI)) and social media platforms (e.g., Facebook page, Twitter account) analysis.

Challenges Met and Lessons Learned from Existing Citizen Science Activities

Citizen science initiatives require strict data management. Since most of the citizen science initiatives take place outside organizational frameworks, special efforts have to be made to guarantee the quality of the obtained data. This can be reached through either controlling the quality of data during their acquisition or subsequently after their acquisition by comparing them with reference data (Goodchild & Li, 2012). The first approach requires the introduction into and training of data collection and/or interpretation methods. Here, participants receive instructions and guidelines (e.g., provision of standardized equipment, instruction sheets, online training, etc.) on the use of the measuring device and how to collect data. In the case of VGI, a certain quality assurance is guaranteed when several participants deliver the same GPS (Global Positioning System) coordinates for the same object. In some occasions, some participants can also be asked to monitor and validate the data collected by participants with less experience, as for example in bird watching activities or monitoring of invasive species. Another example is Wikipedia, where a group of individuals with special rights acts as moderators or gatekeepers to avoid vandalism, remove copyrighted materials and resolving conflicts. In other occasions, it can be valuable to use existing knowledge of citizens to evaluate the validity of the data from volunteers (Grossberndt & Liu, 2016). For example, by using existing knowledge about the fauna in Norway, it is very unlikely that a giraffe will be spotted in the streets of Oslo, Norway (Goodchild & Li 2012, Haklay, 2015).

Next to data issues, there are a number of further challenges for the implementation of public participation. In their review of citizen science and community-based initiatives, Conrad & Hilchey (2011) provided a list of challenges of community-based monitoring. Apart from data issues mentioned above, these challenges include:

1. Lack of interest of the volunteers;
2. Lack of opportunities for networking;
3. Funding difficulties;
4. The participants, inability to get access to appropriate information and expertise;
5. Insufficient experimental design;
6. Utility of community-based monitoring data for, e.g., decision-making.

Another aspect that it is needed to bear in mind is the fact that even though international agencies, such as the EEA, promote approaches of inclusive governance processes, this does not mean that this advice is followed automatically on national, regional or even the local level. Not only the willingness, but also the readiness of decision-makers has to be taken into consideration when designing and implementing any participative initiative. In this context, it should also be mentioned that even if the participative process has been carried out successfully, it should strive to translate the prepared work into action (Litke & Day, 1998).

Although the usage of the citizen science concept is becoming a more common practice for environmental management, challenges still exist regarding how to deal with the quality of the data collected as well as how to use it for environmental policy (Tennessee Water Resources Research Center, University of Tennessee 2004; Goeschl & Jürgens, 2012).

Such challenges suggest that the following areas should receive careful consideration (Liu et al., 2014):

1. **Data Quality** (i.e., accuracy and uncertainty): Especially when comparing crowd-sourced and reference data;
2. **Data Privacy and Security**: Sharing of data and information requires strong ethical and security considerations;
3. **Data Interpretation**: Qualitative indicators such as "quality of life", "well-being", "happiness", etc., should be developed in parallel with more quantitative indicators that are based not only on individual perception, but on an integrated sensor network;
4. **Systematization and Structuring of Citizens-Created Content and Feedback:** Establishing a viable model(s) to support decisions and empower the public (Engelken-Jorge et al., 2014);
5. **Involving and Maintaining a Broad Spectrum of Society**: Implementing various location-specific and target group-tailored tools in recruiting and sustaining citizens' participation in environmental monitoring (Fernandez-Gimenez, Ballard, & Sturtevant, 2008).

In practice, various citizen science programs all share a similar structure, yet in the current status, constructing a new citizen science program for a new type of campaign (e.g., air pollution, mobility patterns of users of public transportation, biodiversity, climate change, etc.) requires that all software infrastructure to be rebuilt from scratch (Liu et al., 2014). The lack of a systematic, easy and reusable methods for setting up new citizen science programs and for defining new campaigns poses an unsurmountable hurdle for communities and organizations as they usually lack the specific technical ICT-skills and programming knowledge to create the necessary server infrastructure and mobile applications. This often forces these organizations to opt for a non-technological approach (i.e., pen and paper) or to spend big chunks of their restricted budget on external ICT-consultants (D'Hondt et al., 2014; Zaman et al., 2014).

SOLUTIONS AND RECOMMENDATIONS

Citizen science has showed its potential to help meet the demands of monitoring the environment. There are several guides exist in the field of biodiversity monitoring, for example, Guide to citizen science (Tweddle, Robinson, Pocock, & Roy, 2012), Choosing and using citizen science (Pocock, Chapman, Sheppard, & Roy, 2014), Guide to running a bioblitz 2.0 (Robinson, Tweddle, Postles, West, & Sewell, 2013), User's guide for evaluating learning outcomes from citizen science (Phillips, Ferguson, Minarchek, Porticella, & Bonney, 2014). However, a generic guide that can offer the scientific community advice on how to develop, implement and evaluate citizen science projects including how to get the most out of citizen science projects in the environmental filed is needed.

In order to ensure a usable citizen science program and to maintain citizens in a citizen science project in practice, Liu et al. (2014) addressed the following development needs:

1. The need to adequately promote the citizen science platform and tools for raising awareness, recruiting and sustaining citizens' participation;
2. The need of a good understanding of citizens' demographics in order to develop the citizen science platform to meet their needs, especially as they change;
3. The need to build a long lasting infrastructure that uses open standards, is easily exploitable through an open Application Programming Interface (API), can be widely accessed, extended and maintained, and is seen as a generic environmental enabler rather than a project specific outcome;
4. The need to address and evaluate Citizens' Voice (Citizen' Views on certain environmental issues and its related environmental actions) and government Accountability (i.e., governments that can be held accountable for their environmental action) in the social and political context in which citizen science programs are embedded (Rocha Menocal & Sharma, 2008; Fernandez-Gimenez et al., 2008), to actively promote the Citizens' Voice and Accountability concepts as important dimensions of good environmental governance (Rocha Menocal & Sharma, 2008), to address citizen science potential role to influence environmental equity and to improve social justice (Kamar, Hacker, & Horvitz, 2012);
5. The need to develop particular channels and mechanisms that can underpin the sound environmental-social-political actions in which citizen science programs are addressed, in a manner that facilitates citizens to influence environmental governing priorities and processes;
6. The need to explore and further develop technologies, which deal with data collection and analysis by building necessary technical capacity and overcoming the 'digital divide' for environmental monitoring, data exchange, visualizing and communicating results back to the broader users (Rocha Menocal & Sharma, 2008; Brabham, 2009), managing and analyzing increasing data volumes, variety and velocity (Zikopoulos, Eaton, & Zikopoulos, 2011); reducing measurement uncertainties; developing reliable and fast Quality Assurance/Quality Control (QA/QC) tools that can work in real-time; and increasing need for interdisciplinary use of data, integration of different types of data.

FUTURE RESEARCH DIRECTIONS

The rapid diffusion of ICTs during the past decades has facilitated the process of data collection and sharing by the public and has resulted in the formation of more and more online environmental citizen observatory networks. There is no doubt that the increasing trend of citizen science programs has the potential to enhance the spatial and temporal resolution of current data streams from terrestrial and areal sensors, as well as being the best approach to increase citizen's environmental awareness.

One of the challenges is defining what is and what not citizen science is. However, due to the boom of citizen science programs that are currently been seeing, there is a need for having a clear global picture, in order to monitor the rapid growth of such initiatives, to catalogue and analyze differences and to synchronize already ongoing activities (Cobweb project, 2016; Ibercivis, 2016). The basic concept is to understand who is doing what and where.

In addition, there is also a need to understand how citizen science is changing the relationship between science and society. Therefore, a set of tools, metrics, shall be further developed and used. Among others, techniques from both humanities and social networks analysis shall be developed and used, to explore whether the key stakeholders are properly engaged or not, and if there is interactive information exchange amongst the key stakeholders within any citizen science program.

Citizen science is a growing worldwide phenomenon thanks in large part to the evolving new technologies that connect people easily and effectively with the scientific community (ECSA, 2016). There is a need to explore what types of new technologies can become a useful tool for citizens to play a more active role in sustainable development, which citizens can contribute valuable information that can be used to develop and deliver policies, improve understanding and respond to many of the challenges facing society today. Furthermore, methods for analyzing and sharing citizen science tools and data are also required (Cobweb project, 2016). There is a need to identify the major requirements for citizen science project repositories and their relation to existing citizen science platforms (Cobweb project, 2016).

Methods for maximizing the value of citizen science data, including data accessibility, availability and quality in support of the sustainable development goals, are also required (Grossberndt & Liu, 2016). Citizen science has evolved over the past four decades. Recent projects place more emphasis on scientifically sound practices and measurable goals for public education. Modern citizen science differs from its historical forms primarily in the access to, and subsequent scale of, public participation. This is, in large part, due to new technology, which is credited as one of the main drivers of the recent explosion of citizen science activity.

There is an emerging trend that indicates that research will focus on cross-cutting computational topics such as optimization, dynamical models, big data, machine learning, and citizen science, applied to sustainability challenges. Advances in computational sustainability will lead, for example, to novel strategies for helping citizens to improve their quality of life.

So far, assistance from the general public is often limited to data collection, as in classical citizen science. Moreover, whether and how people actually perceive the educational outputs of citizen science programs is generally lacking and needs to be addressed more in the future citizen science studies.

In fact, citizen science programs vary from top-down to more bottom-up approaches, depending on how people are involved. Devictor, Whittaker, & Beltrame (2010) argued that such participative citizen science approaches could be even more effective in promoting positive reconnections between the general public and conservation issues. Assessing how citizen science programs are perceived by citizens themselves would also offer opportunities to initiate more challenging levels of participation, and help develop stimulating connections between scientists, authorities and the general public (Devictor et al., 2010).

The authors do believe in the near future, citizen scientists could thus not only help to set surveys and scientific investigations, but also become the best allies of managers and stakeholders in the fields of environmental sustainability. Involving citizen participants directly in monitoring and active management of environmental issues can generate very powerful management efforts, leading to positive, cumulative and measurable impacts on environmental governance.

Although it is increasingly recognized that environmental governance, informed by local knowledge and citizen observations, may be a better option than top-down environmental protection restrictions, a practical means to implement protection strategies using citizen science as a social process is still largely missing (Devictor et al., 2010). In this context, citizen science could be of great help to promote environmental sustainability based on local environmental knowledge in several socio-economic contexts (i.e., not limited to the most developed countries).

CONCLUSION

With this chapter, the authors have provided insight into a topic that creates ambivalent feelings for many, i.e., citizen science in sustainable development. Based on the information from literature and practice, it can be seen that during the last years, policy-makers are beginning to value citizen science approaches as an important tool in their decision-making process. The fact that issues related to environment and health are usually complex and uncertain calls for a paradigm shift in these methods. Integrating knowledge and experience of different citizen groups and other stakeholders will help authorities in their decision-making process.

Citizen science plays a positive role in environmental sustainability. A review of relevant citizen science projects shows that the involvement of volunteers offers high value to research, policy and practice. Two major gaps were identified:

1. A need to compare and contrast the success of citizen science activities which present sound evidence of citizen scientists influencing positive environmental changes in the local environment they monitor;
2. More case studies showing usage of citizen science data by decision-makers, or the barriers to linkages and how these might be overcome.

If new research focuses on these gaps, and on the difference of opinions that exist, the researchers will have a much better understanding of the environmental, social, economic, ecological and cultural benefits of the citizen science role in sustainable development.

As it is well known, not one approach fits all. The choice of both participants and participation methods are always context-dependent and require a thorough preparation phase. Thus, time and personnel resources are crucial preconditions. Additionally, the authors see a need in fostering bottom-up approaches to a larger degree. In this context, the authors see citizen science as an increasingly essential tool for public participation in environmental monitoring. This approach assists citizens in observing and understanding environmental related problems, as well as reporting and commenting on them by help of advanced ICTs.

The authors discovered that different ongoing citizen science related programs in the environmental domain share a common model that is significantly relevant for the outcome of sustainable development. It consists of the following elements:

1. Citizens are playing an active role in observing environment by using novel monitoring-technologies and citizen as monitor approach;
2. Unique virtual places are created to gather and share data from a variety of sources;
3. Extraction and making use of relevant citizens-related data and providing multimodal services for citizens, communities and authorities;
4. Raising citizens' environmental awareness to form a strong public voice; and
5. Enabling dialogue among citizens, researchers, policy-makers and other stakeholders.

In addition, the review reached a number of conclusions about the value of data collected by volunteers:

1. The development of technologies was "revolutionizing citizen science", for example through online recording and smartphone apps;
2. Data quality could be excellent, but was not fully recognized by all researchers or policy-makers;
3. It is a cost-effective way of collecting environmental data;
4. There was potential to make considerably more use of citizen science that currently was the case.

To be successful in a citizen science program, the authors recommend the following sequential steps:

1. Identifying what citizens want and what citizens can offer;
2. Exploring what products and services a citizen science program can provide for the citizens;
3. Recruiting and retaining citizens to participate in and contribute to environmental governance;
4. Providing tools that support citizens to report their observations, inference and concerns; and
5. Supplying tools to access/receive timely information on the environment in a manner that is both easily understood and useful.

With these elements in mind, a successful performance and sustainable implementation should be possible.

Citizen science programs are geographically explicit, standardized and cover large spatial and/or temporal scales. Although citizen science is not a panacea, the development of citizen science programs should be increased and encouraged in the future as they can both be highly valuable for environmental sustainability and promote the reconnection between people and nature and more generally between people and science. Yet, although their number has increased in some countries (specifically the USA and UK) during the last decade, they remain relatively rare in others.

Moreover, the strength of citizen science programs directly relies on the curiosity and pleasure of the volunteers to learn and observe things that they have never noticed in their most familiar places.

REFERENCES

Bastiaensen, J., DeHerdt, T., & D'Exelle, B. (2005). Poverty reduction as a local institutional process. *World Development*, *33*(6), 979–993. doi:10.1016/j.worlddev.2004.09.019

Brabham, D. C. (2009). Crowdsourcing the public participation process for planning projects. *Plan Theory*, *8*(3), 242–262. doi:10.1177/1473095209104824

Buytaert, W., Dewulf, A., De Bièvre, B., Clark, J., & Hannah, D. (2016). Citizen Science for Water Resources Management: Toward Polycentric Monitoring and Governance? *Journal of Water Resources Planning and Management*, *142*(4), 01816002. doi:10.1061/(ASCE)WR.1943-5452.0000641

Buytaert, W., Zulkafli, Z., Grainger, S., Acosta, L., Alemie, T. C., & Bastiaensen, J. et al. (2014). Citizen science in hydrology and water resources: Opportunities for knowledge generation, ecosystem service management, and sustainable development. *Frontiers of Earth Science*, *2*, 1–21.

Ciravegna, F., Huwald, H., Lanfranchi, V., & Wehn de Montalvo, U. (2013). Citizen observatories: the WeSenseIt Vision. In Proceeding of the Infrastructure for Spatial Information in the European Community (INSPIRE 2013). Florence, Italy: INSPIRE 2013.

Cobweb project. (2016). *JRC's Citizen Science: data and service infrastructure meeting - Ispra (Italy)*. Retrieved June 1, 2016, from https://cobwebproject.eu/jrcs-citizen-science-data-and-service-infrastructure-meeting-ispra-italy-jan-2016

Conrad, C. C., & Hilchey, K. G. (2011). A review of citizen science and community-based environmental monitoring: Issues and opportunities. *Environmental Monitoring and Assessment*, *176*(1-4), 273–291. doi:10.1007/s10661-010-1582-5 PMID:20640506

Conrad, T. C., & Daoust, T. (2008). Community-based monitoring frameworks: Increasing the effectiveness of environmental stewardship. *Environmental Management*, *41*(3), 358–366. doi:10.1007/s00267-007-9042-x PMID:18026783

D'Hondt, E., Zaman, J., Philips, E., Boix, E. G., & De Meuter, W. (2014). Orchestration Support for Participatory Sensing Campaigns. In *Proceedings of the 2014 ACM International Joint Conference on Pervasive and Ubiquitous Computing (UbiComp 2014)*. ACM.

Devictor, V., Whittaker, R. J., & Beltrame, C. (2010). Beyond scarcity: Citizen science programs as useful tools for conservation biogeography. *Diversity & Distributions*, *16*(3), 354–362. doi:10.1111/j.1472-4642.2009.00615.x

Dillon, R. (2014). *Engage, empower, energize – Leading tomorrow's schools today*. London: Rowwan & Littlefield.

Dua, A., Bulusu, N., Feng, W., & Hu, W. (2009). Towards trustworthy participatory sensing. In *Proceedings of the 4th USENIX conference on hot topics in security (HotSec'09)*. Berkeley, CA: USENIX Association.

ECSA. (2016). *About Citizen Science and ECSA*. Retrieved June 1, 2016, from http://ecsa.citizen-science.net/about

Engelken-Jorge, M., Moreno, J., Keune, H., Verheyden, W., & Bartonova, A., & the CITISENSE consortium. (2014). Developing citizens' observatories for environmental monitoring and citizen empowerment: challenges and future scenarios. In *Proceedings of the Conference for E-Democracy and Open Government (CeDEM14)*. Danube University.

Fernandez-Gimenez, M. E., Ballard, H. L., & Sturtevant, V. E. (2008). Adaptive management and social learning in collaborative and community-based monitoring: A study of five community-based forestry organizations in the western USA. *Ecology and Society*, *13*(2), 4.

Fore, L. S., Paulsen, K., & O'Laughlin, K. (2001). Assessing the performance of volunteers in monitoring streams. *Freshwater Biology*, *46*(1), 109–123. doi:10.1111/j.1365-2427.2001.00640.x

Goeschl, T., & Jürgens, O. (2012). Environmental quality and welfare effects of improving the reporting capability of citizen monitoring schemes. *Journal of Regulatory Economics*, *42*(3), 264–286. doi:10.1007/s11149-012-9191-6

Goodchild, M. F., & Li, L. (2012). Assuring the quality of volunteered geographic information. *Spatial Statistics, 1*, 110–120. doi:10.1016/j.spasta.2012.03.002

Grossberndt, S., & Liu, H.-Y. (2016 in press). Citizen participation approaches in environmental health. In J. Pacyna & E. Pacyna (Eds.), *Environmental Determinants of Human Health* (pp. 1–25). London, UK: Springer.

Haklay, M. (2015). *Citizen science and policy: a European perspective*. Washington, DC: Woodrow Wilson International Center for Scholars.

Hand, E. (2010). Citizen science: People power. *Nature, 466*(7307), 685–687. doi:10.1038/466685a PMID:20686547

Ibercivis. (2016). *Ibercivis starts to draw the map of citizen science in Spain*. Retrieved June 2, 2016, from http://www.ibercivis.com/ibercivis-starts-draw-map-citizen-science-spain/

Kamar, E., Hacker, S., & Horvitz, E. (2012). Combining human and machine intelligence in large-scale crowdsourcing. In *Proceedings of the 11th International Conference on Autonomous Agents and Multiagent Systems (AAMAS 2012)*. Valencia, Spain: International Foundation for Autonomous Agents and Multiagent Systems.

Kight, C. (2012). *A Brief History of Citizen Science*. Retrieved June 2, 2016, from http://www.science20.com/anthrophysis/brief_history_citizen_science-93317

Kobernus, M., Berre, A. J., Gonzalez, M., Liu, H.-Y., Fredriksen, M., Rombouts, R., & Bartonova, A. (2015). A practical approach to an integrated citizens' observatory: The CITI-SENSE framework. In *Proceedings of the Workshop 'Environmental Information Systems and Services - Infrastructures and Platforms 2013' (ENVIP 2013)*. Neusiedl am See, Austria: CEUR Workshop Proceedings.

Korpela, E. J., Anderson, D. P., Bankay, R., Cobb, J., Howard, A., & Lebofsky, M., … Werthimer, D. (2011). Status of the UC-Berkeley SETI efforts. In *Proceedings of SPIE conference 8152, Instruments, Methods, and Missions for Astrobiology XIV, 815212*. San Diego, CA: Cornell University Library.

Lanfranchi, V., Wrigley, S. N., Ireson, N., Ciravegna, F., & Wehn, U. (2013). Citizens' observatories for situation awareness in flooding. In *Proceedings of the 11th International ISCRAM Conference*. University Park, PA: ISCRAM.

Litke, S., & Day, J. S. (1998). Building local capacity for stewardship and sustainability: The role of community-based watershed management in Chilliwack, British Colombia. *Environments, 25*, 91–110.

Liu, H.-Y., Bartonova, A., Neofytou, P., Yang, A., Kobernus, M. J., Negrenti, E., & Housiadas, C. (2012a). Facilitating knowledge transfer: Decision support tools in environment and health. *Environmental Health, 11*(Suppl 1), S17. doi:10.1186/1476-069X-11-S1-S17 PMID:22759499

Liu, H.-Y., Bartonova, A., Pascal, M., Smolders, R., Skjetne, E., & Dusinska, M. (2012b). Approaches to integrated monitoring for environment and health impact assessment. *Environmental Health, 11*(1), 88. doi:10.1186/1476-069X-11-88 PMID:23171406

Liu, H.-Y., Kobernus, M., Broday, D., & Bartonova, A. (2014). A conceptual approach to a citizens' observatory – supporting community-based environmental governance. *Environmental Health, 13*(1), 107. doi:10.1186/1476-069X-13-107 PMID:25495204

Liu, H.-Y., Skjetne, E., & Kobernus, M. (2013). Mobile phone tracking: In support of modelling traffic-related air pollution contribution to individual exposure and its implications for public health impact assessment. *Environmental Health, 12*(1), 93. doi:10.1186/1476-069X-12-93 PMID:24188173

Miller-Rushing, A., Primack, R., & Bonney, R. (2012). The history of public participation in ecological research. *Frontiers in Ecology and Evolution, 10*(6), 285–290. doi:10.1890/110278

Phillips, T., Ferguson, M., Minarchek, M., Porticella, N., & Bonney, R. (2014). *User's Guide for Evaluating Learning Outcomes in Citizen Science*. Ithaca, NY: Cornell Lab of Ornithology.

Pocock, M. J. O., Chapman, D. S., Sheppard, L. J., & Roy, H. E. (2014). *Choosing and Using Citizen Science: a guide to when and how to use citizen science to monitor biodiversity and the environment*. London: Centre for Ecology & Hydrology.

Robinson, L. D., Tweddle, J. C., Postles, M. C., West, S. E., & Sewell, J. (2013). *Guide to running a BioBlitz*. London: Natural History Museum, Bristol Natural History Consortium, Stockholm Environment Institute York and Marine Biological Association.

Rocha Menocal, A., & Sharma, B. (2008). *Joint evaluation of citizens' voice and accountability: synthesis report*. London: DFID.

Roy, H. (2012, November 23). *Review highlights role of citizen science projects*. Retrieved June 1, 2016, from http://www.bbc.com/news/science-environment-20445296

Roy, H. E., Pocock, M. J. O., Preston, C. D., Roy, D. B., Savage, J., Tweddle, J. C., & Robinson, L. D. (2012). *Understanding Citizen Science & Environmental Monitoring*. London: NERC Centre for Ecology & Hydrology and Natural History Museum.

Salonen, A., & Åhlberg, M. (2013). Towards sustainable society – From materialism to post-materialism. *International Journal of Sustainable Society, 5*(4), 374–393. doi:10.1504/IJSSOC.2013.056846

Science Communication Unit, University of the West of England, Bristol. (2013). *Science for Environment Policy In-depth Report: Environmental Citizen Science*. Retrieved June 1, 2016, from http://ec.europa.eu/science-environment-policy

Scyphers, S. B., Powers, S. P., Akins, J. L., Drymon, J. M., Martin, C. W., & Schobernd, Z. H. et al. (2015). The Role of Citizens in Detecting and Responding to a Rapid Marine Invasion. *Conservation Letters, 8*(4), 242–250. doi:10.1111/conl.12127

Tennessee Water Resources Research Center, University of Tennessee. (2004). *A comparative analysis of water quality monitoring programs in the southeast: lessons for Tennessee*. Retrieved June 2, 2016, from http://isse.utk.edu/wrrc/programsprojects/pdfs/mainbook.pdf

Tweddle, J. C., Robinson, L. D., Pocock, M. J. O., & Roy, H. E. (2012). *Guide to citizen science: developing, implementing and evaluating citizen science to study biodiversity and the environment in the UK*. London: Natural History Museum and NERC Centre for Ecology & Hydrology for UK-EOF.

UNEP. (2009). *Environmental governance*. Retrieved May 29, 2016, from http://www.unep.org/pdf/brochures/EnvironmentalGovernance.pdf

Whitelaw, G., Vaughan, H., Craig, B., & Atkinson, B. (2003). Establishing the Canadian community monitoring network. *Environmental Monitoring and Assessment*, *88*(1/3), 409–418. doi:10.1023/A:1025545813057 PMID:14570426

Wikipedia. (2016). *List of citizen science projects*. Retrieved May 30, 2016, from https://en.wikipedia.org/wiki/List_of_citizen_science_projects

World Commission on Environment and Development. (1987). *Our Common Future*. Oxford: Oxford University Press.

Zaman, J., D'Hondt, E., Boix, E. G., Philips, E., Kambona, K., & De Meuter, W. (2014). Citizen-Friendly Participatory Campaign Support. In *Proceedings of 2014 IEEE International Conference on Pervasive Computing and Communications Work in Progress (PerCom WiP'14)*. Budapest, Hungary: PerCom WiP'14.

Zikopoulos, I. B. M. P., Eaton, C., & Zikopoulos, P. (2011). Understanding Big Data: Analytics for Enterprise Class Hadoop and Streaming Data. New York: McGraw-Hill Professional.

KEY TERMS AND DEFINITIONS

Citizen Engagement: The encouragement of the general public to become involved in the political process and the issues that affect the community.

Citizen Observatory: The citizens' own observations and understanding of environmentally related problems and in particularly as reporting and commenting on them

Citizen Science: A form of research collaboration involving members of the public in scientific research projects to address real-world problems

Environmental Governance: A concept in political ecology and environmental policy that advocates sustainability (sustainable development) as the supreme consideration for managing all human activities—political, social and economic

Information and Communications Technology (ICT): An extended term for information technology (IT) which stresses the role of unified communications and the integration of telecommunications (telephone lines and wireless signals), computers as well as necessary enterprise software, middleware, storage, and audio-visual systems, which enable users to access, store, transmit, and manipulate information.

Sustainable Development: The development that meets the needs of the present without compromising the ability of future generations to meet their own needs

Volunteered Geographical Information (VGI): The harnessing of tools to create, assemble, and disseminate geographic data provided voluntarily by individuals.

Chapter 8
Social Context of Citizen Science Projects

Patricia Tiago
Centre for Ecology, Evolution, and Environmental Changes & Research Center in Biodiversity and Genetic Resources, Portugal

ABSTRACT

This chapter provides a brief history of citizen science in our societies, identifies the main stakeholders involved in projects of this topic, and analyzes the main points to take into consideration, from a social perspective, when designing a citizen-science project: communicating; recruiting and motivating participants; fostering innovation, interdisciplinarity and group dynamics; promoting cultural changes, healthy habits, inclusion, awareness and education; and guiding policy goals and decisions. Different governance structures, and a coexistence of different approaches, are analyzed together with how they suit different communities and scientific studies.

INTRODUCTION

Citizen science engages the general public with scientific research activities, and while not new, is becoming a mainstream approach to collect data on a variety of scientific disciplines (Miller-Rushing, Primack, & Bonney, 2012). The consolidation of citizen science can be perceived from the adoption of a formal name, increased research about the field and formalization of international associations. Citizen science maturity has advanced with technology innovations of recent years.

Societies are facing rapid changes in values, interests and expectations. The growth of social networks and collaborative web projects has implications for the relations between scientists, decision makers and different societal groups. Citizen science is growing to be a mechanism that allows citizens to have an active role in science development and in dealing with important environmental and scientific questions.

Scientists who support the rise in citizen science recognize the benefit of volunteer contribution to science in terms of increased scale, data collection and analysis, outreach capacity, while dealing with budget constraints. Consequently, an increasing number of studies have started to work with volunteer citizens, helped by easily accessible technological tools. Awareness among scientists for these social

DOI: 10.4018/978-1-5225-0962-2.ch008

changes has increased, generally in a gradual way, but faster in countries with a higher tradition of public participation, especially scientific participation (Hess, 2010).

Citizen science can also have a positive impact on society and support sustainable development, by fostering connections between environment, society and economy and overcoming barriers between disciplines (Giddings, Hopwood, & O'Brien, 2002).

Given its collaborative nature, citizen science is characterized by a wide range of stakeholders, whose motivations and interactions can be determinant for the success of a citizen science project and thus should be carefully taken into account on project design.

This chapter provides a brief history of citizen science and identifies the main stakeholders involved in these projects. The chapter then analyzes the main points to take into consideration, from the perspectives of these different stakeholders, when designing a citizen science project.

THE HISTORY OF CITIZEN SCIENCE IN OUR SOCIETIES

For centuries, scientific research was conducted by amateurs (people that were not paid to do science) (Vetter, 2011). Professionalization of science, in the late 19th century, drew those amateurs away from the scientific world and created a big gap between "real scientists" (people that are paid to do science) and citizens interested in those subjects (Vetter, 2011).

John Ray, Alfred Russell Wallace, Gregor Mendel are prime examples of amateurs who produced incredible scientific advances. John Ray published important works on botany, zoology, and natural theology and his classification of plants in *Historia Plantarum*, was an important step towards modern taxonomy (Raven, 1942). Alfred Russel Wallace was a British naturalist, explorer, geographer, anthropologist, and biologist. His best known work was on the theory of evolution through natural selection and his paper on the subject was jointly published with some of Charles Darwin's writings in 1858 (Raby, 2001). Gregor Mendel was a friar who gained posthumous fame as the founder of the modern science of genetics. His pea plant experiments established many of the rules of heredity, now referred to as the laws of Mendelian inheritance (Weiling, 1991). These individuals were largely pursuing research because of an innate interest in particular topics or questions (Vetter, 2011) and were recognized experts in their field, conducting research indistinguishable from today's professional scientists.

On a different level of participation, though not yet called citizen scientists, general people have also been involved in scientific activities on a volunteer basis for centuries, documenting observations of nature. Farmers, hunters and amateur naturalists were some of the activities involved in collecting natural world data (Miller-Rushing et al., 2012). In the 18th century, Carl Linnaeus, collected, with the help of many volunteers, animal, plant, rock and fossils specimens and artifacts from around the world. For 1200 years court diarists in Kyoto, have been recording dates of the traditional cherry blossom festival (Primack, Higuchi, & Miller-Rushing, 2009) and in China citizens and officials have been tracking outbreaks of locust for at least 350 years (Tian et al., 2011).

In some specific science issues, such as weather, astronomy and birds surveys, there is a long history of citizen science, particularly in Anglo-Saxon countries and center and northern European countries such as England, United States of America, Australia, Netherlands or Finland.

The project National Weather Service - Cooperative Observer Program (NWS-COOP) has been collecting basic weather data across United States since 1890 with results supporting much of what we know about variability and directional changes in climate (Miller-Rushing et al., 2012). With a twofold mission of providing observational meteorological data and helping to measure long-term climate changes, the project has more than 8,700 volunteers taking observations in farms, in urban and suburban areas, National Parks, seashores, and mountaintops (NOAA, National Weather Service, 2014).

In the astronomy area, the British government funded, in1874, the Transit of Venus project to measure the Earth's distance to the Sun. This project engaged the admiralty to support data collection all over the globe and recruited the services of amateur astronomers (Ratcliff, 2008).

Ornithology has a long linking history with citizen science. Bird monitoring in Europe goes back to 1749, when amateurs, in Finland, collected data on timing of migration (Greenwood, 2007). Wells Cooke, a member of the American Ornithologists' Union, developed one of the earliest known formal citizen science programs in the United States, in the late 18th century. This project, overtime, transformed into today's North American Bird Phenology Program. Citizens involved collect, on cards, information about migratory bird patterns and population figures. Those cards are being scanned and recorded into a public database for historical analysis (Dickinson, Zuckerberg, & Bonter, 2010). Another example of one of the oldest citizen science programs in the United States, which is still active, is the Christmas Bird Count, sponsored by the National Audubon Society. Since 1900, the organization has sponsored a bird count that runs from December 14 through January 5 each year. An experienced birder leads a group of volunteers as they collect information about local populations of birds. More than 2,000 groups operate across the United States and Canada (Dickinson et al., 2010).

Nowadays the focus of citizen science is changing from the traditional "scientists using citizens as data collectors" to citizens as scientists (Lakshminarayanan, 2007). In this new era of citizen science projects, citizens can participate at the diverse stages of the scientific process from co-creating a project with a scientist, following up all the steps of the project, raising new questions, collecting or analyzing data, producing reports and disseminating findings (Tweddle, Robinson, Pocock, & Roy, 2012). Depending on the desired level of engagement in science, different models of action can be adopted, such as pooling of resources, collective intelligence, grassroots activities, data collection, analysis tasks, serious games or participatory experiment (Socientize, 2014).

Citizen Science has already a long history and has recently begun to evolve into a broad research methodology with new applications and different stakeholders' approaches. Several historical case studies and personalities, involved with this subject, may help us analyze what can be the future direction of citizen science.

WHO ARE THE DIFFERENT STAKEHOLDERS INVOLVED IN CITIZEN SCIENCE PROJECTS?

Citizen science, although in its basic form was viewed as a partnership between volunteers and scientists to answer real world questions (Cohn, 2008), was expanded to a multiplicity of stakeholders, ranging from research scientists, teachers, students, managers, environmental organizations, and politicians (Bonney et al., 2009), due to its potential for educational purposes, raising awareness and driving policy changes, among other reasons. These stakeholders have many different interests in citizen science, and face particular constraints in their involvement.

Despite the considerable amount of stakeholders involved, clustering them into four groups: citizen scientists, scientists, other societal groups and policy makers, allows us to analyze the project design from these four different perspectives. Citizen scientists and scientists are directly involved in the scientific process, while other societal groups and policy makers are more indirectly involved, e.g. using data, promoting education, guiding policy goals and decisions or giving answers to social concerns.

Assuring a good and stable relationship between the interests of these groups is important for the project's success (Figure 1).

Thinking specifically in the citizen scientists, depending on the project aims and activities, target people are diverse and may include hobby and professional groups such as schools, students, scouts, naturalists, tourists, sports enthusiasts, farmers, fishermen, or multiplicity of actors. Engaging these different stakeholders into a shared framework with some common and some specific means of communication are good ways to achieve results. Projects like eBird, iSpot and iNaturalis have in their objectives and strategies specific ways of involving and engaging different groups (Sullivan et al., 2009; Clow & Makriyannis, 2011; Bowser, Wiggins, Shanley, Preece, & Henderson, 2014).

Points to Take into Consideration IN Project Design

Project design is a crucial step in ensuring the effectiveness of the project and the capacity to achieve its goals (Raddick et al., 2009). When designing a project, this will inevitably involve trade-offs, e.g. gathering comprehensive, high quality data according to rigorous scientific protocols, and the ease of data collection (Hochachka et al., 2012). If the data collection is too complex or too time consuming, volunteers may lose their desire to participate and thus, understanding and adapting the program to the skills, expectations and interests of the volunteers is critical (Shirk, Bonney, & Krasny, 2012).

Figure 1. Groups of stakeholders involved in citizen science projects

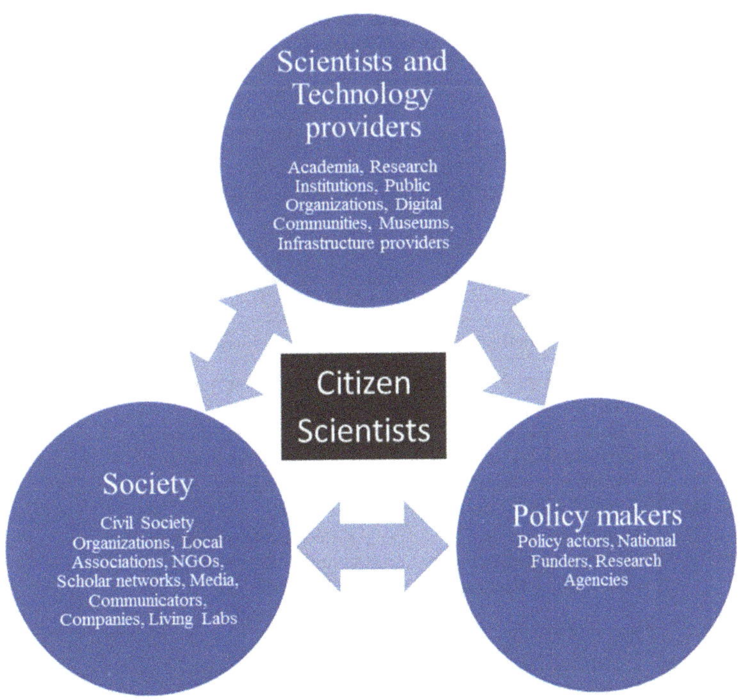

When designing a citizen science project, it is thus important to take into account a social perspective meaning the interactions generated between the different stakeholders, their collective co-existence, regardless of whether they are aware of it or not, and of whether the interaction is voluntary or involuntary.

Although citizen science is nowadays a broad methodology used in many different scientific areas, there are several cross-cutting issues, common to all of them. Highlighting the importance of taking a stakeholder view when designing a citizen science project, table 1 summarizes issues to take into consideration, which will be analyzed in detail below. Some, such as motivation or awareness, are important for several stakeholders, but in very different ways and assuming varying degrees of importance.

Communicating and Recruiting Participants

When designing the project, and after deciding that citizen science is the best methodology to achieve the project's goals (Tweddle et al., 2012), one must identify the groups that might want to be involved, understand the reasons and motivations they have to participate, recruit them and maintain their participation over time. In recent years much has been written on communication and recruiting (Van Den Berg, Dann, & Dirk, 2009; Dickinson et al., 2012; Roy et al., 2012; Pandya, 2012; Tweddle et al., 2012; Silvertown, Buesching, Jacobson, & Rebelo, 2013).

The starting point for recruitment is to determine who the target audience is, which can be from a specific group like students, amateur astronomers, bird watchers, divers or from a broad group like inhabitants of a certain area (Bonney et al., 2009). Knowing who will be the projects' participants is important to decide how to reach them, what will be said, how it will be said, when it will be said, where it will be said and who will say it (Kotler, Wong, Saunders, & Armstrong, 2005). Successful citizen science projects like eBird, Galaxy Zoo, OPAL, attached great importance to the communication approach with the targeted volunteers (Sullivan et al., 2009; Raddick et al., 2010; Tweddle et al., 2012).

Then it is necessary that people who might want to participate get to know that the project exists, to whom it is directed and what are its main objectives (Cohn, 2008).

Table 1. Points to take into consideration for different stakeholder groups, in a citizen science project design

Project design			
Citizen Scientists	**Scientists**	**Other Societal Groups**	**Policy Makers**
• Communicating and recruiting participants • Motivating participants • Promoting education • Giving feedback • Enabling personal recognition and reward • Taking into account work scale preference	• Enabling outputs for scientific studies • Assuring data quality • Sharing open source results • Fostering innovation, interdisciplinarity and group dynamics • Motivating participants • Overcoming reluctance	• Giving answers to social concerns • Promoting healthy habits • Promoting inclusion • Promoting awareness and education • Taking into consideration cultural differences • Overcoming reluctance • Sharing open source results	• Guiding policy goals and decisions • Giving answers to social concerns • Promoting awareness • Overcoming reluctance
Project Evaluation and Governance			

These recruitment efforts may vary, depending on the previous existence – or not - of a community (Robson, Hearst, Kau, & Pierce, 2013). Nowadays there are websites dedicated to host citizen science projects where people can obtain information and enlist as volunteers (Dickinson et al., 2012; Newman et al., 2012). Platforms like Zooniverse, one of the world's largest and most popular platform for people-powered research, covers many disciplines and topics across the sciences and humanities (Reed, Raddick, Lardner, & Carney, 2013). With specific high motivated groups, or in countries with a higher tradition on citizen science, e-mails and newsletters can be sufficient ways of promoting the project (Dickinson et al., 2012). Social networks like Facebook or Twitter provide, nowadays, good opportunities to reach a high range of participants (Pickard et al., 2011; Robson et al., 2013).

In countries facing barriers to public participation, due to lack of tradition in these areas, the actions to promote the project and then to engage people to participate, need to be much more active.

For a more effective recruitment process the use of different types of media is important. Sometimes it is the scale of the project that determines the capacity of the project manager to reach several media channels: print media (newspapers, magazines, direct mail, and specialist publications), broadcast media (radio, television), display media (signs, posters, billboards spread in a country, city, school – depending on the scale of the project), and online and electronic media (websites, social networks) (Kotler et al., 2005).

Organizing a launch event or an event at an existing festival or fair that allows face-to-face contact, can be an important social measure to promote the project (Wiggins & Crowston, 2011), allow citizens to interact directly with the scientists involved and establish a relationship (Tweddle et al., 2012).

Also allowing time for participants to socialize during activities is important for recruitment and retention for longer-time (Silvertown et al., 2013). Word-of-mouth recruitment between peers is one of the most powerful ways of growing a network of contacts. Identifying "influencers" can bring other persons along, increase visibility, credibility and create bandwagon effects (Kotler et al., 2005).

Motivating Participants

Motivations of volunteers and scientists to participate or conduct a citizen science project, have already been the subject of several studies from different authors (Bruyere & Rappe, 2007; Van den Berg et al., 2009; Bramston, Pretty, & Zammit, 2011; Jordan, Ballard, & Phillips, 2011; Silvertown et al., 2013).

Understanding citizen scientists' motivations to contribute may improve the results obtained. These motivations may be different from country to country and for different societal groups or age groups (Dierkes & von Grote,2000; Forte & Lampe, 2013). Cultural differences also influence the reasons to collaborate (Rotman et al., 2014). Some of the main participant motivations highlighted in many of the existing studies include:

- The desire expressed by participants to learn new skills and about the scientific issues behind the project (Bell et al., 2008; Van den Berg et al., 2009; Raddick et al., 2010);
- The desire to see the impact of their work (being able to see and share the efforts undertaken and its further use within a scientific or policy community) (Van den Berg et al., 2009);
- The sense of making a discovery, e.g. finding a new galaxy in Galaxy Zoo project (Raddick et al., 2010);
- The desire to feel as active participants and co-owners of the project (Dickinson et al., 2012; Rotman et al., 2014);

- Gaining recognition for their input, e.g. through feedback and interaction with scientists and peers, and through gaining achievements (Rotman et al., 2014);
- The desire to feel competent in doing a task (Rotman et al., 2014), e.g. progression to expert status or from simple to more complex tasks requiring additional responsibility (Nov, Arazy, & Anderson, 2014);
- The sense of participating in a project that might be relevant to their community (Van den Berg et al., 2009);
- The feeling that they are helping the environment and taking an active conservation action (Van den Berg et al., 2009);
- The enjoyment of developing activities in nature (Bell, et al., 2008; Van den Berg et al., 2009);
- Getting to know other people with similar interests and making new friends (Van den Berg et al., 2009);
- Allowing to explore different career options (Van den Berg et al., 2009);
- The enjoyment of developing team activities that put scientists and citizen scientists working together with a sense of camaraderie, making scientific exploration and discoveries enjoyable (Nov et al., 2011; Newman et al., 2012);
- Competing with other participants. Some projects appeal to the competitiveness of the participants by providing tools for determining the relative status of volunteers (e.g. numbers of species seen) and geographical regions (e.g. checklists per area) (Hochachka et al., 2012).

While most people committed with citizen science projects are likely to belong to an already environmentally aware subsector of the population (Coghlan, 2005), a surprisingly large number of people are motivated by curiosity, tourism motives or because they want to make a new start in life (e.g. after divorce, job redundancy, etc.) (Silvertown et al., 2013), although these motivations may hardly sustain, by themselves, long-term participation.

Studies on motivations, from the perspective of scientists, to participate in citizen science projects are scarce. Some identified motivations for scientists include: professional reasons like further their own professional career, promoting their scientific work in society, the outreach obtained with those projects, advance science, and become more aware of local knowledge and expertise (Carolan, 2006; Rotman et al., 2012).

Providing Education

Some studies show that many skills needed to do research can be obtained by non-experts when they are properly trained (Janzen, 2004; Cohn, 2008). Citizen science projects can benefit greatly from the educational materials provided by scientists, despite some scientists still framing this training and supervising as time wasted away from professional research, rather than a beneficial investment of time (Silvertown et al., 2013).

Most citizen science projects provide volunteers with educational material like training workshops, field lessons (e.g. on species identification, field guides, volunteer manuals or web-based educational tools; Crall et al., 2010). In some cases participants also need to learn how to use maps, technological devices and applications, such as GPS units (Crall et al., 2010).

Protocols used for citizen science should be easy to perform, explainable in a clear and straightforward manner, and engaging for volunteer participants (Bonney et al., 2009). Pilot-testing protocols with naive audiences is important when directed at a wide swath of potential participants. For example, Cornell Lab of Ornithology project designers have tested draft protocols with local bird clubs, school groups, and youth leaders by accompanying participants in the field and observing them as they collect and submit data and tested protocols at distant locations by collecting online feedback. When protocols prove to be confusing or overly complicated, they can be simplified, clarified, or otherwise modified until the participants can follow them with ease (Bonney et al., 2009).

The higher the interaction between scientists and citizen scientists, the higher is the engagement of participants (Dickinson et al., 2012). People tend to participate more when they feel supported by the appropriate expertise while doing the activity. Participants should have all the information about the project and should feel helped by the project team.

Some projects develop learning elements that align with relevant school curriculum standards (Zoellick, Nelson, & Schauffler, 2012). Partnerships between schools and these citizen science projects are likely to become much more important in the future with substantial gains for the projects' multiple goals.

Giving Feedback

Giving participants a rapid feedback and providing regular communication about their contribution and the outcomes of the project, is a powerful way of motivating them and maintaining their participation, since people like to know to what they are collaborating and how is being used the information they are collecting (Devictor, Whittaker, & Beltrame, 2010). Giving feedback is also an important way for people to increase perceived competence and usefulness of their participation.

This feedback can be included by design of the project (e.g. real-time publication and/or validation of the information collected in the project website), or can also be accomplished in many different ways, such as through field events, email, phone, newsletters, blogs, discussion forums and various forms of social media. Organizing a closing event can also be a good way to share results and thank the participants (Tweddle et al, 2012).

Enabling Personal Recognition and Reward

Rewarding citizen scientists, in a number of ways, provides a sense of achievement (Tweddle et al, 2012) and is thus an effective way to encourage and support participation. Volunteers like the idea of knowing that their work is important and that their contributions can help scientists make better and more comprehensive analyses (Musick & Wilson, 2007).

A reward system can be implemented in several different ways such as highlighting the identity of contributors with observations to acknowledge their contributions explicitly (e.g. in Observado, iSpot and iNaturalist; (Clow & Makriyannis, 2011; Bowser et al., 2014), providing participants with certificates of recognition, thanking participants and acknowledging their role (e.g. through organization of a closing event, which can also be used to solicit further inputs and give feedback of project's results) (Tweddle et al, 2012), providing open access to all records in the database, or at least the non-sensitive (Newman et al., 2012), holding a competition (Newman et al., 2012), recognizing the degree of volunteer expertise (e.g. progressing from amateur to expert levels in iSpot; Clow & Makriyannis, 2011), fellowships and sponsorship, symbolic rewards such as badges (Cooper, Dickinson, Phillips, & Bonney, 2010; Clery, 2010).

The eBird website was modified to provide direct rewards to participants and with these modifications participation rapidly increased and eBird has gathered more information in 1 month (almost 3 million observations) than it did during the entire first 2 years of the project (2 million observations) (Hochachka et al., 2012).

Project managers should make an effort to provide easy access to scientific, institutional, managerial and/or legislative information packages produced from project data, in ways of interest to stakeholders. Apart from their immediate value to the target stakeholders, this helps participants understand the value of their contribution. For example, it may not be readily apparent that a few species observations might contribute collectively to e.g. reveal the arrival of invasive / pest species, and eventually promote a policy measure.

Taking into Account Work Scale Preference

Different citizen science projects have different goals and, depending on those, work at different scales such as local, regional, national, continental, global or virtual. Not all citizen scientists like to work at the same scale, preferring to be engage with local questions, helping to give answers to a community problem, while others like broad scale issues like climate change questions or species invasions problems (Cooper et al., 2007; Dickinson et al., 2010; Wiggins et al., 2011; Hochachka et al., 2012).

Enabling Outputs for Scientific Studies

For research scientists, citizen science projects can offer many benefits for their work but usually require a balance between data quality and quantity. The amount of data that can be collected and the geographic scale of these data can give a completely different dimension to a scientific study (Bonney et al., 2009). Funding constrains often limit the amount and scope of these studies and citizen science projects allow them to become a reality (Darwall & Dulvy, 1996; Danielsen, Burgess, & Balmford, 2005). The incorporation of these new sources of data with scientific projects enable them to fill existing gaps e.g. on species distributions (Danielsen et al., 2005).

Some citizen science projects do not produce scientific peer reviewed publications (Theobald et al., 2015) but more awareness publications, in many cases due to a lack of data quality assurance.

To optimize the data quality and quantity provided by participants, researchers must understand which factors affect most their performance (Bueshing & Newman, 2005; Bueshing & Slade, 2012) and then find ways to optimize and mitigate these factors, for instance allocating tasks to the best suited individuals (Mackney & Spring, 2000), which can represent an increase on the quality of data collected.

Another issue to take into account is that some projects, that have specific goals, can have outputs that might be useful for different purposes, not always foreseeable. Flexibility and data open access can increase the projects scientific value.

Assuring Data Quality

Assuring data quality is important in attracting more scientists to use and engage with citizen science projects and become this methodology widely accepted (Dickinson et al., 2010, Bonter & Cooper, 2012). In order to achieve the scientific goals of the projects data collected by citizens should be validated, e.g. checked for errors and entered reliably into databases suitable for further analysis and sharing (Crall et al., 2010).

Statistical approaches should be robust and adapted in order to achieve better data quality (Bird et al., 2014). Protocols need to be developed and later adjusted to any limitation identified (Bonney et al., 2009). Networking among scientists and citizen scientists is an important tool to improve protocols. In this regard a lot can be learnt with computer programmers open source systems (Bonney et al., 2009).

In citizen science projects concerning biology and ecology, the main problems identified in data quality include under or over estimation in species abundance measures (Bray & Schramm, 2001), species misidentification (Brandon, Spyreas, Molano-Flores, Carrol, & Ellis, 2003; Genet & Sargent, 2003; Fitzpatrick, Preisser, Elison, & Elkinton, 2009), protocols too simple that did not produce useful data (Engel and Voshell, 2002). Even though taxonomic identification is a skill that requires years of training the rates of misidentification depend on species rarity or conspicuity (Genet and Sargent, 2003). Many citizen science projects ask people to add information on presence of species (ignoring absences), introducing in those projects problems of bias relating to presence-only data. However, the growing number of participants in citizen science projects can help to reduce uncertainty. A place where large numbers of volunteers submitted presence data for some species, but no data on presence of other species, confidence can be increased that the lack of data on the absent species is due to the true absence of the species, rather than from a lack of sampling effort (Stafford et al, 2010).

Sharing Open Source Results

Nowadays, in citizen science projects the most accepted culture is openness and free data access which is shifting some science paradigms (Newman et al., 2012; Socientize, 2014). However it should still take in consideration the intellectual property rights, fundamental personal data protection rights, ethical standards, legal requirement and scientific data quality (Newman et al., 2012; Socientize, 2014). Information and communication technologies foster open, efficient and agile systems, turning ideas into the actions required to mobilize citizens individual and collective.

There are still some concerns over sharing data due to data sensitivity like species cultural or biological significance (Jarnevich, Graham, Newman, Crall, & Stohlgren, 2007), or private property. To avoid this, many citizen science websites use features that protect those species like year filters (data can be hidden for a certain period) or data will be added with low resolution (Jarnevich et al, 2007).

Fostering Innovation, Interdisciplinarity, and Group Dynamics

When different people with different backgrounds are working together there is all an unpredictable group dynamic that achieve interesting results. Research on collective intelligence indicates that diversity matters and that new leaps of logic, innovation, and invention are more likely to arise when people of different backgrounds and abilities work together toward a common goal (Wooley, Chabris, Pentland, Hashmi, & Malone, 2010).

Comparing to more traditional scientific projects, a citizen science project aims to be more interdisciplinary involving both scientists and citizens with different backgrounds. Apart from the scientific thematic the study is about, a good project design should include people with different skills dealing with technologies available, social and communication aspects (Wiggins et al., 2011; Sullivan et al., 2014).

Creating opportunities for interaction between participants and scientists may also foster innovation and reach useful results (Newman et al., 2012).

Overcoming Reluctance

There is still high reluctance in people from different societal groups (scientists, politicians, decision makers, teachers, students), concerning citizen science projects. Some of them put in question data quality, some do not understand the aims of citizen science projects, the reasons to participate and to use data from these projects (Bonney et al., 2009; Catlin-Groves, 2012).

Project team should be as available as possible to answer questions, to provide any clarification required and to change communication strategies when they are not being effective (Bonney et al., 2009) in order to increase the perceived credibility of the project.

Giving Answer to Social Concerns

Several societal groups take an interest in citizen science due to its ability to give citizens the opportunity to address social concerns and priorities.

In the citizen science process, concerned citizens, government agencies, industry, academia, community groups, and local institutions collaborate to monitor, track and respond to issues of common community concern (Whitelaw, Vaughan, Craig, & Atkinson, 2003). Areas like pollution, public health or species monitoring invasions are sensitive to society in general and thus frequently are subjects for citizen science projects (Cohn, 2008; Bonney et al., 2009; Crall et al., 2011).

Public support for conservation can be increased by building social capital (Schwartz 2006) and this has been measured by increased levels of trust, harmony, and cooperation in communities with scientific engagement (Sultana & Abeyasekera, 2008). This can lead to a more educated community (Pollock & Whitelaw, 2005; Cooper et al., 2007) and a creation of a stewardship ethic (Whitelaw et al., 2003; Cooper et al., 2007).

At the same time, there is evidence that long-term economic and environmental success arrives when people's ideas and knowledge are valued, and power is given to them to make decisions independently of external agencies (Pretty et al., 1995).Citizen science projects also seemed to promote more sustainable communities (Whitelaw et al., 2003).

Communities where citizen science is prevalent tend to be more engaged in local issues, participate more in community development, and have more influence on policy-makers (Whitelaw et al., 2003; Pollock & Whitelaw 2005; Lynam, Jong, Sheil, Kusumanto, & Evans 2007).

Highlighting the social concerns addressed by citizen science may thus be a strong argument when communicating with different society groups, in particular policy makers.

Promoting Healthy Habits

Many citizen science projects promote nature observations in the field. Getting more people into nature is, by itself, an excellent value that may be drawn from citizen science projects (Cohn, 2008). Projects like OPAL have a strong connection with young children and schools promoting a proper childhood development and physical and emotional health once it stimulates interactions with nature which can be quite important (Louv, 2005). The increasing prevalence of childhood obesity has lead policy makers to rank it as a critical public health threat for the 21st century (Koplan, Liverman, & Kraak, 2005). When attending to many citizen science activities, child, young people and adults avoid sedentariness and engage in physical activity. For elder people, participating in citizen science projects is also a way to maintain brain and/or physical activity.

Promoting Inclusion

Citizen science projects aim to be inclusive (Pandaya, 2012). Studies have shown that diversity benefits all participants (Gurin, 1999) and the project itself. Different groups of citizens from different age groups, race, educational level, background, are called directly to take part in some projects, so, it is possible to find projects with specific actions for universities of the third age, minority groups Bushway, Dickinson, Stedman, Wagenet, & Weinstein, 2011) or prisoners (Ulrich & Nadkarni, 2009), with interesting results.

Some communities are frequently excluded from citizen science projects and identifying the barriers to participation is important for finding solutions to widening the community e.g. new technologies may inadvertently create barriers that widen the digital divide between those adopting/having the technology and those avoiding/lacking it (Newman et al., 2012).

Promoting Awareness and Education

Concerns about climate change, species extinctions, land use change, are recurrently discussed in the media but, nevertheless, there is still a widespread scientific illiteracy in society (Miller, 2004). This illiteracy depends on the cultural background of the societies involved (Dierkes & von Grote,2000) and citizen science projects can help promoting citizens' education.

Participating in citizen science projects increases people's awareness in many different areas. Due to its participatory nature, these projects appear well suited to elevating public understanding and support for science, environment and earth stewardship (Shirk, et al., 2012).

The increment of science literacy is a huge benefit of citizen science, giving a personal empowerment to the people involved (Brossard Lewenstein, & Bonney, 2005). Participants that improved their science and technology literacy are better informed to make decisions and can contribute more effectively to society as citizens, workers or consumers (National Science Board, 2008).

Some countries identify the need to put students in an educational environment that instigates them to ask questions, plan and conduct an investigation, use appropriate tools and techniques, think critically and logically about the relationships between evidence and explanations, construct and analyze alternative explanations, and communicate scientific arguments (Natural Research Council, 1996). Citizen science projects can play a role to achieve this objective.

Ideally, citizen scientists will be endowed with knowledge and skills to collect and disseminate this awareness and expertise.

Scientists face nowadays citizen science as a way to connect scientific research to public outreach and education (Lepczyk et al, 2009). It is also true that a public educated on these issues is more likely to fund and support scientific research that seeks to address them (National Science Board, 2008).

Taking into Consideration Cultural Differences

Multiple social and cultural drivers can affect the amount of information available and the efforts required from citizen science projects. To encourage participation in citizen science, project managers should recognize differences across countries, regions, and societal groups.

For example, the number of species observation records per square kilometer is high in countries with high per capita gross domestic product, high proportion of English speakers and high security levels, although those are not necessary countries with higher biodiversity (Amano & Sutherland, 2013). In some countries the value of public participation remains largely unknown to the society so, in these countries, the effort required should be much higher.

Sensitivity to cultural factors will be important to the success of projects that cross boundaries and involve local/traditional ecological knowledge (Dierkes & von Grote,2000; Ballard, Trettevick, & Collins, 2008). Even inside a specific country, significant differences in cultural attitudes towards citizen science may be found between different regions (Rotman, et al., 2014).

The project's strategy should match the target culture, so the two need to be in alignment. Sometimes, focusing on a few critical shifts in behavior may provide best results with the least effort. Measuring and monitoring cultural evolution are also best practices to take into account throughout the project, in order to identify backsliding or correct course (Katzenbach, Steffen, & Kronley., 2012).

Guiding Policy Goals and Decisions

Citizen science projects and their results can guide and influence policy goals and decisions. Community's awareness and literacy attained with these projects may lobby politicians concerning environmental issues.

The data collected by these projects can be used in policy area for management, like natural resource management (Brown, Krasny, & Schoch, 2001) or for environmental regulation (Penrose & Call, 1995). Citizen science projects that focus on biodiversity monitoring, in general, are beneficial to government agencies for several reasons: they offer a cost-effective alternative to government employee monitoring (Whitelaw et al. 2003; Conrad & Daoust 2008), fieldwork can be undertaken over larger areas and during non-office hours (Whitelaw et al. 2003) and they respond to governments' desire to have more stakeholders included in the process (Lawrence & Deagan 2001; Whitelaw et al. 2003). The specific projects of early warning, from pollution to invasive species, enable rapid responses. The inability to avoid invasions and control the existing ones resulted in enormous environmental and economic losses worldwide (Pimentel, 2011) and the costs associated to a false positive identification are much less than the cost of false negatives (Westbrooks, 2004). The need, in this area, of large amounts of data across multiple spatial and temporal scales requires strong collaborations among multiple stakeholders (Lodge et al., 2006). These particular subjects have a good media coverage and projects on these areas attract more participants and have a strong influence in policy decisions.

Conservation of biodiversity has become a major political issue, just like climate change. States are obliged, by international agreements, to implement the Convention on Biological Diversity and several indicators are being developed to achieve the convention's objectives. In France, for instance, the implementation of the indicator "Trends in the abundance and distribution of selected species" is completely dependent on data collected by volunteers, which allows governments to save a significant amount of money (Levrel et al., 2010).

Another example of citizen science projects influencing policy measures comes from the USA. Pond associations from Martha's Vineyard, an island located south of Cape Cod, Massachusetts, had big concerns about water quality, mainly because of local shellfish industry. Numerous dedicated water monitoring initiatives, led by nonprofit organizations, and the partnerships forged with environmental managers in the area, have resulted in policy measures being taken (e.g., pressuring the Board of Health to inspect and replace failed septic systems, address boat related pollution, distributing pamphlets and educating boaters, etc.) and, consequently, in improvements to water quality (Conrad & Hilchey, 2011).

Evaluation

Evaluation of the success of the citizen science initiatives should be taken into consideration to achieve better results in the future (Newman et al., 2012) and contribute to socioecological system resilience (Jordan et al., 2012).

It is quite important to establish evaluation metrics regarding monitoring protocols, to ensure data quality (Engel & Voshell, 2002), to assess the effectiveness of the projects in meeting educational goals (Cohn, 2008; Bonney et al., 2009), to value the scientific outputs that come from the project e.g. scientific publications, and to identify policy goals and decisions influenced by the results.

Evaluating the impacts of citizen science projects on learning can be achieve by selecting appropriate indicators or measures of success, to ensure that the desired outcomes are achieved. Such indicators need to be targeted, feasible, valid, and reliable (Jordan et al., 2012). Though, in some cases there is still lack of effective evaluation mechanisms which can be filled, in educational area, by mechanisms from informal science education (Friedman, 2008).

Governance

The key principles of societal good governance have been categorized by: long-term vision, quality, openness, accountability, effectiveness, and coherence (Socientize, 2014). These principles can be achieved through a strategic commitment of society on citizen science. An urgent need for bottom-up initiatives that address community demands, is important if scientists want a more responsible, proactive and demanding society that uses its rights knowingly. New societies request from the governance area to establish new policies that prioritize science-society-policy interactions, fostering knowledge-based, intelligent and responsible selection choices. So, promoting a democratic governance of science, via public engagement and debate between policy makers, researchers, innovators and the general public in a structured channel for feedback and open criticism is fundamental. All different societal actors should play an important role, adding value to the scientific and social areas of society. Different background knowledge applied to different areas might allow creativity and joint solutions for solving problems.

The way citizens establish their participation commitments was traditionally categorized into top–down and bottom–up governance structures (Conrad & Hilchey, 2011). Lawrence (2006), suggested organizing participation into four forms: consultative (participants contribute with information); functional (participants contribute with information and are also engaged in implementing decisions); collaborative (participants work with governments to decide what is needed and contribute with knowledge) and transformative (participants make and implement decisions with support from experts where needed).

- **Consultative/Functional Governance:** This form of participation is frequently referred as top-down. This case implies that citizen science promoters are asking for help in collecting information or making decisions. The purpose might be to provide early detection of issues of environmental concern, which can then be further investigated by scientific experts. (Whitelaw et al., 2003; Conrad & Daoust, 2008).

An example of the consultative/functional model is the Cornell Lab of Ornithology bird monitoring projects where teams of scientists determine the questions to be answered and decide what segment of the public will be targeted as participants (Ely, 2008a). Most large-scale ecosystem monitoring programs (e.g., bird monitoring programs) tend to be consultative.

- **Collaborative Governance:** In this kind of governance participants might be involved in co-management or adaptive management, if management is part of the goal of the organization (Cooper et al., 2007). In these cases, projects are often governed by a board or group representing as many facets of the community as possible.

An example of a collaborative governance project is the Global Systems Science, which combines advanced information and communications technology with citizen dialogues to understand and shape global systems. This produces evidence, concepts and doubts needed for effective and responsible policies dealing with global systems, engaging citizens into policy processes and process to acquire data. The vision guiding Global Systems Science is to make full use of the progress in information communication technologies to improve the way scientific knowledge can simulate, guide, be used by, and help evaluate policy and societal responses to global challenges like climate change, financial crisis, pandemics, and global growth of cities (http://global-systems-science.eu/).

- **Transformative Governance:** In this case of governance participants make and implement decisions with support from experts where needed. Participants are governed from the "bottom-up", a model often arising out of crisis situations (this may also be called community based, grassroot, or advocacy groups). The group focuses on an issue hoping to initiate government action (Conrad & Daoust, 2008). These type of groups often focus on specific local issues and sometimes have no private sector or government support (Whitelaw et al., 2003). Initiation, organization, leadership, and funding of these groups are provided by the local community (Mullen & Allison, 1999). Emerging alternative funding mechanisms, such as crowdfunding, allow projects to be funded in more direct and democratic way by the public.

The transformative or community-based model has the advantage of involving participants in every stage of the project from defining the problem through communicating the results and taking action. In this case, the role of the scientist is to advise and guide community groups rather than to set their agendas (Ely, 2008b).

Some researchers believe that by transferring authority over decision-making to those most affected by it (the public), better, more sustainable management decisions will be made— thus, making the bottom–up model a desirable type of governance. However, many failures of bottom–up approaches have also been mentioned. These include lack of success due to little organization credibility and capacity (Bradshaw, 2003).

The Global Community Monitor serves as an example of how transformative governance structures can best serve the concerns of a community, although it has evolved into a collaborative framework. It was created to provide community-based tools for citizens to monitor the health of their neighborhoods, with a focus on air quality. One of the organizations in India is the SIPCOT Area Community Environmental Monitors. Villagers have been trained in the science of pollution and have been engaged in environmental monitoring, which over time has led to published scientific reports. This work formed the basis for a Supreme Court order calling for the establishment of national standards for toxic gases in ambient air in State Industries Promotion Corporation of Tamil Nadu (SIPCOT) (Global Community Monitor, 2006).

It is important to take into account that certain governance structures suit different communities and monitoring situations, with collaborative and transformative participation being associated with local scales of participation and consultative and functional participation being more feasible across broader geographic scales (Conrad & Hilchey). Regardless of the fact that different approaches are being held in an exclusive way, some cases of coexistence between different approaches can result in interesting outputs (Lawrence, 2006). Some longer term projects have already changed their typology of governance to better adjust to different scenarios.

RECOMMENDATIONS AND FUTURE RESEARCH DIRECTIONS

Citizen science projects have plenty of social trade-offs that need to be taken in consideration and evaluated when designing. Giving straight information about the goals of the project to stakeholders is fundamental for them not to feel disappointed with the expectations raised. For instance people should know whether the project intends to be rigorous, with a straight scientific approach, or to provide participation from a wide range of volunteers. The outcomes of the project also depend on those trade-offs.

Some of the trade-offs include: deciding the scope and scale of the project, deciding to keep small with local data control, or connect with larger initiatives to benefit data usage, focusing more on guaranteeing data quality and reliability or on the easiness of producing data, with benefits to environmental education and engagement.

Future research should focus on scientists motivations for participating; transdisciplinary relations between stakeholders, from which added benefits may still be further exploited; robustness of data quality and statistical analysis of data; measures of project success, taking into account scientific, policy and social outputs.

CONCLUSION

Citizen science engages the general public with scientific research activities, and while not new, since for centuries scientific research was conducted by amateurs, is becoming a mainstream approach to collect data on a variety of scientific disciplines, much supported by technology advances. Nowadays the focus of citizen science is changing from the traditional "scientists using citizens as data collectors" to citizens as scientists.

Given its collaborative nature, citizen science is characterized by a wide range of stakeholders, ranging from research scientists, teachers, students, managers, environmental organizations, and politicians, whose motivations and interactions can be determinant for the success of a citizen science project and thus should be carefully taken into account on project design.

Despite the considerable amount of stakeholders involved, clustering them into four groups: citizen scientists, scientists, other societal groups and policy makers, allows us to analyze the project design from these four different perspectives, taking a stakeholder view, and identify issues for each group which are common to projects from many different scientific areas.

It is important that the design of a citizen science project takes into consideration issues such as communicating, recruiting and motivating participants, fostering innovation, interdisciplinarity and group dynamics, promoting cultural changes, healthy habits, inclusion, awareness and education, and guiding policy goals, among several others. Analyzing these factors may contribute to the increased success of citizen science initiatives. Some issues, such as motivation or awareness, are important for several stakeholders, but in very different ways and assuming varying degrees of importance.

Around the globe, every day, new citizen science programs are being launched offering:

1. New opportunities for citizen scientists to get involved and increase their scientific literacy
2. New working challenges and opportunities for scientists
3. Chances for rethinking societies and
4. New ways to influence policy makers.

ACKNOWLEDGMENT

P.T. was supported by Portuguese Science Foundation (SFRH/BD/89543/2012).

REFERENCES

Amano, T., & Sutherland, W. J. (2013). Four barriers to the global understanding of biodiversity conservation: Wealth, language, geographical location and security. *Proceedings. Biological Sciences*, *280*(1756), 20122649. doi:10.1098/rspb.2012.2649 PMID:23390102

Ballard, H. L., Trettevick, J. A., & Collins, D. (2008). Comparing participatory ecological research in two contexts: an immigrant community and a Native American community on Olympic Peninsula, Washington. In *Partnerships for empowerment: participatory research for community-based natural resource management*. London, UK: Earthscan.

Bell, S., Marzano, M., Cent, J., Kobierska, H., Podjed, D., Vandzinskaite, D., … Mursic, R. (n.d.). What counts? Volunteers and their organisations in the recording and monitoring of biodiversity. *Biodiversity and Conservation, 17*(14), 3443–3454. http://doi.org/10.1007/s10531-008-9357-9

Bird, T. J., Bates, A. E., Lefcheck, J. S., Hill, N., Thomson, R. J., Edgar, G. J., & Frusher, S. et al. (2014). Statistical solutions for error and bias in global citizen science datasets. *Biological Conservation, 173*, 144–154. doi:10.1016/j.biocon.2013.07.037

Bonney, R., Cooper, C. B., Dickinson, J., Kelling, S., Phillips, T., Rosenberg, K. V., & Shirk, J. (2009). Citizen science: A developing tool for expanding science knowledge and scientific literacy. *Bioscience*, *59*(11), 977–984. doi:10.1525/bio.2009.59.11.9

Bonter, D. N., & Cooper, C. B. (2012). Data validation in citizen science: A case study from Project FeederWatch. *Frontiers in Ecology and the Environment, 10*(6), 305–307. doi:10.1890/110273

Bowser, A., Wiggins, A., Shanley, L., Preece, J., & Henderson, S. (2014). Sharing data while protecting privacy in citizen science. *Interaction, 21*(1), 70–73. doi:10.1145/2540032

Bradshaw, B. (2003). Questioning the credibility and capacity of community-based resource management. *The Canadian Geographer*, *47*(2), 137–150. doi:10.1111/1541-0064.t01-1-00001

Bramston, P., Pretty, G., & Zammit, C. (2011). Assessing environmental stewardship motivations. *Environment and Behavior*, *43*(6), 776–788. doi:10.1177/0013916510382875

Brandon, A., Spyreas, G., Molano-Flores, B., Carrol, C., & Ellis, J. (2003). Can volunteers provide reliable data for forest vegetation surveys? *Natural Areas Journal*, *23*, 254–261.

Bray, G. S., & Schramm, H. L. Jr. (2001). Evaluation of a statewide volunteer angler diary program for use as a fishery tool. *North American Journal of Fisheries Management*, *21*(3), 606–615. doi:10.1577/1548-8675(2001)021<0606:EOASVA>2.0.CO;2

Brossard, D., Lewenstein, B., & Bonney, R. (2005). Scientific knowledge and attitude change: The impact of a citizen science project. *International Journal of Science Education*, *27*(9), 1099–1121. doi:10.1080/09500690500069483

Brown, W. T., Krasny, M. E., & Schoch, N. (2001). Volunteer monitoring of nonindigenous invasive plant species in the Adirondack Park, New York, USA. *Natural Areas Journal*, *21*, 189–196.

Bruyere, B., & Rappe, S. (2007). Identifying the motivations of environmental volunteers. *Journal of Environmental Planning and Management*, *50*(4), 503–516. doi:10.1080/09640560701402034

Buesching, C. D., & Newman, C. (2005). Volunteers in ecological research: Amateur ecological monitors – the benefits and challenges of using volunteers. *Bulletin of the Ecological Society of America*, *36*, 20–22.

Buesching, C. D., & Slade, E. (2012). Citizen science in Wytham Woods: collecting scientific data using amateur volunteers. In D. W. Macdonald & R. E. Feber (Eds.), *Wildlife Conservation on Farmland.* Oxford, UK: Oxford University Press.

Bushway, J., Dickinson, L., Stedman, C., Wagenet, P., & Weinstein, A. (2011). Benefits, motivations, and barriers related to environmental volunteerism for older adults: Developing a research agenda. *International Journal of Aging & Human Development*, *72*(3), 189–206. doi:10.2190/AG.72.3.b PMID:21834387

Carolan, M. S. (2006). Science, expertise and the democratization of the decision-making process. *Society & Natural Resources*, *19*(7), 661–668. doi:10.1080/08941920600742443

Catlin-Groves, C. L. (2012). The citizen science landscape: from volunteers to citizen sensors and beyond. *International Journal of Zoology,* 1–14. http://doi.org/10.1155/2012/349630

Clery, D. (2011). Galaxy Zoo volunteers share pain and glory of research. *Science*, *333*(6039), 173–175. doi:10.1126/science.333.6039.173 PMID:21737731

Clow, D., & Makriyannis, E. (2011). iSpot Analysed: Participatory Learning and Reputation. In *Proceedings of 1st International Conference on Learning Analytics and Knowledge*. doi:10.1145/2090116.2090121

Coghlan, A. (2005). *Towards an Understanding of the Volunteer Tourism Experience.* (Unpublished doctoral dissertation). School of Business, James Cook University, Townsville, Australia.

Cohn, J. P. (2008). Citizen science: Can volunteers do real research? *Bioscience*, *58*(3), 192–197. doi:10.1641/B580303

Conrad, C., & Daoust, T. (2008). Community-based monitoring frameworks: Increasing the effectiveness of environmental stewardship. *Environmental Management*, *41*(3), 356–358. doi:10.1007/s00267-007-9042-x PMID:18026783

Conrad, C. C., & Hilchey, K. G. (2011). A review of citizen science and community-based environmental monitoring: Issues and opportunities. *Environmental Monitoring and Assessment*, *176*(1-4), 273–291. doi:10.1007/s10661-010-1582-5 PMID:20640506

Cooper, C. B., Dickinson, J., Phillips, T., & Bonney, R. (2007). Citizen science as a tool for conservation in residential ecosystems. *Ecology and Society*, *12*, 11–21.

Cooper, S., Khatib, F., Treuille, A., Barbero, J., Lee, J., Beenen, M., & Players, F. et al. (2010). Predicting protein structures with a multiplayer online game. *Nature*, *466*(7307), 756–760. doi:10.1038/nature09304 PMID:20686574

Crall, A. W., Newman, G. J., Jarnevich, C. S., Stohlgren, T. J., Waller, D. M., & Graham, J. (2010). Improving and integrating data on invasive species collected by citizen scientists. *Biological Invasions*, *12*(10), 3419–3428. doi:10.1007/s10530-010-9740-9

Crall, A. W., Newman, G. J., Stohlgren, T. J., Holfelder, K. A., Graham, J., & Waller, D. M. (2011). Assessing citizen science data quality: An invasive species case study. *Conservation Letters*, *4*(6), 433–442. doi:10.1111/j.1755-263X.2011.00196.x

Danielsen, F., Burgess, N. D., & Balmford, A. (2005). Monitoring matters: Examining the potential of locally-based approaches. *Biodiversity and Conservation*, *14*(11), 2507–2542. doi:10.1007/s10531-005-8375-0

Darwall, W. R. T., & Dulvy, N. K. (1996). An evaluation of the suitability of non-specialist volunteer researches for coral reef surveys. Mafia Island, Tanzania – A case study. *Biological Conservation*, *78*(3), 223–231. doi:10.1016/0006-3207(95)00147-6

Devictor, V., Whittaker, R. J., & Beltrame, C. (2010). Beyond scarcity: Citizen science programmes as useful tools for conservation biogeography. *Diversity & Distributions*, *16*(3), 354–362. doi:10.1111/j.1472-4642.2009.00615.x

Dickinson, J. L., Shirk, J., Bonter, D., Bonney, R., Crain, R. L., Martin, J., & Purcell, K. et al. (2012). The current state of citizen science as a tool for ecological research and public engagement. *Frontiers in Ecology and the Environment*, *10*(6), 291–297. doi:10.1890/110236

Dickinson, J. L., Zuckerberg, B., & Bonter, D. N. (2010). Citizen science as an ecological research tool: Challenges and Benefits. *Annual Review of Ecology Evolution and Systematics*, *41*(1), 149–172. doi:10.1146/annurev-ecolsys-102209-144636

Dierkes, M., & von Grote, C. (Eds.). (2000). *Between understanding and trust: the public, science and technology*. Amsterdam: Harwood Academic.

Ely, E. (2008a). Getting volunteer data into science journals. *The Volunteer Monitor*, *19*, 8–11.

Ely, E. (2008b). Volunteer monitoring and the democratization of science. *The Volunteer Monitor, 19*, 1–5.

Engel, S. R., & Voshell, J. R. (2002). Volunteer biological monitoring: Can it accurately assess the ecological condition of streams? *American Entomologist, 48*(3), 164–177. doi:10.1093/ae/48.3.164

Fitzpatrick, M. C., Preisser, E. L., Elison, A. M., & Elkinton, J. S. (2009). Observer bias and detection of low-density populations. *Ecological Applications, 19*(7), 1673–1679. doi:10.1890/09-0265.1 PMID:19831062

Forte, A., & Lampe, C. (2013). Defining, understanding and supporting open collaboration: Lessons from the literature. *The American Behavioral Scientist, 57*(5), 535–547. doi:10.1177/0002764212469362

Friedman, A. J. (Ed.). (2008). *Framework for evaluating impacts of informal science education projects.* The National Science Foundation. Available from: http://insci.org/resources/Eval Framework.pdf

Genet, K. S., & Sargent, L. G. (2003). Evaluation of methods and data quality from a volunteer-based amphibian call survey. *Wildlife Society Bulletin, 31*, 703–714.

Giddings, B., Hopwood, B., & O'Brien, G. (2002). Environment, Economy and Society: Fitting Them Together into Sustainable Development. *Sustainable Development, 10*(4), 187–196. doi:10.1002/sd.199

Global Community Monitor. (2006). Retrieved from www.gcmonitor.org

Greenwood, J. J. D. (2007). Citizens, science and bird conservation. *Journal für Ornithologie, 148*(S1), S77–S124. doi:10.1007/s10336-007-0239-9

Gurin, P. (1999). The compelling need for diversity in education. University of. Massachusetts School of Education Journal, 32, 36-62.

Hess, D. (2010). *Social movements, publics, and scientists. Invited Plenary Lecture.* Tokyo: Japanese Society for Science and Technology Studies.

Hochachka, W. M., Fink, D., Hutchinson, R. A., Sheldon, D., Wong, W. K., & Kelling, S. (2012). Data-intensive science applied to broad-scale citizen science. *Trends in Ecology & Evolution, 27*(2), 130–137. doi:10.1016/j.tree.2011.11.006 PMID:22192976

Janzen, D. H. (2004). Setting up tropical biodiversity for conservation through non-damaging use: Participation by parataxonomists. *Journal of Applied Ecology, 41*(1), 181–187. doi:10.1111/j.1365-2664.2004.00879.x

Jarnevich, C. S., Graham, J. J., Newman, G. J., Crall, A. W., & Stohlgren, T. J. (2007). Balancing data sharing requirements for analysis with data sensitivity. *Biological Invasions, 9*(5), 597–599. doi:10.1007/s10530-006-9042-4

Jordan, R. C., Ballard, H. L., & Phillips, T. (2012). Key issues and new approaches for evaluating citizen-science learning outcomes. *Frontiers in Ecology and the Environment, 10*(6), 307–309. doi:10.1890/110280

Katzenbach, J., Steffen, I., & Kronley, C. (2012). Cultural Change That Sticks. *Harvard Business Review, 90*(7/8), 110–117. PMID:22852451

Koplan, J. P., Liverman, C. T., & Kraak, V. I. (Eds.). (2005). *Preventing Childhood Obesity: Health in the Balance*. Washington, DC: National Academies Press.

Kotler, P., Wong, V., Saunders, J., & Armstrong, G. (2005). *Principles of Marketing* (European Edition). Pearson Education Limited.

Lakshminarayanan, S. (2007). Using citizens to do science versus citizens as scientists. *Ecology and Society, 12. Response, 2*, 1.

Lawrence, A. (2006). No personal motive? Volunteers, biodiversity, and the false dichotomies of participation. *Ethics Place and Environment, 9*(3), 279–298. doi:10.1080/13668790600893319

Lawrence, R. L., & Deagan, D. A. (2001). Choosing public participation methods for natural resources: A context-specific guide. *Society & Natural Resources, 14*, 857–872. doi:10.1080/089419201753242779

Lepczyk, C. A., Boyle, O. D., Vargo, T. L., Gould, P., Jordan, R., Liebenberg, L., & Vaughan, H. et al. (2009). … Vaughan, H. (2009). Citizen science in ecology: The intersection of research and education. *Bulletin of the Ecological Society of America, 90*(3), 308–317. doi:10.1890/0012-9623-90.3.308

Levrel, H., Fontaine, B., Henry, P. Y., Jiguet, F., Julliard, R., Kerbiriou, C., & Couvet, D. (2010). Balancing state and volunteer investment in biodiversity monitoring for the implementation of CBD indicators: A French example. *Ecological Economics, 69*(7), 1580–1586. doi:10.1016/j.ecolecon.2010.03.001

Lodge, D. M., Williams, S., MacIsaac, H. J., Hayes, K. R., Leung, B., Reichard, S., & McMichael, A. (2006). Biological invasions: Recommendations for US policy and management. *Ecological Applications, 16*(6), 2035–2054. doi:10.1890/1051-0761(2006)016[2035:BIRFUP]2.0.CO;2 PMID:17205888

Louv, R. (2005). *Last Child in the Woods: Saving our Children from Nature Deficit Disorder*. Chapel Hill, NC: Algonquin Books.

Lynam, T., Jong, W., Sheil, D., Kusumanto, T., & Evans, K. (2007). A review of tools for incorporating community knowledge, preferences and values into decision making in natural resources management. *Ecology and Society, 12*, 5–18.

Mackney, P., & Spring, N. (2000). *The Earthwatch Perspective on Volunteers and Data Validity*. Oxford, UK: Earthwatch Institute.

Miller, J. D. (2004). Public understanding of, and attitudes toward, scientific research: What we know and what we need to know. *Public Understanding of Science (Bristol, England), 13*(3), 273–294. doi:10.1177/0963662504044908

Miller-Rushing, A., Primack, R., & Bonney, R. (2012). The history of public participation in ecological research. *Frontiers in Ecology and the Environment, 10*(6), 285–290. doi:10.1890/110278

Mullen, M. W., & Allison, B. E. (1999). Stakeholder involvement and social capital: Keys to watershed management success in Alabama. *Journal of the American Water Resources Association, 35*(3), 655–662. doi:10.1111/j.1752-1688.1999.tb03621.x

Musick, M., & Wilson, J. (2007). *Volunteers: A Social Profile*. Bloomington, IN: Indiana Univ. Press.

National Science Board. (2008). *Science and Technology: Public Attitudes and Understanding. Science Engineering Indicators 2008*. National Science Board.

Natural Research Council. (1996). *The National Science Education Standards*. Washington, DC: National Academy Press.

Newman, G., Wiggins, A., Crall, A., Graham, E., Newman, S., & Crowston, K. (2012). The future of citizen science: Emerging technologies and shifting paradigms. *Frontiers in Ecology and the Environment, 10*(6), 298–304. doi:10.1890/110294

NOAA, National Weather Service. (2014). Retrieved from http://www.nws.noaa.gov/os/coop/what-is-coop.html

Nov, O., Arazy, O., & Anderson, D. (2011). Dusting for science: motivation and participation of digital citizen science volunteers. In *Proceedings of the 2011 iConference- Association for Computing Machinery*. doi:10.1145/1940761.1940771

Nov, O., Arazy, O., & Anderson, D. (2014). Scientists@Home: What Drives the Quantity and Quality of Online Citizen Science Participation? *PLoS ONE, 9*(4), e90375. doi:10.1371/journal.pone.0090375 PMID:24690612

Pandya, R. (2012). A framework for engaging diverse communities in citizen science in the US. *Frontiers in Ecology and the Environment, 10*(6), 314–317. doi:10.1890/120007

Penrose, D., & Call, S. M. (1995). Volunteer Monitoring of Benthic Macroinvertebrates – Regulatory Biologists Perspectives. *Journal of the North American Benthological Society, 14*(1), 301–306. doi:10.2307/1467735

Pickard, G., Pan, W., Rahwan, I., Cebrian, M., Crane, R., Madan, A., & Pentland, A. (2011). Time-Critical Social Mobilization. *Science, 334*(6055), 509–512. doi:10.1126/science.1205869 PMID:22034432

Pimentel, D. (Ed.). (2011). *Biological invasions: economic and environmental costs of alien plant, animal, and microbe species*. CRC Press. doi:10.1201/b10938

Pollock, R. M., & Whitelaw, G. S. (2005). Community based monitoring in support of local sustainability. *Local Environment, 10*(3), 211–228. doi:10.1080/13549839.2005.9684248

Pretty, J. (2003). Social capital and the collective management of resources. *Science, 302*(5652), 142–148. doi:10.1126/science.1090847 PMID:14671287

Primack, R. B., Higuchi, H., & Miller-Rushing, A. J. (2009). The impact of climate change on cherry trees and other species in Japan. *Biological Conservation, 142*(9), 1943–1949. doi:10.1016/j.biocon.2009.03.016

Raby, P. (2001). *Alfred Russel Wallace: A Life*. Chatto & Windus.

Raddick, M. J., Bracey, G., Carney, K., Gyuk, G., Borne, K., Wallin, J., & Jacoby, S. (2009). *Citizen science: status and research directions for the coming decade*. Astro2010: The Astronomy and Astrophysics Decadal Survey, Position Papers, no. 46.

Raddick, M. J., Bracey, G., Gay, P. L., Lintott, C. J., Murray, P., Schawinski, K., & Vandenberg, J. et al. (2010). Galaxy Zoo: Exploring the motivations of citizen science volunteers. *Astronomy Education Review*, *9*(1), 18. doi:10.3847/AER2009036

Ratcliff, J. (2008). *The Transit of Venus Enterprise in Victorian Britain*. London: Pickering & Chatto.

Raven, C. E. (1942). *John Ray, Naturalist: his life and work*. Cambridge, UK: Cambridge University Press.

Reed, J., Raddick, M. J., Lardner, A., & Carney, K. (2013) An Exploratory Factor Analysis of Motivations for Participating in Zooniverse, a Collection of Virtual Citizen Science Projects. In *Proceedings of the Hawaii International Conference on Systems Sciences* (pp. 610–619). doi:10.1109/HICSS.2013.85

Robson, C., Hearst, M., Kau, C., & Pierce, J. (2013). Comparing the use of social networking and traditional media channels for promoting citizen science. In *Proceedings of the 2013 conference on Computer Supported Cooperative Work Companion* (pp. 1463-1468). doi:10.1145/2441776.2441941

Rotman, D., Hammock, J., Preece, J., Hansen, D., Boston, C., Bowser, A., & He, Y. (2014). Motivations affecting initial and long-term participation in citizen science projects in three countries. In Proceedings of iConference (pp. 110–124). http://doi.org/ doi:<ALIGNMENT.qj></ALIGNMENT>10.9776/14054

Rotman, D., Preece, J., Hammock, J., Procita, K., Hansen, D., Parr, C., … Jacobs, D. (2012) Dynamic changes in motivation in collaborative citizen-science projects. In *Proceedings of the Association for Computer Machinery Conference on Computer Supported Cooperative Work* (pp. 217-226). doi:10.1145/2145204.2145238

Roy, H., Pocock, M. J. O., Preston, C. D., Roy, D. B., Savage, J., Tweeddle, J. C., & Robinson, L. D. (2012). *Understanding citizen science & environmental monitoring*. NERC Centre for Ecology & Hydrology and Natural History Museum.

Shirk, J., Bonney, R., & Krasny, M. E. (2012). Public Participation in Scientific Research: A framework for intentional design. *Ecology and Society*, *17*(2), 29p. doi:10.5751/ES-04705-170229

Silvertown, J., Buesching, C. D., Jacobson, S. K., & Rebelo, T. (2013). Citizen science and nature conservation. In Key Topics in Conservation Biology (vol. 2, pp. 127-142). John Wiley & Sons, Ltd. doi:10.1002/9781118520178.ch8

Socientize. (2014). *White Paper on Citizen Science for Europe*. European Commission.

Stafford, R., Hart, A. G., Collins, L., Kirkhope, C. L., Williams, R. L., Rees, S. G., & Goodenough, A. E. et al. (2010). Eu-Social Science: The Role of Internet Social Networks in the Collection of Bee Biodiversity Data. *PLoS ONE*, *5*(12), e14381. doi:10.1371/journal.pone.0014381 PMID:21179423

Sullivan, B., Aycrigg, J. L., Barry, J. H., Bonney, R. E., Bruns, N., Cooper, C. B., & Kelling, S. et al. (2014). The eBird enterprise: An integrated approach to development and application of citizen science. *Biological Conservation*, *169*, 31–40. doi:10.1016/j.biocon.2013.11.003

Sullivan, B. L., Wood, C. L., Iliff, M. J., Bonney, E., Fink, D., & Kelling, S. (2009). eBird: A citizen-based bird observation network in the biological sciences. *Biological Conservation*, *142*(10), 2282–2292. doi:10.1016/j.biocon.2009.05.006

Sultana, P., & Abeyasekera, S. (2008). Effectiveness of participatory planning for community management of fisheries in Bangladesh. *Journal of Environmental Management*, *86*(1), 201–213. doi:10.1016/j.jenvman.2006.12.027 PMID:17291678

Theobald, E. J., Ettinger, A. K., Burgess, H. K., DeBey, L. B., Schmidt, N. R., Froehlich, H. E., & Parrish, J. K. et al. (2015). Global change and local solutions: Tapping the unrealized potential of citizen science for biodiversity research. *Biological Conservation*, *181*, 236–244. doi:10.1016/j.biocon.2014.10.021

Tian, H., Stige, L. C., Cazelles, B., Kausrud, K. L., Svarverud, R., Stenseth, N. C., & Zhang, Z. (2011). Reconstruction of a 1910-y-long locust series reveals consistent associations with climate fluctuations in China. *Proceedings of the National Academy of Sciences of the United States of America*, *108*(35), 14521–14526. doi:10.1073/pnas.1100189108 PMID:21876131

Tweddle, J. C., Robinson, L. D., Pocock, M. J. O., & Roy, H. E. (2012). *Guide to citizen Science: Developing, implementing and evaluating citizen science to study biodiversity and the environment in the UK.* Natural History Museum and NERC Centre for Ecology & Hydrology for UK-EOF.

Ulrich, C., & Nadkarni, N. M. (2009). Sustainability research and practices in enforced residential institutions: Collaborations of ecologists and prisoners. *Environment, Development and Sustainability*, *11*(4), 815–832. doi:10.1007/s10668-008-9145-4

Van Den Berg, H. A., Dann, S. L., & Dirk, J. M. (2009). Motivations of adults for non-formal conservation education and volunteerism: Implications for Programming. *Applied Environmental Education and Communication*, *8*(1), 6–17. doi:10.1080/15330150902847328

Vetter, J. (2011). Introduction: Lay participation in the history of scientific observations. *Science in Context*, *24*, 259–280. doi:10.1017/S0269889711000093 PMID:21797076

Weiling, F. (1991). Historical study: Johann Gregor Mendel 1822–1884. *American Journal of Medical Genetics*, *40*(1), 1–25. doi:10.1002/ajmg.1320400103 PMID:1887835

Westbrooks, R. G. (2004). New approaches for early detection and rapid response to invasive plants in the United States. *Weed Technology*, *18*(sp1), 1468–1471. doi:10.1614/0890-037X(2004)018[1468:NAFEDA]2.0.CO;2

Whitelaw, G., Vaughan, H., Craig, B., & Atkinson, D. (2003). Establishing the Canadian Community Monitoring Network. *Environmental Monitoring and Assessment*, *88*(1/3), 409–418. doi:10.1023/A:1025545813057 PMID:14570426

Wiggins, A., & Crowston, K. (2011). From conservation to crowdsourcing: a typology of citizen science. In *Proceedings of the 44th Annual Hawaii International Conference on Systems Sciences* (pp. 1–10). doi:10.1109/HICSS.2011.207

Wooley, A. W., Chabris, C. F., Pentland, A., Hashmi, N., & Malone, T. W. (2010). Evidence for a collective intelligence factor in the performance of human groups. *Science*, *330*(6004), 686–688. doi:10.1126/science.1193147 PMID:20929725

Zoellick, B., Nelson, S. J., & Schauffler, M. (2012). Participatory science and education: Bringing both views into focus. *Frontiers in Ecology and the Environment*, *10*(6), 291–297. doi:10.1890/110277

Chapter 9
Citizen Observatories as Advanced Learning Environments

Josep M. Mominó
Universitat Oberta de Catalunya (UOC), Spain

Jaume Piera
Institut de Ciències del Mar (ICM-CSIC), Spain

Elena Jurado
1000001 Labs, Spain

ABSTRACT

Citizen Observatories are the technological platforms where a diverse range of tools are developed, such as web portals, smartphone apps, electronic devices, that allow the development of citizen science projects, particularly those with the principal objective of large scale participation of the people, covering large geographical areas and long periods of time. These new observatories integrate the latest Information and Communication Technologies (ICT) to connect the citizens digitally, improve their observational capabilities and provide information flows. The concept of Citizen Observatories offers great possibilities as an educational experience, precisely due to the opportunities offered by the participation of the people, with different levels and roles and therefore, it is assumed in terms of active collaboration of the citizens, in shared processes of knowledge creation. This is especially clear when we pay attention to the complexity of the challenges education must face today, within the framework of a society of knowledge like ours.

WHAT ARE CITIZEN OBSERVATORIES?

Citizen science promotes public participation in the collection of large quantities of observations of a very diverse nature (from the identification of new stars or comets to the detection of cancerous cells and the presence of invasive species, to mention a few examples). These observations provide data in a wide variety of settings and during long periods of time. Citizen science projects have had notable

DOI: 10.4018/978-1-5225-0962-2.ch009

success in the advancement of scientific knowledge and the contributions of the people are providing large quantities of data in different scientific disciplines. An essential requirement for the functioning of citizen science is the participation of the people, in an extensive and sustained manner. Citizen Observatories are the research infrastructures (i.e. the technological platforms where a diverse range of tools are developed such as web portals, smartphone apps, electronic devices) that allow the development of citizen science projects, particularly those with the principal objective of large scale participation of the people, covering large geographical areas and long periods of time. These new observatories integrate the latest information and communication technologies (ICT) to connect the citizens digitally, improve their observational capabilities and provide information flows.

The Citizen Observatories therefore offer an environment where people can participate in citizen science projects. According to an original idea by Arnstein (1969) different levels of participation can be defined, considering to the social and cognitive involvement of the people. In an adaptation of this idea, within the context of citizen science, Haklay (2011) defined the hierarchy of participation with four levels of social and cognitive involvement:

- **Crowdsourcing:** At this level, people generate information passively with little cognitive involvement. For example, people are invited to wear a sensor for a time, after which it is returned to the organisers of the experiment. The information generated is subsequently analysed by specialists in the subjects being studied.
- **Distributed Intelligence:** At this level, the central resources of the project are the cognitive abilities of the people. Most current citizen science projects fall within this level. The people who participate often receive training, which may be in person or using didactic resources accessible on the web. People who can provide information through observations made or through the interpretation of existing information (for example, validating observations made by others).
- **Participative Citizen Science:** This level is characterised by the problem being defined by the community itself. In some cases, it may be derived from an evolution of the projects from the previous level, when the people that have participated have acquired sufficient expertise in the collection and analysis of data, as well as the ability to think up new questions to solve.
- **Collaborative (or Extreme) Citizen Science:** This is the most integrated level where the people participate at all levels. It is not only possible to think up new questions, like at the previous level, but the people can also participate in the design of the methods for acquiring the observations and analysing the information.

Aside from the levels of participation, the Citizen Observatories also offer an environment where the people can collaborate with different roles of participation (see Figure 1):

- **Makers:** These are people with abilities and interests in the technological field. In this role, the people participate by designing new observation instruments and tools, often with low-cost material that is easy to acquire, promoting the concept of *Do-It-Yourself* (DIY).
- **Observers:** These are the people who provide observations, either using existing technology or technology designed by the makers.
- **Analysers:** In this role, the people participate by interpreting or validating existing information. With this analysis it is possible to improve the quality of the information provided by the observers.

Figure 1. The different roles that can be associated with the design, use and exploitation of the data from the DIY KdUINO buoy, designed within the CITCLOPS project.

The roles in Citizen Observatories: An example in the European project CITCLOPS. *One of the first European Citizen Observatories was developed within the context of the project CITCLOPS (Citizens' Observatory for Coast and Ocean Optical Monitoring). In this project, the creation of an observatory was promoted to evaluate water quality, using observations provided with the participation of the people measuring simple optical parameters such as the colour, transparency or fluorescence of the water (Wernand, et al. 2012). One of the devices designed to measure water transparency was the KdUINO buoy (Bardaji, et al. 2016). This easy-to-build DIY buoy made it possible to develop the idea of the different roles of participation. People who wanted to participate in the role of Makers could develop their own KdUINO buoy and modify it to improve its features, as the system was developed in a fully open environment (open hardware and software). In the role of Observers, anybody who, due to their frequent contact with the sea (whether leisure or professional activities), would like to take on the installation and maintenance of a KdUINO buoy can participate. The data collected by these buoys is stored in open servers and is accessible anyone interested in using it. It is in this latter context where the role of the Analysers comes into play. The people participating in this role can verify that the data being obtained is correct or can be used (for example in school workshops) to see the differences in the water transparency at different locations and environments of the coast (to ask for example "Is the water transparency different when the coast is rocky or sandy?") or analyse the changes that occur throughout the year in a specific location ("how and why does the water transparency change in the different seasons of the year?").*

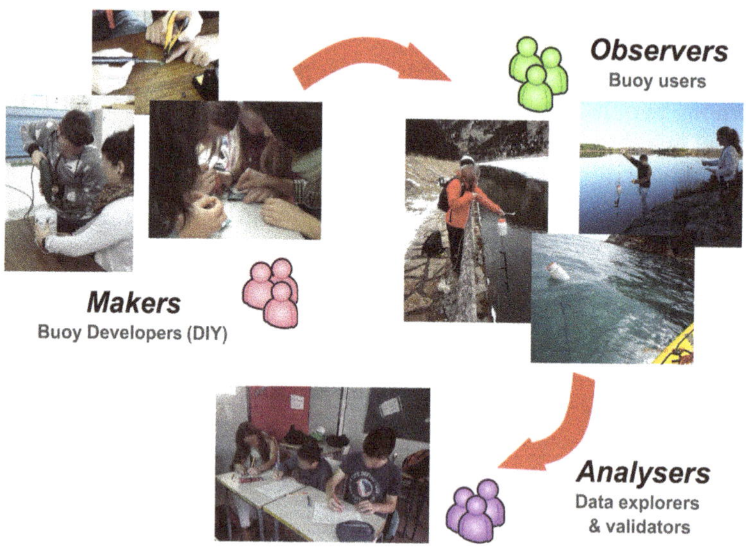

This concept of science offers great possibilities as an educational experience. This potential for education comes from the opportunities offered by active collaboration of the citizens, with different levels and roles in shared processes of knowledge creation. This is especially clear when we pay attention to the complexity of the challenges education must face today, within the framework of a society of knowledge like ours.

RETHINKING THE EDUCATIONAL SCENARIOS, IN THE SOCIETY OF KNOWLEDGE

We live in a world that is transforming profoundly at a very high speed. The parameters that industrialisation established are being left further and further behind. Our societies are finding new methods of organisation and cultural expression, and new mechanisms of economic production. This reconfiguration, the *network society* (Castells, 2000) builds its structure through information and communication technology (ICT) and is driven by a new form of economy, based on the production of knowledge, with the ability to learn as a primary resource. Innovation and creativity (Florida, 2004), finally, have become the essential fuel for the engine driving this process.

In a globalised world, the governments understand more and more that the dynamics of the economy of knowledge depend more than ever on effective organisation of education (Dale, 2005) that, consequently, must be contemplated as a strategic priority, in order to ensure sustainable growth and social cohesion. The question is that, in the current climate, the dynamics of the organisations and, ultimately, the work of the people is converted, more and more, into the resolution of complex problems with, consequently, the ability to generate and handle information with sophisticated procedures with dexterity when participating in flexible and open structures of collaboration in networks. It also consists of forming relevant questions and learning to resolve them, through forms of systematic reasoning, making use of the enormous potential provided by ICT for the resolution of these processes and, finally, having a high level of initiative, autonomy and critical spirit in decision-making.

The need to find models with the effective ability to form the learners of the new millennium (OECD, 2009) and give them the skills the XXI Century requires (Partnership for the 21st Century Skills, 2005) goes hand in hand with a strong demand for transformation into the distinctive ranges of the traditional education systems. There is a requirement of "rethinking education" (UNESCO, 2015), reconsidering its purpose and knowing how to provide the most appropriate educational scenarios. The need for change refers to the fundamental principles with which it is necessary to design the learning environments (Dumont et al, 2010). The revision must pay attention to what needs to be taught and to the type of skills that must be provided nowadays for a person's integral education. It must also be translated, consequently, into an update of the way they are taught. For this advanced concept of education, finally, education centres have to be able to provide appropriate organisational formulas. This organization should be able to promote and facilitate the development, in their everyday activity, of the teaching formulas and, finally, the learning experiences required by a society like ours.

The measurement of a challenge that considers a change of this type in schools has to be calibrated taking into account the complexity of the factors involved. The education centres gradually change when updating forces them to change the DNA that sustains the way the activity takes place in their classrooms on an everyday basis (Cuban, 1986; 2001; 2012). The traditional concept of teaching practices is structured on a "school grammar" that, in this regard, acts in a similar way to the way it would in language regulation (Hargreaves, 2000; Tyack and Tobin, 1994). The ability to update must always be measured taking into account the significant pitfall that must be overcome to act outside of this "grammar" and leave behind organisational methods and practices that were strongly established in the preceding educational models. Finding the appropriate strategies to make headway in this process of "creative destruction" (Kozma, 2012) often constitutes the main challenge for essential change. Our intention is to show that, within the concept of the citizen science, schools can find an interesting alternative when embarking on this trajectory. The condition is that they are able to approach this concept making the most of the potential of participation in the Citizen Observatories as a new educational scenario. The opportunity isto advance towards a new "ecology of learning" (Barron, 2006; Coll, 2013, 2016) that occurs in different parameters from those that set out traditional teaching practices.

CITIZEN OBSERVATORIES: AN OPPORTUNITY TO RETHINK EDUCATION FOR SUSTAINABLE DEVELOPMENT

The educational priorities in the political agendas are closely linked to the social, cultural, environmental and economic needs of each era. Industrial society promoted the model of a transmissive and academicist school and established a marked dichotomy between the cognitive, emotional and ethical dimensions in the processes of teaching and learning. The current society of knowledge is facing the challenge of promoting a more holistic model of teaching, with a humanist focus. This perspective is aimed at "Sustaining and enhancing the dignity, capacity and welfare of the human person, in relation to others and to nature" (UNESCO, 2015, 36). From this angle, education can be seen as a process that goes beyond the mere acquisition of knowledge and aims to facilitate the development of skills for life.

The integrated, humanist focus of education that was proposed in the influential report by Delors (1996) conceptualised the purpose of education and the organisation of learning into four fundamental pillars: learning to know, learning to do, learning to be and learning to live together. The same report already underlined that learning to know and, to a lesser extent, learning to do, are the pillars that have traditionally received more attention in formal teaching. The other two types of learning depend more on random circumstances and they are often considered a natural extension of the first two. For an advanced education, however, each one of the four pillars of education must warrant the same consideration so that education is a global experience for the person, providing the integral training necessary to be able to participate actively as responsible citizens in the society of knowledge when faced with the complexity of a globalised world.

Currently, fully immersed in the society of knowledge, in a world that is highly interdependent, interconnected and under pressure, especially in matters of sustainability, these four pillars remain valid, but the need to reinterpret them has arisen. This needs to "rethink education" in accordance with the latest reflection of UNESCO (2015) emphasises sustainability as a worrying concern for global development that has to be incorporated into an integrated, humanist focus of education. The growing concern for sustainability fully affects the reinterpretation of one of the four pillars of the Delors report: learning to live together must go further than the social and cultural dimensions of human interactive, to include a concern for the relationship of society with the natural environment.

The current dominant model of economic development entails the degradation of vital natural resources such as water and the loss of biodiversity. Similarly, the growing superpopulation of cities has consequences for the natural environment, and climate change has brought an increase in natural disasters. It is, therefore, essential to educate citizens to be responsible for economic growth guided by responsible environmental management and social justice. This should be a fundamental aim of education in the 21th century (UNESCO, 2015).

In a context of active concern for sustainable development, Citizen Observatories can provide a valuable educational platform for obtaining life skills. There the citizens have the opportunity to actively participate in the process of scientific research and work for sustainability, collecting, interpreting or analysing data. At the same time, these technological platforms offer a good opportunity to go beyond the traditional concept of the digital divide with a more complex overview about the use of technologies by young people. From this perspective, participation in Citizens Observatories can be understood as an experience that can contribute to reducing digital inequalities (DiMaggio and Hargittai, 2001). The incorporation of these platforms for educational purposes offers an opportunity to pay attention to the multiple factors that affect the process of appropriation of technology in everyday life of young people. Reviewing the limitations of a simple division between users and non-users (Lenhart & Horrigan, 2003; Steyaert, 2002) allows to adopt a broader view of the social implications of the use of technology (Robinson, DiMaggio, & Hargittai, 2003 ; van Dijk and Hacker 2003; DiMaggio et al. 2004; Warschauer 2003). The incorporation of these platforms allows the school to influence the multiplicity of variables involved in the unequal appropriation of technology (van Dijk, 2005; DiMaggio, Hargittai, Celeste & Shafer, 2004). In this sense, participation in Citizen Observatories can be seen as an opportunity to open a space in the school for compensation of digital inequalities, paying attention to the unequal conditions of access, motivations, abilities, and purposes of use and contribute, from these platforms, to the digital inclusion of young people (Meneses & Mominó, 2010).

Citizen Observatories provide a bridge connecting learners with their environment, promoting active participation. They are also platforms that enable the cultivation of values such as solidarity and shared responsibility for our joint future. From this point of view, it is possible to understand the interest in seeing these observatories as environments for teaching and learning that actively contribute to educating citizens concerned about sustainable development, competent in using tools to characterise the current situation of the ecosystems and their future evolution, and able to collaborate with other learners and with entities involved in knowledge generation.

CITIZEN OBSERVATORIES: A PLATFORM FOR AN ADVANCED CONCEPT OF TEACHING AND LEARNING

The concept of citizen science to which we refer offers favourable ground when promoting this humanist vision of education. This conception of science enables connections to be made with a holistic representation of education in which knowledge is understood as a joint process and the people who participate in this collective construction can learn bringing significance to the experience, processing information and recombining it, but above all deepening understanding and, at the same time, developing skills, and acquiring attitudes and values. The Citizen Observatories can offer a platform for the generation of this integrated way of understanding knowledge. Those who contribute can learn to use it, collaborating from diverse positions, developing the necessary skills to apply this knowledge in complex contexts, such as those provided by the citizen science project, to respond to specific, relevant demands.

Anyway, the ability of the Citizen Observatories to provide an advanced environment for learning does not depend solely on the potential we have just mentioned. The result of the educational experiences carried out with these platforms ultimately depend on the ability to guide the participation of the different parties involved taking into account the knowledge accumulated on the way people learn. For this reason, the pedagogic design of the activity carried out in these observatories, within the framework of the citizen science project, ends up being fundamental. The research by the OECD (2010) on the nature of learning synthesised some of the principles that, from this point of view, should enable support for the configuration and dynamics of learning environments in the XXI Century. The revision of these principles, applied to Citizen Observatories, must allow us to rethink the structure and functioning of these platforms, with the aim of getting maximum effectiveness in educational terms from them.

- *Citizen Observatories: A Learning Experience with the Students in the Lead Role*

The socio-constructivist concept (Bransford et al., 2006) that, today, prevails when interpreting how learning occurs reveals that in this construction process, nobody can replace the learner. Therefore, the students have to be at the centre of these platforms, it must be possible to find the possibility of being actively involved in projects that connect with their interests, stimulating their commitment and promoting the active exploration of knowledge. Appropriate activity design, in the observatories, must be able to offer the opportunity to develop metacognitive habits, related to the processing of information, and the analysis, evaluation and diffusion of knowledge. It must facilitate the acquisition of mechanisms for establishing and monitoring personal objectives linked to the project in which they are participating, for their own time management and organisational ability. Ultimately, the Citizen Observatories have to provide an environment that provides the gradual growth of the students' ability to self-regulate and, in short, their autonomy.

- *Citizen Observatories: A Space for Learning through Cooperation*

This is especially valuable when we take into account that, from the same socio-constructivist concept to which we have referred and in keeping with the contributions of neuroscience in particular, we can underline the fact that learning is a shared construction process that occurs through social interaction. Thus, the fact that participation, through different levels of social and cognitive implication, as we have highlighted in the first section of this chapter, is a defining aspect of these observatories is a fundamental factor for the construction of these platforms as an advanced learning environment. The effectiveness of cooperative work, however, with regard to obtaining knowledge, but also on a level of behaviour, emotions and development of values, ends up depending on the pedagogic design of the activity. The dynamics of cooperative work has to be well-organised in order to provide learning opportunities for all of the students, establishing not only group, but also individual, objectives and evaluating them.

- *Citizen Observatories: Motivation and Emotions in Generating Knowledge*

This is essential for learning (Pekrun et al., 2007). The experience of participating in citizen science is a magnificent opportunity for young people when identifying the relevance of what they want to learn. This is especially valuable in an environment such as sciences in which, all too often, schools find themselves with difficulties connecting the activity that takes place in the classroom with life skills. Participation in Citizen Observatories helps the students to make sense of their learning, connected to specific purposes of participation in citizen science. The real value that young people can attribute to the data collected and the perception of their effective contribution to the body of knowledge facilitates the interaction between motivation, cognition and emotion that ends up being critical for the quality of the learning.

- *Citizen Observatories: A Space Open to Participation of All Students*

The individual differences that we find in the classrooms are very important and cannot be ignored when the aim is learning that can be significant for all students, not just a few. Young people do not only have different interests and motivations, they also start with diverse prior knowledge, skills and even cultural contexts. Participation in citizen science projects, however, allow on one hand different levels of participation that, as we have seen, provide different levels of social and cognitive implication. On the other hand, participation in these observatories also enables those involved to adopt alternative roles for collaboration, able to adapt to the differences to which we refer. From this point of view, versatility is another of the Citizen Observatories' strengths as a learning environment.

- *Citizen Observatories: A Challenge for Each Student.*

The range of options that can be provided by these platforms, when attending to the differences we have just mentioned, also have to be able to be deployed in the form of diverse challenges. Each goal should be adapted to the characteristics of each student, linked to their interests and thus, finally, stimulating their implication and personal commitment when achieving their own objectives. The potential of the Citizen Observatories must also be seen from this point of view. It is about paying attention to the flexibility that these environments can provide when setting appropriate milestones for each student that allows them to gradually self-regulate their participation and gain a degree of personal autonomy in a motivating, and at the same time demanding, context of collaboration with classmates and supervision by the teachers.

- *Citizen Observatories: A Process of Ongoing Assessment*

The quality of the activity carried out at these observatories is closely linked to the assessment mechanisms incorporated. In this regard, the teaching staff's role is fundamental. Beyond the definition of the objectives and expectations for each student, as we have just mentioned, the contribution of the teachers providing feedback at all stages of this project is fundamental in two ways. On one hand, feedback on the activity of each student must provide an essential formative feedback to be able to gradually adapt their actions, gain autonomy and develop the skills necessary to contribute to the activity of the observatories more and more efficiently. On the other hand, this ongoing assessment process that must be carried out by the teachers must also result in a review of the observatory's activity as a whole. The assessment of this dynamic must allow the design to be adjusted in such a way that provides better performance in terms the learning results and, at the same time, quality in the contribution of the observatory to the citizen science project.

- *Citizen Observatories: A Permeable Environment*

Finally, we must bear in mind that the activity of the Citizen Observatories, by their very nature, cannot be developed in a closed environment. The very concept of citizen science, based on ideas of extensive, large-scale participation, advocates forms of activity based on collaboration in a network of diverse participants with different roles and even geographical locations. Thus, the incorporation of Citizen Observatories into school activity can effectively contribute to the promotion of schools. This is a fundamental challenge for some institutions that, all too often, have functioned as a closed site, if they aim to offer education able to respond to the challenges that arise in the knowledge society, in a globalised world. Furthermore, the ability to be open that Observatories can also bring must be able to be contemplated in an internal sense, paying attention to the opportunities offered by participation in the citizen science projects for the horizontal connectivity between diverse areas of knowledge. Understanding that the ability to establish this type of transverse connections helps to develop more profound forms of knowledge enables the potential of Citizen Observatories as a learning environment to be seen once more.

Ultimately, it is important to note that it is the combination of these principles, more than the emphasis on one of them, that gives these learning environments their power (OECD, 2010, 2012). Similarly, when setting them in motion, Citizen Observatories can be used from different pedagogic focuses. Let's look at some:

CITIZEN OBSERVATORIES: AN OPPORTUNITY FOR LEARNING BASED ON RESEARCH

Research-based learning occurs when the learners participate in projects using methods inherent in scientific research. Thus, the learners question, investigate, consider hypotheses, test, analyse and communicate. This type of research-based project provides the opportunity to participate in real-world situations, which has shown its positive effect in terms of depth and meaning of the learning that can be obtained (Barron & Darling Hammond, 2012). Citizen Observatories have the potential to provide environmental data from the users, which can benefit the sustainability of their environment. This data, duly filtered, can provide decisive information for the construction of models that may allow the evolution of ecosystems to be predicted. Thus the users of the Citizen Observatories can feel empowered through their participation in the fabric of scientific research and, to a certain extent, of environmental government.

Ideally, and with the attentive guidance of an educator, the learners learn to consider questions and projects that are relevant to them. The result may be learner-based learning that can take into account the individual differences, as the design of these projects can contemplate the possibility of participation respecting different paces and different methods and paying attention to different contents. The active intervention in this type of authentic experiences provides better opportunities for more in-depth and meaningful learning due to the importance that young people may attribute to these processes and the involvement they get out of them, in emotional and motivational terms. In this regard, the participation in Citizen Observatories as a research-based learning experience brings into play the principles that come together in the nature of learning (OECD, 2010).

Participation in citizen science projects, through these Observatories, with a research-based learning focus, provides a magnificent opportunity for the development very diverse skills, including independence, creativity, critical thinking, communication, collaboration and persistence when faced with uncertainty (Dumont et al., 2010). Research-based learning, similarly, requires students to know how to develop the skills to filter and discriminate an avalanche of often unreliable information that can be accessed on the Internet. It is important to bear in mind that all of these types of learning can be incorporated into what we have come to call XXI Century skills (Partnership for the 21st Century Skills, 2005). From this perspective, research-based learning enables the students to perceive the creation of knowledge as an open process that is not only in the hands of the teachers, but also that it is built collectively and acquired throughout life in diverse situations.

Citizen Observatories, when conceived as environments for research-based learning, incorporate mechanisms for cooperation and have ICT tools that enable communications within the framework of the centre where they are located, but also beyond their limits. Thus collaboration between learners is sought, but also with various other parties involved, linked to research, paying attention, specifically, to the social nature of learning and the process of knowledge generation. In the same way, this type of environment must provide a high potential for the development of strategies of systematic work and analysis, of metacognitive skills over the learning process itself, but also for creativity. In short, the participation in research-based environments allow the deployment of in-depth forms of reasoning that give young people the opportunity to develop higher-level skills in knowledge generation.

This is how, in schools, with resources and an appropriate pedagogic design, Citizen Observatories can provide suitable environments for developing research-based learning. The strength of the observatory as a learning environment is built with a sufficient degree of structuring, with suitable mechanisms for support and guidance for the research and analysis, able to guarantee desirable results and ambitious products. The design and development of these environments require a high level of involvement, the ability to manage resources and, principally, educational evaluation instruments able to identify the range of skills set into motion in these research-based environments, aiming to go further than what can be grasped using traditional evaluation systems. In this regards, another fundamental factor for the correct functioning of these experiences comes to light: the teaching staff who must design and promote these environments must have the necessary training to be able to achieve the best yield in educational terms.

CITIZEN OBSERVATORIES AS A LEARNING ECOSYSTEM

Immersed as we are in the society of knowledge, the new economic and social stage has brought with it a paradigm shift in the way people learn. Many of the factors on which learning processes have traditionally been based are gradually becoming blurred. The time and place we learn, how we act during this process, the parties involved that we interact with, and even what we learn and the purpose we give to learning has taken on a new configuration. It is more and more palpable that learning occurs and will occur throughout life, but also that it extends, in a transverse sense, across the breadth of life, in new learning scenarios that are ever more modelled by digital ICT. This technology and in particular mobile technology such as smartphones, laptops and tablets with wireless connections are playing a decisive role in the definition of new contexts of activity that provide new opportunities and resources when learning. ICT plays a role in both in the support of new niches of learning and in the reinforcement of the more tradition contexts of learning and, ultimately, they are paying a decisive role in the mediation of the learning process itself. All of these substantial modifications situate us in what some authors have called a new ecology of learning (Barron, 2006; Coll, 2013)

These great lines of change in the way people learn require new interconnections between parties involved in education, new ways to adapt to the new configuration of the ecosystem of learning. Before the complex challenge schools must face in these circumstances, Citizen Observatories can provide the types of educational experience, in line with the paradigm shift to which we are referring and, in this way, they can provide learning environments able to respond to the education challenges posed by the society of knowledge. Citizen Observatories are built on technological platforms, through vertical spaces, making use of mobile technology. Thus, they redefine the possibility of collaborating with the community and cultural or leisure institutions for learning through a process of shared knowledge construction. It is in this regard that these Observatories, accommodated in or closely linked to schools, may be especially relevant within the contemporary context of reconsidering the interconnections of schools with their environment. At a time when many schools still base their educational model on the traditional dynamic classroom-centred and teacher-student knowledge acquisition, there is a need for distributed educational models able to contemplate the extent of the network of learning contexts through which the students will pass and integrate others into non-traditional learning environments. This change of focus responds to the vision of education as ongoing, where formal education centres can interact more closely with other, less formal, educational experiences. This opening in the network allows schools to get the best out of their social and cultural capital with a more permeable position to network collaboration.

Furthermore, in a scenario like this, in which people are faced with learning needs in more and more diverse moments and contexts, the acquisition of the types of skills that enable learning in this diversity of situations has special importance. Citizen Observatories, in this regard, can offer a favourable platform for the acquisition of this type of skills, stimulating the ongoing learning of the students and, at the same time, other essential capabilities such as creativity, critical thinking, independence and persistence, among others. In these spaces, more than the product created, attention should be directed at the process and the skills required for the creation of engines for obtaining environmental data or for the observation or analysis of this data. Furthermore, participation in the Observatories through different learner profiles (makers, observers and analysers) provides the opportunity for everyone to go a step further in their learning paths, according to their own pace and trajectory. The confluence of these learning principles in the participation of the Observatories in the learning ecosystems again show the contribution they can make for the configuration of advanced learning environments.

The possibility of participating in the Observatories through diverse profiles makes it possible to connect with another of the effects of the modifications of the learning ecosystem to which we refer: the diversification of the learning trajectories adhering to the ever wider range of contexts and experiences that people go through in their learning trajectory. The needs to recognise these individual trajectories and rethink the personalisation of this learning (Coll, 2013) is a significant challenge for formal education institutions. When moving the focus of attention towards this diversity of trajectories, Citizen Observatories can again provide an appropriate platform, to the extent that through their design the emphasis is also placed on knowing how to accompany, in personalised manner, diverse itineraries and ways of participation to educate competent people, able to learn to learn in an ubiquitous manner, from different situations and positions. From this perspective, schools have to be able to maintain their central role, interconnecting the learning contexts through which the individual students pass.

Also from this point of view it can be understood how it is important to properly integrate Citizen Observatories into schools, through appropriate collaboration formulas. This process of integration does not happen easily; it requires an effort in the design, organisation and management. In contrast, it is easy for the Observatories to end up acting as independent centres, with an autonomous dynamic disconnected from the everyday activity of the centre as a whole. In this case, we have an annex that provides little added value to the school. The potential of Observatories, in these circumstances, is wasted. To boost the interconnections necessary, it is important that the projects carried out are interdisciplinary and favour transverse connections with the content of the curriculum in the different areas of knowledge. For this reason, the education centres need to be organised with sufficient versatility, able to approach these Observatories properly to promote good use for an advanced concept of the educational processes. The participation of the teaching staff in this process is fundamental and requires and education that allows them to properly guide the activity of these environments with educational purposes. Fundamental guidance is necessary when promoting the investigative spirit and research skills in these spaces. Ultimately, the pedagogic design of this process also has to incorporate appropriate education evaluation methods that reveal and contribute to the development of the whole range of skills that the Observatories put into motion, monitoring the activity as a whole to readjust it and appropriately redirect the individual learning trajectories.

It is in this case that Citizen Observatories can provide an educational experience adapted to the new ecology of learning, able to be deployed beyond the walls of the school, through projects that capture the interest of the students with problems that are relevant to their surroundings.

CITIZEN OBSERVATORIES: AN OPPORTUNITY FOR LEARNING AND SERVICE

In a world where technification, urbanisation and, in general terms, the models of economic consumption and production are helping to distance citizens from their environment, weakening the institutions before the large economic corporations, the purpose of education must be to guide towards the value of sustainable human and social development once more (UNESCO, 2015). As a response to the ecological tension, it reveals the importance of responsible action of the people towards the environment and active participation of young people in projects that improve the social, economic and environmental conditions of their surroundings. Projects working towards sustainability are especially important, in circumstances like those we are experiencing in which the degradation of natural resources is a fact and environmental awareness is a necessity. In this regard, it is essential to educate powerful citizens with an active role in environmental governance.

The projects of learning and services are designed so that learners undertake actions for the good of the community, using knowledge to participate in authentic situations and complex social problems. These projects can entail a wide range of social questions (environmental, health, human needs, multicultural issues, etc.), and a service of direct action in the community or indirect action (performing a study to improve a community problem, for example). It is however a requirement that in this type of projects there is equilibrium between learning academic knowledge and serving the community. If it were only a case of undertaking projects for the good of the community it would be more voluntary work than learning and service that should benefit both participants and end-users.

In participating in projects of learning and service, the students are involved in the problems of their surroundings. The community gives them the magnificent opportunity to participate in authentic learning situations that allow them to put into motion their skills, to design and apply solutions to real-life problems, linked to the immediate context or to society in a more general sense. Participation in these projects often results in greater empowerment and awareness of the possibilities of active participation in social problems and, definitively, a greater degree of commitment in civic and political matters (OECD, 2015).

Beyond what it contributes through the acquisition of these values, participation in this type of projects, with the appropriate design, can also lead to the confluence of the principles involved in significant learning. The students must find the opportunity to leave behind their traditional role of passive receiver to adopt a leading role as a creator, designer and active supplier of specific actions (Cairn and Kielsmeier, 1991) that can be as relevant for the community as for the students themselves. With this socioconstructivist concept, knowledge is constructed or reconstructed by the subject itself that learns through the action and social interaction, in a building process shared and guided by the teaching staff. In keeping with the significant range of skills that these processes put into motion, it is not unexpected that, currently, participation in initiatives of learning and service are gaining weight in education systems all over the world, in both the context of the OECD and beyond (Furco, 2010).

Experiences of learning and service can be generated through very diverse disciplines. They can incorporate students at very different educational levels and can be linked to a wide range of social questions, including sustainability and environmental tensions, without doubt, a challenge for participation from an educational setting. Citizen Observatories, from this point of view, can offer a valuable alternative when responding to this challenge. Participation in citizen science projects allows the Observatories to adopt this concept of learning and service and, from these platforms, design learning environments in which school activities can be linked to specific social problems, for example those related to maintaining the balance of aquatic ecosystems to which we referred to previously. In this case, it is about participating in the evaluation of the problems that threaten them, such as pollution or overuse of resources, and that impact the measures that can be carried out in order to develop more sustainable activities.

The link of the Observatories to this type of project generates a type of activity that must be deployed transversally, through different disciplines. The interaction that occurs in these environments, when linked to authentic problems that have real consequences for the community and for the complexity inherent in these situations, must be possible to expand beyond the limits of the education centres. Thus it has been possible to establish links for collaboration with diverse entities and parties involved, through various types of networks, for the configuration of which ICT is an essential ally. Participating in these processes, students can deepen their knowledge, discovering the link between concepts and procedures in diverse areas of knowledge and, same time, they can develop attitudes and values of civic commitment, regarding problems that affect the dynamics of daily life in their surroundings. Participation in these projects makes it possible for learning to occur, making sense of the experience, in a process in which the link between the cognitive and emotional aspects are particularly highlighted.

The adoption of Citizen Observatories in education centres for the development of learning and services projects can incorporate the guidance of the teaching staff, necessary for the design of the learning environment, for the correct implementation of the project and, in short, to be able to get the most out of it in terms of education and community service. For this reason, however, it is fundamental for the teaching staff to be able to design and apply the appropriate evaluation mechanisms to measure the learning that participation in these projects allows us to contemplate in its holistic sense. Ultimately, the possible performance of Citizen Observatories, linked to this type of project, also depends on the flexibility of their internal organisation for interdisciplinary collaboration and, at the same time, on the ability to open up education centres for collaboration with the environment and making the most of its social and cultural capital.

CITIZEN OBSERVATORIES: A PROPOSAL FOR COOPERATIVE LEARNING

As sustained, extensive participation is a fundamental component of citizen science and taking into account that this participation in Citizen Observatories is classified into different levels and possibilities of the social and cognitive involvement of the participants, it is important to pay attention to the possibilities provided, in educational terms, by this dynamic of collaboration that is consubstantial to these projects. Primarily, it is important to note that the possibilities of interaction between people and the group work of the different parties involved, in themselves, do not entail any benefit in educational terms. The characteristics taken on by this working dynamic directly affect the quality of the results in terms of learning. In this regard, solid evidence enables us to ensure that these positive results are only obtained in certain circumstances that, only when they occur, allow us to speak of "cooperative work". In fact, the knowledge available (Slavin, 2010) on the effective contribution of this form of collaboration in the construction of knowledge, but also for motivation and social cohesion, does not correspond sufficiently, in general terms, with the level of implementation it has in the everyday activity of education centres. Participation in Citizen Observatories provides a magnificent opportunity to benefit from the possibilities of collaboration between the different participants, if this working dynamic is designed and implemented effectively in terms of cooperation.

Thus it is important to understand that participation in citizen science projects, through the different levels possible (Arnstein, 1969), requires the activity to be correctly designed so that it properly guides the interaction of the different parties involved, taking into account their different roles: makers, observers or analysers. The responsibility of the teaching staff is fundamental in this regard, when achieving effective collaboration of students in cooperative terms. For this purpose, it is essential for the activity to be considered so that each of the participants has to be aware that the result of their own work is important, but no more so than that of the rest of their colleagues with whom collaboration is established (Webb, 2008). In this sense, the activity must be designed and guided by the teaching staff in such a way that the objective cannot be deemed completed simply by having completed a certain joint observation task. The aim must also refer to what the team is able to learn together. Thus the success of the team must be linked to the result obtained by the group as a whole, and also what each of the members has learned. The design of the evaluation tools that adequately contemplate this double aspect is a necessary mechanism to guarantee the functioning of cooperation in the Observatories, the performance of the collaborators and, ultimately, the quality of the contribution to the citizen science projects.

Furthermore, it is also important to bear in mind that cooperative work is only effective when all members of the teams can find opportunities to collaborate. The activity design must contemplate this requirement and make the most of the different roles from which it is possible to participate in the Observatories to attend to the individual differences. With appropriate guidance by the teaching staff, in keeping with their interests and learning style, each one can find their place as makers, observers and analysers and interact in an interdependent manner. The methodology of the cooperative working established by the teaching staff has to set clear and ambitious objectives and for the teams, as well as personalised objectives for each of the participants, in keeping with their different roles. Ultimately, to guarantee effective cooperation, the evaluation criteria must establish that the success of the team collaborating with the Observatory depends on the individual learning of each of the participants.

Cooperation, furthermore, should not only be established among diverse roles, the activity of the Observatories must also be able to be designed so that the work of those who collaborate within the same role is also carried out in collaboration. In the case of makers, for example, their cooperation must occur in the process of designing new observations tools and instruments. The teaching staff will have to facilitate cooperation in keeping with the interests, often of a technological nature, of these makers. Their guidance must consist of promoting, on one hand, the concept of *Do-It-Yourself (DIY)*, providing autonomy, self-regulation and entrepreneurship for each person. At the same time, however, the teaching staff must also ensure the cooperation is able to lead to a veritable process of co-creation. With a well-defined work strategy and the appropriate evaluation instruments, Citizen Observatories can offer cooperative environments in which *do-it-yourself* (DIY) can give way to *do-it-together* (DIT) (Hagel, Brown, & Davison, 2010).

The opportunities that Citizen Observatories provide as an environment of cooperative working are not reduced to the dynamic established within the framework of the education centre itself; cooperation can also extend beyond the limits of the school. The information generated from the technological platforms in the Observatories are disseminated via the Internet and is used by other Observatories. Thus, collaboration can be established between different education centres and, similarly, with other types of public or private institutions that, with a well-designed work strategy, can participate in effective processes of cooperation in networks. Thus different Citizen Observatories and different schools can learn together, acting collaboratively in joint projects and establishing effective links in terms of social cohesion. This extensive, complex concept of cooperation that allows participation in Citizen Observatories constitutes another aspect of these environments that enable us to harness the potential to obtain the type of skills needed in a society as complex, interconnected and interdependent as ours.

CITIZEN OBSERVATORIES: LEARNING WITH TECHNOLOGY

Citizen Observatories, by definition, incorporate a wide range of diverse technology types that enable us to improve the skills of the participants, in both acquisition and transmission of information. One of these general objectives, shared by all citizen science projects, is that both quantity and the quality of the observations reported by the citizen scientists gradually improve. There is an intrinsic interest in the people participating acquiring, over time, better abilities in the process of obtaining information needed. The technology linked to Citizen Observatories therefore contributes to the creative learning environments from which the Observatory itself benefits, as the people participating in the project become more expert.

It is necessary to analyse, therefore, the different technology that can be used in the educational processes, within the context of the observatories. In studies carried out previously to evaluate the various technological options for learning, Graesser and colleagues (Graesser et al., 2008; Graesser & King, 2008) suggest ten genres of technology-based learning environments, which we will analyse below within the context of Citizen Observatories:

- **Computer-Based Training:** In this case, learning occurs through classes, experiments and comments presented on a computer screen. This is often done in a format of progressive learning in which the student moves on to the next section after passing a test verifying that the knowledge presented in the current section has been assimilated.
- **Multimedia:** In this field, training is given through images (for example, illustrations, photographs, animation and/or video) and words (such as printed or spoken text). This format enables information of a certain complexity to be transmitted in a format that facilitates understanding. It can be very useful in the context of the Citizen Observatories, for example, to explain to the participants with the role of *Makers* the different steps to build a DIY instrument using videos that explain the process. It can also be useful in other roles: images and explanations on how to distinguish, for example, a species of interest, in the case of the *Observers*, or how to validate an observation or apply an analysis method, in the case of the role of *Analysers*.
- **Interactive Simulation:** In this type of simulation, the student has some degree of control for example establishing input parameters and observing what happens. It may be of special interest in the case of the *Analysers* who can present predictions regarding the variation of the results in different user-controlled, simulated scenarios.
- **Hypertext and Hypermedia:** In this field, the educational material is presented in a format with which the student can have a certain control over the flow of information, navigating through links and references.
- **Smart Tutorials** *:* These learning systems monitor the knowledge of the student and adapt the contents presented in accordance with the results. With this system, the diversity of knowledge between the students is taken into account, so the training is made more flexible and adapted according to the degree of expertise of the users.

- **Obtaining Information Through Searches:** This is currently one of the most widely used systems for accessing new sources of information, in an environment as dynamic as the web.
- **Animated Teaching Agents:** In this case, characters are implemented that appear on the screen and help guide the student through a computer-based lesson. In the context of Citizen Observatories, these characters can be designed related to the target topic of study.
- **Virtual Environments with Agents** *:* In general, these are realistic environments that simulate interaction with real characters, often using natural language. Sklar and Richards (2010) identify three types of agents in virtual environments:
  - *Pedagogic agents*, that are often presented as narrators that offer voice-over explanations without appearing in the scene and that are activated when the students (directly or indirectly) indicate they need help.
  - *Peer-learning agents* that are presented and interact with the students as if they were classmates, tutors or instructors.
  - *Demonstrative agents*, very different to the other two types, these agents are used above all in the context of learning by doing, one of the contexts most used by *Makers*.

For example, demonstrative agents allow the students to program using simple commands and verifying graphically, at that moment, the effects of their code. This does not only allow the students to learn the concepts of programming, but it also provides them with lessons of a wide-reaching scope in process modelling.

- **Serious Games:** This term is used to define games aimed at incorporating an educational function. Bowser, Hansen and Preece (2013) reviewed the different options of gamification in citizen science projects, signalling the potential of games in both learning and complementary systems to increase participation.
- **Learning Supported by Collaboration Technology:** In this field, the groups of students work together on a task communicating through computers. This is one of the fields that enables the integration of activities developed for the different roles (*Makers*, *Observers* and *Analysers*) that participate in the observatories. It is in this environment where it is possible to develop the new concept of Collaborative Science of *Do-It-Together* (DIT) (Hagel, Brown, & Davison, 2010) integrating the expertise of each one of the roles.

Despite the range of tools that can favour educational processes in the context of Citizen Observatories, it is important to bear in mind some of the barriers and challenges that can affect some of these learning processes.

On one hand, the motivation and preparation of the students. There is a growing concern regarding the preparation of enough students, teachers and professionals in the areas of technology, engineering and mathematics (STEM) (Fairweather, 2008; Osborne & Dillon, 2008). A large majority of secondary school students do not manage to achieve the skills in mathematics and science, in a context in which these subjects are often given by teaching staff who do not have sufficiently up to date knowledge to teach these disciplines.

On the other hand, it is important to bear in mind the fact that the ability to participate actively in technological activities requires time, knowledge and the economic resources to access or purchase equipment, etc. These aspects can mean that some people are excluded due to a lack of confidence in their skills, in the educational models or in the resources available. These potential problems, however, can be considered a challenge to change in the future. Citizen Observatories can constitute innovative platforms that act as catalysts and favour the incorporation of the transverse principles that must guide the creation of the learning environments of the XXI Century.

CITIZEN OBSERVATORIES: THE CHALLENGE OF RENEWING EVERYDAY EDUCATION PRACTICES

Ultimately, however, it would be naïve to wait for our expectations of the enormous potential of Citizen Observatories, as an environment for the development of the skills required by the XXI Century to give results through the mere incorporation of these platforms into primary and secondary schools. Demanding that impact, as if it were a switch that could be activated in any circumstances, can only be based on an absolute lack of knowledge of the functional dynamics of education centres and teaching itself. In reality, making use of all that these observatories can provide, in terms of the effectiveness of the education action and finally the improvement in the nature of the learning that the students can obtain, is almost never a straightforward process. The difficulty lies in the complexity that, all too often, education centres find in adapting themselves to these Observatories with an advanced vision of what the school has to teach and the young people have to learn in this society of knowledge.

Thus, the effect Citizen Observatories can have on the everyday activity of schools must be linked to their ability to deploy education centres when building an appropriate context for the parameters in which the educational action must be situated within a society such as ours. The construction of this space, without doubt, requires more and more flexible structures with an open capacity for collaboration in networks in projects with a variable extent and configuration, such as those generated around the Citizen Observatories. Similarly, it requires greater opportunities for the initiative and creativity of all the participants: students, teaching staff and other parties involved, whatever their role level of participation. The challenge consists of knowing how to link collaboration in citizen science projects with of holistic concept of education, able to break the dichotomy between cognitive, emotional and ethical aspects. It is therefore about promoting scenarios where the students can put into motion the skills they need for life, participating in situations that may be relevant, such as those these projects can offer.

Subordinating the organisation of schools to this concept of educational aims is a complex undertaking. Generating the conditions that must enable the promotion of advanced learning environments such as those that can be generated from the Citizen Observatories have into account at least three conditions of possibility (OECD, 2015):

1. **Formulas of School Organisation at the Service of the Nature of Learning:** It would not make sense for this adaptation to occur the other way around, even though in reality in the case of schools this logic is contradicted all too frequently. The organisational formulas of the education centres must be subordinate to the needs considered by the design, development and, let's not forget, sustainability and, therefore, the continuity of the learning experiences proposed by the participation in Citizen Observatories. It is about rethinking the dynamics between the core elements of teaching practices to optimise them according to the knowledge available, incorporating advanced formulas for a formative evaluation of the skills in play, versatile formulas for the organisation of

the teaching staff and students, more flexible structures for the use of space and time that facilitate the establishment of the necessary transverse connections for solid learning.

2. **Leadership for Learning:** Promoting the development of these advanced environments requires forms of leadership (Pont et al., 2008) with a high level of commitment to facilitate the best conditions possible for the learning of all the students. Thus, the effective incorporation of citizen science projects into schools also depends on ability of the education centres when developing this vision. This capacity should, ultimately, result in the deployment of the strategies necessary to advance towards schools, conceived as learning organisations (Senge, 2006). In this case, the centres have a shared vision of the teaching staff as a whole and a high capacity for teamwork when designing and sustaining the dynamics of advanced learning environments such as those that can be developed from the Citizen Observatories.
3. **Schools as Open Organisations:** Schools often carry out their activity with limited contact with their environment. This closed position has led to a certain separation between the activity occurring in the classrooms and the everyday life of the young people outside of school. In the network society, however, the spatial limits of educational activity are gradually being blurred (Davidson & Goldberg, 2010). In the new social order, the distinction between those who have the knowledge, those who generate it and those who transmit it are harder and harder to perceive. A profusion of diverse players build their artefacts and cultural products into the network. This context offers great opportunities for schools to access a significant range of environments and learning resources and also for the possibility of collaborating with diverse institutions and players The challenge of schools and their professionals is to reverse their traditional position of closure, know how to decode the potential of the environment and the players to generate different collaboration projects that can result in the improvement of the education quality. In this regard, it is necessary to understand collaboration in Citizen Observatories as a magnificent opportunity to use and expand their social, cultural and professional capital through collaboration with the diverse parties involved and institutions linked to these projects.

Schools need to know how to meet these challenges, not just to be able to harness the enormous potential that Citizen Observatories provide as a learning environment, but also to be able to properly position itself as an organisation able to respond to the challenges posed by education in the XXI Century and, at the same time, able to offer a reference that may still be valid for young people, within the framework of the society of knowledge.

REFERENCES

Arnstein, S. R. (1969). A ladder of citizen participation. *Journal of the American Institute of Planners*, *35*(4), 216–224. doi:10.1080/01944366908977225

Bardaji, R., Sánchez, A. M., Simon, C., Wernand, M. R., & Piera, J. (2016). Estimating the Underwater Diffuse Attenuation Coefficient with a Low-Cost Instrument: The KdUINO DIY Buoy. *Sensors (Basel, Switzerland)*, *16*(3), 373. doi:10.3390/s16030373 PMID:26999132

Barron, B. (2006). Interest and self-sustained learning as catalysts of development: A learning ecologies perspective. *Human Development*, *49*(4), 193–224. doi:10.1159/000094368

Barron, B., & Darling Hammond, L. (2010). Prospects and challenges for inquiry-based approaches to learning. In H. Dumont (Ed.), *The Nature of Learning. Using research to inspire practice*. Paris: OECD. doi:10.1787/9789264086487-11-en

Bowser, A., Hansen, D., & Preece, J. (2013, April). Gamifying citizen science: Lessons and future directions. In *Workshop on Designing Gamification: Creating Gameful and Playful Experiences.*

Bransford, J., Vye, N., Stevens, R., Kuhl, P., Schwartz, D., Bell, P., & Sabelli, N. et al. (2006). Learning Theories and education: Toward a Decade of synergy. In P. A. Alexander & P. H. Winne (Eds.), *Handbook of Educational Psychology* (2nd ed.; pp. 209–244). Mahwah, NJ: Lawrence Erlbaum Associates.

Cairn, R., & Kielsmeier, J. (1991). Growing Hope: A Sourcebook on Integrating Youth Service into the School Curriculum. National Youth Leadership Council.

Castells, M. (2000). *The rise of the Network Society*. Malden, MA: Blackwell Publishers.

Coll, C. (2016). La educación formal en la nueva ecología del aprendizaje: tendencias, retos y agenda de investigación. In J.M. Mominó & C. Sigalés (Eds.), El impacto de las TIC en educación. Más allá de las promesas. Barcelona: Editorial UOC.

Cuban, L. (1986). *Teachers and machines. The classroom use of technology since 1920*. Teachers College Press.

Cuban, L. (2001). *Oversold & Underused. Computers in the classroom*. Cambridge, MA: Harvard University Press.

Cuban, L. (2012). *Dilemes polítics i docents de l'ús de les TIC a l'aula. El cas dels Estats Units*. Barcelona: Fundació Jaume Bofill. Retrieved from http://www.debats.cat/sites/default/files/debats/pdf/dilemes-politics-docents.pdf

Dale, R. (2005). Globalisation, knowledge economy and comparative education. *Comparative Education, 41*(2), 117–149. doi:10.1080/03050060500150906

Davidson, C. N., & Goldberg, D. T. (2010). *The future of thinking. Learning institutions in a digital age*. Cambridge, MA: The MIT Press. Retrieved from https://mitpress.mit.edu/books/future-thinking

Delors, J. (Ed.). (1996). Educació: hi ha un tresor amagat a dins. Informe de la UNESCO de la comissió internacional sobre Educació per al s. XXI. Barcelona: Centre UNESCO de Catalunya.

DiMaggio, P., & Hargittai, E. (2001). *From the "digital divide" to "digital inequality": Studying Internet use as penetration increases*. Working Paper 15. Center for Arts and Cultural Policy Studies. Retrieved from http://www.princeton.edu/~artspol/workpap15.html

DiMaggio, P., Hargittai, E., Celeste, C., & Shafer, S. (2004). From unequal access to differentiated use: A literature review and agenda for research on digital inequality. In K. M. Neckerman (Ed.), *Social inequality* (pp. 355–400). New York: Russell Sage Foundation.

DiMaggio, P., Hargittai, E., Celeste, C., & Shafer, S. (2004). From unequal access to differentiated use: A literature review and agenda for research on digital inequality. In K. M. Neckerman (Ed.), *Social inequality* (pp. 355–400). New York: Russell Sage Foundation.

Dumont, H., Istance, D., & Benavides, F. (2010). *The Nature of Learning. Using research to inspire practice*. Paris: OECD.

Fairweather, J. (2008). *Linking evidence and promising practices in science, technology, engineering, and mathematics (STEM) undergraduate education*. Washington, DC: Board of Science Education, National Research Council, The National Academies.

Florida, R. (2004). *The Rise of The Creative Class*. New York: Perseus Books Group.

Furco, A. (2010). The community as a resource for learning: an analysis of academic service-learning in primary and secondary education. In H. Dumont (Ed.), *The Nature of Learning. Using research to inspire practice*. Paris: OECD. doi:10.1787/9789264086487-12-en

Graesser, A. C., Chipman, P., & King, B. G. (2008). Computer-mediated technologies. In Handbook of research on educational communications and technology, (pp. 211-224). Academic Press.

Graesser, A. C., & King, B. (2008). Technology-based training. *Human Behaviour in Military Contexts*, 127-149.

Hagel, J., Brown, J. S., & Davison, L. (2010). *From Do It Yourself to Do It Together*. Retrieved February 3, 2016, from https://hbr.org/2010/02/from-do-it-yourself-to-do-it-t.html

Haklay, M. (2011). *Citizen Science as Participatory Science*. Retrieved February 3, 2016, from https://povesham.wordpress.com/2011/11/27/citizen-science-as-participatory-science/

Hargreaves, A. (2000). Four ages of professionalism and professional learning. *Teachers and Teaching: History and Practice, 6*(2), 151-182.

Kozma, R. B. (2012). *Les TIC i la transformació de l'educació en l'economia del coneixement*. Barcelona: Fundació Jaume Bofill. Retrieved from http://www.debats.cat/sites/default/files/debats/pdf/kozma.pdf

Lenhart, A., & Horrigan, J. B. (2003). Re-visualizating the digital divide as a digital spectrum. *IT & Society, 1*(5), 23–39.

Meneses, J. & Mominó, J.M. (2010). Putting Digital Literacy in Practice: How Schools Contribute to Digital Inclusion in the Network Society. *The Information Society: An International Journal, 26*(3), 197-208. 10.1080/01972241003712231

OECD. (2009). *21st Century Skills and Competences for New Millennium Learners in OECD Countries*. OECD Education Working Papers, No. 41. OECD Publishing.

OECD. (2010). *The nature of learning. Using research to inspire practice*. OECD Publishing. Retrieved from http://www.oecd.org/edu/ceri/thenatureoflearningusingresearchtoinspirepractice.htm-3

OECD (2012). *Better Skills, Better Jobs, Better Lives: A Strategic Approach to Skills Policies*. OECD Publishing. 10.1787/9789264177338-en

OECD. (2015). *Innovative Learning Environments: Implementation and change*. OECD Publishing. Retrieved from http://www.oecd.org/edu/ceri/The ILE project.pdf

Osborne, J., & Dillon, J. (2008). *Science education in Europe: Critical reflections* (Vol. 13). London: The Nuffield Foundation.

Partnership for the 21st Century Skills. (2005). *A Report on the Landscape of 21st Century Assessment.* Washington, DC: Partnership for the 21st Century Skills.

Pedró, F. (2006). *The New Millennium learners: Challenging our Views on ICT and Learning*. Paris: OECD – CERI.

Pekrun, R., Renzel, A. C., Goetz, T., & Perry, R. P. (2007). Theoretical perspectives on emotion in education. In P. Schutz, R. Pekrun, & G. Phye (Eds.), Emotion in Education. Academic Press.

Pont, B., Nusche, D., & Moorman, H. (2008). *Improving School Leadership: Policy and Practice*. OECD. Retrieved from http://www.oecd.org/edu/school/44374889.pdf

Robinson, J. P., DiMaggio, P., & Hargittai, E. (2003). New social survey perspectives on the digital divide. *IT & Society*, *1*(5), 1–22.

Senge, P. (2006). *The Fifth Discipline. The Art and Practice of the Learning Organization*. New York: Currency Doubleday.

Sklar, E., & Richards, D. (2010). Agent-based systems for human learners. *The Knowledge Engineering Review*, *25*(02), 111–135. doi:10.1017/S0269888910000044

Slavin, R. E. (2010). Co-*operative* learning: what makes group-work work? In H. Dumont (Ed.), *The Nature of Learning. Using research to inspire practice*. Paris: OECD. doi:10.1787/9789264086487-9-en

Steyaert, J. (2002). Inequality and the digital divide: Myths and realities. In S. F. Hick & J. G. McNutt (Eds.), *Advocacy, activism and the Internet* (pp. 199–211). Chicago: Lyceum Press.

Tyack, D., & Tobin, W. (1994). The grammar of Schooling: Why has it been so hard to change? *American Educational Research Journal*, *31*(3), 453–480. doi:10.3102/00028312031003453

UNESCO. (2015). *Rethinking Education - Towards a global common good?* París: UNESCO.

van Dijk, J. A., & Hacker, K. (2003). The digital divide as a complex and dynamic phenomenon. *The Information Society*, *19*(4), 315–326. doi:10.1080/01972240309487

van Dijk, J. A. G. M. (2005). *The deepening divide: Inequality in the information society*. Thousand Oaks, CA: Sage.

Warschauer, M. (2003). *Technology and social inclusion: Rethinking the digital divide*. Cambridge, MA: MIT Press.

Webb, N. (2008). Co-operative Learning. In T.L. Good (Ed.), 21st Century Education: A Reference Handbook. Sage.

Wernand, M. R., Ceccaroni, L., Piera, J., & Zielinski, O. (2012). Crowdsourcing technologies for the monitoring of the colour, transparency and fluorescence of the sea. In Proceedings of Ocean Optics XXI (pp. 8-12).

Chapter 10
The Role of Citizen Science in Environmental Education:
A Critical Exploration of the Environmental Citizen Science Experience

Ria Ann Dunkley
Cardiff University, UK

ABSTRACT

Citizen Science is increasing in popularity and used by many academics, community groups and Non-Governmental Organizations in scientific data collection. Despite this, little is known about the motivations and experiences of those who contribute to citizen science projects, nor about the impacts of involvement in citizen science upon the individual. Moreover, few have considered the pedagogic process that individuals undergo as they participate in these activities. Citizen science practitioners and program developers stand to benefit from increased understanding of these experiences in terms of their capacity to enhance environmental education. Such increased understanding of the implications of citizen science may also promote the development of sustainability education. This chapter synthesizes insights from existing literature, policy documents and practical projects to explore the pedagogic potential of the convergence of citizen science and environmental education. The chapter concludes that progressive evaluation approaches are needed to complement what is an emergent field.

INTRODUCTION

This chapter will explore the role of citizen science within environmental education or education for sustainable development, as it is also know. On the one hand, it will examine the motivations of scientists for developing environmental citizen science programs. It will also address what they perceive the motivations of those who contribute to citizen science projects to be. On the other hand, it considers the motivations of individuals who become involved in environmental citizen science programs. This chapter will explore the place of citizen science initiatives within the lives of those who choose to participate

DOI: 10.4018/978-1-5225-0962-2.ch010

within them. This is a rarely considered topic within the field of citizen science studies. However, this is perhaps unsurprising given the fact that studies of citizen science are a relatively recent research development. Within this chapter, it is argued that considering the motivations and experiences of the individuals who contribute to environmental citizen science projects is essential to understanding the role of citizen science within sustainable development, as Irwin (1995) originally set out to achieve.

THE APPEAL OF INVOLVING PUBLICS IN SCIENTIFIC RESEARCH

Citizen Science projects have grown rapidly since the mid-1990s. Involving publics in research, through citizen science, enables scientific institutions to expand their scientific endeavors. Twenty-first century technological advances are seen as tools to enable collaborative projects to be ever more ambitious. The current emphasis within science and society on 'big data', which involves collecting data across spaces and time spans previously unthinkable, means that there are ever more opportunities to contribute to global and significant research projects. Individuals can contribute, for example, to online projects like E-bird (http://ebird.org/content/ebird/) an online citizen science initiative. E-bird is an ornithology program, launched in 2002 by the Cornell Lab of Ornithology and National Audubon Society. The project receives over five-million contributions per month (Bonney, Shirk, Phillips, Wiggins, Ballard, Miller-Rushing and Parrish, 2014). Online, citizen science projects, such as E-bird, can be engaged with regardless of geographical location. They are, therefore, able to include a limitless number of participants as contributors, due to the technological advances of the latter half of the Twentieth and Twenty-First Century. Such projects become part of the expansion of scientific endeavor, which is portrayed as a benefit to all human beings, due to the capacity to 'do' science, at ever-larger scales. Therefore, technological innovations are often considered a driver of citizen science within the present day.

Nevertheless, the appeal to scientists of involving publics in scientific research predates the emergence of the internet. Indeed, environmental citizen science has evolved within disciplines that have traditionally depended upon contributors to help facilitate research processes. These include, for example, ornithology, paleontology and atmospheric science (Bonney et al., 2014). Currently, citizen science is regarded as incredibly important to environmental conservation research (Dickinson, Zuckerberg, & Bonter, 2010; Dickinson and Bonney, 2012 and Dickinson, Shirk, Bonter, Bonney, Crain, Martin, & Purcell, 2012 and Johnson, Acton, Popovici, Karanth, & Weinthal, 2014). It contributes to the study of a diverse range of ecological fields ranging from macro-ecology to landscape ecology and forest ecology to urban ecology, while land managers and conservationists, policy makers and activists widely use the results of such studies in practical settings (Bonney, Cooper, Dickinson, Kelling, Phillips, Rosenberg and Shirk, 2009).

Furthermore, for some, citizen science goes beyond merely being a method of collecting data. For these individuals, it is a revolutionary activity, capable of affecting how the environment is managed. For example, Cooper, Dickinson, Phillips and Bonney (2007) suggest that harnessing citizen science represents a 'new frontier to advance the theory and practice of conservation in residential ecosystems' (p. 8). They suggest that this is possible because of the scale upon which citizen science makes it possible to operate. Far from being an activity once reserved for English Gentlemen who considered natural history as a hobby, twenty-first century citizen science is regarded to be open to all amateur observers, irrespective of knowledge, background or social status.

CLASSIFYING CITIZEN SCIENCE EXPERIENCES: AN ORGANIZATIONAL APPROACH

Wide varieties of organizations are involved in the running of citizen science initiatives. Table 1 is an attempt to represent these various organizations. In doing so, it is possible to represent the nuances between various kinds of citizen science projects, including the drivers for their initial set-up. It also helps to provide a basis upon which it is possible to suggest a range of motivations and experiences of individuals who participate in different kinds of project. This typology builds upon others conceptual frameworks presented within the literature that have sort to typify citizen science. These include, that proposed by Wiggins and Crowston (2011: p. 2) whose typology describes citizen science projects 'by primary goal orientation and degree of virtuality'. More recently, Shirk, Ballard, Wilderman, Phillips, Wiggins, Jordan, and Bonney (2012) defined citizen science projects based on how involved contributors were in the entire process of research within a citizen science project. A final example is a typology proposed by Haklay (2013). Within this typology, Haklay (2013) describes four levels of participation in citizen science. These levels include:

- **Level One**: Crowd-sourcing,
- **Level Two**: Distributed intelligence,
- **Level Three**: Participatory science and
- **Level Four:** Extreme citizen science.

Table 1. Types of environmental citizen science initiatives

Types of Environmental Citizen Science Initiatives	Examples of Citizen Science Initiative
School or college -based citizen science projects	• Journey North's Tulip Test Gardens (https://www.learner.org/jnorth/tulip/index.html) • Seeds in Space (https://schoolgardening.rhs.org.uk/news/News-results/National/2015/May/rocket-science) • Microverse (http://www.nhm.ac.uk/take-part/citizen-science/microverse.html)
Citizen science projects based at eco-attractions	• Projects hosted by the Natural History Museum includes: Orchid Observers (https://www.orchidobservers.org/) • Bioblitz, run by the National Park Service in the US (https://www.nps.gov/subjects/biodiversity/national-parks-bioblitz.htm).
Citizen science projects run by conservation charities	Projects run by the Woodland Trust, UK: • Natures Calendar include: (http://www.naturescalendar.org.uk/) • and • The Big Bluebell Watch (http://www.woodlandtrust.org.uk/visiting-woods/bluebell-watch/) • Project run by the Royal Society for the Protection of Birds in the UK, for instance, The Big Garden Bird Watch (https://ww2.rspb.org.uk/discoverandenjoynature/discoverandlearn/birdwatch)
Citizen science projects organised by Non-governmental organisations	An example of such an international charity is Earthwatch (http://eu.earthwatch.org/scientific-research/our-approach-to-research-citizen-science)
Local programs organized by the community, the council or other entities	• Projecte Rius (http://www.projecterius.cat/) • Citizen Crane (http://www.cranevalley.org.uk/projects/citizen-crane.html)
University led programs citizen science programs	• YardMap (Cornell Lab of Ornithology and funded by the National Science Foundation (NSF)) (http://content.yardmap.org/) • Project Splatter, Cardiff University, UK (https://projectsplatter.co.uk/report-some-data-to-project-splatter/)

As participants progress through the citizen science projects that they are involved in, the degree to which they progress may increase. They may, for instance, become more interested in being involved in the entire scientific process as time passes.

In contrast to these pre-existing frameworks, Table 1 presents an alternative way to categorize citizen science projects. This approach is based upon the organizational context within which such projects take place. In the section that follows, each of the categories will be described in turn. It is important to note that the categories are not mutually exclusive. There are, for instance, many instances of hybrid projects, which involve many organizations in their running. Examples include the OPAL project (http://www.opalexplorenature.org/), which involves contributors in conducting ecological surveys that include a range of species, including lichens and earthworms. The project places community scientists in museums across the UK to engage wider publics, but primarily large numbers of schoolchildren. The Opal Project is made up of a consortium of several partners, including Imperial College London, University of Birmingham, Natural History Museum and The Met Office. Another example, of such a hybrid project is Track-a-Tree (http://trackatree.bio.ed.ac.uk/). Track-a-Tree is a phenology project that looks for the first signs of spring in trees and flowering plants. It is a collaboration of the University of Edinburgh; the Woodland Trust; the British Ecological Society and is supported through funding from the Natural Environment Research Council (NERC). Individuals may also contribute to projects within more than one category, simultaneously. Alternatively, they may become involved in differing citizen science projects at different stages of their lives.

The first category presented in Table 1 is school-based citizen science projects. Students of all ages, at all levels of study, from pre-school to college level may be involved. Schoolchildren may be formally required to participate, as part of their curriculum. However, some individuals at later stages of their education, for example, at a post-compulsory level may be motivated by the desire to supplement formal education. The typical primary emphasis of such projects is science education and engagement. Thus, projects may vary in the extent to which the scientific data gathered is used within formal scientific research. An example of a school based citizen science project is *Seeds in Space* (https://schoolgardening.rhs.org.uk/news/Newsresults/National/2015/May/rocket-science). This UK-based mass observation exercise compares how seeds grow in space, as opposed to on school grounds. Astronaut Tim Peak launched this project to coincide with his recent space mission. It is a collaborative project between the Royal Horticulture Society, the UK Space agency and thousands of schools across the UK.

The second category of citizen science project detailed in Table 1 concerns projects based at eco-attractions. Eco-attractions include museum, arboretums, botanical gardens, national parks, urban parks and zoos (Dunkley, 2016). They are organizations that are involved in both engaging the public in learning experiences and encouraging appreciation of the natural world (Davis 1996). They often have large public engagement programs and host vast numbers of school, college and community groups, through specialist programs. Individuals who participate in citizen science through eco-attractions may do so to supplement their learning through informal, experiential learning (Kolb, 1984). Eco-attraction-based citizen science is inclusive of individuals of all ages, from those involved in formal education to life-long learners, who might perhaps be in their retirement. These individuals may also be motivated to interact with organizations that they respect and want to support in their conservation efforts. They may also be motivated by that fact that collecting citizen science data for these organizations enables them to enjoy the natural surrounds of these sites. An example, an eco-attraction-based citizen science initiative is the *Orchid Observers* project (https://www.orchidobservers.org/), run by the Natural History Museum, in London, UK. The project asks contributors to photograph wild orchids. It also asks contributors to help classify 300 years of orchid records that the museum holds. This specific project contributes to climate change research. Participants may consider the projects role in monitoring environmental change to be a worthy endeavor. They may want, therefore, to participate in a program on this basis.

A third category of organization involved in running citizen science projects are conservation charities. Organizing citizen science initiatives may enable conservation charities to reach their aims on a number of levels. For example, developing citizen science programs may help them to gather data on ecological threats to the species that they seek to protect. This will assist them in gathering wider support from both publics and policy makers. Citizen science initiatives may also be looked upon by such organizations as an effective means of engaging publics. An example of a citizen science project, run by a conservation charity in the UK, would be *Natures Calendar* (http://www.naturescalendar.org.uk/). *Natures Calendar* is a longstanding citizen science program, run by the Woodland Trust. Similarly to its sister project listed above (*Track-a-Tree*), it is a phenology project. It, therefore, looks at the effects of climate upon natural phenomena.

A fourth category of citizen science utilizing organizations includes Non-government organizations (NGOs). Such organizations may be concerned with environmental issues that have global implications. Examples might include climate change, deforestation and biodiversity loss. Setting up citizen science initiatives enables NGOs to collect data at vast scales, in a cost effective manner. Thus, this may help them to achieve their goals, in that the research insights gained through citizen science projects may help to support the claims they make in lobbying government. Individual contributors may seek to participate within such projects because they agree with the values that the particular NGO seeks to promote. An example, of such an NGO is *EarthWatch* (http://eu.earthwatch.org/scientific-research/our-approach-to-research-citizen-science), an organization that seeks to support local-level citizen science projects. They seek to effect environmental change at a global level by supporting local people all over the world to tackle local environmental issues through environmental monitoring.

A fifth category concerns programs coordinated by communities, councils, or other entities that are stakeholders within their local environment. Such projects may emerge from a concern for the local environment. For example, they may, be developed in response to a pollution breach. This was the case for the *Citizen Crane* (http://www.cranevalley.org.uk/projects/citizen-crane.html), a citizen science project run by the Crane River Partnership in London. The project monitors pollution levels in a river tributary. A further example, is *Projecte Rius* (http://www.projecterius.cat/) in Spain, a project that organizes contributors for monitoring pollution in local rivers. Individuals who choose to participate in these projects may be motivated to contribute to scientific data collection on local issues. They may be driven to do so because of their commitment to the places within which they live.

The sixth and final category presented within Table 1 is that of university-led programs. Such programs seek to answer specific research questions, developed by an academic or university-based research group. Such initiatives may be driven primarily by the needs of academics to create peer-reviewed journal publications. In addition, with the increasing focus within universities on creating societal impact, citizen science initiatives may be developed by academics who see the methodology as a means to ensure the relevance of their research. An example of such a project is *Project Splatter* (https://projectsplatter.co.uk/), at Cardiff UK University. Academics within the College of Biomedical and Life Sciences developed this citizen science project to record sightings of wildlife roadkill within the UK. The aim of the project is to explore the impact of roads upon wildlife. As well as publishing findings, researchers behind this project also hope it might be possible to influence policy maker's decisions about road construction through presenting evidence of some of the negative impacts of roads on biodiversity.

THE MOTIVATIONS OF CITIZEN SCIENCE CONTRIBUTORS: THE SCIENTIFIC ORGANIZERS PERSPECTIVE

It is suggested within the citizen science literature that there are various reasons why vast numbers of people participate in citizen science globally. Miller-Rushing, Primack & Bonney (2012) argue that participation in citizen science enhances scientific literacy. Moreover, it was stated in a recent UK government PostNote (2014) that participating in citizen science enables contributors 'to learn new skills, often with value for future employment' (p. 3). There is also a sense that people take part in citizen science projects in order to work with scientists. Dickinson et al. (2012) state that: 'collecting data for use by professional scientists is highly motivating [and], fosters scientific knowledge' (p. 295). A related potential reason for participation in citizen science, recently recognized by scientists in the field, focuses on the social context of citizen science. Price and Lee (2013) argue that the 'social component of the project' (p.795-796) is its most important dimension. Citizen science projects often allow individuals to track their progress and compare their performance to that of other citizen science participants through, for example, contests, games and challenges. This may also be encouraged by rewarding the participants with certificates. This might also include coverage in the media, on project blogs and newsletters (Dickinson et al., 2012). All this, together with the prestige associated with being involved in scientific processes, means that citizen science may subsequently enable participants to enhance their social capital, by enabling those who take part to construct a desired personal identity.

Along with the thrill of the competition, academics suggest making friends drives contributors. Internet forums often support citizen science projects. Examples include those available via citizen science platforms, such as Zooniverse (https://www.zooniverse.org/) and iRecord (http://www.brc.ac.uk/irecord/). Bonney et al. (2009) argue that such online forums increase participant's visibility at ever-larger scales. Dickinson, Crain, Reeve & Schuldt, (2013) have discussed the benefits of online social networking. One of these benefits, they suggest is that participants can appreciate their role in the collection of large data sets. As a result, they argue that citizen science could create 'massive shifts in pro-environmental behavior and social norms' (p.1). Therefore, citizen science may nurture collective action, by appealing to pro-social sensibilities. Involvement in large-scale research projects, they argue, may lead to awareness of group efficacy. This may combat individual feelings of helplessness in confronting environmental issues (MacNaghten, 2003).

Such discourses of learning and contribution are well established in the citizen science literature. The democratization of science is thought to be a concern for organizers of citizen science (Johnson et al., 2014). This involves increasing opportunities for science learning, as well as a preoccupation with the expansion of scientific endeavor in both scale and scope. Bonney et al (2009), for example, describe citizen science as a process that 'enlightens the public' (p.977). Simultaneously, the citizen scientist emerges as a value-driven individual, willing to contribute their time, skills and efforts for the good of science. In return, it is often suggested that citizen scientists expect to learn about scientific methodologies and to benefit from interacting with scientists. In sum, the non-scientist gains the support of the scientist, while participating in vast and scientific endeavors. It is suggested that both stand to make positive contributions towards environmental citizenship. Dickinson et al. (2012), thus, argue that citizen science is a 'shareable public good' (p. 291) in both its processes and its outcomes. These authors argue that citizen science contributes to 'public participation and earth stewardship'.

PROBLEMATISING EXISTING SUGGESTION FOR THE MOTIVATIONS OF CITIZEN SCIENCE CONTRIBUTORS

The suggested motivators discussed above are plausible reasons for individuals to choose to become involved in a wide variety of environmental data monitoring activities. Yet, existing literature on citizen science motivations may overemphasize the role of citizen science organizers. This is perhaps not surprising given that those operating citizen science programs have, to date, conducted most program evaluations for the field. Within this context, the scientist is in a position to influence participant's motivations. Moreover, as scientists guide participants in scientific processes, it is, thus, their role to enhance the participant's science literacy. It is the scientist and organizing team also who make decision on what interactive components to include within a citizen science project. For example, they decide whether to include gamification aspects and online networks as components of the projects they set-up. As a result of this, they are, therefore, able to facilitate social interactions. Existing literature, therefore, portrays citizen science organizers as facilitators of learning and social networking.

Furthermore, Cooper et al (2007) suggest that environmental citizen science offers hope, not only in terms of its ability to engage citizens with environmental issues, but also for conservation more broadly. This, they suggest, is because 'it operates over such large scales by drawing on spatially dispersed participants'. Therefore, the data generated through citizen science 'can be used to create a new frontier to advance the theory and practice of conservation in residential ecosystems' (p. 8). Thus, they see citizen science not only as a research and monitoring tool, but also as a tool for conservation. Cooper et al (2007) propose that citizen science can be used in the 'adaptive management' of residential habitat. To this end, they suggest a new approach calling it 'adaptive citizen science'. This approach, they suggest, is a 'an effective means of organizing citizens, residents and habitat management activities to achieve cumulative, positive impacts on biodiversity in research landscapes' (p.1).

It would seem then that the scientific organizers of citizen science projects have a crucial role to play in mobilizing the public in scientific data collection. On the other hand, however, contributors to citizen science projects give their time freely to be involved in what could be regarded as often quite mundane tasks. Such tasks may include measuring and counting plant and animal species or collecting water samples. These observations are often passed on to scientists who have the specialist expertize to analyze the data that emerges from them. They then use this data within research projects, the results of which the participant may never see. The likelihood of contributing participants being involved in reflections upon citizen science data is perhaps decreased, not only by the fact that these individuals are unlikely to possess the skills to interpret the data, but also by a focus on major outcomes of citizen science studies. For example, while reporting the success of citizen science projects, there is often a focus on the significant findings that emerge following participation. For example, the RSPB *Big Garden Bird Watch* (https://ww2.rspb.org.uk/discoverandenjoynature/discoverandlearn/birdwatch) reports that over half a million people counted eight and a half million birds across the UK in 2015. Under this scheme, a contributor submits her data via a smart-phone, iPad or a web-portal. Following this, all the records are collated and the results are reported online. Yet contributors rarely receive feedback within macro-level-schemes on the specific records that they have contributed. They are, thus, unable to contextualize their local results in a wider context. It is, of course, likely that some participants may keep their own records. This would enable them to compare their data to the national result. However, often there is little feedback from citizen science organizers. It could be argued, therefore, that the greatest benefits from citizen science go to scientists. The scientists behind citizen science make new discoveries at ever-vaster scales through these programs. They frequently acknowledge that these new discoveries are only made possible by the recording efforts of contributors. Yet, it is rare for participants to be included as co-authors on the peer-reviewed publications upon which scientific careers are built (Dickinson et al., 2012 and Venkatraman, 2010).

Further, there is also a political implication in the expansion of citizen science. The Parliamentary Office of Science and Technology (POST) in the UK, regards citizen science as key to scientific progress. For instance, they draw attention to the fact that citizen science informs the Biosecurity Action Plan and the National Pollinator strategy. In a climate of austerity, environmental citizen science is looked upon by POST (2014) as a 'cost effective' means of gathering data' (p. 3). More cynically then, public participation in scientific research could be regarded as a process that capitalizes on the labor of contributors.

Problematizing the scientific narrative concerning the value of citizen science to the public leads us towards a deeper exploration of the citizen science experience. As appealing as such scientific explanations may be, little evidence supports the assertions concerning the value of the process to participants. This places reliance upon narratives that posit the value of citizen science to those involved, posed by those who arguably stand to gain most from citizen science as a process. It is not proposed within this chapter, that scientists are over-claiming the significance of citizen science. Rather, the chapter seeks to highlight the implications of the reliance upon citizen science organizers to provide insights into the citizen science experience. It is argued here that in order to understand the role of environmental citizen science in addressing sustainability problems it is necessary to understand how such processes affect the individual person, involved in citizen science. Their role involves data collection, and less frequently, data analysis and interpretation on the ground. Gaining insights into the meanings of such experiences for participations is key to the field's development. Yet, as a relatively young field, this is yet to be explored deeply.

ENVIRONMENTAL EDUCATION PERSPECTIVES ON CITIZEN SCIENCE: A CALL FOR FURTHER EXPLORATION

When an environmental education lens is applied in order to study the citizen science experience, a different story of what motivates citizen science project contributors emerges. This perspective offers a social, political and emotional understanding rather than a scientific understanding of the role of citizen science. As citizen science has gained in popularity, its conjunction with science education and environmental education has been seen as a major opportunity by leaders within the field of Education for Sustainable Development (ESD). In a recent paper published in Science, Wals, Brody, Dillon and Stevenson (2014) stated that a convergence of efforts would be feasible for these fields. They also argue that it would be effective given that both fields seek to address sustainability challenges. These authors characterize citizen science as a process through which phenomena are classified and monitored. They portray Environmental Education as an educational field concerned with identifying causes and solutions. For Wals et al. (2014) citizen science delivers synergy between science education and environmental education. These fields, they argue have historically been regarded as 'distant, competitive, predatory and host-parasite' (p. 583). Citizen science is, thus, a particularly promising development for these environmental educationists. This is because they believe that 'citizen science enables for people to engage with science on relevant environmental issues in collaboration with scientists working in local contexts' (p. 584).

TROUBLING THE APPLICATION OF AN ENVIRONMENTAL EDUCATION LENS TO CITIZEN SCIENCE

Nevertheless, it is also necessary to problematize the application of an environmental education lens to citizen science. Applying an environmental education lens to the analysis of the citizen science experience may be, for example, particularly problematic because of the controversies surrounding discourses of Education for Sustainable Development (ESD). There are many who have criticized the effectiveness of Education for Sustainable Development and the related United Nations Decade for Education for Sustainable Development (UNESD). For instance, Richard Kahn (2010) has suggested that Education for Sustainable Development can only serve to reinforce the status-quo because of its ideological underpinnings, which are based upon current neo-liberal values, currently held by Western Society. To this end, Kahn (2010) states:

the next decade will ultimately decide whether ESD is little more than the latest educational fad or, worse still, turns out to be a pedagogical seduction developed by and for big business-as-usual in the name of combating social and ecological catastrophes (p. 16).

Kahn (2010) and others may argue, therefore, that applying an environmental education or Education for Sustainable Development lens to any endeavor that seeks to raise ecological awareness and sensibility, may be ineffective. This is because looking at citizen science through such a lens may bring to it an anthropomorphic perspective. It may do so, by considering how humans benefit from environmental citizen science. This could happen at the expense of exploring citizen science from the angle of its significance for biodiversity, as arguably conservation biologists would. In analyzing the existing developments within the field of Education for Sustainable Development, Kahn (2010) calls for wider reform of education systems, based upon an 'eco-pedagogy'. This, he argues, would involve centralizing environmental concerns within education systems, so that humans gain a better appreciation of their place as part of such systems.

Using an environmental education lens to study of citizen science may be problematized on a second basis. Focusing on the benefits of participation may result in the ignorance of interactions with 'nature' that are perhaps lost as a result of participation. In a different context, Pergams and Zaradic (2006) have argued that a love of technology drives individuals away from experiencing the natural world. It is possible to make the case, therefore, that the use of smartphones in environmental citizen projects may distract people from directly observing and experiencing the natural world. It is also possible to envisage that participating in citizen science might mask other more significant motives. For instance, in the case of internet-based projects, the desire to participate could be a consequence of the love for technology that Pergams and Zaradic (2006) speak of. There is also yet a third basis, upon which it is possible to problematize the study of citizen science through the lens of environmental education. To date, there has been little consideration of the inequality of opportunity in terms of who gets to engage with citizen science. Some individuals, for example, may lack chances to become involved due to absence of access to technology, or project unavailability in the areas within which they live.

Nevertheless, although it is crucial to be mindful of these problematics, the involvement of environmental educators, or perhaps eco-pedagogues, in debates about citizen science is key. This will make it possible to recognize the benefits to individuals involved. It will also allow exploration of how environmental citizen science may address sustainability problems. To achieve this, it is necessary to understand what drives contributors to become involved in citizen science projects. To date, many studies have focused on the effectiveness of citizen science in facilitating traditional scientific processes. Yet the meaning of citizen science for the individual has been regarded as less significant. The impact of involvement upon the individual, the local school or community groups appear to have been considered as secondary.

THE NEED FOR FURTHER RESEARCH CONCERNING EXPERIENCES OF CITIZEN SCIENCE

As noted in the preceding section, it is necessary to seek the perspectives of wider disciplines, beyond those who are responsible for the design of citizen science programs. There are interesting stories about the contexts in which participation takes place and about the making of citizen science data that currently go unreported. Understanding who collects the data and why they do so is crucial to understanding the value of citizen science. From an environmental education perspective, such insights are as important as understanding scientific outcomes.

Aside from a few notable recent studies (Jones, Riddell, Morrow, 2013; Johnson, Hannah, Acton, Popovici, Karanth, & Weinthal, 2014 and Rotman, Preece, Hammock, Procita, Hansen, Parr, & Jacobs, 2012), there has been a lack of focus on the motivations of those who participate in citizen science and on the citizen science experience itself. This means that little is known about who might and might not participate and for what reasons. Furthermore, while the ancient roots of citizen science are known, there has been no consideration of what appeared to motivate those early contributors. The short number of motivations that emerge from the scientific literature may also be the result of the fact that there has been a tendency to focus upon current citizen science programs for understanding. There are many historical examples of individuals engaging with citizen science. Lighthouse keepers, for instance, began collecting data concerning bird strikes as long ago as 1880 (Bonney et al., 2009), while there has been no shortage of past citizen science projects. For example, the National Weather Service Cooperative Observer Program (http://www.nws.noaa.gov/om/coop/) began in 1890, while the National Audubon Society Christmas Bird Count (http://www.audubon.org/conservation/science/christmas-bird-count) began in the 1900's and the Breeding Birds Survey of the British Trust for Ornithology (BT0) (https://www.bto.org/volunteer-surveys/bbs) was established in the 1930's. It would be possible to conduct a study of the motivations of the early participants in citizen science programs. If such a study was conducted, a different picture might emerge of some of the central drivers for the individuals who choose to participate in citizen science.

Moreover, it would also be interesting to conduct studies of how the motivations and experiences of contributors to citizen science projects vary in accordance to the types of citizen science program, within which they participate. Using Table 1 as a starting point it might be interesting, for instance, to explore the experiences of those who participate in citizen science programs run by NGO's as opposed to those run by schools, or those run by eco-attractions. A particularly interesting and underexplored instance of citizen science experiences includes projects that are considered community-led, perhaps even more activist projects. Such projects are often considered to be less connected to the endeavors of 'grand science' (McQuillan, 2014). Thus, conducting studies that involve an exploration of the experiences of individuals involved within them may provide different comprehensions of the motivators that drive those who contribute to citizen science projects.

The few existing studies that analyze the practices of those individuals who contribute their time to become involved in citizen science provide insights into the variety of reasons that influence individual decisions to become involved and the impacts of their experiences. Both areas, motivations and experiences, will be socially and culturally constructed. Yet, the social and cultural factors that drive individuals to contribute to citizen science are often not clearly visible. It is necessary, therefore, to explore such drivers in depth, drawing upon a range of disciplines, including history, geography and sociology. Doing so, may reveal the underlying factors that influence individuals to contribute to citizen science. It may also uncover the important effects of participation upon individuals.

The ability of citizen science to expand the scale and scope of data collection appears to be what makes it appealing to scientists, as well as to the popular imagination. Yet it is clear from the preceding discussion that it is possible to consider the benefits of citizen science from an alternative perspective. This involves beginning at the scale of the individual participant, considering what motivates them to become involved and how participants experience the process. The outcomes of involvement for the individual are also important to understanding the significance of citizen science within contemporary society. Therefore, as well as considering citizen science at a scale that is relevant to scientific progress, it is also worthy of consideration at the scale of personal progress for the individual. Such individuals may learn about and see and sense novel surroundings through their participation, which could have key implications for their everyday lives.

It may be desirable, therefore, to consider the nature of the connections that individual contributors have to citizen science. This may include, for instance, thinking about how long they participate in projects for and whether they are involved in multiple projects. It may also be valuable to consider the backgrounds of individual contributors. It would then be possible to explore how such personal histories, together with present circumstances, influence decisions to contribute to citizen science projects. Doing so, may provide an appreciation of citizen science and its role within increasing environmental literacy. An appreciation of how such initiatives expand individuals' sense of care for the environment, as well as, their willingness to act in pro-environmental ways may also be gained. This would go some way in providing understanding of how environmental citizen science initiatives might contribute to solving 'sustainability problems' through social responses at a personal and collective community level. Such studies would therefore be useful to those working within the field of sustainability science. Providing understandings of what makes individuals get involved and stay involved in citizen science programs would also be useful to those conservation biologists looking to harness citizen science as a research method.

BROADER INTERPRETATIONS OF THE MOTIVATIONS OF CITIZEN SCIENCE CONTRIBUTORS: EXPLORING THE TOPOPHILIA HYPOTHESIS

It may be the case then that the language of education, social cooperation and technological innovation may be inadequate for understanding the motives of individuals who contribute to environmental citizen science. Indeed, it is possible to argue that environmental citizen science differs from other citizen science programs, such as those focused upon human health, from this point of view. In addition to those motivations suggested by previous authors, it might be beneficial to adopt a language of 'philos' – that is a language of love and affection to understand what might also inspire individuals to contribute to environmental citizen science projects. In 'A Sand Country Almanac', Leopold (2001) suggested that the development of an ecological consciousness is very much dependent upon an 'individual's internal emphasis, loyalties, affections and convictions' (p. 174). Leopold (2001) argued that 'we can be ethical only in relation to something we can see, feel, understand, love, or otherwise have faith in' (p.179). It has been argued in the preceding section that by considering the deeper, affective drivers that may be present within citizen science, it is possible to gain a greater appreciation of the values, perceptions and emotions that bring individuals to participate in citizen science activity. The remainder of this chapter, therefore, seeks to explore a hypothesis concerning how a particular phenomenon, 'topophilia' (Tuan, 1974) may influence citizen science experiences. In doing so, this chapter demonstrates how underlying ways of seeing (Berger, 1972) might affect people's desires to become involved and continue to participate in citizen science.

Topophilia describes the phenomenon of having a deep knowledge of a place and love of such a place (Tuan, 1974). It is comparable to the notion of biophilia, which indicates a love of life. It is proposed here that topophilia might drive individuals to become involved in environmental citizen science. Furthermore, topophilia may be a desirable outcome of participating in citizen science, for individuals. These individuals may become connected to their locale through participating in local environmental monitoring initiatives. For instance, *Citizen Crane* is a citizen science project that asks local contributors to monitor a stretch of river near to their homes. The contributors of Citizen Crane have worked in groups of two or three to conduct the river monitoring surveys once a month for the past three-years. They may be motivated to do so because of a love of the place within which they live. Topophilia is not necessarily, however, a phenomenon that connects participants to their home environments. Participants may be, for example, encouraged to develop affective bonds with places that they consider special through their participation in a citizen science projectwhich may be considerable distance from their homes.

Sampson (2012) argues that individuals form such bonds with place through 'both an attraction to place and sense of place-based history' (p. 41). Building upon these early theorizations of topophilia, he proposes the 'topophilia hypothesis', suggesting that the ability to bond with local place is an evolutionary adaptation, which enabled human beings to learn specific place-based skills required to adapt to and thrive within the particular places to which they found themselves bound. For Sampson (2012), 'topophilia' is something that is innate to human beings, but has been lost within modern societies. He argues that 'the proposed affective connection with place that characterized humans during the bulk of their history has been largely severed today in industrialized societies' (p. 35). Within the present day, therefore, human beings are failing to recognize their dependencies upon local places, which results in a 'dysfunctional human-nature relationship at the heart of the ecocrisis' (Sampson 2012: p.42). Yet, he suggests there is cause for hope in that 'the human brain is genetically wired to incorporate knowledge through local place' (Sampson 2012: p.38). Sampson (2012), therefore, calls for efforts to reinstate the bonds between people and places. He suggests that place-based education, beginning at the earliest possible stage would be one way to approach this.

Nevertheless, it is also possible to contest the idea that such topophilic bonds to local places are not experienced by individuals within the present day. Both Welsh and German cultures have expressions with no direct translations that describe the bond and attachments that an individual can have to their local place. The Welsh word 'hireath' is used to describe the bond that one feels to the land to which they were born and their connection to their culture. It is often thought to be felt as homesickness by those who are displaced. This term is still pervasive within Welsh language and culture. Indeed, it has been adopted within popular culture and has even been used in tourist campaigns to market to the Welsh diaspora (Morgan, Prichard and Pride, 2003). Similarly, the German concept of heimat is used to refer to a communion with place. Thus, a healthy skepticism might be maintained regarding the nostalgic overtones of such concepts. Yet, the concept of affective bonds to place does seem to persist within cultures of the present day. It is also possible to argue that within a modern industrialized world, towns and cities have adapted so that gardens, parks and nature reserves offer intrinsically valuable opportunities to reconnect with the natural world within urban settings. Over the last half a century, the importance of natural spaces within urban contexts has been increasingly acknowledged (Goode, 2011), while urban initiatives, including community-based science programs and nature festivals that address disconnections between the human and natural world have been established (Goode, 2014), including many citizen science programs.

Regardless of whether we feel topophilia is present in contemporary society or in the need of reawakening, it is clear that the presence of affective bonds to place, at a local or indeed global scale, may act as a motivator for those who choose to contribute to citizen science projects. Understanding what drives individuals to engage with place-based environmental citizen science is, therefore, fundamental to our appreciation of participation. If we follow the topophilia hypothesis, as Sampson (2012) proposes, it could be posed that individuals may be attracted to participate in citizen science initiatives because of an innate desire to preserve, protect or restore places that they value. Such initiatives often occur within urban spaces. Environmental citizen science initiatives provide a means through which individuals can become attentive to urban nature. Furthermore, while many environmental citizen science projects rely upon web content, email and postal means to report data, they are still place-based in their collection of records. There are also environmental science initiatives that are being organized through partnerships with local museums, science centers and local organizations. The Open Air Laboratories (OPAL) Network (http://www.opalexplorenature.org/), for example, is a citizen science project that is operated by a team of community scientists, based regionally within museums across the UK. These community scientists then work with local schools, community and social groups to involve young people and communities in citizen science. Environmental citizen science in such contexts may well, therefore, offer the opportunity to bond with nature within urban settings, as Sampson (2012) puts it, 'as our foraging forbearers bonded with savannahs, rainforests, tundra and deserts' (p.39). This may well be part of the appeal of environmental citizen science for those who contribute. Therefore, a consideration of 'topophilia' as playing a role in citizen science motivation and experiences warrants further exploration.

An awareness of the fact that participating within local citizen science initiatives might have a positive effect on an area that an individual cares about could well be a motivation for that individual. For example, Cooper et al. (2007) highlight the capacity of citizen science projects to have a positive effect on residential areas that are regarded as important in terms of ecosystem services and biodiversity support. They argue that an 'adaptive citizen science model' could be an effective 'means of organizing citizens, residents and habitat management activities to achieve cumulative, positive impacts on biodiversity in research landscapes' (Cooper et al., 2007: p.1). The ability to contribute to the restoration of local areas would undoubtedly be seen as rewarding by many who choose to contribute to environmental citizen science, who may arguably increase their 'environmental stewardship' through 'active participation in research and subsequent informal (i.e. not classroom based) science education' (Cooper al, 2007: p.7). Thus, the experience is assumed meaningful within the everyday lives of individual participants, as well as beneficial for scientists.

In many cases, those who contribute to citizen science are choosing to pursue their own projects with their own research questions enabling them to produce evidence to protect local areas and influence local decision making as collectives, often working with scientists in this process. Dickinson et al. (2012) suggests that learning outcomes are 'more robust among volunteers who explore their own questions'

(p. 295), while such groups benefit from increasing their social networks and their opportunities for social learning. In this context, there is perhaps also an activist motivation for citizen science contributors. McQuillan (2014) has recently noted the activist potential of citizen science projects that arise out of disorder rather than order as a scientific endeavor. He argues that though those who participate in citizen science 'rarely characterize themselves as countercultural' (p. 1), usually aiming instead to create 'orthodox scientific knowledge', citizen science shares counter-cultural resonance. It possesses such resonance, he argues, because of its focus on 'participatory experimentation and the principles of environmental sustainability and social justice'. For some, citizen science is, therefore, seen as a means of democratizing science, involving people in caring for and making decisions about local landscapes, and in cases where citizen science projects are 'grassroots' challenging received wisdom about the state of the environment (McQuillan, 2014). Citizen science offers opportunities to challenge through mass data collection. Through collecting large amounts of data on current issues, it is possible to support counter-cultural movements (McQuillan, 2014). In responding to the destruction of places and life forms that they see to be at risk, citizen science contributors may become involved in projects that seek to halt environmental damage or change. Examples of such community-led citizen science initiatives include *Grupo Tortuguero* (https://www.oceanfdn.org/projects/international-partner-project/grupo-tortuguero), which has through research, helped to establish marine projected areas and sustainable fisheries. Furthermore, the *West Oakland Environmental Indicators Project* (http://www.woeip.org/) in California has helped empower individuals in disadvantaged communities to gather evidence of air quality and health data, while Bonney et al. (2014) discuss a project, based at University College London, which documents poaching and illegal logging in the Democratic Republic of the Congo. Such projects: 'use science to address community-driven questions [which involves] attentiveness to diverse interests including why and how members of the public would even want to be involved (p. 1437).

The implications of affective bonding to place are also further reaching than they first appear. Indeed, within a global context, it may be possible that citizen science helps individuals to make connections between the global and the local. For example, they could make such connections by considering how the data they collect on a local scale links to data that is collated at an international level. Sampson (2012) argues that 'achieving sustainability at higher levels (state, nation, biosphere) will be realized only through iterative accumulation of sustainable societies in local places' (p. 45). In this sense, it is not only the outcomes of specific citizen science projects that will help to address environmental change, but crucially the effect of participation on the individual participant in terms of being generative of a sense of care for place through establishing affective bonds to place. If topophilia is an outcome of citizen science participants, it may emerge that such experiences have even wider implications for enabling societies to tackle both local and global environmental crises.

CONCLUDING THOUGHTS

It is clear that citizen science is a field capable of capturing the imaginations of a large number of people, while also having vast media appeal. There is great potential within the phenomenon to advance sustainable development through links with Environmental Education. Yet, this chapter has sought to highlight that the motivations and experiences of those who choose to contribute to citizen science projects need to be explored in greater depth than occurs at present. This chapter is likely to be valuable to citizen science providers given its emphasis on finding novels ways of enabling understanding of the individuals with whom they engage. Crucially, expanding knowledge concerning the effects of citizen science will be achieved through harnessing progressive methods of exploring the meanings and impacts of individual citizen science experiences. As Bonney et al. (2009) states, 'the full potential of citizen science is just beginning to be understood' (p. 983). This applies perhaps as much to its potential as a form of environmental learning and public engagement, as it does to the exploration of its ability to contribute to rigorous, peer-reviewed science.

In-depth qualitative study would lend themselves to such analysis through their ability to elicit meaningful insights into experience. Narrative is central to our way of making sense of our lived experiences. Therefore, a possible way of gaining such insights would be through narrative inquiry (Riessman, 1993). This approach involves listening to the stories that participants tell about why they became involved, why they stay involved or leave projects and why they choose to spend their time on environmental citizen science, as opposed to other means of interacting with the environment. Through employing narrative techniques, we may gain insight into emotional, social and cognitive experiences. Such studies would provide insights into how participants feel their involvement in citizen science shapes their identity. This future research would enable us to gain an in-depth understanding of the citizen science experience for contributors within the context of their histories, political leanings, communities and their views on environmental change. Furthermore, through understanding the nuanced stories of involvement, we gain insight into the multiple meanings of citizen science within the lives of different individuals, while taking into account their diverse backgrounds and circumstances. In going through this process, we are likely to encounter conflicts and contradictions in how individuals who choose to contribute to citizen science projects come to understand their involvement in scientific endeavor. Yet, doing so would enable us to move away from the grand narratives provided by science, concerning the effectiveness of citizen science. In providing opportunities for such insights into the educative and cooperative process, we stand to gain a richer understanding that acknowledges that citizen science is part of a longer history, as an activity that can play a prominent role within the everyday lives of participants. This may help conservation biologists to appreciate, in turn, how the stories they tell motivate individuals to become part of their scientific endeavors.

REFERENCES

Berger, J. (1972). *Ways of Seeing*. London: Penguin.

Bonney, R., Cooper, C. B., Dickinson, J., Kelling, S., Phillips, T., Rosenberg, K. V., & Shirk, J. (2009). Citizen science: A developing tool for expanding science knowledge and scientific literacy. *Bioscience*, *59*(11), 977–984. doi:10.1525/bio.2009.59.11.9

Bonney, R., Shirk, J. L., Phillips, T. B., Wiggins, A., Ballard, H. L., Miller-Rushing, A. J., & Parrish, J. K. (2014). Next steps for citizen science. *Science*, *343*(6178), 1436–1437. doi:10.1126/science.1251554 PMID:24675940

Brossard, D., Lewenstein, B., & Bonney, R. (2005). Scientific knowledge and attitude change: The impact of a citizen science project. *International Journal of Science Education*, *27*(9), 1099–1121. doi:10.1080/09500690500069483

Cooper, C. B., Dickinson, J., Phillips, T., & Bonney, R. (2007). Citizen science as a tool for conservation in residential ecosystems. *Ecology and Society*, *12*(2), 1–11.

Davis, P. (1996). *Museums and the natural environment: the role of natural history museums in biological conservation*. Leicester, UK: Leicester University Press.

Dickinson, J. L., Zuckerberg, B., & Bonter, D. N. (2010). Citizen science as an ecological research tool: Challenges and benefits. *Annual Review of Ecology Evolution and Systematics*, *41*(1), 149–172. doi:10.1146/annurev-ecolsys-102209-144636

Dickinson, J. L., & Bonney, R. (Eds.). (2012). *Citizen Science: Public participation in environmental research*. Ithaca, NY: Cornell University Press.

Dickinson, J. L., Shirk, J., Bonter, D., Bonney, R., Crain, R. L., Martin, J., & Purcell, K. (2012). The current state of citizen science as a tool for ecological research and public engagement. *Frontiers in Ecology and the Environment*, *10*(6), 291–297. doi:10.1890/110236

Dickinson, J. L., Crain, R. L., Reeve, H. K., & Schuldt, J. P. (2013). Can evolutionary design of social networks make it easier to be 'green'? *Trends in Ecology & Evolution*, *28*(9), 561–569. doi:10.1016/j.tree.2013.05.011 PMID:23787089

Dunkley, R. A., (2016). Learning at Eco-attractions: Exploring the Bifurcation of Nature and Culture through Experiential Environmental Education. *The Journal of Environmental Education*, 1-9.

Goode, D. (2011). Planning for nature in towns and cities. In I. Douglas, D. Goode, M. C. Houck, & R. Wang (Eds.), *The Routledge Handbook of Urban Ecology* (pp. 84–92). Abingdon, UK: Routledge.

Goode, D. (2014). *Nature in Towns and Cities*. London: HarperCollins.

Haklay. (2013). *Citizen Science and Policy: A European Perspective*. Commons Lab. Retrieved May 16, 2016, from: https://www.wilsoncenter.org/sites/default/files/Citizen_Science_Policy_European_Perspective_Haklay.pdf

Irwin, A. (1995). *Citizen science: A study of people, expertise and sustainable development*. Abingdon, UK: Routledge.

Johnson, M., Hannah, C., Acton, L., Popovici, R., Karanth, K., & Weinthal, E. (2014). Network environmentalism: Citizen scientists as agents for environmental advocacy. *Global Environmental Change*, *29*, 235–245. doi:10.1016/j.gloenvcha.2014.10.006

Jones, M., Riddell, K., & Morrow, A. (2013). *The impact of citizen science activities on participant behavior and attitudes: Project Report 2013: The Conservation Volunteers*. Retrieved May 16th 2016, from: http://www.environment.scotland.gov.uk/media/58143/Impact_Of_Citizen_Science_Activities_On_Participant_Behaviour_And_Attitude.pdf

Kahn, R. (2010). *Critical pedagogy, ecoliteracy, & planetary crisis: The ecopedagogy movement*. New York, NY: Peter Lang.

Kolb, D. A. (1984). *Experiential Learning: Experience as the source of learning and development*. Prentice-Hall.

Leopold, A. (2001). *A Sand County Almanac: Outdoor Essays and Reflections.* New York: Ballantine.

Macnaghten, P. (2003). Embodying the environment in everyday life practices. *The Sociological Review, 51*(1), 63–84. doi:10.1111/1467-954X.00408

McQuillan, D. (2014). The Countercultural Potential of Citizen Science. *M/C Journal, 17*(6), 1-9.

Miller-Rushing, A., Primack, R., & Bonney, R. (2012). The history of public participation in ecological research. *Frontiers in Ecology and the Environment, 10*(6), 285–290. doi:10.1890/110278

Morgan, N., Pritchard, A., & Pride, R. (2003). Marketing to the Welsh diaspora: The appeal to hiraeth and homecoming. *Journal of Vacation Marketing, 9*(1), 69–80. doi:10.1177/135676670200900105

Pergams, O. R., & Zaradic, P. A. (2006). Is love of nature in the US becoming love of electronic media? 16-year downtrend in national park visits explained by watching movies, playing video games, internet use, and oil prices. *Journal of Environmental Management, 80*(4), 387–393. doi:10.1016/j.jenvman.2006.02.001 PMID:16580127

Parliamentary Office of Science and Technology. (2014). *Environmental citizen science.* Parliamentary Office of Science and Technology (POST), Houses of Parliament, POSTNOTE 476, August 2014. Retrieved May 18, 2016, from http://researchbriefings.parliament.uk/ResearchBriefing/Summary/POST-PN-476

Price, C. A., & Lee, H. S. (2013). Changes in participants' scientific attitudes and epistemological beliefs during an astronomical citizen science project. *Journal of Research in Science Teaching, 50*(7), 773–801. doi:10.1002/tea.21090

Riessman, C. K. (1993). *Narrative Analysis.* London: Sage.

Rotman, D., Preece, J., Hammock, J., Procita, K., Hansen, D., Parr, C., & Jacobs, D. (2012). Dynamic changes in motivation in collaborative citizen-science projects. In *Proceedings of the Association for Computer Machinery 2012 conference on Computer Supported Cooperative Work* (pp. 217-226). New York, NY: Association for Computer Machinery. doi:10.1145/2145204.2145238

Sampson, S. D. (2012). The topophilia hypothesis: Ecopsychology meets evolutionary psychology. In P.H. Kahn & P.H. Hasbach (Eds.), Ecopsychology: Science, totems, and the technological species (pp. 23-53). Cambridge, MA: MIT Press.

Shirk, J. L., Ballard, H. L., Wilderman, C. C., Phillips, T., Wiggins, A., Jordan, A., & Bonney, R. (2012). Public participation in scientific research: A framework for deliberate design. *Ecology and Society, 17*(2), 29. doi:10.5751/ES-04705-170229

Tuan, Y. F. (1974). *Topophilia.* Englewood Cliffs, NJ: Prentice-Hall.

Venkatraman, V. (2010). Conventions of scientific authorship. *Science.* Retrieved May 18, 2916, from http://www.sciencemag.org/careers/2010/04/conventions-scientific-authorship

Wals, A. E., Brody, M., Dillon, J., & Stevenson, R. B. (2014). Convergence between science and environmental education. *Science, 344*(6184), 583–584. doi:10.1126/science.1250515 PMID:24812386

Wiggins, A., & Crowston, K. (2011). From conservation to crowdsourcing: A typology of citizen science. In *Proceedings of the 2011 44th Hawaii international conference on system sciences* (pp. 1-10). Washington, DC: Institute of Electrical and Electronics Engineers Computer Society. doi:10.1109/HICSS.2011.207

Chapter 11
Citizen-Driven Geographic Information Science

Thomas J. Lampoltshammer
Danube University Krems, Austria

Johannes Scholz
Graz University of Technology, Austria

ABSTRACT

This chapter shows how global environmental changes put society in front of new challenges, and how immediate and intense actions have to be undertaken in order to foster necessary progress in global sustainability research. The technological infrastructure has reached a status of ubiquitous computing and virtually unlimited data availability. Yet, the dynamic nature of the global environment makes continuous and in-situ monitoring challenging. Citizen-driven geographic information science can bridge this gap by building on inputs, observations, and the wisdom of the crowd, represented by the citizens themselves. This chapter argues for the important role of citizen science in geographic information science, presents its position in current research, and discusses future potential research streams, based on the participation by and collaboration with citizens. In particular, the chapter sheds light on three major pillars of the future of citizen-driven geographic information science, namely: big geo-data; education; and open science.

INTRODUCTION

Global environmental changes put society in front of new challenges. According to Craglia et al. (2012), immediate and intense actions have to be undertaken in order to foster necessary progress in global sustainability research. According to them, five major research challenges have to be addressed:

1. **Observation Systems:** To monitor environmental changes on all geographic scales (local, regional, and global)
2. **Forecasts:** Have to be improved in order to react timely regarding future changing environmental conditions and related direct and indirect consequences

DOI: 10.4018/978-1-5225-0962-2.ch011

3. **Key Thresholds:** Have to be identified in order to act properly on rapidly changing conditions or the occurrence of abrupt phenomena
4. **Impact Factors:** Have to be identified in a transdisciplinary approach to cover institutional, economic, and behavioral aspects in order to reach global sustainability
5. **Encouraging Innovation:** To boost the development and application of new technologies, as well as political and social progress; always paired with solid evaluation methods

In order to be able to realize solutions towards these presented challenges, new ways of digitalization and networking on a global scale throughout society have to be put in place. Former U.S. Vice President Al Gore first presented such an overall concept back in 1998 titled "Digital Earth" (Gore, 1998). At the time of being presented, the concept was criticized as not being realistic due to problems such as interoperability issues of existing geographic information systems, data accessibility, or overall Internet connectivity and available bandwidth (Craglia et al., 2008). However, major improvements have been made since then and the currently available technological infrastructure is ready to take a big step forward towards making the vision once expressed a reality.

Geographic Cyberinfrastructure

The first time that the term cyberinfrastructure appeared was in 1998. It should describe a generic information infrastructure that is able support actions such to collect, archive, share, analyze, visualize, and simulate data throughout all scientific areas. While each scientific field features its own kind of common types of data, data with included or attached geographic references can largely be found throughout all disciplines (Yang, Raskin, Goodchild, & Gahegan, 2010). A cyberinfrastructure dedicated to the resulting challenges is called geographic cyberinfrastructure (see Figure 1). These challenges relate, for example, to the necessity of specific methods and tools for the data to be processed, due to their inherent spatial characteristics (see, e.g., de Smith, Goodchild, & Longley, 2007). The required calculations can be quite demanding as spatial dimensions increase from 2D to 3D and even beyond, if time-base analysis has to be considered as well. But the newly available infrastructure presents more than just pure computational resources.

Based on the underlying technology, especially the distributed networking capability and the high grade of interconnectivity, knowledge exchange between various stakeholders, working on the cyber-infra-structure, becomes possible. Knowledge exchange can be performed, e.g., via the application of ontologies (Gruber 1993), describing environmental phenomena and their spatial and temporal dimension as fundamental cornerstone to ensure semantic interoperability (Harvey, Kuhn, Pundt, Bishr, & Riedemann, 1999; Klien, Lutz, & Kuhn, 2006). Taking the next step, these shared concepts can then be linked together (Heath & Bizer, 2011) in order to provide a web of knowledge for interdisciplinary and transdisciplinary exchange. Maybe the most important addition to the technical part of the cyberinfrastructure comes in form of the community. This essential way of contribution comes in two forms. The first form relates to users providing additional services on the cyberinfrastructure. With the establishment of standards, e.g. for web services, users can set-up their own services and offer them to be integrated in other platforms. Furthermore, communities can use the cyberinfrastructure for exchanging ideas and concepts, which can be immediately linked with data, interpretations, and visualization on the very same platform. The second form is for users to take up the role as data providers. They can act as proxies for sensors (Goodchild, 2007) and therefore collect data, in situ and in a dynamic way, arrays of sensors

Figure 1. Geographic cyberinfrastructure framework cube
Adapted from Yang et al., 2010

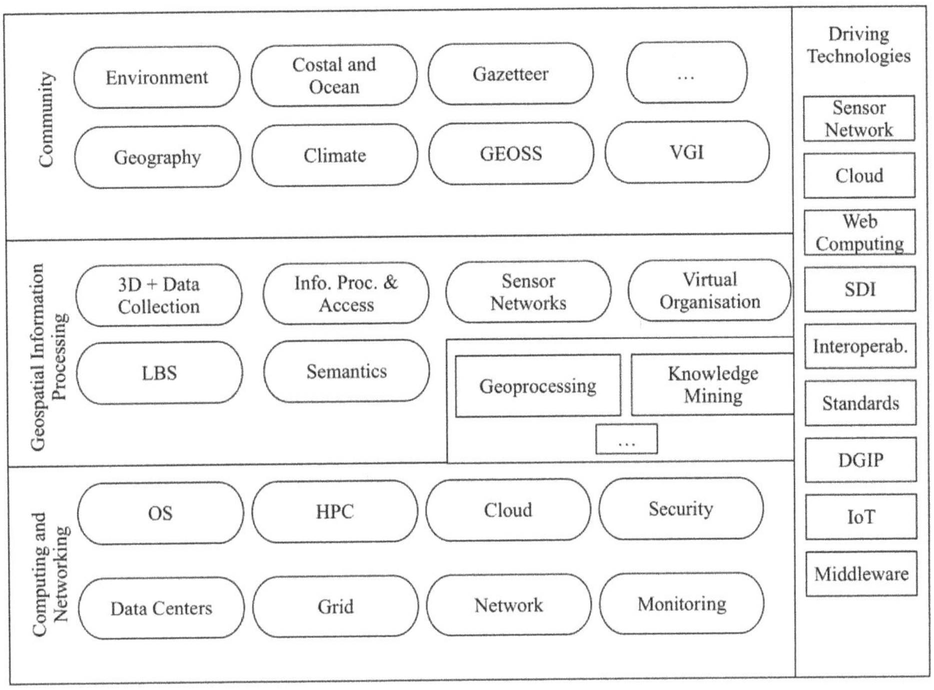

are not able to. This concept of data contribution and active participation of citizens in research areas is called citizen science, with a special peculiarity called volunteered geographic information, each of which will be explained in detailed throughout the chapter.

THE CONCEPT OF CITIZEN SCIENCE

The concept of Citizen Science can be described as civilians acting as researchers in a scientific/research context (Kruger & Shannon, 2000). Going alongside this definition, Carr (2004) describes the joint actions of individuals towards a research project as community science. According to Whitelaw, Vaughan, Craig, & Atkinson (2003), this community of citizen scientists is able, due to them joining forces, to collaborate with stakeholder from various interest groups such as public administration and agencies, industry, and academia. Yet, the authors of the paper at hand argue that the term "collaboration" has to be scrutinized as it implies a level of interaction that is not necessarily provided in all cases. Collaboration implies, from the authors' point of view, to work on eye level with someone else. However, levels of participation in citizen science are under discussion within the scientific community (Conrad & Hilchey, 2011). Lawrence (2006) suggests a literature-derived approach, defining four major forms of participation: i) consultative (public contributes information to a central authority), ii) functional (public contributes information and is also engaged in implementing decisions), iii) collaborative (public works with government to decide what is needed and contributes knowledge), iv) and transformative (local people make and implement decisions with support from "experts" where needed).

The actual level of participation is not only determined by the cooperating/involved other parties, also the motivation of participants plays an important role. As can be inferred from other fields, the actual motivation (or maybe also hidden agenda) is as manifold as the involved individuals (Lerner & Tirole, 2002). When considering contributions to the open source community as an example, people contribute, as they want to disseminate their achievements, improve their own skills, or even due to pure altruism (Bogers & West, 2012).

The Evolution of Citizen Science

The concept of citizen science, although not existent in its current form and understanding, has significantly changed over the period of the last few centuries. In the very beginning, it was understood of people from the upper-class (e.g., aristocrats) conducting research and providing sufficient evidence in order to prove their results, of which the very least was to give their word as gentlemen (Cho, McGee, & Magnus, 2006). Newman et al. (2012) provide a comprehensive overview of the evolution of citizen science since then. The phenomenon of citizen science started as individualists/hobbyists worked together out of common interest in a certain scientific discipline on a local, small geographic scale. The research questions that were pursued were purely following a top-down approach. The process of gathering data strictly followed a monitoring protocol, established by experts/scientists and the resulting data collection was available in paper-based forms only, not to mention the absence of any form of real-time availability or access. The performed analyses as well as the related publication of results were – again – performed by scientists. Furthermore, the impact that was or was not triggered by the project, was not of concern that time. The main driving force for conducting the particular research was mostly based on individual interest in the related sciences field, triggered by, e.g., personal observations of the individual's surroundings. The technological level at that time was rather limited and reached only towards basic instruments regarding data collection.

Newman et al. (2012) describe current movements to cover a much broader audience in terms of group sizes and local coverage. Groups of contributors cooperate through emerging national and international projects. While these projects still primarily focus on top-down defined research questions, bottom-up approaches are on the upgrade. The contribution of acquired data is now online-based, which significantly increases accessibility of these data, as well as possibilities to cover aspects regarding data integration as well as data quality. In addition, the facets covered by the acquired data have changed and are now much more complex as they cover spatial and temporal aspects. The associated analysis and interpretation is again performed by scientists/experts. The results are commonly distributed via publications by scientists, yet there is an increasing trend to also make the data and results available online. While the participants of current projects are globally distributed, the evaluation of the results of the performed analyses is still restricted towards the context of the project. This is due to the limited transferability of developed key performance indicators. The currently available socio-demographic data suggests that there is still space for improvement regarding the composition of research groups. The main motivation of taking part in such projects extended from pure individual interests towards the social benefits and interaction with groups of common interests. The current technological advances are not only positively influencing data availability, but also allow better integration of results of other projects into own research endeavors, mainly through social media channels such as community boards or blogs.

Newman et al. (2012) further foresee a paradigm shift regarding the extensive use of social media and viral marketing approaches in the science area to motivate individuals to participate and collaborate in new project ideas. Large virtual communities will emerge, backed up by the availability of a technologically evolved cyberinfrastructure. Research questions will be mainly data-driven and therefore based on bottom-up approaches, supported by real-time visualizations. The collected and provided data will be of high quality and are available for the global community to intensively interconnected databases and networks. High performance computing solutions within the existing cyberinfrastructures will enable mixed approaches between social sciences and natural sciences in a seamless way. The dissemination of results is performed via social media channels within the virtual community, which allows a peer-review process beyond closed knowledge circles. It is due to these new community-based solutions that evaluation processes will be more standardized and methods as well as indicators can be used across projects as well as across disciplines. It is via this mixture of real-life and virtual communication and collaboration that traditional and local knowledge can be exchanged and therefore bridging currently existing cultural, societal, and geographical boundaries. As these communities grow, people will start to compete (in a positive way) and will be rewarded through acknowledgement from the community. As all emerging projects will be part of this community, technology adaption will be fostered, as projects cannot afford to lack behind in order to be compatible with other existing ventures.

Supporting Actions for Citizen Science in the European Union

The before-described evolution of citizen science would have not been possible without substantial funding and establishment for new citizen science-based approaches, as well as the creation of a positive, tolerant, and open environment for such endeavors. Therefore, the authors present in the following section a selection of research calls to provide an overview of on-going research funding activities by the EC in the area of citizen science. The list is in no means comprehensive or complete:

- **SC5-18-2017: Novel In-Situ Observation Systems:** Current earth observation systems based on remote sensing technologies are not able to always provide the required resolution when it comes to societal observation tasks. In order to fill the existing data and knowledge gaps, calibrate and validate existing remote sensing based insights, further extended and improved in-situ-based technologies and methodologies are required to be established. Yet, existing in-situ technologies are not suitable to serve as persuasive solutions as they are often too bulky, and expensive in order to be probably used in a large scale monitoring concept. Therefore, the challenge arises to develop new technology concepts to provide cheap as well as easy solutions regarding deployment and maintenance. Via these new technologies, existing gaps in earth observation systems can potentially be closed. Furthermore, these new approaches can also be used in less developed countries to be able to contribute as well towards the deeper understanding of our planet. The envisioned research in this program should focus on the development of new technologies and in-situ application with low-energy sensors, costs-effective, easy to maintain sensor technologies. Concepts to be covered regarding the demonstration of the proposed solutions should include disposable sensors, unmanned platforms, and citizens' observatories.
- **SC5-19-2017: Coordination of Citizens' Observatories Initiatives:** Community-based environmental monitoring and information systems, also known as citizens' observatories, focus on the use of portable or mobile devices to take part in earth observation applications. Due to their mobility, ubiquitous information can be collected in-situ, providing important insights relevant, e.g., for environmental policy making while complementing existing environmental monitoring

systems. The emerging co-operations between involved stakeholders such as NGOs, citizens, and public administration offers new opportunities for SMEs in related technological fields. In order to establish a basis across research fields, efforts have to be made to elude replications and enable interoperability, which will after all foster sustainability. As the number of citizen-involved projects is increasing rapidly, coordination efforts have to be realized in order to manage these projects on numerous levels of scale (local, regional, and global).

- **ICT-11-2017: Collective Awareness Platforms (CAPS):** There exists a huge unexploited potential regarding capitalization of participatory innovation across Europe. In order to overcome this gap, new and additional models and approaches are required to utilize the power of collective intelligence in key areas. The challenge herein lays in leveraging interconnected technologies in order to establish a level of sustainability, which in the long run should lead to mass adaption, together with a significant globally-recognized impact. A possible way towards this achievement is Collective Awareness Platforms (CAPs) that make use of bottom-up (virtual) social collaboration. These emerging communities are envisioned to make heavily use of open data and open knowledge, was well as open hardware and open software, pushing crowdsourcing approaches to the next level.

Considering these calls, earth observation, in situ sensing, and associated collaboration platforms play a leading role from the point of view of the European Commission. All of these aspects have a high level of association towards geo data and gained georeferenced information. Therefore, the next section will introduce the equivalent concept of citizen science in geographic information science, namely Volunteered Geographic Information.

VOLUNTEERED GEOGRAPHIC INFORMATION

The term Volunteered Geographic Information (VGI) was coined by Goodchild (2007) and describes the activities of volunteers to collect and share spatial data. VGI arises from a number of Web 2.0 technologies (e.g., Sui, Elwood, & Goodchild, 2013), such as social media, wikis, blogs, or others. Hence, VGI can range from less serious activities, like geo-located holiday photographs, to more serious data collection for disaster relief (Haklay, 2013). Therefore, VGI is more than just a new data source or a new data type. VGI changes the paradigm for spatial research and in detail for monitoring of behaviors, opinions and social interactions of societies in urban environments (Jiang & Thill, 2015). The value of VGI can be described as follows. First, citizens can act as sensors and participate in e.g. decision processes or the development of new data sources (e.g., travel logs). Hence, individuals can be equipped with sensors (e.g., smartphones) and monitor their environment – by taking photos, recording noise or detecting air quality. In addition, humans can also directly use their senses and share their observations – e.g. what does an individual see, feel, hear, or smell. The second value of VGI is based on the value for the community, as the data are shared at no cost. Hence, citizens may have alternative data sources at hand for, e.g., planning a hiking trip or for avoiding potential dangerous areas in a city. Generally, the value of VGI can be easily justified when looking at the VGI project with a high societal impact: OpenStreetMap. OpenStreepMap has become one of the most complete and up-to-date street data collections for urban environments. Haklay (2010) reports on the quality of OpenStreetMap data, by a comparison with data originating from Ordnance Survey. The comparison shows that VGI can deliver a compelling data quality that is sufficient for a number of applications. In addition, Ostermann & Spinsanti (2011) conclude that data quality in VGI might be an important issue but data their contribution to close the gap between science and the public is even more valuable. Furthermore, VGI can play a vital role in emergencies, crisis management, as individuals can collect data, and report on the situation ahead of them before official data sets satellite data become available or disaster relief forces reach the spot. Hence, with VGI it is possible to get a quick and accurate overview of the spatial phenomena present – e.g. roads, forest fires, bird sightings. This can be justified by Craglia et al. (2012), who describe the role of citizens as the main contributors of data. Due to the fact that there are approximately 10 billion social network accounts in 2010, it seems obvious that the potential of these digital volunteered data for a variety of applications – like emergency management, quality of life and environmental monitoring should be utilized.

Current Research Trends in VGI

Of particular interest in VGI are data representing how humans perceive the world using their sensing organs – as sensors cannot measure such phenomena. Humans are able to detect approximately 1 trillion smells (Bushdid, Magnasco, Vosshall, & Keller, 2014), but such olfactory data are difficult to record, analyze and map. Nevertheless, the relationship between smell and space is shown in Quercia, Schifanella, Aiello, & McLean (2015), McLean (2016), as well as Henshaw (2013). MacDonald, Cummins, & Macintyre (2007) elaborate on the relationship between odor and socio-economic boundaries. Hence, small maps originating from volunteered data may serve as additional data source for spatial segregation simulation. This fact is subject to a European research project investigating the impact of open data and volunteered data sources – including olfactory data – for spatial segregation simulation in urban environments.

A certain level of Quality of Life (QoL) of citizens is a target of spatial planning in cities and encompasses ecological, social, and economic aspects of living (Haslauer, Delmelle, Keul, Blaschke, & Prinz, 2015). In order to assess QoL. it is possible to use subjective, individual perceptions of citizens or objective secondary data sources. Although Haslauer et al. (2015) show that there is a strong match between objective and subjective data, they stress that there is a certain spatial heterogeneity in residential QoL perceptions. Another example of citizen science in QoL includes the integration of qualitative contextual data to identify the contextual factors that strongly influence asthma (Keddem et al., 2015). Hence, in QoL studies, VGI contributes as on the one hand as robust and reliable data source and on the other hand as data source for validation purposes.

The term Emotional Mapping is used to describe the approach to map how an individual or a group of individuals perceive space – i.e. to map their emotions with respect to the urban context. Here highlight two approaches are highlighted: EmoMap (Klettner, Huan, & Schmidt, 2011) makes use of a Smartphone application to collect volunteered data on individual emotions with respect to the spatial and temporal context. In contrast, a sentiment analysis from twitter feeds applies natural language processing, computational linguistics and text analysis to extract information. Frank, Mitchell, Dodds, & Danforth, (2013) use a collection of 37 million geo-located tweets to characterize the movement of 180000 individuals together with their happiness – which is expressed in their twitter feeds. Frank et al. (2013) conclude that the expressed happiness increases logarithmically with the distance from the average location.

Other examples of citizen science encompass the usage of VGI for movement analysis purposes. Hawelka et al. (2014) conclude that geo-located twitter feeds may be regarded as proxy for global mobility patterns. They validate their hypothesis by comparing their results with global tourism statistics and other mobility patterns published. Similar to Hawelka et al. (2014), Sagl, Delmelle, & Delmell (2014) utilize mobile phone to evaluate human activity in an urban environment. In addition, social media, e.g. Twitter data, have not only been used as proxy for mobility, but also to analyze the influence on the spread level of crime news in the public (Lampoltshammer, Kounadi, Sitko, & Hawelka, 2014) as well socio-demographic analysis of cities to get a better understanding of the tangible and intangible social infrastructure (Hofer, Lampoltshammer, & Belgiu, 2015).

FUTURE RESEARCH DIRECTIONS

Citizen-driven Geographic Information Science has already achieved a significant impact on the overall participatory science movement. Yet, these achievements are only the beginning as technology evolves, so does societal thinking. In the following, the authors present, from their point of view, three major pillars that will support the sustainably of citizen science in the area of geographic information science (see Figure 2).

The first pillar is represented by the concept of *Big Geo Data* (Miller & Goodchild, 2015). While all challenges relevant to the concept of Big Data apply here as well, there are some specific aspects that have to be considered, when dealing with geographic data in the Big Data context. Firstly, the geographic aspect enables not only a topic-wise or semantic analysis of networks contained within the "big data pool", but also regarding their geographic location. By doing so, local, regional, and global aspects

Figure 2. The pillars of future citizen-driven geographic information science

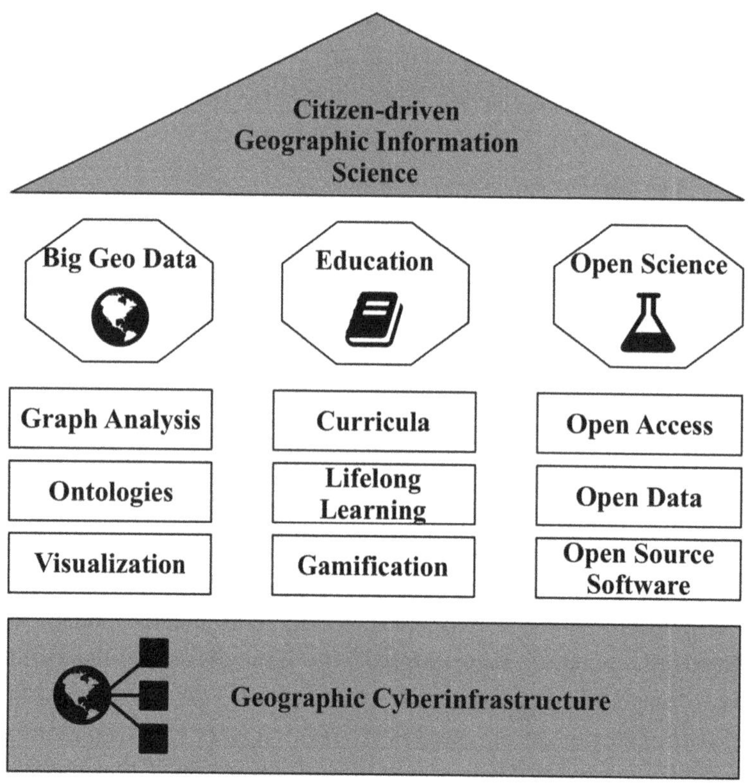

of the data can not only be discovered, but also be set into relation with other observed phenomena at the same place or somewhere else. One way of achieving this kind of analysis can be found in form of graph databases. These databases, as the name implies, structure their entire data as graphs, with nodes representing data sets or entities and (weighted) edges as the relationships between them. Such databases have already been successfully employed in the area of classifying remote sensing data (Lampoltshammer & Wiegand, 2015) based on a priori-modeled knowledge and could potentially be extended towards coverage of spatial and temporal aspects as well. Another area that has the potential to handle very large amounts of georeferenced data are ontologies. Although ontologies are already a fundamental tool to model and share expert knowledge, there still exist issues regarding classifications of large data sets in real-time. Furthermore, there is still the existing lack of a common methodology regarding the design, creation, and maintenance of ontologies (Neuhaus et al., 2013). Although there do exist efforts towards the solution of this issue (e.g., Lampoltshammer & Heistracher, 2014) the issue has not been solved yet. Besides the before-mentioned challenges, knowledge harmonization is a big topic, providing the necessary means for semantic interoperability (Janowicz, 2009). Finally, visual data exploration is key in order to get an overall idea of the data at hand, as well can it support individuals during hypothesis generation (Kehrer et al., 2008) and results dissemination.

The second pillar is dedicated to *Education*. The concept of citizen science in terms of participation should be already included in curricula early as elementary school in order to get pupils familiar with the idea of participating in the task of better understanding of their environment. But also secondary schools should include and further elaborate on such concepts, as discussed by Jekel, Koller, & Strobl (2011). A step in the right direction can be found, e.g., in the Austrian grant program called "Sparkling Science" (SparklingScience, 2016). The idea of this program by the Federal Ministry of Science, Research and Economy (BMWFW) started way back in 2007 and presents a unique opportunity to foster young scientists. Up to 260 projects have been funded so far, with young people working closely together with scientists and experts. They take an active part by working on their own on distinct facets and therefore contributing to the overall goal of the particular project. But they are not only executing certain tasks, they are also – from the very beginning – involved in the design of the project. Furthermore, they are also disseminating their results and the results of the entire project on various events airing at schools, universities, up scientific conferences. The funding program is not limited to a particular field of science and therefore opens up possibilities for nearly every interest group. The topics range from biology, acoustics, informatics, and literature, up to art and migration research. While involvement during the first, second, and even third educational phase is important, also lifelong learning-oriented aspects have to be part of future endeavors as the demographic curve is shifting. An important factor for the success of any kind of contributed work is intrinsic motivation. To foster this kind of motivation, the concept of gamification offers a high potential. Gamification is the idea to trigger behavioral outcomes like in games as well as associated motivational and emotional states (Huotari & Hamari, 2012; Hamari, 2013). Hamari, Koivisto, and Sarsa (2014) have demonstrated according to literature that gamification-based concepts indeed work; yet it is a multi-faceted environment and other important influential factors may not be neglected. A successful and citizen science-relevant adoption of gamification is presented in the work of Martella, Kray, and Clementini (2015). They introduced a gamification framework, particularly designed for volunteered geographic information. As this framework is rather new, more projects have to actually make use of it in order to foster the evolution of the framework.

The third pillar covers the concept of *Open Science*. One important aspect is the idea of Open Access. In the academic area, this is mostly related to the open availability of publications. This availability is important for researchers, as it was demonstrated that open access publications tend to be cited more often (Lawrence, 2001; Harnad & Brody, 2004), which in turn represent one measure of knowledge distribution. As quantitative metrics such as citations counts are part of measuring the success of researchers (Brody, 2013), this has a significant impact on further grant allocations and therefore the availability of projects in all areas, including citizen science-oriented projects. While there are possibilities to publish results via open access (e.g., by self-archiving or open access journals, see Björk, Laakso, Welling, & Paetau, 2014), this approach has some caveats that may not be neglected. On the one hand side, these open access offerings are not coming for cheap and researches and institutions respectively have to pay several thousand Euros for the article to be released under an open access license. While this may not be an issue for big institutions, it is an issue for smaller organizations and individual researchers without huge financial backing. This leads to an imbalance, which also affects citizen scientists, as it makes the situation even more difficultly for them to publish their results. One the other hand, the self-archived version may come for free, but they often do not present the final results and could include potential errors, which have not been corrected yet. This in turn can affect citizen scientists that may not have the particular experience to identify these potential pitfalls when relying on the previous work of others. Furthermore, if the data are available towards a larger, geographically extended community, errors can spread throughout numerous projects quickly, having a significant impact on the overall sustainability of results of these projects. While many research works have already addressed the issue of information and data quality (Lee, Strong, Kahn, & Wang, 2002; Pipino, Lee, & Wang, 2002), current approaches specifically address requirements and needs from the Open Data community (Umbrich, Neumaier, & Polleres, 2015; Höchtl & Lampoltshammer, 2016), a particularly important group for citizen science. These efforts have to be strengthened even more in the future, as open data are key when it comes to the core elements of many citizen science projects. Finally, contributions out of the open source community have to increase, as software licensing costs are a major issue for individuals and volunteered workers. As the availability of open software increases, the newly gained possibilities can support not only educational programs (e.g., Steiniger & Hunter, 2010), but also push beyond existing closed mindsets to foster innovations (Lakhani & Panetta, 2007) in transdisciplinary ways, involving stakeholders from the public, companies, and the public administration.

CONCLUSION

Citizen Science has come a long way from a limited circle of privileged individuals towards a phenomenon that enables virtually everybody to participate in the deeper understanding of the global environment. Yet, the lack of availability of technology and compatible social and political structure in some countries of the world still present major hurdles that have to be taken in order to enable truly a global citizen science culture. Volunteered Geographic Information as particular form of citizen sciences offers the possibility to collaborate with citizens to gather dynamic, in situ data about environmental phenomena. This kind of data acquisition is much faster as common sensor arrays and can, in addition, build on human logic and intuition, both attributes that are not yet fully replaceable by artificial intelligence. Yet there are challenges that have to be overcome, in order to foster this movement. As the authors have demonstrated throughout the chapter, data fusion is a critical point, as VGI is not the only form of data that is processed on geographic cyberinfrastructures. Yet, data fusion of quantitative and qualitative data can be complex and context-aware solutions have to be created to overcome this impediment. It is furthermore important to create awareness of potential errors or biases in VGI as it is distributed throughout the community. Thus, new ways of quality insurance will be necessary to limit error propagation. Finally, funding is a severe issue. Classical research projects struggle already with the current national and international funding strategy, which puts even more pressure on voluntary projects by citizens, even if they are conducted together with professional scientists. As a change of this situation is not on the horizon, its is up to the local public administrations to seek out towards the citizens and to join forces in order to foster sustainability on a local, regional – and on the long run – global level.

REFERENCES

Björk, B. C., Laakso, M., Welling, P., & Paetau, P. (2014). Anatomy of green open access. *Journal of the Association for Information Science and Technology*, *65*(2), 237–250. doi:10.1002/asi.22963

Bogers, M., & West, J. (2012). Managing distributed innovation: Strategic utilization of open and user innovation. *Creativity and Innovation Management*, *21*(1), 61–75. doi:10.1111/j.1467-8691.2011.00622.x

Brody, S. (2013). Impact factor: Imperfect but not yet replaceable. *Scientometrics*, *96*(1), 255–257. doi:10.1007/s11192-012-0863-x

Bushdid, C., Magnasco, M. O., Vosshall, L. B., & Keller, A. (2014). Humans Can Discriminate More than 1 Trillion Olfactory Stimuli. *Science*, *343*(6177), 1370–1372. doi:10.1126/science.1249168

Carr, A. J. L. (2004). Why do we all need community science? *Society & Natural Resources*, *17*(9), 841–849. doi:10.1080/08941920490493846

Cho, M. K., McGee, G., & Magnus, D. (2006). Lessons of the stem cell scandal. *Science*, *311*(5761), 614–615. doi:10.1126/science.1124948

Conrad, C. C., & Hilchey, K. G. (2011). A review of citizen science and community-based environmental monitoring: Issues and opportunities. *Environmental Monitoring and Assessment*, *176*(1-4), 273–291. doi:10.1007/s10661-010-1582-5

Craglia, M., de Bie, K., Jackson, D., Pesaresi, M., Remetey-Fülöpp, G., Wang, C., & Woodgate, P. et al. (2012). Digital Earth 2020: Towards the vision for the next decade. *International Journal of Digital Earth*, *5*(1), 4–21. doi:10.1080/17538947.2011.638500

Craglia, M., Goodchild, M. F., Annoni, A., Camara, G., Gould, M., Kuhn, W., & Parsons, E. (2008). Next-generation digital earth: A position paper from the vespucci initiative for the advancement of geographic information science. *International Journal of Spatial Data Infrastructures Research*, *3*(6), 146–167.

De Smith, M. J., Goodchild, M. F., & Longley, P. (2007). *Geospatial analysis: a comprehensive guide to principles, techniques and software tools*. Leicester, UK: Troubador Publishing Ltd.

Frank, M. R., Mitchell, L., Dodds, P. S., & Danforth, C. M. (2013). Happiness and the patterns of life: A study of geolocated tweets. *Scientific Reports*, *3*(2625), 1–9.

Goodchild, M. F. (2007). Citizens as sensors: The world of volunteered geography. *GeoJournal*, *69*(4), 211–221. doi:10.1007/s10708-007-9111-y

Gore, A. (1998). The digital earth: understanding our planet in the 21st century. *Australian Surveyor*, *43*(2), 89-91.

Gruber, T. R. (1993). A translation approach to portable ontology specifications. *Knowledge Acquisition*, *5*(2), 199–220. doi:10.1006/knac.1993.1008

Haklay, M. (2010). How good is volunteered geographical information? A comparative study of OpenStreetMap and Ordnance Survey datasets. *Environment and Planning. B, Planning & Design*, *37*(4), 682–703. doi:10.1068/b35097

Haklay, M. (2013). Citizen science and volunteered geographic information: Overview and typology of participation. In D. Sui, S. Elwood, & M. Goodchild (Eds.), *Crowdsourcing geographic knowledge* (pp. 105–122). Dordrecht: Springer Netherlands. doi:10.1007/978-94-007-4587-2_7

Hamari, J. (2013). Transforming homo economicus into homo ludens: A field experiment on gamification in a utilitarian peer-to-peer trading service. *Electronic Commerce Research and Applications*, *12*(4), 236–245. doi:10.1016/j.elerap.2013.01.004

Hamari, J., Koivisto, J., & Sarsa, H. (2014). Does gamification work? - a literature review of empirical studies on gamification. In *47th Hawaii International Conference on System Sciences (HICSS)*, 2014 (pp. 3025-3034). IEEE. doi:10.1109/HICSS.2014.377

Harnad, S., & Brody, T. (2004). Comparing the impact of open access (OA) vs. non-OA articles in the same journals. *D-Lib Magazine*, *10*(6).

Harvey, F., Kuhn, W., Pundt, H., Bishr, Y., & Riedemann, C. (1999). Semantic interoperability: A central issue for sharing geographic information. *The Annals of Regional Science*, *33*(2), 213–232. doi:10.1007/s001680050102

Haslauer, E., Delmelle, E. C., Keul, A., Blaschke, T., & Prinz, T. (2015). Comparing Subjective and Objective Quality of Life Criteria: A Case Study of Green Space and Public Transport in Vienna, Austria. *Social Indicators Research*, *124*(3), 911–927. doi:10.1007/s11205-014-0810-8

Hawelka, B., Sitko, I., Beinat, E., Sobolevsky, S., Kazakopoulos, P., & Ratti, C. (2014). Geo-located Twitter as proxy for global mobility patterns. *Cartography and Geographic Information Science*, *41*(3), 260–271. doi:10.1080/15230406.2014.890072

Heath, T., & Bizer, C. (2011). Linked data: Evolving the web into a global data space. *Synthesis Lectures on the Semantic Web: Theory and Technology*, *1*(1), 1-136.

Henshaw, V. (2013). *Urban smellscapes: understanding and designing city smell environments*. Oxford, UK: Taylor & Francis Group Ltd.

Höchtl, J., & Lampoltshammer, T. J. (2016) ADEQUATe - Analytics and Data Enrichment to Improve the Quality of Open Data. In Parycek, P.; Edelmann, N. (Eds.), *Proceedings of the International Conference for E-Democracy and Open Government CeDEM16*, (pp. 27-32). Krems: Edition Donau-Universität Krems.

Hofer, B., Lampoltshammer, T. J., & Belgiu, M. (2015). Demography of Twitter Users in the City of London: An Exploratory Spatial Data Analysis Approach. In J. Brus, A. Vondrakova, & V. Vozenilek (Eds.), *Modern Trends in Cartography* (pp. 199–211). Cham: Springer International Publishing. doi:10.1007/978-3-319-07926-4_16

Huotari, K., & Hamari, J. (2012). Defining gamification: a service marketing perspective. In *Proceeding of the 16th International Academic MindTrek Conference* (pp. 17-22). ACM. doi:10.1145/2393132.2393137

Janowicz, K. (2009). Semantic Interoperability. In B. Warf (Ed.), *Encyclopedia of Geography*. SAGE Publications.

Jekel, T., Koller, A., & Strobl, J. (2011). Research – education cooperations for GI in secondary education. In *International perspectives on GIS in secondary education* (pp. 27-36). New York: Springer.

Jiang, B., & Thill, J. C. (2015). Volunteered Geographic Information: Towards the establishment *of a new paradigm. Computers, Environment and Urban Systems, 53*(September), 1–3. doi:10.1016/j.compenvurbsys.2015.09.011

Keddem, S., Barg, F. K., Glanz, K., Jackson, T., Green, S., & George, M. (2015). Mapping the urban asthma experience: Using qualitative GIS to understand contextual factors affecting asthma control. *Social Science & Medicine, 140*(September), 9–17. doi:10.1016/j.socscimed.2015.06.039

Kehrer, J., Ladstadter, F., Muigg, P., Doleisch, H., Steiner, A., & Hauser, H. (2008). Hypothesis generation in climate research with interactive visual data exploration. *Visualization and Computer Graphics. IEEE Transactions on, 14*(6), 1579–1586.

Klettner, S., Huang, H., & Schmidt, M. (2011). EmoMap – considering emotional responses to space for enhancing LBS. In *Advances in Location-Based Services Proceedings 8th International Symposium on Location-Based Services*. Springer.

Klien, E., Lutz, M., & Kuhn, W. (2006). Ontology-based discovery of geographic information services—An application in disaster management. *Computers, Environment and Urban Systems, 30*(1), 102–123. doi:10.1016/j.compenvurbsys.2005.04.002

Kruger, L. E., & Shannon, M. A. (2000). Getting to know ourselves and our places through participation in civic social assessment. *Society & Natural Resources, 13*(5), 461–478. doi:10.1080/089419200403866

Lakhani, K. R., & Panetta, J. A. (2007). The principles of distributed innovation. *Innovations, 2*(3), 97–112. doi:10.1162/itgg.2007.2.3.97

Lampoltshammer, T. J., & Heistracher, T. (2014). Ontology evaluation with Protégé using OWLET. *Infocommunications Journal, 6*(2), 12–17.

Lampoltshammer, T. J., Kounadi, O., Sitko, I., & Hawelka, B. (2014). Sensing the public's reaction to crime news using the 'Links Correspondence Method'. *Applied Geography (Sevenoaks, England), 52*, 57–66. doi:10.1016/j.apgeog.2014.04.016

Lampoltshammer, T. J., & Wiegand, S. (2015). Improving the Computational Performance of Ontology-Based Classification Using Graph Databases. *Remote Sensing, 7*(7), 9473–9491. doi:10.3390/rs70709473

Lawrence, A. (2006). No personal motive? Volunteers, bio- diversity, and the false dichotomies of participation. *Ethics Place and Environment*, *9*(3), 279–298. doi:10.1080/13668790600893319

Lawrence, S. (2001). Online or invisible. *Nature*, *411*(6837), 521–523. doi:10.1038/35079151

Lee, Y. W., Strong, D. M., Kahn, B. K., & Wang, R. Y. (2002). AIMQ: A methodology for information quality assessment. *Information & Management*, *40*(2), 133–146. doi:10.1016/S0378-7206(02)00043-5

Lerner, J., & Tirole, J. (2002). Some simple economics of open source. *The Journal of Industrial Economics*, *50*(2), 197–234. doi:10.1111/1467-6451.00174

Macdonald, L., Cummins, S., & Macintyre, S. (2007). Neighbourhood fast food environment and area deprivation—substitution or concentration? *Appetite*, *49*(1), 251–254. doi:10.1016/j.appet.2006.11.004

Martella, R., Kray, C., & Clementini, E. (2015). A Gamification Framework for Volunteered Geographic Information. In F. Bacao, M. Y. Santos, & M. Painho (Eds.), *AGILE 2015* (pp. 73–89). Cham: Springer International Publishing. doi:10.1007/978-3-319-16787-9_5

McLean, K. (2016). Smellmap: Amsterdam—Olfactory Art & Smell Visualisation. *Leonardo*.

Miller, H. J., & Goodchild, M. F. (2015). Data-driven geography. *GeoJournal*, *80*(4), 449–461. doi:10.1007/s10708-014-9602-6

Neuhaus, F., Vizedom, A., Baclawski, K., Bennett, M., Dean, M., Denny, M., …, , & Obrst, L. (2013). Towards ontology evaluation across the life cycle. The Ontology Summit 2013. *Applied Ontology*, *8*(3), 179–194.

Newman, G., Wiggins, A., Crall, A., Graham, E., Newman, S., & Crowston, K. (2012). The future of citizen science: Emerging technologies and shifting paradigms. *Frontiers in Ecology and the Environment*, *10*(6), 298–304. doi:10.1890/110294

Ostermann, F. O., & Spinsanti, L. (2011). A conceptual workflow for automatically assessing the quality of volunteered geographic information for crisis management. In *Proceedings of 14thAGILE International Conference on Geographic Information Science*.

Pipino, L. L., Lee, Y. W., & Wang, R. Y. (2002). Data quality assessment. *Communications of the ACM*, *45*(4), 211–218. doi:10.1145/505248.506010

Quercia, D., Schifanella, R., Aiello, L. M., & McLean, K. (2015). *Smelly maps: the digital life of urban smellscapes*. arXiv preprint arXiv:1505.06851

Sagl, G., Delmelle, E., & Delmelle, E. (2014). Mapping Collective Human Activity in an Urban Environment based on Mobile Phone Data. *Cartography and Geographic Information Science*, *41*(3), 272–285. doi:10.1080/15230406.2014.888958

SparklingScience. (2016). Retrieved May 09, 2016, from http://www.sparklingscience.at/en

Steiniger, S., & Hunter, A. J. S. (2010). Teaching GIScience with Free and Open Source Software? A first assessment. In *6th International Conference of GIScience*.

Sui, D., Elwood, S., & Goodchild, M. (Eds.). (2013). *Crowdsourcing geographic knowledge: volunteered geographic information (VGI) in theory and practice*. Springer Science & Business Media. doi:10.1007/978-94-007-4587-2

Umbrich, J., Neumaier, S., & Polleres, A. (2015). Quality Assessment and Evolution of Open Data Portals. In *Future Internet of Things and Cloud (FiCloud), 2015 3rd International Conference on* (pp. 404-411). IEEE. doi:10.1109/FiCloud.2015.82

Whitelaw, G., Vaughan, H., Craig, B., & Atkinson, D. (2003). Establishing the Canadian Community Monitoring Network. *Environmental Monitoring and Assessment*, *88*(1), 409–418. doi:10.1023/A:1025545813057

Yang, C., Raskin, R., Goodchild, M., & Gahegan, M. (2010). Geospatial cyberinfrastructure: Past, present and future. *Computers, Environment and Urban Systems*, *34*(4), 264–277. doi:10.1016/j.compenvurbsys.2010.04.001

KEY TERMS AND DEFINITIONS

Big Geo Data: Big data sets similar to Big Data with all their associated challenges but with the additional complexity of space and local context.

Digital Earth: A concept that sees the entire planet earth being represented as a digital globe to foster understanding of its inner processes and the human role within them.

Geographic Cyberinfrastructure: An infrastructure that possess high performance computing capabilities, a high level of extensibility and accessibility, as well as multiple data input channels (including data from citizens) in order to perform spatio-temporal analysis operations.

Open Science: The concept of everybody being able to participate in scientific projects as well as resulting data and information are made again freely available for everybody.

Quality of Life: A fusion of qualitative and subjective as well as quantitate data regarding variables that impact living quality.

Volunteered Geographic Information: A particular form of citizen science where citizens act as proxies to gather dynamic, in situ data about environmental phenomena.

Chapter 12
Can Citizen Science Seriously Contribute to Policy Development?
A Decision Maker's View

Colin Chapman
Welsh Government, UK

Crona Hodges
University of Aberystwyth, UK

ABSTRACT

This chapter considers the potential for citizen science to contribute to policy development. A background to evidence-based policy making is given, and the requirement for data to be robust, reliable and, increasingly, cost-effective is noted. The potential for the use of 'co-design' strategies with stakeholders, to add value to their engagement as well as provide more meaningful data that can contribute to policy development, is presented and discussed. Barriers to uptake can be institutional and the quality of data used in evidence-based policy making will always need to be fully assured. Data must be appropriate to the decision making process at hand and there is potential for citizen science to fill important, existing data-gaps.

INTRODUCTION

The research presented in this chapter has contributed to the European Union FP7 project COBWEB (EU FP7 reference number: 308513). COBWEB is a multidisciplinary citizen observatories project that aims to develop a data collection and sharing platform for crowdsourced or citizen science data, using standards and interoperability principles (Leibovici *et al.* 2015; Hodges *et al.* 2014). Contributors are mostly non-expert and can use this platform to customise their own data collection projects or campaigns, communicate with other contributors, and share and find data that are of interest to them. An overarching goal of the project is to enable the collection of data that ultimately can be sourced and accessed

DOI: 10.4018/978-1-5225-0962-2.ch012

and used by policy makers; this data must be suitable, robust and of a known quality for evidence-based policy making. As such, the Welsh Government has been involved since the start of COBWEB and contributes the views of government and decision makers to the development of the project. The views presented in this chapter are based on the involvement of the authors in the COBWEB project, rather than a comparison of a range of structure data collection methods, and draws on personal experience in the policy development arena.

With respect to policy development, governments are interested in the quality of data, knowing where they came from, why and, whether or not the data are fit for purpose. Collecting environmental datasets on the geographical and temporal scales required to make informed decisions and to develop a broad and robust evidence base can be an expensive and resource costly exercise when employing traditional methods of data collection. Budgets are becoming more and more limited, and governments are increasingly open to innovative and cost-effective solutions to source reliable data (Haklay *et al.* 2014). Concurrently, mobile devices are becoming increasingly ubiquitous and more powerful, enabling ordinary citizens and volunteers to contribute more and more data about their local environments than ever before. Decision makers are acutely aware of the potential for these developments to significantly contribute to the 'data gap' and, increasingly, government papers are calling for more and more volunteer data (POSTnote 2014).

The importance of stakeholder engagement as a central precept to policy development and decision making cannot be overstated although it has been argued that the quality of decisions made through stakeholder participation is strongly dependant on the nature of the process leading to them (Reed 2008). Directly working with stakeholders, COBWEB presents the opportunity to explore and demonstrate effective methods of engagement with different stakeholder groups that are organising high volume environmental data collection projects. This is an opportunity for Welsh Government to learn effective measures to strengthen the interface between policymakers, citizens and scientists with the aim of educating, informing and communicating. The potential for citizen science as a method to stimulate local participation in environmental governance in line with the key themes of sustainable development is also an area of great interest to government.

The question that is being tested here is simply whether or not there is real potential for volunteered data on the environment, on the whole collected by the non-expert, to contribute to evidence-based policy making. If this is the overall goal of this work (and it is ongoing) then there are a number of objectives that contribute to answering this:

- To identify the main barriers to the uptake of projects like COBWEB.
- To develop strategies that can be applied to overcome these barriers.
- To incorporate effective stakeholder engagement to better align the objectives and applications of citizen science projects with those of the policy/decision makers.

BACKGROUND

The term 'citizen science' is used within this chapter to denote the collection of spatially referenced data and information by persons involved in on-the-ground environmental projects that use the internet to upload, manage or contribute that data to the wider public and other end-users for the greater good and knowledge of others. The spatial component in this respect is directly related to the term 'volunteered geographic information' (VGI) of which there is a large body of scientific literature (Goodchild 2007). In many ways, citizen science data that are of a spatially referenced nature is indeed VGI (Haklay 2013).

In order to understand the value of citizen science data for policy development and decision making in government, the processes and concepts that underpin decision making or policy development in government must be understood. These processes not only provide a contextual backdrop to understand the value of citizen science but also provide insight into how the suitability of a piece of data or information can be assessed and considered to support or inform a policy or decision. Policy making can be a complex and dynamic process that is reliant on a range of factors or processes that seek to ensure that any decision, policy or intervention is based upon the best available supporting data, information or 'evidence.' The idea of evidence-based decision making is central to the concept of policy development, formulation, implementation and evaluation and in order to understand the potential for citizen science to add value to the decision making process, the mechanisms of how policy is created, developed and monitored must be understood.

WHAT IS 'EVIDENCE'

The Oxford English Dictionary defines evidence as: 'the available body of facts or information indicating whether a belief or proposition is true or valid' (Oxford Dictionaries 2013). This definition is at the heart of policy making within the contemporary governments of the UK and stresses the importance of providing supporting information or material in support of assertions, actions or statements. England's Department of for Environment, Food and Rural Affairs (DEFRA) suggests that evidence can have a broader definition:

'Evidence is the best available information used to support decisions in developing, implementing and evaluating policy, operations and services.' (DEFRA 2006)

With specific regard to environmental policy making within government, evidence can comprise a number of diverse types and forms, although the underlying theme that unites these elements is the use of the appropriate scientific methods in their collection, creation and synthesis. The following list provides an overview of the types of activities that are generally considered credible sources of evidence that are suitable to inform policy creation and development (DEFRA 2014):

- Research & Development
- Monitoring & Surveillance
- Secondary Analysis and Synthesis

This short list is not exhaustive and, of course, does not account for individual quality of research, analysis or monitoring exercises. Of note is the absence of unsynthesised data in a raw or primary state – this provides a broad but useful example of the separation of 'raw data' from the scientific collation, analysis and interpretation of the same. In other words, the production of evidence is more complex than the collection of data or information about observations or phenomena; the informed interpretation, analysis and contextualization of those data is where the real value for policy lies. These points must be considered where it is the case that citizen science aspires to contribute to policy development and the *evidence base*.

EVIDENCE BASED POLICY IN EUROPE

In 1999 the Modernising Government white paper noted the importance of robust and fit-for-purpose evidence in informing the creation of policies in the UK. It stated that Government 'must produce policies that really deal with problems that are forward-looking and shaped by evidence rather than a response to short-term pressures, that tackle causes not symptoms' (DEFRA, 2006). Since this declaration there has been a considerable shift of emphasis towards evidence based (or evidence informed) policy in the UK public sector. This can be seen in the creation of evidence strategies such as the Department for Environment, Food & Rural Affairs (DEFRA) Evidence & Investment Strategy 2010-2013 and the subsequent DEFRA One Evidence programme. Within Wales, and particularly within the environment sector, the evidence agenda has gathered considerable momentum and is of paramount concern in the creation, refinement and evaluation of policy and management of natural resources.

Within the European Union, a single unified concept of evidence based policy is less well defined. Evidence based policy has been defined as 'peculiarly British' (Solesbury 2001) but the concepts and elements involved are represented within guidance and policies created by the Commission and these are becoming increasingly prevalent in policy making discussions. For example in June 2014 the 2nd European Risk Summit titled 'Living in an uncertain world' focused on how EU and national policy-makers should communicate risk and formulate evidence-based risk communication in the health, food, environment and pharmaceutical policy areas. *Members of the European Parliament should take a closer look at how European agencies are handling risks and ensure that EU legislation is evidence based and risk informed* (Summit Chair Prof. Ragnar Löfstedt). Similarly, the 7th Environmental Action Programme (Decision No 1386/2013/EU) included as its fifth priority objective *to improve the knowledge and evidence base for Union environment policy*. The emphasis on the role of science and science based evidence as a way of increasing the effectiveness and accountability of EU legislation and the policies of member states is also supported by associations of professional scientists and researchers such as EuroScience (see www.euroscience.org).

In relation to environmental policy in Europe specifically, the need for scientific data relating to the state of natural resources is well documented. Projects such as the European Biodiversity Observation Network (EUBON) demonstrate the intention to more fully integrate biodiversity data into decision making at a range of scales. EUBON is explicitly linked to the 'Strategic Plan for Biodiversity 2011-2020' (see: www.cbd.int/doc/strategic-plan/2011-2020/Aichi-Targets-EN.pdf), in particular to the strategic goal: *By 2020, knowledge, the science base and technologies relating to biodiversity, its values, functioning status and trends, and the consequences of its loss, are improved, widely shared and transferred and applied* (Hoffman *et al.* 2014). The application and integration of these data into environmental policies and legislation are key cornerstones of addressing biodiversity loss. They would form the basis of any measurement of trends, positive or otherwise and allow monitoring of strategies and the creation of milestone targets.

From these statements, the importance of the availability and collection of evidence as a driving force in the creation of strategies and policies is clear. Whilst the process may be less overt, and arguably have a less structured form than as laid out in UK policy development, the consideration of evidence, scientific advice and abstraction, trend analysis and effective communication are shared concerns. Whereas it should be noted that the evidence based policy process can be seen as a conceptual ideal it does ensure the consideration of key factors that surround and influence the uptake of research, data and information as sources of evidence by policy makers. These factors are common to any consideration of the use of evidence to support decision making in the UK or further afield.

THE RELATIONSHIPS BETWEEN CITIZEN SCIENCE AND EVIDENCE-BASED POLICY DEVELOPMENT

Good evidence comes initially from good data. In its simplest case data are the measurements and observations from which evidence is abstracted. Data collection exercises themselves need to be robust and appropriate and the survey design needs to be configured to answer the question in mind. Difficulties for the decision makers come when the only data available is not quite appropriate to the policy question at hand. The use of data collected for another purpose raises questions about their application and use outside of the context and reason for which it was originally collected. That, coupled with sometimes significant issues around data currency can add ambiguity, or negatively affect confidence levels of evidence provided.

How does citizen science derived data fit into this framework of good evidence and add value to policy and decision making? Certainly data, when viewed in isolation, are not a suitable source of evidence for use in decision making. It is not appropriate to think that data collection in and of itself adds value to decision making. It is clear that, in order to realise the value of citizen science in a policy context, any data collection should be scientific and adhere to the principles of good study design (Wiersma 2010) and, in addition, that the data collection method and its subject will be appropriate to fit a pre-defined policy question or need.

The appraisal of evidence and whether or not specific types of information are indeed 'evidence', has to be regarded within a social and political context (Nutley *et al.* 2013). A wide range of variables exist that can affect any judgement of whether evidence is appropriate for policy making outside of study design. Evidence that is persuasive or actionable in a practical way can be attractive to a policy maker (Cameron *et al.* 2011) and similarly, implications of real world considerations such as timescale or availability of resources can influence the adoption of different evidence gathering activities; the latter factor being particularly relevant across all member states and even globally at the time of writing. This leaves considerable opportunity for innovation and development within this area both technologically and conceptually.

Although the *commissioning* of evidence creating activity (be it research, monitoring or evaluation or other analytical processes) needs to demonstrate appropriate degrees of scientific rigor, there remains considerable scope for innovation to transform traditional concepts of *good* evidence and to further contribute to developing an understanding of the theme which the evidence or research is investigating (Sanderson 2009). That is not to say that the large scale uptake of citizen science derived datasets into the decision-making process would be a straightforward process. Evidence derived from these sources would need to convince policy makers, assessment panels or evidence specialists that it was appropriately conceived, designed, implemented and analysed and that it was suitably aligned with the intended purpose for its use or policy outcome. Each project would need to instil sufficient confidence that the findings or data are suitable for use. The greater the confidence in the evidence under consideration, then the greater the chance that it will be drawn upon to support decision making and policy development.

Technological, methodological and conceptual innovation are all vital to driving the collection of new data. For the policy maker or decision maker the ultimate benefit of such innovation would be the subsequent production of more robust, timely and appropriate evidence. The need for new data is increasing and the reality of a dynamic and shifting policy landscape adds additional pressure on the ability to gather data on new themes, in new and efficient ways. It is into this situation where citizen science offers real potential, especially when delivered by utilising internet and mobile technologies as proposed by COBWEB (see also http://www.citizen-obs.eu/). The adoption of any new technology or approach however, will bring with it a set of issues and conditions that need to be considered in order for it to be of value to policy and decision making. This potential needs to be balanced against the primacy of the use of citizen science approaches to answer predefined questions and framed within effective study design as part of a scientific exercise (a research project or a monitoring programme for example). It is when the potential of approaches like citizen science are framed within the context of 'good' evidence where they offer most value.

INITIAL CONCERNS AND CHALLENGES

In order to increase accuracy and confidence in data collection exercises the use of validation and quality assurance methods are widely considered important, if not essential to the validity of citizen science projects (SCU 2013; Delaney *et al.* 2008, Gollan *et al.* 2012). It follows that this advice aligns with evidence assurance practices too, where data quality plays an important part. Therefore an appreciation of the occurrence of bias in the data must be addressed. This is not to state that any dataset will be expected to be free of error or inaccuracies but it is important that there is an awareness or understanding of types of issues to enable specialists and decision makers to take into account these potentially misleading factors.

Appropriate study design can mitigate or address bias in data; detailed and comprehensive metadata becomes a key factor in contextualising datasets for evidence assurance exercises to reduce the risk of inappropriate usage of data, particularly here where a failure to recognise bias in reporting could lead to the misapplication of effort or resource and affect any subsequent evaluation or measurement of impact of that effort. In other words, it is as important for policy and decision making that data is of a *known quality* as it is for any scientific subject matter. The documentation of quality assurance practices and the supply of data quality statements will add value to citizen science derived data and improve its uptake and use/re-use especially in light of the reluctance to use data of unknown provenance by scientists (Alabri *et al.* 2010).

The GPS functionality that is available in mobile devices offers great potential for the collection of spatially referenced information by the relatively non-expert user. With many available applications or 'apps' this can be captured with the observation in an automated way without input from the user. This process reduces the risk of error in the transcription or recording of abstract coordinate references, especially in the case of non-expert users. There can be inconsistency in the quality of GPS derived spatial references however, and this can have subsequent impact on the accuracy of a particular observation (Meek *et al.* 2015). If information on the 'known quality' of observations can be recorded along with the observation itself this can have profound implications on the reuse of those data and the potential for its use in the policy and decision making process. This is an area of research that is being explored as part of the COBWEB project as well as the incorporation of other quality assessments that will assist the end-user in judging whether or not that dataset might be fit for purpose with respect to the policy question or evidence requirements being considered (Wiemann *et al* 2015, Meek *at al* 2014).

Quality has long been a contentious issue with regards to citizen science derived or VGI datasets with numerous references in the available literature (Flanagin & Metzger 2008). The main thrust of discussion is whether observations and recordings made by non-expert citizens are as accurate as those made by specialist or professional recorders. With regards to evidence informed decision making this issue is particularly pertinent; generalised statements about the quality of *citizen science data* on the whole cannot be made due to the diversity in project approaches and subjects addressed however it is acknowledged that *Citizen science can yield high quality, policy relevant information* (Haklay 2015). Certainly a goal in COBWEB is the process of attributing a known confidence to the observation made or the set of observations collected by any one survey as well as to filter observations based on this or any given parameter such as an acceptable numerical range; it is believed that given the ability to effectively interrogate the observations and/or to access the observations and dataset *after* they have already been conflated with authoritative data sources for example, will result in uptake of those data for use in policy making.

The complexity of the survey or data collection process or protocol is another area that is of a concern. This can greatly influence the quality of observations and their recording as well as influence the popularity or uptake of the survey. Pocock *et al.* (2014) rightly note that there appears to be a relationship between complexity of the survey protocol and the number of participants that may be motivated to take a project, this has also been seen during the 'co-design' phase in the COBWEB project (see below). This can impact on the statistical validity of a given data set if too few records are submitted, or potentially threaten the longevity of a project if the protocol is so demanding as to become a barrier to participation. It also follows that projects that require complex observations of particular phenomena may be associated with a higher level of erroneous recording by non-specialists (Dickinson *et al.* 2010). This can be mitigated to a degree by the production of supporting materials or professional specialist support in identification exercises.

STAKEHOLDER ENGAGEMENT: THE BENEFITS OF GOOD ENGAGEMENT PRACTICES

Embracing the concept of government engaging with citizens through citizen science projects not only raises the potential of enriching the evidence-base with fit for purpose data of a known and acceptable quality, but there are also additional intangible benefits. The importance of engagement, participation in governance and the communication of knowledge and development of skills are important aspects of sustainable development (SD). Informed and politically active citizens can play a crucial role in the success of local, national and global policies by playing a significant part in the development of policies (Irvin and Stansbury 2004). In the literature this type of VGI where the public can contribute to the decision making process of government is often referred to as 'public participation GIS' or PPGIS (Kingston *et al.* 2010, Ganapati 2011) though there are some notable differences between the two (Tulloch 2008). Citizen science provides this potential for government to engage with citizens in a practical and meaningful way, contributing to knowledge and skill development and providing the potential for *additional value* to be gained from being involved in these early stages of evidence gathering.

Shirk *et al.* (2012) suggest the need for 'deep collaboration between stakeholders in the early stages [of project design] to achieve environmental policy and management goals.' This deep collaboration should ideally allow for the balancing of the aims, objectives and motivations of those commissioning/

designing the project, be they scientists, policy makers or motivated individuals, and those who will be participating in the project. This deep collaboration also aligns well with the political need to engage meaningfully with citizens and to encourage and value input and participation in issues of interest to them. Of course other factors play key roles, such as trust building and the development of shared and committed understanding (Ansell and Gash 2008). In the case of COBWEB, stakeholders (project designers, scientists, citizens and policymakers) aim to look at the creation of a project together, in order to make it more effective. Effectiveness in this sense comes from the balancing of scientific, educational, motivational and practical elements. Keys to this process are the establishment and agreement of a shared vision and of increased levels of mutual understanding between all stakeholders that take into consideration the different perspectives and expectations of those involved.

Co-design and co-creation of projects is a growing theme in the research community in the consideration of citizen science and digital innovation. It is referred to frequently in the Socientize Green Paper on Citizen Science (Citizen Science for Europe: Towards a better society of empowered citizens and enhanced research see: www.socientize.eu/?q=eu) where it is noted that co-design in citizen science bridges the gap not only between policy and citizens but also increases understanding between scientists/research community, citizens and policymakers as a whole. This increased understanding has the potential to not only increase the relevance of research to policy makers by strengthening the science-policy interface but also to increase the value and engagement with citizens and society with the aim of influencing the research agenda to maximise societal value of research outputs. This cross-sectoral dialogue provides a compelling opportunity to embody core principles of SD in policy creation through the use of citizen science projects, particularly in relation to the environment as well as to help target the effectiveness and sensitivity of policies and research at local (and wider) scales.

The co-design process can be considered key to ensuring the sustained motivation of stakeholder involvement. The success of a citizen science project may rely on having a critical mass of recorders (human sensors) to collect, analyse or review an appropriate amount of data. Motivation can be effected by many factors within both the design and implementation of a citizen science project. UK Environmental Observation Framework's 'Guide to Citizen Science' (Tweddle *et al.* 2012) note the importance of having motivated, interested and keen participants as a primary factor in any citizen science exercise. It follows that any such project should have, as a core objective, an effective mechanism or strategy to motivate and retain participants. Failure to successfully secure a group of enthused, motivated and able (or enabled) participants can have a direct impact on the quality, quantity and longevity of the data or data series collected and ultimately, of course, the scientific success of the project itself which in turn, will have an impact on the potential for that data to be reused in policy development.

Citizen science can positively effect and influence more sustainable behaviours or attitudes in citizens and communities and allow a more informed dialogue with stakeholders as part of the democratic process. Furthermore, the potential for citizens to participate in projects that generate data that could well contribute positively to local, national and global issues, e.g. biodiversity loss or the spread of invasive/non-native species. Gollan *et al.* (2012) suggest 'citizen science has the potential to bring society closer to science and to nature, bringing about a sense of ownership and helping create the kind of society that works to protect its natural environment' (SCU 2013).

If citizen science can affect behaviour positively then its usage may provide a meaningful way not only to engage with citizens but also to empower and enthuse them through participation in specific projects and/or events. The citizen engagement elements stressed under SD principles can be supported through the use of citizen science type activities not only through establishing a novel and engaging method of communication, education and knowledge exchange between public authorities (or the scientific community) but also through positive action in the participation of environmental science to contribute to the understanding and management of our natural resources. Access to and distribution of these data online provide a tangible presence of the effort of individuals or groups to provide evidence to scientists and policy makers to ensure that they have access to the best available data and information to support their respective research or responsibilities.

The extent to which citizen science can alter behaviour, effect societal change and increase awareness is hard to measure (SCU 2013). The education value from citizen science can vary depending on the project, the complexity of the protocol and the level of resource available to provide contextual and educational material. However the ability of citizen science to encompass a means to provide an engagement mechanism, to encourage participation in science and enable the collection of data to support and effect policy development align well with the concepts of effective policy making and will require further scrutiny as citizen science continues to develop.

MAKING DATA MORE AVAILABLE

The ability to understand the context and make up of a project, its objectives and aims and the provenance and quality of its data and outputs is captured in the metadata associated with the data created by that project. Structured and standardised metadata allows a rich test environment to determine the suitability of a dataset for communicating key concepts, metrics and narratives. The role of metadata for communicating essential contextual information for spatial information is a primary factor that underpins the establishment of an infrastructure for spatial information through the European driven INSPIRE Directive (see: http://inspire.ec.europa.eu/). The definition below is taken from the Access to European Law website (See: http://eur-lex.europa.eu/legal-content/EN/ALL/?uri=CELEX:32008R1205).

The definition of a set of metadata elements is necessary in order to allow identification of the information resource for which metadata is created, its classification and identification of its geographic location and temporal reference, quality and validity, conformity with implementing rules on the interoperability of spatial data sets and services, constraints related to access and use, and organisation responsible for the resource. Metadata elements related to the metadata record itself are also necessary to monitor that the metadata created are kept up to date, and for identifying the organisation responsible for the creation and maintenance of the metadata.

The importance of metadata, for the policy maker, is its potential to communicate this contextual information effectively and concisely to improve the assessment of its suitability for any subsequent use of the data. Consequently, this will increase any confidence levels that are appropriate in design and output. The use and potential extension of current metadata standards tailored to citizen science projects could provide additional value here to allow the often complex spatial and temporal variability in citizen science projects, concepts and records to be made more understandable and transparent to the user. The use of metadata and established spatial data infrastructure (SDI) initiatives to increase accessibility, openness and support the use and uptake of data across a multiplicity of scenarios also provide additional value to decision makers. The use of controlled vocabulary and terminology and a standardised approach to recording and displaying metadata to published standards allows the assessment of key project details in an informed way. This can be seen as a method of further contextualisation to allow consideration of utility of the data in a familiar and rich digital environment.

The many variables at play within any given citizen science project can greatly affect the quality and accuracy of the outputs. Metadata should provide valuable insight into these variables which can lead to an increase in the likelihood for that data to be reused appropriately and with a measured level of confidence. The varying ways that quality can be affected (and potentially defined) at record, dataset and project level and at the time of capture (user experience and on-site verification to name just two) present a complex set of situations which may need to be understood. If these situations can be summarised to reduce ambiguity and support the notion of a *known quality* then the data derived has an increased chance of being used appropriately and effectively within policy development and decision making.

The publication of data with appropriate metadata and additional information that contextualises the quality of that data not only has the potential to stimulate its uptake within government (and government commissioned projects and research) but also to stimulate its use within the wider context of the environment sector. Increased access to more inclusive or accurate environmental data provides countless opportunities to positively stimulate targeted and informed research, activities and technology. Release of information at required quality levels can improve local, national and wider reporting effort, making any derived outputs more representative of real world phenomena.

The ability to publish standardised data of known quality digitally and potentially across established key environmental data repositories improves access to data for re-use and also increases transparency of the reasoning behind decisions and the evidence supporting those decisions. The democratic benefit to increased understanding of how and why decisions have been made and the evidence upon which those decisions have been based are potentially great and strengthen the concepts of openness espoused under sustainable development principles. An important point here may be the use of standards to share and discover data held within these repositories. The COBWEB project has worked on the development of a profile of the relevant OGC standards to maximise interoperability known as 'swe4citizenscience 2015' (further information on github: https://github.com/opengeospatial/swe4citizenscience/wiki/What's-this-all-about). This has resulted in progressing a vision of a common data model, to which data can be published using OGC web services; with sufficient community support, citizen science type data can be published to this open standard. This will then increase the immediate usefulness of these data and allow the myriad of potential users of such data to exploit existing standards based tooling and develop new standards based solutions (Higgins *et al.* 2015).

The use of digital infrastructures to publish and promote access to environment data, concepts and methods could facilitate or support more sharing of data and methods especially when supported by online digital communities that allow collaboration of stakeholders across a range of geographical areas. This improved picture will in turn benefit future and present decision making by allowing a more representative picture of the environment or issue to be presented. This may be used to improve the effectiveness of lobbying or be incorporated and captured in dialogue with stakeholders as part of the policy creation process.

FUTURE RESEARCH DIRECTIONS

The need for practical and effective links between research and policy to effectively tackle key environmental and social issues is a growing area of interest in the research community: 'Close collaboration is essential between the scientific community and stakeholders across the public, private and voluntary sectors to encourage scientific innovation and address policy needs' (Future Earth, 2013). The overall aim of this effort toward closer collaboration is to target research outputs to reflect societal and environmental needs and to support and inform policy development to incorporate the latest science and knowledge.

The involvement of Welsh Government in the COBWEB research project can be seen as an example of the strengthening of this science/policy interface. Given that the aims and objectives of the researchers and policy makers within the consortium can be quite different, the closer working and improved dialogue between these parties helps concentrate effort and activities to communicate the potential of new ideas and science and to test or ideally realise that potential to support the tackling of real world problems or concerns. For COBWEB, the agreed use of a co-design model to develop, refine and test technology and systems, assess their policy applicability and usefulness in the real world is a further benefit and strengthens not only the science/policy interface but also the interface between these two parties and citizens, community groups and volunteer organisations (real world actors). This collaboration supports the potential for this research project to provide valuable project outputs to stimulate and support local community, research and policy agendas.

If datasets gathered from citizen science projects are to realise their full potential as a source of evidence for policy and decision making there are a range of potential issues that need to be examined that relate to the expenditure of *resource* (in time, effort and financially). Co-design is certainly a promising way to bridge gaps between science, community and governments but it does not come without a cost. Full cost-benefit analysis studies in this area would be useful information for government departments in their planning of involvement with citizen science projects particularly where there is a co-design element. Input will be required in terms of time and expertise, resources and other material, possibly publications can be required and there will be overall management costs in these interactions. These costs can be far outweighed by the potential tangible and intangible benefits of government involvement in these projects but justifications will have to be made.

As citizen science can be seen as the involvement of volunteers in science (Roy *et al.* 2012) the source of risk or uncertainty in a project's longevity or quality can rely on a range of complex factors that revolve around a project's ability to recruit and retain a team of citizen scientists. The success or failure of a citizen science project can depend on this potentially unknown variable and it should not go unmentioned. This uncertainty could reduce the desire in the commissioning of citizen science projects specifically by policy makers if the need for representative levels of high quality data is a primary factor in the delivery of a policy. This may be especially the case for contributory or collaborative projects (terminology from Bonney *et al.* 2009) that may have disconnect from potential participants. Further understanding of ensuring the long term viability of these projects is an important area for research and an area where government would like reasonable assurance that any investment made will make a positive and effective return and will be a good use of public money.

CONCLUSION

The value of citizen science projects to contribute to the 'evidence-base' for decision making and policy development lies in its ability to harness the effort and knowledge of contributors to produce data and information. The process of decision making and policy development requires the careful consideration of the *suitability* of potential sources of evidence. Confidence in that data is a key component that will determine whether or not that data or information will feed into the policy development process. Data or abstracted information from a project that is well conceived, designed and implemented and produces data of suitable quality and quantity has a greater chance theoretically of being considered as a serious potential source of evidence. Geographical and temporal scales are also variables that need to be considered. Challenges arise when there is any degree of mismatch between the data that are required to fulfil a specific policy need and the data that are available through a vast number of very different citizen science projects.

The use of co-design as a method of deep collaboration with stakeholders involved in a citizen science project presents opportunities to further add value and confidence to the data collected for use in decision making. Through co-design the different elements of a project can be discussed and agreed and it is at this point where a mismatch can be avoided with respect to what data are required and what data are produced, in other words, better aligning project design and outputs to policy needs and/or priorities. In addition, this approach may help reduce risk to delivery through the valuing and incorporation of local knowledge and motivations, factors noted as important to effective and successful projects. Understanding motivation as a barrier and opportunity for better and more effective engagement is an area of research in citizen science that would be welcomed by anyone who wishes to invest time and resources into this area.

Pressure on resources and the drive to gather evidence to inform new and increasingly complex environmental scenarios and management systems brings the value of citizen science based environmental data collection into sharp focus. The concept of citizens' observatories is an important advancement in recent years as undoubtedly there is vast potential to use these rich data sources in the area of policy development and decision making. Working directly with groups contributing to these citizens' observatories however can be resource costly and it does seems unlikely that for every policy that requires data, appropriate resourcing can be targeted to coordinate and manage co-design projects with interested stakeholders. It is expected that as the concept of citizens' observatories develops, more and more citizen science projects with a co-design element that is supported by the public sector will be seen. It will only be a matter of time that government will utilise this data but it will always need to demonstrate sufficient quality and relevance to the policy or decision making process in hand.

Technological and conceptual innovation can add value to citizen science by utilising a range of sensors present within mobile devices (or from the sensor web) to measure and collect data about environmental conditions to enhance or assure observations interpreted by human sensors. This kind of data will no doubt increasingly contribute to citizens' observatories. Effective software design can assist standardised data capture routines that reduce the risk of transcription error and help facilitate automated quality measurement exercises. These processes can contribute to identifying *known quality* and communicate contextual information through the production of meaningful metadata. Conflation with authoritative datasets and attributing citizen science data with measures of quality will no doubt increase confidence levels in the uptake of such data as suitable sources of evidence and increase the likelihood of this data contributing to the policy development process. The area of data quality with respect to citizen science data is a known barrier to uptake in government but it is expected that as methods are developed to automatically or semi-automatically assign known levels of confidence to that data, potential users will be more informed of the suitability and reliability of that dataset; with known and acceptable levels of quality will come greater and broader uptake.

Data quality will always need to be assured and this is an ongoing area of research that will be closely watched by those hoping to use citizen science data more prolifically. Data accessibility and availability will also be a key factor in how extensively this kind of data will be used in government, the data must be easy to find and ideally be made available to data streams already accessed by those sourcing data that will contribute to the *evidence-base*. Our expectations is that citizen science data *can* contribute to the policy development process but in order to do so it will always have to assure the user that it can be fit for purpose, robust and timely. As momentum gathers in the development of citizens' observatories and as opportunities for collaboration are increasing (Haklay 2014), uptake of this sort of data in government and amongst decision makers and policy makers will no doubt increase. Undoubtedly there will be institutional cultures that will need to adapt (Brabham 2013) and change always takes time to develop fully but there are plenty of practical benefits associated with engaging with the development of citizen projects, particularly if these are on a large scale. In addition are the intangible benefits associated with empowering citizens, better environmental stewardship and fostering understanding between science, community and government.

REFERENCES

Alabri, A., & Hunter, J. (2010). Enhancing the quality and trust of citizen science data. In *e-Science (e-Science), 2010 IEEE Sixth International Conference on* (pp. 81-88). IEEE. doi:10.1109/eScience.2010.33

Ansell, A., & Gash, A. (2008). *Collaborative Governance in Theory and Practice. J Pblic Adm Res Theory*, *18*(4), 543–571. doi:10.1093/jopart/mum032

Bonney, R., Ballard, H., Jordan, R., McCallie, E., Phillips, T., Shirk, J., & Wilderman, C. C. (2009). *Public Participation in Scientific Research: Defining the Field and Assessing Its Potential for Informal Science Education. A CAISE Inquiry Group Report.*

Brabham, D. (2013). *Using Crowdsourcing in Government. Collaboration Across Boundaries.* IBM Center for The Business of Government. Available at: http://www.businessofgovernment.org/report/using-crowdsourcing-government

Cameron, A., Salisbury, C., Lart, R., Stewart, K., Peckham, S., Calnan, M., & Thorp, H. et al. (2011). Policy makers' perceptions on the use of evidence from evaluations. *Evidence & Policy*, *7*(4), 429–447. doi:10.1332/174426411X603443

DEFRA. (2006). *Evidenced Based Policy Making.* Available at: https://www.wbginvestmentclimate.org/uploads/modgov.pdf

DEFRA. (2014). *Making the most of our evidence: A strategy for Defra and its network.* Available: https://www.gov.uk/government/uploads/system/uploads/attachment_data/file/318610/evidence-strategy-defra.pdf

Delaney, D., Sperling, C. D., Adams, C. S., & Leung, B. (2008). Marine invasive species: Validation of citizen science and implications for national monitoring networks. *Biological Invasions*, *10*(1), 117–128. doi:10.1007/s10530-007-9114-0

Dickinson, J. L., Zuckerburg, D., & Bonter, D. N. (2010). Citizen Science as an Ecological Research Tool: Challenges and Benefits. *Annual Review of Ecology Evolution and Systematics*, *41*(1), 49–72. doi:10.1146/annurev-ecolsys-102209-144636

Earth, F. (2013). *Future Earth Initial Design: Report of the Transition Team.* Paris: International Council for Science (ICSU). Available at: http://www.icsu.org/news-centre/future-earth/media-centre/relevant_publications/future-earth-initial-design-report

Flanagin, A. J., & Metzger, M. J. (2008). The credibility of volunteered geographic information. *Geo-Jounal, 72*(3-4), 137–148. doi:10.1007/s10708-008-9188-y

Ganapati, S. (2011). Use of Public Participation Geographical Information Systems. *Applications in e-Government, 71*(3), 425-434.

Gollan, J., De Bruyn, L. L., Reid, N., & Wilkie, L. (2012). Can volunteers collect data that are comparable to professional scientists? A study of variables used in monitoring the outcomes of ecosystem rehabilitation. *Environmental Management, 50*(5), 969–978. doi:10.1007/s00267-012-9924-4

Goodchild, M. F. (2007). Citizens as sensors: The world of volunteered geography. *GeoJournal, 69*(4), 211–221. doi:10.1007/s10708-007-9111-y

Haklay, M. (2013). Citizen Science and Volunteered Geographic Information – overview and typology of participation. In D. Z. Sui, S. Elwood, & M. F. Goodchild (Eds.), *Crowdsourcing Geographic Knowledge: Volunteered Geographic Information (VGI) in Theory and Practice* (pp. 105–122). Berlin: Springer; doi:10.1007/978-94-007-4587-2_7

Haklay, M., Antoniou, V., Basiouka, S., Soden, R., & Mooney, P. (2014). *Crowdsourced geographic information use in government Report to GFDRR.* London: World Bank.

Higgins, C., Williams, J., Leibovici, D., Simonis, I., Davis, M. J., Muldoon, C., & O'Grady, M. (2015). Citizen OBservatory WEB (COBWEB): A Generic Infrastructure Platform to Facilitate the Collection of Citizen Science Data for Environmental Monitoring. In *Proceedings of the Workshop "Environmental Infrastructures and Platforms 2015 - Infrastructures and Platforms for Environmental Crowd Sensing and Big Data" for the European Citizen Science Association General Assembly 2015* (ECSA GA'2015).

Hodges, C. J., Leibovici, D. G., & Ties, S. F. (2014). Building confidence in crowdsourced data for biological monitoring and policy making as part of FP7 Project COBWEB. Role of Volunteered Geographic Information in Advancing Science: Effective Utilization (Workshop). In *Proceedings of GIScience 2014 Eight International Conference on Geographical Information Science.*

Hoffmann, A., Penner, J., Vohland, K., Cramer, W., Doubleday, R., Henle, K., & Häuser, C. et al. (2014). Improved access to integrated biodiversity data for science, practice, and policy - the European Biodiversity Observation Network (EU BON). *Natureza & Conservação, 8*, 49–65. doi:10.3897/natureconservation.6.6498

Evidence. (2013). In Oxford Dictionaries. Available at http://www.oxforddictionaries.com/definition/english/evidence

Irvin, R. A., & Stansbury, J. (2004). Citizen participation in Decision Making: Is it Worth the Effort? *Public Administration Review, 64*(1), 55–65. doi:10.1111/j.1540-6210.2004.00346.x

Kingston, R., Carver, S., Evans, A., & Turton, I. (2000). Web-based public participation geographic information systems: An aid to local environmental decision-making. *Computers, Environment and Urban Systems, 24*(2), 109–125. doi:10.1016/S0198-9715(99)00049-6

Leibovici, D. G., Jackson, M., & Higgins, C. (2015). *Geospatial Data Curation and Interoperability in the COBWEB project: citizen science & crowdsourcing for environmental policy*. Paper presented at RDA (Research Data alliance) symposium Plenary 5, joint meeting of IG Geospatial & IG Big Data, San Diego, CA.

Meek, S., Jackson, M., & Leibovici, D. G. (2014). *A flexible framework for assessing the quality of crowdsourced data*. Paper presented at the AGILE conference, Castellon, Spain.

Meek, S., Jackson, M., & Leibovici, D. G. (2015). Addressing the quality assurance challenge for location-based crowd-sourced data through workflow composition of OGC web services. *Computers & Geosciences*, (Jan): 2015.

Nutley, S., Powell, A., & Davies, H. (2013). *What counts as good evidence? Provocation paper for the Alliance of Useful Evidence*. Retrieved from http://www.alliance4usefulevidence.org/assets/What-Counts-as-Good-Evidence-WEB.pdf

Pocock, M. J. O., Chapman, D. S., Sheppard, L. J., & Roy, H. E. (2014). A Strategic Framework to Support the Implementation of Citizen Science for Environmental Monitoring. Final Report to SEPA. Wallingford, UK: Centre for Ecology & Hydrology. Retrieved from http://www.ceh.ac.uk/products/publications/documents/hp1114final_5_complete.pdf

POSTnote. (2014). *Environmental Citizen Science*. POSTnote number 476. Houses of Parliament. Retrieved from http://researchbriefings.parliament.uk/ResearchBriefing/Summary/POST-PN-476

Reed, M. S. (2008). Stakeholder participation for environmental management: A literature review. *Biological Conservation, 141*(10), 2417–2431. doi:10.1016/j.biocon.2008.07.014

Roy, H. E., Pocock, M. J. O., Preston, C. D., Roy, D., Savage, J., Tweddle, J. C., & Robinson, L. D. (2012). *Understanding Citizen Science and Environmental Monitoring*. Final Report on behalf of UK-EOF. NERC Centre for Ecology & Hydrology and Natural History Museum. Available at: http://www.ceh.ac.uk/news/news_archive/documents/understandingcitizenscienceenvironmental monitoring_report_final.pdf

Sanderson, I. (2009). Intelligent Policy Making for a Complex World: Pragmatism, Evidence and Learning. *Political Studies, 57*(4), 699–719. doi:10.1111/j.1467-9248.2009.00791.x

SCU. (2013). *Science Communication Unit, University of the West of England, Bristol (2013). Science for Environment Policy In-depth Report: Environmental Citizen Science*. Report produced for the European Commission DG Environment. Available at: http://ec.europa.eu/environment/integration/research/newsalert/pdf/IR9_en.pdf

Shirk, J. L., Ballard, H. L., Wilderman, C. C., Phillips, T., Wiggins, A., Jordan, R., & Bonney, R. et al. (2012). Public participation in scientific research: A framework for deliberate design. *Ecology and Society, 17*(2), 29–48. doi:10.5751/ES-04705-170229

Simonis, I. (2015). Standardised Information Models to Optimise Exchange, Reusability and Comparability of Citizen Science Data. A Specialised Approach. In *Proceedings of the Workshop "Environmental Infrastructures and Platforms 2015 - Infrastructures and Platforms for Environmental Crowd Sensing and Big Data" for the European Citizen Science Association General Assembly 2015* (ECSA GA'2015).

Solesbury, W. (2001). *ESRC UK Centre for Evidence Based Policy and Practice: Working Paper 1*. Retrieved from http://eur-lex.europa.eu/legal-content/en/TXT/?uri=celex:52001DC0428

Tulloch, D. L. (2008). Is VGI Participation? From Vernal Pools to Video Games. *GeoJournal*, *72*(3–4), 161–171. doi:10.1007/s10708-008-9185-1

Tweddle, J. C., Robinson, L. D., Pocock, M. J. O., & Roy, H. E. (2012). *Guide to Citizen Science: Developing, Implementing and Evaluating Citizen Science to Study Biodiversity and the Environment in the UK*. London, UK: Natural History Museum and Centre for Ecology & Hydrology for UK-EOF.

Wiemann, S., Meek, S., Chapman, C., Leibovici, D. G., Jackson, M., & Lars, B. (2015). *Service-based combination of quality assurance and fusion processes for the validation of crowdsourced observations*. Paper presented at AGILE 2015 conference, Lisbon, Portugal.

Wiersma, Y. F. (2010). Birding 2.0: citizen science and effective monitoring in the Web 2.0 world. *Avian Conservation and Ecology, 5*(2), 13. Available at: http://www.ace-eco.org/vol5/iss2/art13/

Chapter 13
Smart Activation of Citizens: Opportunities and Challenges for Scientific Research

Maria Gilda Pimentel Esteves
Universidade Federal do Rio de Janeiro, Brazil

Alexandre Prestes Uchoa
Universidade Federal do Rio de Janeiro, Brazil

Jano Moreira de Souza
Universidade Federal do Rio de Janeiro, Brazil

Carla Viana Pereira
Empresa de Tecnologia e Informações da Previdência Social – DATAPREV, Brazil

Marcio Antelio
Universidade Federal do Rio de Janeiro, Brazil

ABSTRACT

This chapter focuses on how, by "activating" the citizen's engagement in the research process, the scientific community has a smart way to benefit from the wisdom of the "crowd". There are countless success stories in which citizens participate, contributing with their knowledge, cognitive capacity, creativity, opinion, and skills. However, for many scientists, the lack of familiarity with the particular nature of citizen participation, which is usually anonymous and volatile, turns into a barrier for its adoption. This chapter presents a problem-based typology for citizen-science projects that aims to help scientists to choose the best strategy for engaging and counting on citizen participation based on the scientific problem at hand; and some examples are included. Moreover, the chapter discusses the main challenges for researchers who intend to start involving the citizens in order to solve their specific scientific needs.

INTRODUCTION

Historically, scientific research has been based on integrity, objectivity, truth-seeking and autonomy. This autonomy has led to the creation of a boundary between academia and society, thus dictating that science should be conducted only by scientists and acknowledged by their peers, for the benefit of society. However, new technologies and the popularity of Internet have led to the establishment of a new collaboration paradigm. The combination of crowdsourcing together with the advancement of mobile technologies opens up huge potential benefits for science, society, and the environment.

DOI: 10.4018/978-1-5225-0962-2.ch013

One of the greatest challenges of modern science is to transform the former border between academia and society into a place where ideas and interests can encounter and collide. Citizen science transforms this border into a permeable boundary that allows the union and exchange of different knowledge, skills and interests with benefits for all participants: scientists, citizens, and partners.

Although there are numerous citizen science projects in various parts of the world, only few studies have addressed the specific managerial aspects of citizen engagement in scientific domain and the dimensions that should be evaluated before and during its adoption. A better understanding of these aspects and related mechanisms can provide the "perfect experience" for the citizen scientist. To attract and retain citizens willing to collaborate with science, assure the quality of the contributions and the attendance to standards, and support a large number of contributors and contributions, are some of the challenges faced by managers of citizen science projects.

This chapter will present an overview of different opportunities for smart collaboration between citizens and scientists. It presents a problem-based typology in order to explore some representative examples of citizen science projects. Additionally, different types of projects will be grouped according to a pushed or pulled data approach adopted by the scientist. A smart activation decision tree is proposed in order to: help project managers assess which types of problems they need to solve, identify if the particular scientific objective of the project is compatible with the use of citizen science, and select which category of solution is best suited to the problem. An assessment of the main challenges for the design and management of these projects, as well as the challenges related to motivational aspects and quality control, will be presented. We believe this chapter will serve as a guide for scientists to advance towards this new paradigm and achieve the benefits associated with the smart activation of citizens in modern science.

BACKGROUND

In the new age of modern science, which is increasingly global, interconnected, and involves more international collaboration ("The Royal Society", 2011), citizen science has emerged as a form of crowdsourcing in which geographically distributed members of the crowd are invited to collaborate with scientists by applying some human cognitive ability on a large scale. This new paradigm has been studied by many authors, including Haklay (2013, 2014), Wiggins and Crowston (2010, 2012), Dickinson et al. (2010), Nov et al. (2010), Alabri and Hunter (2010), and Bonney et al. (2009) to name just a few. In accordance to this paradigm, members of the general public are promoted to the role of citizen scientists, in the stages of real scientific research and, therefore, collaborating to the creation of scientific knowledge.

New scientific methods are being created with the support of the Internet and mobile technology, thus allowing scientists to expand their network of collaborators beyond the limits of institutions. Ubiquitous and pervasive technology has broken the barriers of time and space, allowing a greater and more diverse number of collaborators to be engaged in scientific activities. The use of crowdsourcing platforms is making possible the participation of large groups to perform tasks that were once confined to small groups of experts. Recent innovations in information, communication, and technology — from smartphone apps to real-time crowdsourcing — are undoubtedly making citizen engagement far easier than ever before in history.

Currently, crowdsourcing is considered to be an umbrella or generic term, since it embraces a variety of approaches that exploit the labor force and cognitive potential of a large and open crowd of people

(Geiger, 2011a). Crowdsourcing for science can be characterized either by an *open call* — which is also described as "self-identification of contributors" and allows anyone, who is interested and capable, to participate (Howe, 2009) — or a *restricted call* (pre-selection of contributors), which is concerned with restrictions regarding the group of potential contributors (Geiger et al., 2011a). In the second case, the interested participant must possess certain qualifications (e.g., specific skills or knowledge) or represent a specific context or ethnography, such as geographic distribution, social class or education, age, and gender, among others.

Citizen science projects go beyond the simple use of a citizen workforce. They promote opportunities for entertainment, education, and quality of life improvements, since many projects focus on local issues related to the everyday life of the citizens. Involving citizens in authentic research provides participants with valuable experience and the opportunity to make significant contributions to scientific research. Furthermore, it can promote behavioral change and increased environmental awareness, if also designed to educate citizens through their participation (Dickinson et al., 2012; Newman et al., 2012 & Bonney et al., 2009). On the other hand, for the scientist, citizen participation adds value to the scientific process bridging the gap between science and society (Pfeffer & Wagenet, 2007). The involvement of partners — such as civil society organizations, local associations, scholar networks and non-governmental organizations — is common in citizen science.

What distinguishes collaborative projects with citizen participation from the conventional mode of collaboration in scientific research is precisely the lack of a formal agreement and commitment to work, which leads to more flexibility in performing the tasks. For Haklay et al. (2014), this can be a challenge to professional scientists who are used to working only with their peers, in a top-down manner.

Citizen science projects require "hybrid" management that allows a balance between leadership rigidity, hierarchical organizational structures, and formal working relationships, as opposed to the engagement flexibility of the amateur scientist; that is, a contributor without a formal working relationship who has different motivations (entertainment, altruism, seeking new skills and knowledge, etc.). Therefore, a tailored management approach is required to increase the chances that both contributors and scientists achieve their expectations and goals (Uchoa et al., 2013).

Figure 1. Main differences between conventional scientific projects and citizen science projects

	Agreement to work	**Group Characteristics**
Conventional Science	• *Formal* Scientists and institutions define needs and services according to a scientific project or a contract	Pre-selected contributors Small groups Experts only Integration of distributed teams Strong ties Formalization of participation
Citizen Science	• *Flexible* Scientists, institutions and/or bottom-up practice define scientific activities and invite external contribution of lay public and amateur scientist volunteers	Open contribution Large groups Not only experts Integration of distributed volunteers Weak ties Flexible participation Anonymity

Despite the knowledge acquired with many successful citizen science projects, the integration of such projects with more conventional scientific research activities is still challenging. It involves culture changes in research institutions and more rigorous control of data quality. Ensuring reliable inputs by citizens depends on the good design of the task to be performed by the citizen, which includes the choice of technology and best management strategy for mobilization, participation, and communication, as well as the definition of quality control methods, which must be taken into account from the beginning of the project design up until the final stage of validation and approval of the contribution.

Citizen science projects previously conducted in local communities and in teaching initiatives, now benefit from: the infrastructure of the Web; the popularity of mobile device usage; and a large number of potential contributors connected to the Internet, who are ready to be activated on scientific projects of common interest.

CITIZEN SCIENCE OPPORTUNITIES

The opportunities for citizens to participate and help solve real-world scientific problems grow every day. These opportunities can be grouped into two broad categories, according to a pushed or pulled data approach adopted by the scientist. These two approaches in citizen science describe the exchange of data or information between citizens and scientists. It should be guided by the scientists' needs for new data and information, or due to the demand of massive data analysis or processing, depending on the scientists' points of view:

1. **Pushed Approach (Data Analyses)**: Scientists ask citizens to collaborate with science through classifying and analyzing large datasets that neither computer nor individuals alone can deal with. Scientists "push" the data toward citizens in order to use their collective or individual cognitive capacity.
2. **Pulled Approach (Data Collection)**: It is based on scientific demand for new data or information, and is especially used for surveys, investigations, and monitoring. For instance, when demand is higher than the scientists' ability to collect data from the physical environment, the scientists "pull" the data they need from citizens and ask them to collaborate with their capacity to observe and collect environmental data.

The pushed data approach should be used in projects that require the analysis and classification of large volumes of digital data obtained automatically and continuously by sensors, telescopes, and other electronic devices. Digitalizing information of old analog data collections such as the weather observations of ship's logs or herbarium specimens are also part of this approach. These type of citizen science projects were labeled as "virtual" by Wiggins & Crowston (2011) and "volunteer thinking" by Ponciano et al. (2014). These projects aim to gather citizen scientists, who can contribute by executing human computation tasks. Participants collaborate via the Internet through games or tools that assist scientists in solving important scientific problems such as: recognizing patterns in images, sounds, and videos; text transcription; and geocoding, among other activities which also include scientific discovery outcomes. Contributors follow a sequence of activities, predefined by scientists dealing with a scientific computational problem.

This approach takes advantage of distributed human computation, which is described by Quinn & Bederson (2009) as a kind of task used to resolve problems that cannot be solved just by computers or just by humans, but for which a solution can be reached if both work together.

On the other hand, the pulled approach is usually associated with the collection of large amounts of data across large areas and/or for long periods of time. These collection tasks are usually performed outdoors and the data obtained are shared with projects in the form of text, images, sounds, and/or videos. This approach, which is also referred to as "citizens as sensors", has been widely discussed by Burke et al. (2006) and Goodchild (2007). Most citizen science projects belong to this category and there are many different examples in the literature: Conservation and Investigation (Wiggins & Crowston, 2011), Volunteer Monitoring (EPA, 2012), Volunteered Geographic Information (Goodchild, 2007; Haklay et al. 2013), and Volunteer Sensing and Participatory Sensing (Cuff, 2008; Estrin, 2010). Also belonging to this category are projects — generally related to the areas of psychology and medicine — aimed at collecting personal information.

There is also a third approach that does not fit properly in neither of the two data oriented categories above:

3. **Ideation Approach**: includes those projects that use the citizen participation for the generation of new ideas or for the solution of complex problems. The main issue of these projects is not exactly weather to use the pulling or pushing of data, but rather to develop entirely novel ideas with the direct help of the citizen.

These three approaches were used to create a decision tree for the smart activation of citizens, in order to help scientists determine which type of approach is most useful for each specific scientific problem. Figure 2 illustrates these logical steps.

Figure 2. Decision tree for the smart activation of citizens

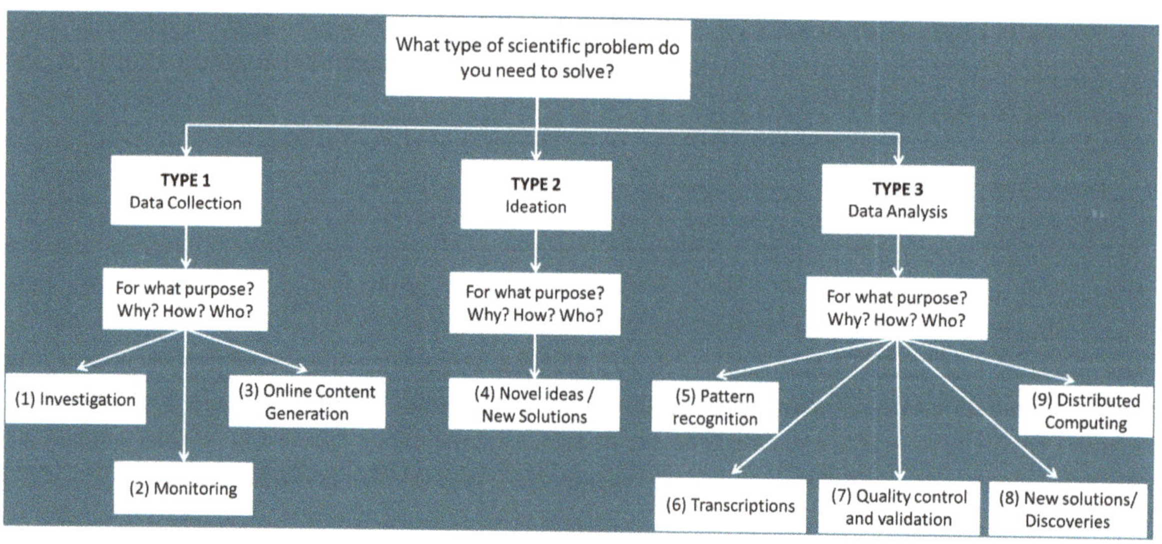

First, the scientist needs to determine whether the type of scientific challenge is a data management problem; that is, whether it involves:

1. Management of data collection (pushed data approach);
2. Analyzing existing data (pulled data approach); or
3. An ideation problem.

After this decision, the next question to answer is: What is the purpose of the citizen collaboration? Understanding the purpose and the "why" and "how" of the external contributions can accelerate the conventional science is a first step towards deciding whether or not to engage citizen participation.

Depending on the type of problem, a specific target public must be defined and mobilized. The profile of "who" will be activated correlates strongly with the characteristics of the scientific activities to be delegated to the participants. The precise identification of the skill set, abilities, knowledge, and individual characteristics of potential participants is as important as the design of appropriate task.

Like any other scientific research, before starting a project, the scientist needs to clearly define the problem and the corresponding solution. Defining the problem to be investigated and the question to be answered will help in describing the tasks and defining what type of scientific contribution the citizen can make to the project.

It is also important to discover and evaluate what others have done in similar projects. This helps verify the need to start a new project or to extend an existing one.

The most common opportunities for activating citizen participation in scientific research were grouped into 9 different categories. These citizen science categories are not exclusive, and in some projects they are combined to enhance the qualities of each other. Depending on the strategic management approach and the question to be solved, the best category of task to be given to the citizen scientists may be one of the following:

Category 1: Investigation or Surveillance

Most scientific opportunities for engaging citizens with science belong to this category. The projects are particularly focused on ecology, environmental sciences, and related fields that require data collection from the physical environment (Wiggins & Crowston, 2011; McKinley et al., 2015), and they depend on the geographic location and the volunteer's use of mobile devices to contribute with data and observations. According to Nichols and Williams, (2006), in this category of project, scientists are not guided by a priori hypotheses and their corresponding models. For example, in research areas such as ornithology (eBird) and marine biology (JellyWatch), the citizen scientist contributes with the sighting and collecting of data on birds, and jellyfish and other marine organisms (e.g., man-of-war, squid, mammals and algal blooms), respectively. The citizen scientist follows standardized data collection protocols to increase the amount of data, in time and space, for one or more biological or physical parameter. These data are used for various purposes, and combining data analysis with other information allows scientists to discover new patterns and trends.

On the other hand, scientific research can benefit from crowdsourcing as a means of obtaining personal information. For instance, the Animal Ownership Interaction Study, which was recently launched by the Center for Canine Behavior Studies, recruits citizen scientists to fill out an online form and take part in surveys collaborating with science. The main objective of this research is to understand how owner personality affects pet behavior.

These projects require large amounts of data and, in general, participants do not need to have expertise to contribute. Often, participants join the project due to having a personal interest in the project's theme or because they already perform the activity as a hobby. For Bonney and Dickinson (2012), what motivates citizen scientists is a strong interest in the organisms being studied, a curiosity about the world around them, and a desire to advance the field of science. The benefits obtained by the managers of this category of project are: increased amount of data collected, expansion of geographic coverage of the surveyed area, increased period of collection, dissemination of knowledge, and increased participant awareness in relation to their environment (Tweddle et al., 2012) and health issues.

Category 2: Monitoring

Many scientific research institutions are developing "volunteer monitoring" programs, which involve volunteers, members of the general public, and a wide range of community groups in collecting and reporting personal or environmental data.

Projects in this category, according to Nichols & Williams (2006) are designed and implemented based on a priori hypotheses and associated models of system responses to management.

This type of smart activation helps in: bridging the gap between science and society, reinforcing public confidence in science, and allowing direct involvement of the general public in the generation of data. As a result, decision-making may become more democratic (Pfeffer & Wagenet, 2007), thus giving it clear relevance in education and policy.

Projects of this category are aimed at the management and monitoring of public health and natural resources, and they can be classified into two sub-categories:

1. Ongoing data collection for scientific investigation or environmental conservation; and
2. Immediate data collection for monitoring and decision-making on epidemics or catastrophic events.

Long Term Scientific Investigation

This sub-category of projects seek to educate citizens on important issues related to science and to the environment, by contributing to projects involving water and air quality monitoring, global warming, biodiversity, global health, and many other scientific opportunities.

For instance, CoralWatch is a citizen science project managed by the University of Queensland, which aims to improve the extent of information on coral bleaching events and coral bleaching trends (Reid et al., 2009).

The BudBurst project draws on the voluntary cooperation of citizens to help scientists understand climate change by making regular observations of the occurrence and phenology (red, flowering, fruiting, and leaf loss) of various plant species. Participation was designed to be performed in six steps:

1. Register for an account;
2. Choose a plant;
3. Download Datasheet;

4. Locate/Describe site;
5. Make observations;
6. Report observations online.

Volunteer contributors can choose the plants they want to observe from a decision tree. Five groups of plants are available at the project website: wild flowers and herbs (110); deciduous trees and shrubs (105); evergreen trees & shrubs (39); grasses (18); and conifers trees (10). These groups and their respective species were selected because they are easy to identify and they occur widely in the United States. For each of the five groups there is educational material explaining the phenophases.

The citizen scientist can also contribute to science by providing personal information that helps scientists monitor public health. For instance, the Flu Survey project[1] aims to collect personal information about flu-like symptoms experienced during the winter months. These data have been used by researchers at the London School of Hygiene and Tropical Medicine and also in United Kingdom's Tropical Medicine and Public Health (NHS) since 2009 to monitor influenza trends in the UK. The data provided by citizens are analyzed and displayed on a Web map updated every three minutes. Currently, more than 6,000 citizens, from all over the UK, have contributed to this project.

Crisis, Emergency, and Disaster Monitoring

The goal of these projects is to allow citizens to report relevant information during disasters or emergencies. This information is relevant to many scientific research areas such as biology, seismology, climatology, geology, and public health, among others.

According to Okolloh (2009), "Information in a crisis is a patchwork of sources. You can only hope to build up a full picture by having as many sources as possible" (p.66). The Ushahidi[2] platform was built to gather geolocation data reported by large groups of volunteers and to facilitate the sharing of information through visualization and interactive mapping. Scientists can use these data to monitor and understand catastrophic events, such as the magnitude 7.0 earthquake in Haiti on January 12, 2010. With the same goal, the "Did You Feel It?" (DYFI) [3] of the U.S. Geological Survey harnesses the potential benefits of citizen science for monitoring earthquake events.

The growing use of smartphones, which are increasingly accessible, and the pervasive connectivity and consolidated data in monitoring and surveillance projects are working together to create a public that can objectively record, analyze, and discover a variety of patterns that are both important in their lives and also contribute to scientific research. In the literature, new terms like volunteered geographic information (VGI) and participatory sensing have come to describe this new potential for participation. These two new terms, which are not mutually exclusive, are described below (both are applied to the categories 1 and 2 discussed above).

Volunteered Geographic Information (VGI)

First defined by Goodchild (2007), VGI "has the potential to be a significant source of a geographers' understanding of the surface of the Earth".

Since the vast majority of citizen science is geographic, that is, it requires a location on the Earth, the overlapping of citizen science with VGI can be seen as a way of updating geographical databases

(Goodchild 2007; Haklay et al., 2013). Notwithstanding, the opportunities of using VGI for science go beyond collection and database storage. VGI enhances the collective creation of georeferenced dynamic maps. There are huge opportunities for citizen science to build collaborative maps for various objectives and interests, as described above.

Maps are not static, because information is added and it changes over time and space. The citizen's data collection activities can be valuable for scientists, because these activities enable the generation of a greater volume of data, as well as increasing the accuracy and quality of information.

Participatory Sensing

For Estrin (2010), participatory sensing is the process by which individuals and communities are increasingly using mobile devices and cloud services to collect and analyze data systematically for several purposes.

The use of humans as a network of sensors is an approach that has already been discussed in the literature and it has been employed in various citizen science projects for data collection (Burke et al., 2006; Goodchild, 2007).

A new collective capacity is emerging through the use of sensors built into mobile phones and connected to web services (e.g., cameras, motion sensors, and GPS). For Goldman et al. (2009) participatory sensing allows people to participate in activities in which they can detect and analyze aspects of their lives that were previously invisible. An example of the discovery of this invisible aspect is the Noise Tube project.

NoiseTube is a research project started in 2008 at the Sony Computer Science Lab in Paris, which is currently maintained by the Software Languages Lab at the Vrije Universiteit in Brussels. The main goal of this project is to enable citizen scientists to measure the personal exposure to noise resulting from using mobile phones. The geolocation and the measurement results are shared online in collective noise mapping of cities (Maisonneuve, 2009).

Category 3: Online Content Generation

Projects classified as category 3 aim at creating collective online content. In many cases, citizen collaboration are often supported by Wiki technology. Like other citizen science activities, it is only necessary to register a username and password to join the online community of citizen scientists. Effective community participation is what guarantees the content and the quality of the data. Included in this category are the Wikiflora, WikiAves and Polimathy projects, just to name a few.

The WikiAves project[4] is a Brazilian citizen science project that is one of the largest repositories on the distribution and abundance of birds. Unlike other projects that request a species list of birds to be filled in, WikiAves only requests that citizens send an image or sound record, as well as a description and location of the bird sighting. Known as the "Facebook for birds", it currently has 1,858 pages — one for each Brazilian species — and all pages contain relevant information published collaboratively by volunteers. This number of pages is almost equivalent to the total of 1,901 species known and registered on the official list of the Brazilian Ornithological Records Committee. Up until March, 2016, the total number of participants was of 23,558. The number of records included 1,578,557 images and 93,365 sound records.

Category 4: Ideation: Creating Novel Ideas and New Solutions

Such projects promote great challenges and competitions involving issues related to scientific research. The coordinators of this category of projects can create their own platforms or use marketplace platforms designed exclusively to manage the tasks. InnoCentive, Climate Collab, and Citizen Sky are examples of this type of opportunity, which brings citizens together to collectively create novel ideas and solutions.

Today's scientific projects are no longer restricted by the ability and skills of professional teams. They can now appeal to open and varied audiences, thus allowing different skills, knowledge, and interests to be added to projects. Climate Collab[5], for example, is a research project of the Massachusetts Institute of Technology, and its goal is to harness the collective intelligence of thousands of people worldwide to address global climate change. The participant can create and submit proposals for what should be done about climate change, or collaborate with people worldwide to outline ideas that will help reduce climate change impacts on the planet.

Open competitions discuss topics in different areas of knowledge such as low-carbon energy, building efficiency, energy supply, land use, and waste management, among many other topics.

Category 5: Pattern Recognition

The increasing scientific need for analyzing and processing large datasets, which neither small groups of scientists nor computers can solve alone, opens up a variety of opportunities for citizen collaboration in web-based citizen science projects. Generally, the tasks to be performed require some type of pattern recognition, which can only be performed by the human brain.

Pattern recognition is used to classify massive online datasets of images, sounds, or videos. Citizens participate by following protocols with standards and guidelines previously set by scientists. Activities are executed exclusively online, and results have proven that the quality of such activities is similar to that produced by professional scientists (Canfield, Jr. et al., 2002; McKinley et al., 2015). One of the most successful examples is the Galaxy Zoo project, which is a citizen science project on astronomy (Lintott et al., 2008). Planet Hunters[6] is a similar project that features a quick and easy-to-assimilate tutorial that gives credits for the volunteer work performed. The tutorial is always available at any stage of task execution. The task is simple, and in a few steps the collaborator can finish the classification of an image.

Some projects stimulate task completion through games and competition. For instance, "Citizen Sort" is a platform where there are many games to help scientists with classification tasks about species of insects, animals and plants. One of these games is "Happy Match" a taxonomic classification game, in which the taxonomic titles vary according to the pictures that must be classified. The game design includes a question, pictures that must be classified, and pictures that represent the possible classifications. Thus, the collaborators must compare the images that need classification with the classification options. From the images that must be classified, at least two of them have already been classified by experts and they are used to compute a collaborator's accuracy (score). The scores are used by Citizen Sort to maintain a ranking of the collaborators, who are invited to perform more tasks (or play more games) and enhance their position in the ranking (Crowston & Prestopnik, 2013).

Category 6: Document Transcription

Like the previous category, the science community makes use of distributed human computation to benefit from the cognitive ability of millions of minds connected to the Internet to perform simple tasks related to digital document transcription.

These projects are aimed at completing tasks that are easily performed by humans but which computers alone cannot yet perform. Tasks that are part of such projects are simple and do not require specific skills. Digitalizing information of old analog data collections such as the weather observations of a ship's log - Old Weather project (Eveleigh et al., 2013) or herbarium specimens - herbaria@home projects (Groom et al., 2014) are examples of this approach.

Category 7: Quality Control and Data Validation

Citizens participate in these projects by helping to improve data quality. In the RiverWatch project, participants try to identify out-of-range observations. They copy their observations into a worksheet, sort them by ascending or descending numerical attributes, and then compare observations with each other. After this, it is possible to identify duplicated observations or out-of-range values that were mistyped and must be corrected or removed (Sheppard & Terveen, 2011).

Hutt et al. (2013) investigated the best task design for obtaining annotations for microscopic images, in order to determine how clumpy an image is. Three task designs were proposed: classification, scoring, and ranking. In classification tasks, collaborators must classify an image as either clumpy or not clumpy by clicking on the corresponding button. The scoring tasks ask collaborators to give a score for an image. Finally, the ranking tasks show three images for collaborators, who are required to order them from the least clumpy to the clumpiest, in which the least clumpy image must appear to the left and the clumpiest image to the right.

Category 8: New Solutions/Discoveries

This type of opportunity is related to scientific problems that require large-scale human-computer interaction for solving unpredictable problems. Typical examples of success cases include the Foldit and EteRNA projects. These projects use GWAP, human computing, crowdsourcing, and social computing in order to attract common citizens to perform scientific tasks. Such projects seek to encourage the participation of people who have an interest in science, but who are also looking for entertainment and leisure.

Foldit was developed by the Center for Game Science at University of Washington (UW) in collaboration with that university's Department of Biochemistry. It is a crowdsourcing computer game that enables citizens to contribute to solving protein-folding problems (Cooper et al., 2010). Moreover, it is a multiplayer online game that uses the knowledge and intuition of non-specialists to solve protein-folding problems by using a host and tools provided. Since its release, Foldit has gained over 100,000 players. The best Foldit players have little to no prior exposure to biochemistry (http://fold.it/).

"Foldit showed that it is possible to effectively 'crowdsource' human problem solving in order to solve very hard scientific problems, and that the gaming environment is capable of turning novices into highly skilled researchers. The goal of the Center for Game Science[7] is to generalize and expand the success of Foldit to a wider range of problems in science, education, and beyond.

Category 9: Distributed Computing

In this category of project, the scientific result does not depend on direct human activity, but rather on the computational processing of the machine provided by the volunteer. This type of project has been classified as citizen science, despite there being no human-computer interaction in this activity. Nevertheless, volunteer computation is helping scientists to save time and also speed up the processing of scientific information through distributed computation. These are projects aimed at using the idle time of participants' computers to process scientific research data. Volunteers contribute by simply "lending" their computers, without doing any additional activity. The computer network formed by collaborators produces a huge network of distributed computing, which is a way of overcoming the technological limitations a project may face. These projects typically use the idle time of computers connected to the internet to: find extraterrestrial intelligence (SETI@Home), study samples of solid matter collected outside the solar system (Stardust@home), or help determine three-dimensional shapes of proteins (Rosetta@Home).

Every day, new forms of citizen participation in scientific research arise and, consequently, the volume of data that are generated, processed, or recombined is increasing. There are numerous current possibilities for the use of this data in the present, and huge potential for its use in the future. Therefore, more and more data are being stored, even if it does not have value today. After all, besides preserving raw data, what really matters for scientists and decision makers is the added value; that is, the ability to recombine and transform raw data into useful information that is available and easily accessible.

Grouping citizen science projects is a difficult task, due to the wide variety of potential opportunities in scientific research. However, if we consider that what binds the scientist to the citizen are the data and/or the knowledge to be acquired or transmitted, grouping these projects into a pushed or pulled data approach seems appropriate and cohesive with the new era of data-intensive scientific discovery.

CITIZEN SCIENCE CHALLENGES

Citizen science refers to contexts in which most individuals are not part of the academic community, so they engage in scientific projects, waiting or not for a reward for their effort (Tweddle et al., 2012). The use of amateurs, as opposed to scientists, can reduce confidence in the results if the task has not been designed to prevent or minimize errors. Well-defined criteria should be developed at the design stage and accompanied during the performance of the task and the delivery of the contribution. Motivation, quality control, and management of a large number of contributors and contributions are the main challenges faced by managers of citizen science projects. These three topics are discussed below.

Motivation

Often what motivates the citizen is the satisfaction of being part of a real research project, with the possibility for learning, leisure, activism, altruism, fun, recognition, as well as becoming acquainted with new people, places, and socio-environmental contexts.

If, on the one hand, citizen science enables data collection and analysis on a larger scale than the conventional method, on the other hand, it brings challenges associated with motivating individuals to engage in task performance.

The reasons associated with why citizens participate in scientific projects have been the focus of various scholars' research: (Wenger, 2002; Raddick et al., 2010b; Nov et al., 2011; Crowston & Prestopnik, 2011). Depending on the context of the project and the interest of the participant, the selection process may be triggered by one of the following factors: the research issues may arouse interest or they are already part of the person's day-to-day activities as a hobby; participants may have a vested interest in the outcome of the research and thus contribute to its goals; some projects make use of games and entertainment approaches, thus leading people to participate for leisure; or the reason for participation may be altruistic in nature.

Raddick et al. (2010) identifies four social groups as the main beneficiaries of successful citizen science projects: the volunteer contributors of the projects, the researchers behind projects, educators, and society. For the public, projects provide knowledge, entertainment, and experiences, which can ensure their motivation and lasting engagement. For researchers, they ensure that scientific activity, which cannot be solved in a short amount of time by a small team of professional scientists, will be executed quickly and cost-effectively within required quality standards. For educators, this kind of research helps them professionally by providing new opportunities for education. And for the affected communities, citizen science projects generate results and concrete benefits, which can bring scientific practice and society closer together. Thus, when designing experiments and products that will attract and retain the participants, the organization of citizen science projects should take into account the specific needs and concerns of the social groups that these projects interrelate with. In order for this to happen, identifying and knowing them seem to be essential requirements for the design of appropriate and effective projects.

Pant (2009) believes that a way for an organization to recognize potential participants is by paying attention to the way they act in the virtual communities in which they participate, including those that form around the research subjects of citizen science projects. Thus, social tools such as forums, blogs, wikis, microblogs, and chats — all of which are employed by Galaxy Zoo[8] and similar projects such as Foldit[9]— offer great potential for promoting the exchange of knowledge and experiences.

According to Nov (2007), it is important to understand the factors that lead people to freely share their time and knowledge with others, in order to increase the number of contributions and improve the user-generated content.

Wikipedia, the Web encyclopedia created by users, is a successful example of collaboration in the form of online content generation. The motivational factors of contributors appear to be critical for the maintenance of Wikipedia, as well as for other similar initiatives, since they depend on the contribution of volunteers who offer their time and talent without monetary reward. Therefore, in order to understand what is behind the contribution, it is necessary to understand what motivates participants and identify the motivations that are associated with high or low levels of contribution (Nov, 2007).

Several aspects of the project are directly related to the motivation of regular citizen to participate, including the complexity of the tasks delegated to them, and the importance they attach to their goals. Nov et al. (2011) conducted an experiment which aimed to study possible motivational factors in crowdsourcing systems. In this experiment, volunteer contributors formed two activities — the first was related to solving small parts of a large task and the second was related to image classification. Results showed that the motivational factors associated with the second activity were higher for all motivation indicators. Among the reasons influencing participants of the experiment, the major indicators were related to: the importance to be part of something collaborative; and due to intrinsic reasons such as altruism, entertainment, reciprocity, intellectual stimulus, and a sense of obligation to contribute.

Although compelled by a scientific need, citizen science is also able to meet the public's desire and provide them with many benefits, including knowledge, entertainment, and the opportunity to take part in scientifically authentic experiences. In the Galaxy Zoo project, for instance, this experience can be considered to be an authentic involvement with scientific practice, since the data analyzed, and the publications to which volunteers contribute, are exactly the same for professional scientists. Citizen science projects often provide the public with some form of education or development in their understanding of science (Wiggins & Crowston, 2010), and the resources used usually vary from texts, publications, sources of virtual content, and manuals, to online courses and interactive games that can be accessed on websites. These instructional resources provide benefits for both the participants and the projects themselves, because they simultaneously enrich the participatory experience of the volunteer contributors and improve the quality of their contribution.

However, the main motivator for effective engagement seems to be the pleasure associated with the participatory experience (Nov et al., 2011). For Wenger (2002), nothing replaces the vividness of participating as the main attraction for the engagement and retention of citizen scientists. The ability to generate interest, relevance, and value among participants, to the point of leading them to work in favor of scientific research is one of the main success factors in citizen science projects.

Data Quality

Data quality is one of the major concerns in scientific research. In particular, if the project is designed for collaboration with non-experts or amateurs, the chance of introducing errors may increase (Jordan et al., 2012). As a result, part of the scientific community considers data coming from citizen science projects to be unreliable to be used in conventional scientific research (Alabri & Hunter, 2010). However, the literature has shown that citizen science projects, when properly designed and conducted, can produce results as good as those produced by conventional science (Canfield, Jr. et al., 2002; McKinley et al., 2015), and also do it quickly and cost-effectively (Bowser & Shanley, 2013). Errors are expected, but collectively, the participation of nonprofessional scientists can generate knowledge and reliable results for scientific research (Soares, 2011).

Data quality is a key aspect in citizen science projects and requires further study. A study conducted in a project involving the mapping of invasive marine species revealed a high value placed on the accuracy of observations made by high school students. This study also found that the motivation had a positive impact on the completeness of the data set (Delaney et al., 2008).

Tweddle et al. (2012) emphasize the importance of training the participants to collect and analyze reliable data. Besides training, it is also important to provide user support, as well as create forms, handouts, field guides, or even a direct communication channel with the project team, in order to facilitate the participant's activity and minimize the complexity of the task.

The quality of data in monitoring programs should be evaluated from two perspectives: external and internal (Conrad & Hilchey, 2010). Nicholson et al. (2002) conducted a statistical comparison of environmental monitoring data collected by professionals and volunteers, and from this they concluded that the data quality is comparable to certain parameters.

Alabri & Hunter (2010) formulated a hypothesis that the use of trust and reputation metrics (as used to provide recommendation services in online social networks such as eBay and Netflix) can be applied to citizen science data. Trust models can provide simple and effective mechanisms for filtering unreliable data. Additionally, combining trust/reputation metrics with data validation services can significantly improve the quality and reliability of the data generated in the community, thus allowing its safe re-use by the scientific community. Among the various aspects evaluated to identify a set of criteria and attributes for measuring the confidence of the citizen science data, the authors suggested as criteria: the role and qualifications of the contributor (e.g., primary or secondary school student, doctoral student, volunteer, scientist, etc.); the quality and amount of data contributed; whether or not the contributor has had some kind of training; the frequency and the period of the contribution; and the classification of the contributor by other members (inferred or calculated). The authors developed a simple tag system in which members of the network can assign a degree of confidence to other contributors in the network. The added value of community confidence in a member of the network is calculated by considering both the direct trust value assigned by the network members and the indirect trust value which is inferred by additional attributes.

Antelio et al. (2013) proposed a collaborative framework named Qualitocracy to improve data quality in Citizen Science projects. By associating data quality dimensions to scientists through a voting network, the authors aimed to create a continuous process for data quality validation. Another project that employs experts' validation is FeederWatch Project – a citizen science program whose objective is monitoring the distribution and abundance of birds during winter. The system may require additional information or expert analysis regarding the data uploaded by participants. In this case, experts (project staff or regional biologists) may either accept observations or request additional information and photographic documentation from participants. When the extra information is uploaded by the participant, observations are either confirmed and accepted by the platform or non-confirmed and considered invalid and thereof discarded from data analysis (Bonter & Cooper, 2012). Thus, consulting experts permits to identify outliers that are mistakes or even rare observations, reducing the occurrence of errors in final data.

Management

Although there are different types of tasks, each project basically follows the same general structure: citizen scientists follow workflows and specific protocols, perform online tasks or collect data, and make observations of the real world that are later sent to the project's website via the Internet. The team behind the project validates, analyzes, and organizes the information sent by the contributor, and also publishes the results, not only in the scientific literature, but also in a variety of more accessible forums, ranging from websites, blogs, social networks, and wikis, to newsletters and emails. Participants are then able to see their contributions, compare them with the contributions of peers, and understand how their data help science.

Nevertheless, although the execution of citizen science projects seems simple, the design and management of these projects is not a plug-and-play solution. Their implementation requires a tailored approach for each type of task in order to increase the chances that all involved parties meet their expectations and achieve their goals. Uchoa et al. (2013) proposed a conceptual framework known as Mix4Crowds to help citizen science enterprises conceive their crowdsourcing strategy and design their collaboration systems. It departs from the basic principles of traditional marketing models and incorporates the specific requirements of citizen science projects, structured according to a four-stage process, with each stage aimed at gathering, analyzing, and defining the most relevant features, criteria, and requirements related to one of the following four dimensions: crowds, collaboration, communication, and platform. The goal is to produce a coherent, integrated, and balanced mix of features and requirements through the identification and weighting of the design aspects related to each of these four dimensions.

Figure 3. Generic model of citizen science projects

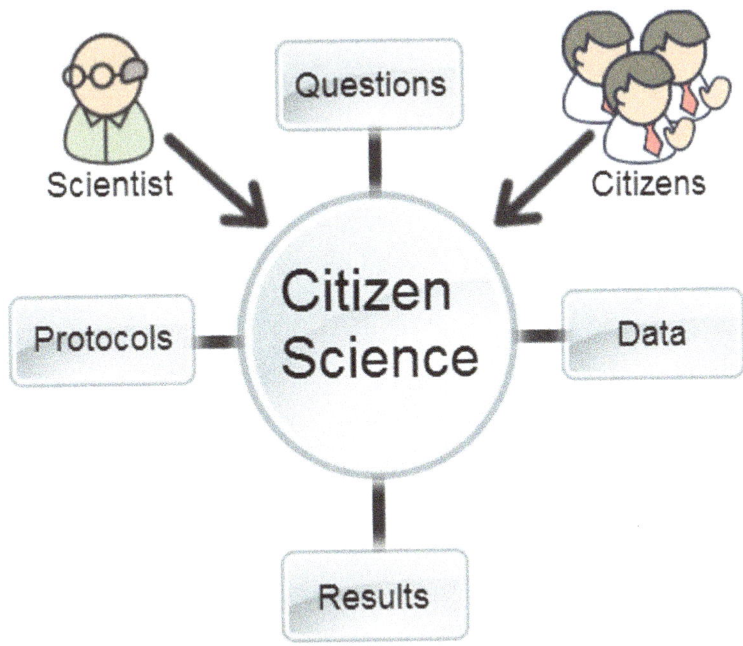

Marketing strategies are widely used by business organizations to develop their relationships with their public. When choosing the appropriate dimensions for developing its marketing mix, an organization usually adopts a predominant producer- or consumer-oriented perspective, each of which will have its benefits, shortcomings, and challenges. For example, product and service features, price policies, promotion, and distribution are the four most common dimensions used by traditional consumer goods organizations when formulating their marketing mix, while consumer capacity, interests and needs, cost, convenience, and communication channels are consumer-oriented dimensions commonly adopted by service organizations. The Mix4Crowds framework resembles the traditional marketing mix model with its four dimensions; however, its approach adopts neither a pure producer nor a pure consumer perspective. Along with the adjustment of the four dimensions, the scientist should also consider the needs and requisites of the three main groups involved: managers (or the team behind the project), volunteer contributors, and institutional partners. For this reason, in order to choose the most appropriate design, it is important to take into consideration the volunteer contributor's point of view — both as producer (e.g., collecting data or transcribing documents) and consumer — about what the project has to offer. In citizen science projects, participants have different motivations, which end up generating additional challenges for researchers who wish to make use of this new collaboration paradigm.

In citizen science projects, each volunteer contributor has a twofold role. When participants begin contributing to a project, they become part of the research team, adding skills and expertise, sharing with other participants and researchers the same goals, methods, and protocols. Nevertheless, each participant is also a consumer of the benefits that the project has to offer, which include: knowledge, experience, social interaction, acknowledgment, entertainment, and leisure. Participants and projects become linked in a way that resembles a provider-consumer or "prosumer" relationship. Prosumers, as they are known

Figure 4. Twofold role of volunteer contributors in citizen science projects

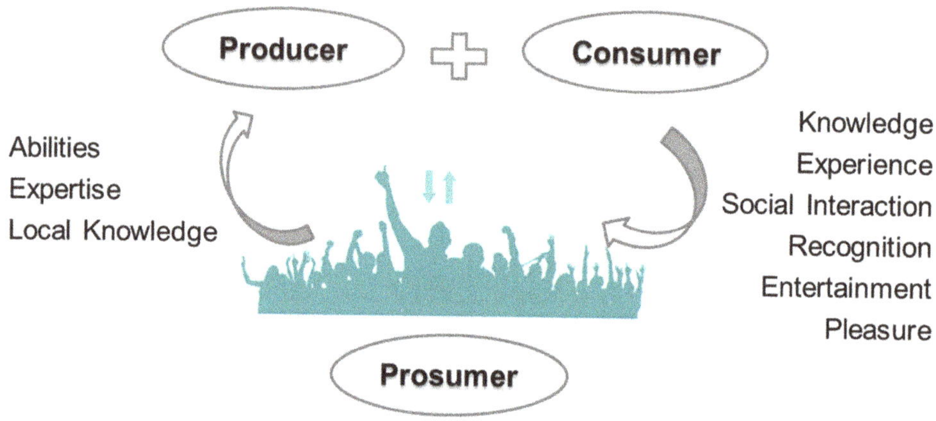

in the marketing area, need their own strategy so that their membership is acknowledged and their connection with the project preserved.

Add to this the fact that citizen science projects depend on the ability to attract and retain volunteer contributors, usually by creating the perception of value and relevance in many levels among participants. Nov et al. (2007, 2011), describe the importance of delivering a "participatory experience" through something that has the power to attract new members and motivate their lasting and productive collaboration.

Additionally, citizen science projects have characteristics that differ from conventional scientific projects used for managing and motivating employees in an organization, in which there are rigid deadlines to suit more formal and hierarchical environments. In citizen science projects, scientists need to make management more flexible due to the fluidity of participation, which is sometimes anonymous and volatile. Different to what happens in conventional scientific collaboration projects, citizen science projects do not spontaneously organize as a result of the interaction of their participants (Preece, 2002). In citizen science, the results depend much more on the individual participation of volunteer contributors than on their mutual interaction, and this participation is subject to the supervision of scientists or managers, as well as the hierarchical structures of command and power previously established (Wiggins & Crowston, 2012). These are structures that can be compared to those of business organizations (Wiggins & Crowston, 2010). After all, citizen science involves projects that do not rarely extend for long periods of time, and achieve large-scale participation and geographical distribution. This requires physical structures, materials, financial and human resources, communication efforts, and organized processes so that a continuous, durable, and reliable operation is guaranteed.

Just as with companies, citizen science projects must seek ways to increase the chances of achieving goals. For this to happen, it seems essential to have a form of organization and management that reconciles the rigidity of aims, timelines, and scientific methods with the flexibility and lack of formality appropriate for volunteer participation. Designing systems to support this type of scientific collaboration requires tailored organizational and task design to ensure scientifically valid results and sustainable contribution (Wiggins & Crowston, 2010; Uchoa et al., 2013). The authors believe that citizen science projects, especially those based on the web, can also benefit from the adoption of a marketing approach when designing their crowd engagement systems.

CONCLUSION

If, on the one hand, smart action to engage citizens with science brings benefits to modern research, by allowing collaboration on a larger scale than conventional methods and reducing time and cost, on the other hand, it brings challenges in terms of keeping citizens motivated and continuously engaged with the project.

Motivation, quality control, and management of a large number of volunteer contributors and contributions are the main challenges. It is important to bear in mind the citizen vision both as a producer (e.g., collecting and classifying data) and as a consumer of what the project has to offer. Often what motivates citizens is the pleasure of being part of a real scientific research project, the possibility of learning, combined with entertainment, as well as the opportunity of getting to know new people, places, and different social and environmental contexts. Thus, social tools, such as forums, blogs, wikis, microblogs and chats, offer great potential for promoting the exchange of knowledge and experience.

Citizen participation has been accelerated by the use of the Internet and new mobile technologies, and there are many successful cases. However, one of the main challenges of citizen science lies not in technology, but in management, quality control, and maintenance of the engagement and motivation of the participants. Constant campaigns, simplicity in user-interfaces, tutorials, acknowledgment, and feedback mechanisms including communication channels and publication of the results are the keys to success of such projects. It is important to provide support and a direct communication channels to ensure continuous feedback between project managers and volunteer contributors.

In order to make data reliable, it is important that the project assesses the needs and mechanisms for training participants to collect trusted data. The use of reliable and reputable metrics can provide simple and effective mechanisms for filtering unreliable data. The greater flexibility of participation, which is sometimes anonymous and volatile, requires the design of a strategy for management, communication, and the use of technology, in order to maximize quality, engagement, and maintenance of voluntary participation, but it also needs to generate benefits for all involved; that is, managers, volunteer contributors and institutional partners.

We believe the typology and concepts presented here can represent a guide to the opportunities and challenges for scientific research and inspire the creation of new initiatives for the smart activation of citizens.

REFERENCES

Alabri, A., & Hunter, J. (2010). Enhancing the Quality and Trust of Citizen Science Data. In *Proceedings of 2010 IEEE Sixth International Conference on e-Science (e-Science)*. doi:10.1109/eScience.2010.33

Antelio, M., Esteves, M. G. P., Schneider, D., & Souza, J. M. (2012), Qualitocracy: A data quality collaborative framework applied to citizen science. In *Proceedings of the International Conference on Systems, Man and Cybernetics (IEEE SMC 2012)*. doi:10.1109/ICSMC.2012.6377847

Bonney, R., Cooper, C. B., Dickinson, J., Kelling, S., Phillips, T., Rosenberg, K. V., & Shirk, J. (2009). Citizen Science: A Developing Tool for Expanding Science Knowledge and Scientific Literacy. *Bioscience*, *59*(11), 977–984. doi:10.1525/bio.2009.59.11.9

Bonney, R. I. C. K., & Dickinson, J. L. (2012). *Overview of citizen science. Citizen Science: Public Participation in Environmental Research*. New York: Cornell University Press.

Bonter, D. N., & Cooper, C. B. (2012). Data validation in citizen science: A case study from Project FeederWatch. *Frontiers in Ecology and the Environment*, *10*(6), 305–307. doi:10.1890/110273

Bowser, A., & Shanley, L. (2013). *New Visions in Citizen Science*. Washington, DC: Woodrow Wilson International Center for Scholars.

Burke, J. A., Estrin, D., Hansen, M., Parker, A., Ramanathan, N., Reddy, S., & Srivastava, M. B. (2006) Participatory sensing. WSW'06 at SenSys '06.

Canfield, D. E. Jr, Brown, C. D., Bachmann, R. W., & Hoyer, M. V. (2002). Volunteer lake monitoring: Testing the reliability of data collected by the Florida LAKEWATCH program. *Lake and Reservoir Management*, *18*(1), 1–9. doi:10.1080/07438140209353924

Conrad, C., & Hilchey, K. (2010). A review of citizen science and community-based environmental monitoring: Issues and opportunities. *Environmental Monitoring and Assessment*, 1–19.

Cooper, S., Khatib, F., Treuille, A., Barbero, J., Lee, J., Beenen, M., & players, F. et al. (2010). Predicting protein structures with a multiplayer online game. *Nature*, *466*(7307), 756–760. doi:10.1038/nature09304

Crowston, K., & Prestopnik, N. R. (2011). Gaming for (Citizen) Science: Exploring Motivation and Data Quality in the Context of Crowdsourced Science through the Design and Evaluation of a Social-Computational System. In *IEEE Seventh International Conference on e-Science Workshops (eScienceW)*.

Crowston, K., & Prestopnik, N. R. (2013). Motivation and Data Quality in a Citizen Science Game: A Design Science Evaluation. In *46th Hawaii International Conference on System Sciences (HICSS)*. doi:10.1109/HICSS.2013.413

Cuff, B. D., Hansen, M., & Kang, J. (2008). Urban Sensing: Out of the Woods. *Communications of the ACM*, *51*(3), 1–33. doi:10.1145/1325555.1325562

Delaney, D., Sperling, C., Adams, C., & Leung, B. (2008). Marine invasive species: Validation of citizen science and implications for national monitoring networks. *Biological Invasions*, *10*(1), 117–128. doi:10.1007/s10530-007-9114-0

Dickinson, J. L., Shirk, J., Bonter, D., Bonney, R., Crain, R. L., Martin, J., & Purcell, K. et al. (2012). The current state of citizen science as a tool for ecological research and public engagement. *Frontiers in Ecology and the Environment*, *10*(6), 291–297. doi:10.1890/110236

Dickinson, J. L., Zuckerberg, B., & Bonter, D. N. (2010). Citizen Science as an Ecological Research Tool: Challenges and Benefits. *Annual Review of Ecology Evolution and Systematics*, *41*(1), 149–172. doi:10.1146/annurev-ecolsys-102209-144636

EPA - United States Environmental Protection Agency. (2012). *Starting Out in Volunteer Monitoring*. EPA 941-F002-004.

Estrin, D. (2010). Participatory sensing: Applications and architecture. *IEEE Internet Computing*, *14*(1), 12–42. doi:10.1109/MIC.2010.12

Eveleigh, A., Jennett, C., Lynn, S., & Cox, A. L. (2013). 'I Want to Be a Captain! I Want to Be a Captain!': Gamification in the Old Weather Citizen Science Project. In *Proceedings of the First International Conference on Gameful Design, Research, and Applications*. doi:10.1145/2583008.2583019

Geiger, D., Rosemann, M., & Fielt, E. (2011b). *Crowdsourcing Information Systems– A Systems Theory Perspective*. Presented at the 22nd Australasian Conference on Information Systems, Sydney, Australia.

Geiger, D., Seedorf, S., Schulze, T., Nickerson, R., & Schader, M. (2011a). Managing the crowd: Towards a taxonomy of crowdsourcing processes.*Proceedings of the Seventeenth Americas Conference on Information Systems*.

Goldman, J., Shilton, K., Burke, J., Estrin, D., Hansen, M., Ramanathan, N.,... West, R. (2009). *Participatory Sensing: A citizen-powered approach to illuminating the patterns that shape our world*. Foresight & Governance Project, White Paper.

Goodchild, M. F. (2007). Citizens as sensors: The world of volunteered geography. *GeoJournal*, *69*(4), 211–221. doi:10.1007/s10708-007-9111-y

Groom, Q. J., O'Reilly, C., & Humphrey, T. (2014). Herbarium specimens reveal the exchange network of British and Irish botanists, 1856–1932. *Nordic Journal of Botany*, *4*(2), 95–103. doi:10.1179/2042349714Y.0000000041

Haklay, M. (2013). Citizen science and volunteered geographic information: overview and typology of participation. In D. Z. Sui, S. Elwood, & M. F. Goodchild (Eds.), *Crowdsourcing geographic knowledge: volunteered geographic information (VGI) in theory and practice* (pp. 105–122). Berlin: Springer. doi:10.1007/978-94-007-4587-2_7

Haklay, M., Antoniou, V., Basiouka, S., Soden, R., & Mooney, P. (2014). *Crowdsourced geographic information use in government, Report to GFDRR*. London: World Bank.

Howe, J. (2009). *Crowdsourcing: Why the Power of the Crowd Is Driving the Future of Business*. Crown Publishing Group.

Hutt, H., Everson, R., Grant, M., Love, J., & Littlejohn, G. (2013). How clumpy is my image? Evaluating crowdsourced annotation tasks. In *UK Workshop on Computational Intelligence (UKCI)*.

Jordan, R., Ballard, H., & Phillips, T. (2012). Key issues and new approaches for evaluating citizen-science learning outcomes. *Frontiers in Ecology and the Environment*, *10*(6), 307–309. doi:10.1890/110280

Lintott, C. J., Schawinski, K., Slosar, A., Land, K., Bamford, S., Thomas, D., & Vandenberg, J. et al. (2008). Galaxy Zoo: Morphologies derived from visual inspection of galaxies from the Sloan Digital SkySurvey. *Monthly Notices of the Royal Astronomical Society*, *389*(3), 1179–1189. doi:10.1111/j.1365-2966.2008.13689.x

Maisonneuve, N., Stevens, M., Niessen, M. E., Hanappe, P., & Steels, L. (2009). Citizen Noise Pollution Monitoring. In *Proceedings of the 10th International Digital Government Research Conference*.

McKinley, D.C., Miller-Rushing, A.J., Ballard, H.L, Bonney, R., Brown, H., Evans, D.M., … Soukup, M.A. (2015). Investing in Citizen Science Can Improve Natural Resource Management and Environmental Protection. *Issues in Ecology,* 19.

Newman, G., Wiggins, A., Crall, A., Graham, E., Newman, S., & Crowston, K. (2012). The future of citizen science: Emerging technologies and shifting paradigms. *Frontiers in Ecology and the Environment*, *10*(6), 298–304. doi:10.1890/110294

Nichols, J. D., & Williams, B. K. (2006). Monitoring for conservation. *Trends in Ecology & Evolution*, *21*(12), 668–673. doi:10.1016/j.tree.2006.08.007

Nicholson, E., Ryan, J., & Hodgkins, D. (2002). Community data - where does the value lie? Assessing confidence limits of community collected water quality data. *Water Science and Technology*, *45*, 193–200.

Nov, O. (2007). What Motivates Wikipedians? *Communications of the ACM*, *50*(11), 60–64. doi:10.1145/1297797.1297798

Nov, O., Arazy, O., & Anderson, D. (2010). Crowdsourcing for science: understanding and enhancing SciSourcing contribution. In *ACM CSCW 2010 Workshop on the Changing Dynamics of Scientific Collaborations*.

Nov, O., Arazy, O., & Anderson, D. (2011). Technology-Mediated Citizen Science Participation: A Motivational Model. In *Proceedings of the AAAI International Conference on Weblogs and Social Media (ICWSM 2011)*.

Okolloh, O. (2009). Ushahidi or 'testimony': Web 2.0 tools for crowdsourcing crisis information. Participatory Learning and Action. *IIED Publications Database*, *59*, 65–70.

Pant, R. B. (2009). *The Social Media Marketing Mix (The 4 "P's")*. Retrieved January, 26, 2012 from http://ritubpant.com/social-media-marketing-mix/

Pfeffer, M. J., & Wagenet, L. P. (2007). Volunteer Environmental Monitoring, Knowledge Creation and Citizen-Science Interaction. In J. Pretty, A. Ball, T. Benton, J. Guivant, D. R. Lee, D. Orr, & P. H. Ward et al. (Eds.), *The SAGE Handbook of Environment and Society*. SAGE. doi:10.4135/9781848607873.n16

Ponciano, L., Brasileiro, F., Simpson, R., & Smith, A. (2014). Volunteers' Engagement in Human Computation Astronomy Projects. *Computing in Science Engineering*.

Preece, J. (2002). Supporting Community as Building Social Capital. *Communications of the ACM*, *45*(4), 37–39. doi:10.1145/505248.505269

Quinn, A. J., & Bederson, B. B. (2009). *A Taxonomy of Distributed Human Computation*. University of Maryland.

Raddick, M. J., Bracey, G., Carney, K., Gyuk, G., Borne, K., Wallin, J., & Jacoby, S. (2010). Citizen Science: Status and Research Directions for the Coming Decade. In astro2010: The Astronomy and Astrophysics Decadal Survey.

Raddick, M. J., Bracey, G., Gay, P. L., Lintott, C. J., Murray, P., Schawinski, K., & Vandenberg, J. et al. (2010b). Galaxy Zoo: Exploring the Motivations of Citizen Science Volunteers. *Astronomy Education Review*, *9*(1). doi:10.3847/AER2009036

Reid, C., Marshall, J., Logan, D., Kleine, D., & Dean, A. (2012). *Coral Reefs and Climate Change. The guide for education and awareness. Published by CoralWatch*. The University of Queensland.

Sheppard, S. A., & Terveen, L. (2011). Quality is a Verb: The Operationalization of Data Quality in a Citizen Science Community. In *Proceedings of the 7th International Symposium on Wikis and Open Collaboration*. doi:10.1145/2038558.2038565

Soares, M. D. (2011). *Employing Citizen Science to Label Polygons of Segmented Images*. (doctoral dissertation), National Institute For Space Research.

The Royal Society. (n.d.). *Knowledge, Networks and Nations: Global scientific collaboration in the 21st century*. Author.

Tweddle, J.C., Robinson, L.D., Pocock, M.J.O., & Roy, H.E. (2012). *Guide to citizen science: developing, implementing and evaluating citizen science to study biodiversity and the environment in the UK.* Natural History Museum and NERC Centre for Ecology & Hydrology for UK-EOF.

Uchoa, A. P., Esteves, M. G. P., & Souza, J. M. (2013). Mix4Crowds - Toward a framework to design crowd collaboration with science. In *2013 IEEE 17th International Conference on Computer Supported Cooperative Work in Design (CSCWD)*.

Wenger, E., Mcdermott, R., & Snyder, W. M. (2002). *Seven Principles for Cultivating Communities of Practice*. HBSWK Pub.

Wiggins, A., & Crowston, K. (2010). Developing a Conceptual Model of Virtual Organizations for Citizen Science. *International Journal of Organizational Design and Engineering*, *1*(1/2), 148–162. doi:10.1504/IJODE.2010.035191

Wiggins, A., & Crowston, K. (2011) From Conservation to Crowdsourcing: A Typology of Citizen Science. In *Proceedings of the 2011 44th Hawaii International Conference on System Sciences*. doi:10.1109/HICSS.2011.207

Wiggins, A., & Crowston, K. (2012). Goals and Tasks: Two Typologies of Citizen Science Projects. In *Proceedings of the Fourth-fifth Hawaii International Conference on Systems Sciences* (HICSS-45). doi:10.1109/HICSS.2012.295

KEY TERMS AND DEFINITIONS

Participant: In general, is an unpaid person who takes part in a project by helping to define its scope, gathering or analyzing data and contributing with new ideas and solutions – a 'citizen scientist'.

Partner: An organization or group of people with a common interest relevant to a citizen science project (e.g. local communities, school groups, governmental agencies or Non-Governmental Organizations (NGOs); others scientific institutions, or members of a natural history group).

Project: It is a citizen science activity. We use this term to incorporate the full range of citizen science including crowdsourcing, long-term monitoring and scientific investigations.

Scientists: Leading or participating in the team behind the citizen science projects are primarily interested in the scientific outputs. They may be students, amateur scientists, professional scientists or research group coordinators.

Smart Activation: An intelligent opportunity for scientists to activate the citizen's engagement in the steps of scientific process. The project management requires a tailored approach in order to increase the chances that both participants and scientists achieve their expectations and goals.

ENDNOTES

http://flusurvey.org.uk/
2 https://www.ushahidi.com/
3 http://earthquake.usgs.gov/data/dyfi/
4 http://www.wikiaves.com.br/
5 http://climatecolab.org/
6 https://www.planethunters.org/
7 http://centerforgamescience.org/
8 http://www.galaxyzoo.org/
9 http://www.fold.it/

Chapter 14
Surface Water Information Collection:
Volunteers Keep the Great Lakes Great

Mark Gillingham
Hermit's Peak Watershed Alliance, USA

ABSTRACT

This chapter's starting premise is that for decades the United States Environmental Protection Agency region subsuming most of the Great Lakes watershed has been partially monitored by private citizens, but collected data have been underutilized by water managers, scientists, and policymakers. Today, citizens with only a smartphone can dramatically increase our understanding of surface water, help managers and policymakers, and educate the general public about the quality of water. The US Clean Water Act and National Strategy for Civil Earth Observations have helped to coordinate citizen scientists and direct funds to surface-water monitoring. And more contributors are being solicited and trained to help with the enormous task of monitoring lakes and streams. At the same time, technology allows citizens with a smartphone to accomplish what previously required experts in a lab: to act for clean water!

INTRODUCTION

The Great Lakes watershed of the United States serves 51 million people with drinking water (United States Environmental Protection Agency, 2016, May 29). For decades the Great Lakes have been partially monitored by private citizens, but these data have been underutilized by scientists and policymakers. However, today, citizens with only a smartphone can dramatically increase our understanding of surface water, help policymakers, and educate the general public about the quality of water in the region. Trained volunteers are able to perform important water quality measurements and use increasingly sophisticated phone applications. With governmental budgets cut, what is needed is a call for citizens to act for clean water!

DOI: 10.4018/978-1-5225-0962-2.ch014

PAST CITIZEN EFFORTS TO MONITOR THEIR GREAT LAKES WATERSHED

In the upper Midwest of the United States are five very large fresh-water lakes including Superior, Michigan, Huron, and Erie and tens-of-thousands of smaller lakes and streams in its six states. This area, designated by the Environmental Protection Agency as Region 5, includes the states of Ohio, Michigan, Indiana, Illinois, Wisconsin, and Minnesota (from east to west). Individuals have been monitoring these local lakes for as many as 8 decades. Each of these states began lake and stream monitoring programs at different times through various agencies but mostly as a consequence of the Clean Water Act of 1972. The following describes each state's lake and stream monitoring programs individually.

Ohio

Ohio law ensures that citizen volunteers are qualified to sample and measure lake and stream water. The state's Environmental Protection Agency requires volunteers to use approved study plans and become Qualified Data Collectors, who are authorized to submit data to the state's site where the data can be used by all interested parties and combined with data from other states (Ohio Environmental Protection Agency, 2016).

Ohio has three levels of credible data. Level 1 was designed with educators in mind and targets conservation districts, parks, health departments, and the general public. The purpose of Level 1 is primarily to promote public awareness and education about surface waters of the state.

The Level 2 group was designed with watershed groups in mind and appears to be closer to other Region 5 states' first tier or basic level. Level 2 information can be used to evaluate the effectiveness of pollution controls, to conduct initial screening of water quality conditions, and to promote public awareness and education about surface waters. Level 2 groups monitor long term surface water quality trends in a watershed.

Level 3 provides the highest level of scientific rigor and incorporates methods which are equivalent to those used by Ohio Environmental Protection Agency personnel. By Ohio law, only Level 3 information can be used for regulatory application (Ohio Environmental Protection Agency, 2016).

Of all the states in Region 5, Ohio rejects most volunteer data collection for research purposes. Whereas most states rely on project coordinators, who are usually government agency employees, for training, Ohio depends on nonprofit and educational organizations to train volunteers and collect data. For instance, the Ohio Watershed Network within Ohio State University Extension (Ohio Watershed Network, 2016).

Michigan

The state of Michigan has been using citizen volunteers to monitor lakes since 1974 and streams since 1992. Since then, the combined program has been called Cooperative Lakes Monitoring Program and more recently has been subsumed under the Michigan Clean Water Corps (MiCorps), which is Michigan's volunteer surface water monitoring network. Like other states, Michigan's volunteers monitor lake and stream water over time, reduce costs to the state, and provide important baseline data. Volunteers also educate their communities and build public support of clean and ecological practices near lakes (Michigan Department of Environmental Quality, 2016).

Indiana

Indiana has a two-tiered volunteer approach. Tier 1 volunteers sample lake water for transparency using the Secchi method in which a disc is lowered into the water until it is just visible at which point the depth is measured. Tier 2 volunteers also perform chemical analyses of lake water. The state understands that without volunteers there would not be enough financial support to monitor all of the lakes in Indiana (Indiana Clean Lakes Program, 2016). Volunteers can become credible information monitors and earn a certificate in watershed leadership from Purdue University.

Illinois

The state of Illinois created the Volunteer Lake Monitoring Program within its Environmental Protection Agency in 1981. Volunteers help gather more information about Illinois lakes than professionals alone could, which adds more information with less cost and allows better lake management (Illinois Environmental Protection Agency, 2016a). The Illinois program also creates an environment for citizen education of the local environment and lake ecosystems.

Illinois has basic and advanced volunteer programs. Basic volunteers monitor lake transparency using the Secchi method twice per month from April to October and they also monitor Zebra Mussels. Advanced volunteers add chemical analysis tasks by collecting lake samples and sending them to a state lab for analysis (Illinois Environmental Protection Agency, 2016b).

Wisconsin

Wisconsin's Citizen Lake Monitoring Network has 1000 citizen volunteers who for 40 years have been working with the Wisconsin Department of Natural Resources to monitor 40,000 sites. The Network's goals are to collect high quality information, to educate and empower volunteers, and to share this information and knowledge with the public (University of Wisconsin Extension, 2016).

Wisconsin provides volunteers with equipment and training to conduct water monitoring. The data are shared with the Wisconsin Department of Natural Resources and university biologists and researchers, university extension offices, and the interested public through a state site (University of Wisconsin Extension, 2016). Wisconsin volunteers measure transparency of lakes using the Secchi method. Some volunteers are also trained to monitor water chemistry, aquatic life, ice on and off a lake, and native aquatic plants (Wisconsin Department of National Resources, 2016).

Minnesota

In 1973, the state of Minnesota was the first to train citizens to use the Secchi method (Minnesota Pollution Control Agency, 2008). Soon, the neighboring states in Region 5 followed suit. Eventually each state created an agency to oversee training, monitoring, and information storage. Minnesota's Citizen Lake-Monitoring Program claims to be the longest running volunteer lake monitoring program in the United States and, like other states, has spawned a stream monitoring program, too. Citizen Lake-Monitoring Program requires their 1200 volunteers to conduct regular seasonal lake sampling. The state supports volunteers with training, discs, and information storage. Training consists of learning to use a Secchi disc, choosing locations, recording information, and sharing it with the state.

The volunteers contribute to the goal of improving water quality by adding to the Minnesota Pollution Control Agency and Environmental Protection Agency information stores and by learning about water quality and lake degradation.

An Example of Volunteer Water Monitoring Data: Minnesota Pollution Control Agency

The Minnesota Pollution Control Agency oversees volunteer water monitoring for the state of Minnesota. Once data are collected, they can be viewed on a public site with common information for each lake or stream. For instance, the information for Upper Prior Lake in Scott County in the Lower Minnesota River watershed (site 70-0072-00, http://cf.pca.state.mn.us/water/watershedweb/wdip/details.cfm?wid=70-0072-00) and Elk Lake in Sherburne County in the upper Mississippi watershed (site 71-0141-00, [cf.pca.state.mn.us/water/watershedweb/wdip/details.cfm?wid=71-0141-00]) is provided by both volunteers and professionals. The volunteers are members of the Citizen Lake Monitoring Program or Advanced Citizen Lake Monitoring Program groups.

The Minnesota Pollution Control Agency (2016b) provides the public with lake data of various types (e.g., transparency, trophic state, chlorophyll-a, and Phosphorus). For instance, one can find a statewide view of transparency data for monitored lakes in Minnesota (see Figure 1). Figure 1 shows lakes that have increased, decreased, and remained the same transparency. Arrows that point up show lakes that have increased transparency (22%) and those with arrows that point down show lakes that have decreased transparency (10%). Circles show lakes with no significant change in transparency (68%). Lake managers can use this information to quickly understand which lakes need more attention.

This information has mostly been collected by volunteers over the years (c.f., [http://cf.pca.state.mn.us/water/watershedweb/wdip/details.cfm?wid=71-0141-00]). Volunteers have been very important in documenting many years of data which are needed to detect trends. The baseline data collected by volunteers helps professionals know when to do further testing and recovery planning and action.

IS VOLUNTEER-COLLECTED INFORMATION OF ANY USE TO SCIENTISTS?

Volunteers are important to the United States federal and state agencies, which have begun using more volunteer-collected information. The United States has initiated a wide-ranging use of volunteer-collected information throughout agencies including the Bureau of Land Management, Department of State, National Archives, National Science Foundation, National Oceanic and Atmospheric Agency, United States Geological Survey, and United States Environmental Protection Agency (Office of Science and Technology Policy, 2014; Office of Science and Technology Policy, 2015). As part of the effort to increase citizen innovation, the U. S. Government issues challenges within agencies with prizes (c.f., Challenge.gov). Many of the challenges are directed to students, school teachers, and college and university faculty. For instance, after it was determined that farm runoff was the major contributor to the algal bloom that closed Toledo Ohio's water supply in August, 2014, the United States Environmental Protection Agency developed a challenge for college students to create an easy way to visualize and predict algal blooms using existing data (Challenge.gov, 2016). United States Government agencies clearly see the value of citizen help with its water quality challenges.

Figure 1. Statewide view of transparency data for monitored lakes in Minnesota (Minnesota Pollution Control Agency, 2016b)

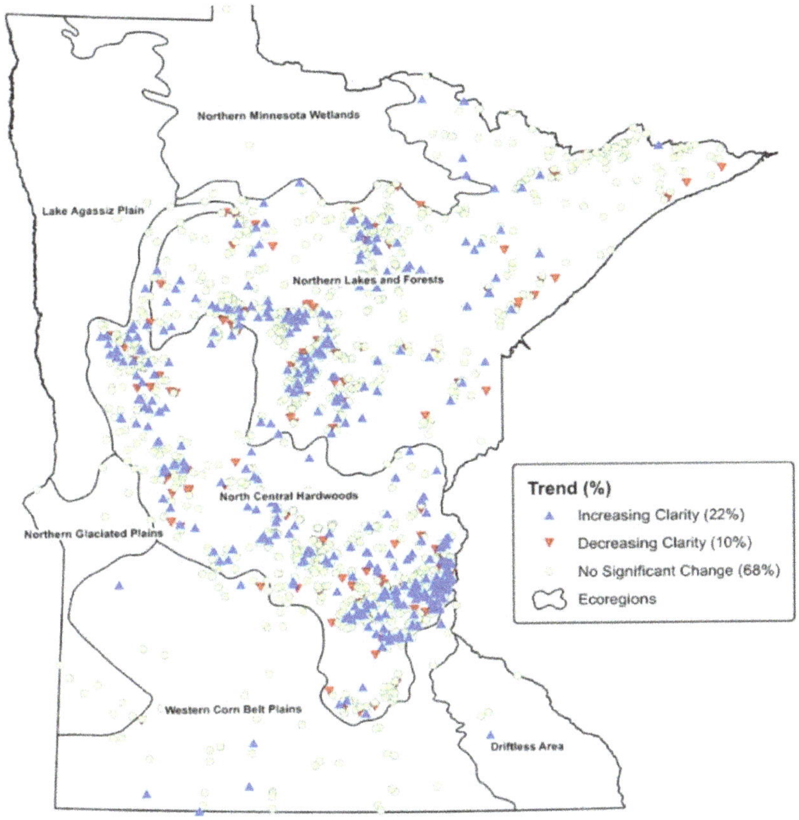

Use of Volunteers in Minnesota

The Minnesota Pollution Control Agency, along with other state government agencies and organizations, has been making greater use of volunteers to collect lake and stream information.

Volunteer-collected information is the main source of data used in state assessments. The Minnesota Pollution Control Agency uses information collected by volunteers of the Citizen Lake Monitoring Program, which monitors lake transparency with the Secchi method, combined with nutrient data to determine the health of lakes and identify impairments. Since 2000, a second-level monitoring program, Advanced Citizen Lake Monitoring Program has been collecting nutrient data on lakes and providing this information to the Minnesota Pollution Control Agency. In 2006, Minnesota Pollution Control Agency began using tubes to assess the turbidity of streams, primarily with the assistance of volunteers. The tubes are a modification of the Secchi method in which water from a stream is poured into a long tube with a small Secchi disc at the bottom. When the disc is just visible, the height of the water in the tube is measured (see Figure 2).

Finally, the volunteer-collected Secchi-based transparency information is the principal source for lake transparency trend analysis in Minnesota. Data for nearly every lake in the state is viewable by the public including lake transparency trends (Minnesota Pollution Control Agency, 2011).

Figure 2. Secchi tube
(pca.state.mn.us/sites/default/files/tube1.jpg)

Example of Elk Lake. The Minnesota Pollution Control Agency (2012) publishes watershed reports periodically to inform the public of the condition of surface water. Elk Lake (site 71-0141-00, [*cf.pca.state.mn.us/water/watershedweb/wdip/details.cfm?wid=71-0141-00*]) is a medium-sized shallow lake in the farmland of central Minnesota northwest of Minneapolis and in the Mississippi watershed. Lake quality information has been reported since 1978 (see Figure 3). Over this period of 4 decades, volunteers have collected transparency measures every year. In addition, chemical analyses have been performed and reported for 15 of those years. Elk Lake is being carefully watched by the state because it is rated as hypertrophic (containing excessive nutrients). Citizens are cautioned against eating fish or bathing in Elk Lake.

COMPARATIVE ANALYSIS OF PROFESSIONAL AND VOLUNTEER-COLLECTED INFORMATION

Studies have shown that information collected by volunteers is robust—nearly indistinguishable from information collected by experts. Ashley Shelton (2013) found that volunteer-collected water quality

Figure 3. Description of Elk Lake in the Upper Mississipi Watershed, Minnesota (Missesota Pollution Control Agency [cf.pca.state.mn.us/water/watershedweb/wdip/details.cfm?wid=71-0141-00])

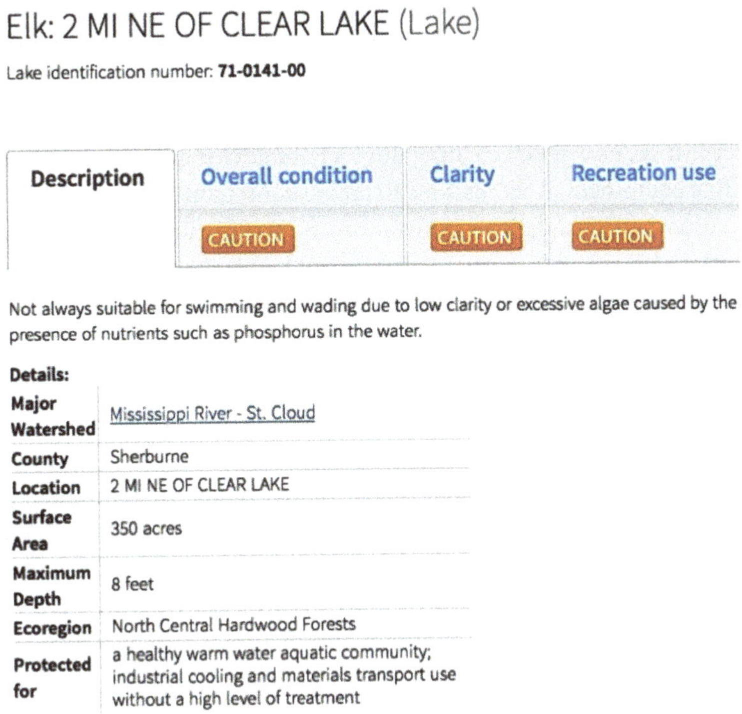

Elk: 2 MI NE OF CLEAR LAKE (Lake)

Lake identification number: **71-0141-00**

Description	Overall condition	Clarity	Recreation use
	CAUTION	CAUTION	CAUTION

Not always suitable for swimming and wading due to low clarity or excessive algae caused by the presence of nutrients such as phosphorus in the water.

Details:

Major Watershed	Mississippi River - St. Cloud
County	Sherburne
Location	2 MI NE OF CLEAR LAKE
Surface Area	350 acres
Maximum Depth	8 feet
Ecoregion	North Central Hardwood Forests
Protected for	a healthy warm water aquatic community; industrial cooling and materials transport use without a high level of treatment

parameters were indistinguishable from that of a skilled professional. Shelton was careful to reduce variability that was not attributable to data collectors, for instance sampling at the same location at the same time and using the same equipment and calibration technique. More specifically, Shelton found four of the five water quality parameters to be indistinguishable between the volunteers and the professional including water temperature, pH, conductivity, and discharge measurements. Only dissolved oxygen was different between the volunteers and professional and may not be a suitable parameter for volunteers to measure. Shelton speculated that measuring dissolved oxygen in the field is prone to error from several variables including water temperature, plant growth, field procedures, and characteristics of the water source (p. 76).

Other studies in a variety of settings have found that when given proper materials and training, volunteer information collection is comparable to professionals (Au, et al., 2000; Canfield, Brown, Bachmann, & Hoyer, 2002; Fore, Paulsen, & O'Laughlin, 2001; Obrecht, Milanik, Perkins, Ready, & Jones,1998). As budgets have been cut, volunteers have been asked to do more to monitor lakes and streams in both the United States and Canada. Each government provides volunteers with training and procedural guidelines (United States Environmental Protection Agency, 2016; Oceans and Fisheries Canada, 2015). Europe is also organizing volunteer citizens to collection information on water and other resources (*European Citizen Science Association, 2016*).

To summarize, with adequate training and simple procedures, volunteer monitors can collect credible information. Volunteers also need the structure and guidance that a governmental body or organization can provide. Also, volunteers must understand the importance of following procedures that they have been taught and to be vigilant during the collection process so that their data can enhance the work of professionals.

Even without much human intervention, permanent on-site electronic monitors can collect flawless data constantly for days and weeks at a time. It may be possible to give similar equipment to volunteers to add more sampling locations. The question is to ascertain how simple procedures and equipment need to be to collect good information? Is a $2000 electronic water meter that requires careful calibration suitable for volunteers? If so, how much training does a volunteer require? Considering that training can be more expensive than equipment, it seems wise to create equipment that can perform the most difficult processes and calculations automatically. For instance, an instrument that adjusts for water temperature while measuring another parameter (e.g., dissolved oxygen). Yet, there are important procedural aspects that cannot be built-into a device such as locating an appropriate water site.

Examples of Water Monitoring Processes by World Water Monitoring Challenge and EyeOnWater

Below are two examples of entry-level water-monitoring processes. The first is a portable lab created through Philippe Cousteau's EarthEcho International called *World Water Monitoring Challenge* (EarthEcho International, 2015). The second is a phone app (EyeOnWater) which was created through the work of the Citizens' Observatory for Coast and Ocean Optical Monitoring (Citclops, 2016a).

World Water Monitoring Challenge (EarthEcho International, 2015; Scientific American, 2013) grew out of a single day in 2012 in which organizers challenged volunteers around the world to collect water samples using their portable test kit (c.f., http://map.monitorwater.org). Now, the *World Water Monitoring Challenge* (EarthEcho International, 2015) is a program used by governmental agencies, school programs, clubs, and organizations for all ages around the world. Participants can purchase a test kit at a reasonable cost to measure basic water quality parameters including temperature, acidity (pH), transparency, and dissolved oxygen. Each kit contains instructions and supplies for 50 tests (Figure 4). Results can be added to an international database (c.f., http://map.monitorwater.org) and participants can share videos of their work with one another around the world.

Figure 4. World water monitoring challenge kit
(EarthEco International, 2015)

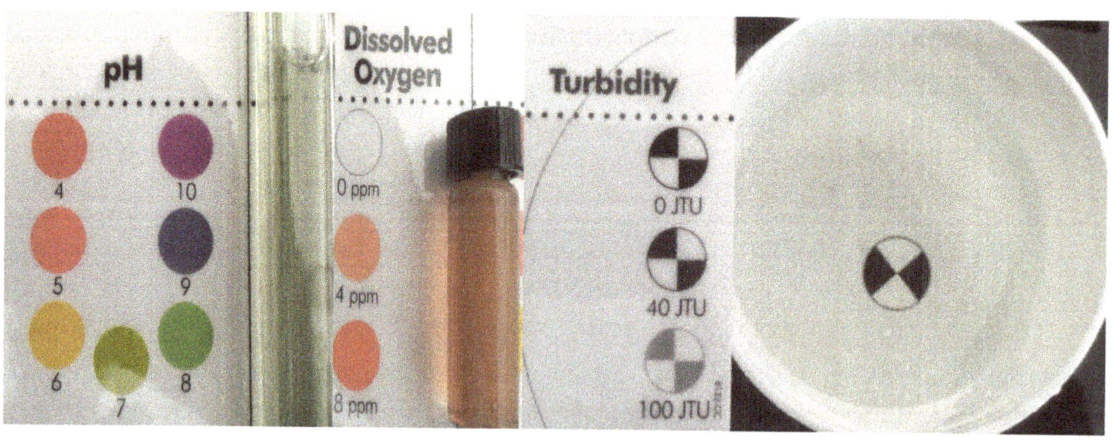

The World Water Monitoring Challenge Kit is a great way to teach novices about water quality parameters but is prone to error because of its many physical and chemical measurements in the field. Each measure requires collecting a water sample and taking several readings with it. A collection container is used to retrieve water from a lake or stream at a depth of 1 foot (⅓ meter). If not cleaned well, this container may have contaminants from prior collections that interfere with transparency or chemical measurements (e.g., acidity and dissolved oxygen). Also, tests must be performed quickly to maintain the original temperature of the sampled water. Transparency is measured with a highly modified Secchi procedure in which a small black and white disc is affixed to the bottom of the collection container, and the water color is compared to a standard (see Figure 4). Other errors can be introduced by contaminants in the collection vials, which are used to measure acidity and dissolved oxygen (see example of vial in Figure 4). Another point of possible error is when one must compare colors in the vials to a chart to estimate the values. Finally, the supplied strip thermometers can be difficult to read. Two are supplied with the kit: one for lower water temperatures and one for higher air temperatures. Both the water and air temperatures are to be recorded, so having high and low reading thermometers is important. Considering all of these error points, one must use the information carefully and perform quality control tests on it.

The Forel-Ule method is a simple means of approximating water quality based on water color. Created in the 1890s by Swiss scientist and medical professor, Francois Alphonse Forel and German geologist and limnologist, Willi Ule (Wernand & van der Woerd, 2010), the Forel-Ule method compares the colors in 21 glass tubes to the current state of the surface of a body of water. This simple handheld method correlates very highly with optical water quality, trophic level state, other measures of color (i.e., colored dissolved organic material), and can be used as a proxy for these measurements (Garaba, Friedrichs, Voß, & Zielinkski, 2015). Also, the method can be used to evaluate streams in addition to lakes and seas. In addition, Forel-Ule can be approximated using satellite imagery, fixed automated stations, and phone apps (Wernand, M., & van der Woerd, H., 2010).

The *EyeOnWater* phone app (eyeonwater.org) incorporates the Forel-Ule method into an easy to use app that has built-in protections against collecting poor information which makes it a good tool for novices and students. To use the *EyeOnWater* app, one takes a photo of the surface of a body of water and compares the processed image of the photo with the 21 Forel-Ule colors. Once the collector chooses a matching color strip (see Figure 5), the collected information is immediately sent to an international database where other data integrity processing is performed. *EyeOnWater* only measures surface water color, but it has few of the drawbacks of the World Water Monitoring Challenge Kit because water does not have to be collected nor tests processed and recorded. In addition, the phone app can record the geolocation of each measurement and allow a contributor to remove a bad data point, further reducing error. Because a Secchi disc is not required, boaters can collect information underway. For instance, sailors on the Barcelona World Race have collected information around the world from their sailboats (Ceccaroni, et al., 2014).

EyeOnWater is easy to use, but there is still a learning curve. One must take a good initial image and then select a good color match from one of the 21 standard colors to the processed image. For instance, my initial attempts had to be discarded. In the first attempt to photograph the Pecos River in New Mexico, the photo had too much shore grass because the river was high with snow melt from the Sangre de Cristo Mountains. The photo should have been taken from a nearby bridge so the center of the river was in view. One can check on observations sent to EyeOnWater by looking at the online map (map.

Figure 5. EyeOnWater application screen to compare processed photo color strip with Forel-Ule color strip (Author)

eyeonwater.org, see Figure 6). The first observation depicted in Figure 6 (labeled "Can see bottom") had a calculated Forel-Ule value of 19 and I entered a value of 18 (all observations in Figure 6 are from the Gallinas Creek, Las Vegas, New Mexico). EyeOnWater calculates a Forel-Ule value from the uploaded image so that each entry has an entered value and a calculated value. For instance, I entered Forel-Ule values ranging between 4 to 16 and the calculated values were from 9 to 20. Obviously, the creek was not changing color so much from one observation to the next within the same month. What was I doing wrong? Taking poor photos and making poor comparisons. Light and shadow effects the observation a great deal. Different light and shadow conditions can be seen in the observations depicted in Figure 6. According to EyeOnWater directions, one should place the sun behind and over ones left or right shoulder and hold the phone between 0 and 30 degrees to the surface of the water (Citclops, 2016b). However, each of my photos had objects from the shore or bridge casting shadows or overhanging tree limbs causing a reflection. Obviously, training is important even for a simple-to-use phone app.

How is Volunteer-Collected Information Being Used?

Volunteer-collected information is increasingly being used to monitor watersheds in Region 5. After decades of volunteer lake monitoring with Secchi discs, this information has finally been added to databases and used for management and planning by state and national agencies. As the technology to store and upload information to central and distributed repositories has improved, states have increased volunteer training. At the same time the technology to test water quality has improved.

Figure 6. Uploaded image data from phone application, EyeOnWater, entrees by the author on Gallinas Creek, Las Vegas, New Mexico during March, 2016
(eyeonwater.org)

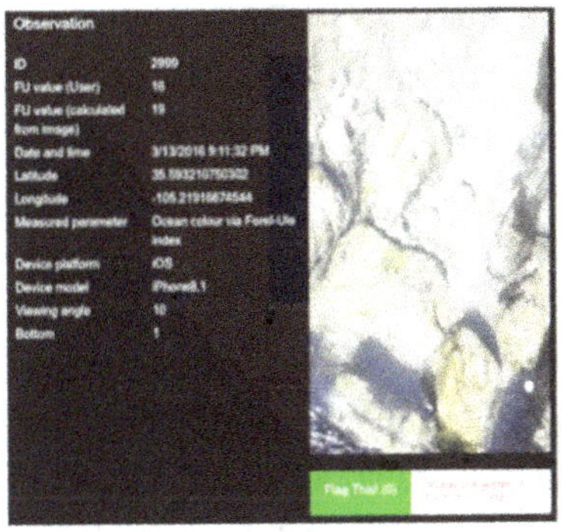

Can see bottom.

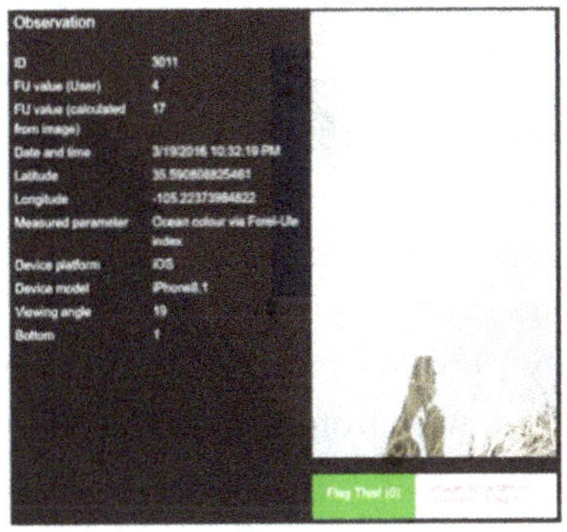

Can see shadow of phone and surface glare.

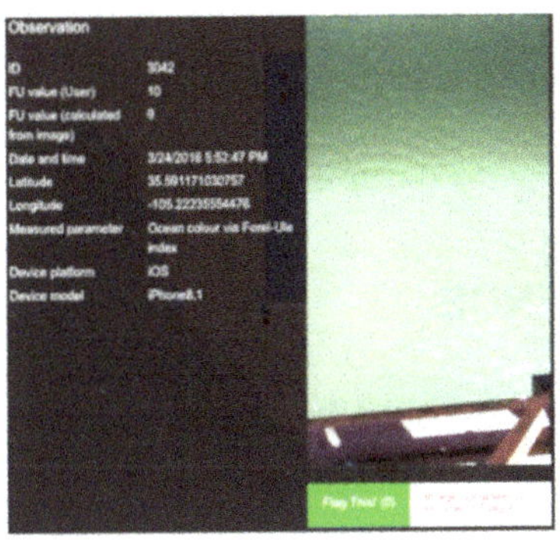

Can see shaded water.

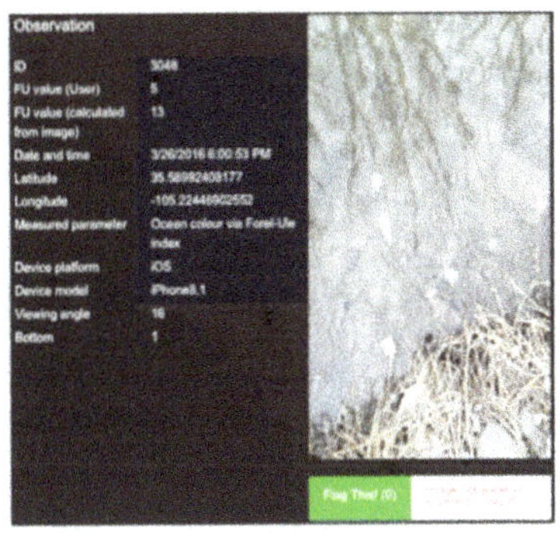

Can see shore and reflected trees.

Lake Trend Analysis

The most common use of volunteer information in Region 5 is for analyzing lakes over long periods of time. Like the example of Elk Lake (above), many lakes have decades of information from which to create an analysis. If a lake's transparency begins to decrease, other measures (e.g., chemical and nutrient) can be taken by volunteers or professionals to create and execute an abatement plan.

Calibration of Satellite Measurement Devices

Since Landsat 7 was launched in 1999 to take photos of the Earth, states have had access to useful ground images suitable for observing water color. Satellite imagery allows a state to cover remote and large lakes that are not easy to access. However, the satellite images have no meaning until they are calibrated with known information about the transparency of some of the lakes in the area. Olmanson and his colleagues (Olmanson, Bauer, & Brezonik, 2002) calibrated images of Minnesota lakes taken from Landsat 7 by comparing them to transparency information that was collected by volunteers on the ground.

How Much Data is Available?

The United States Environmental Protection Agency and the United States Geologic Survey have been adding information to a common store since the 1960s (i.e., STORET, United States Environmental Protection Agency, 2016, February 10). Since 2000, volunteers who have appropriate credentials have added information to this store, which allows the public to access 80 million water quality records. Federal and state agencies access STORET to allow the public to access information about lakes and streams sorted by state or watershed (e.g., United States Geological Survey WaterWatch, Minnesota Pollution Control Agency).

The Water Quality Portal (www.waterqualitydata.us) is a service of the United States Geological Survey, the United States Environmental Protection Agency, and the National Water Quality Monitoring Council. Information collected by over 400 state, federal, tribal, and local agencies is available (www.waterqualitydata.us). The information store contains over 265 million records from over 2.2 million monitoring locations beginning in the 1960s.

The Water Quality Portal allows the public to view how much information has been collected in a state or watershed over the past 1 or 5 years by either the Environmental Protection Agency or United States Geological Survey (National Water Quality Council, 2016). This information combines volunteer and professional contributions.

THE RISE OF ORGANIZED CITIZEN SCIENCE

Government and Nonprofit Organization Support in US and Europe

Monitoring is necessary to ensure that our waters can continue to support the many different ways we use these resources and to track whether protection and restoration measures are working. The information gained from monitoring helps with prioritizing the issues to be addressed and choosing the geographic areas in which to concentrate, thus helping to ensure cost-effective water-resource management (United States Geological Survey, 2016).

The United States Environmental Protection Agency is in charge of the United States water quality. During the 1970s and 1980s, the number of employees increased dramatically from circa 4,000 to over 16,000. This coincided with the passage of the Clean Water Act in 1972. The crystallizing event of that time occurred in 1969 when the Cuyahoga River in Cleveland, Ohio burned because a passing train cast a spark that set the oil on the surface of the river ablaze. Subsequent to Cuyahoga River fire, the number of employees of the United States Environmental Protection Agency increased steadily until it peeked in 1999 at 18,110. Today, the number of employees has dropped to about 15,000, but demand for water monitoring continues to rise (United States Environmental Protection Agency, 2016). Agencies find they must reduce the number of lakes and streams they sample. For instance, in Minnesota, goals for monitoring surface water were set shortly after the Clean Water Act was passed in 1972, but the goals were never reached. By 2006 few lakes had been monitored so a new strategy was created that included incorporating more volunteer monitors (Minnesota Pollution Control Agency, 2016b).

In addition to governmental agencies, a nonprofit organization, CitSci.org, seeks to assist citizen scientists because most researchers find it difficult to use volunteer-collected information unless it follows some scientific etiquette. CitSci.org aims to collect and generate diverse public datasets. CitSci.

org helps volunteers document their process and, most importantly, add other information (metadata) to the content to allow others to use the information, too.

The European Citizen Science Association is an association supported by organizations from over 17 EU and other countries (European Citizen Science Association, 2016) to create an outline and procedures to help the countries in Europe collect and share volunteer-collected information. The European Citizen Science Association purpose includes the following statement:

Citizen science is defined as organised research where the balance between scientific, educational, societal and policy goals varies across projects. It is a growing worldwide phenomenon recently invigorated by evolving new technologies that connect people easily and effectively with the scientific community. New technology provides a valuable tool for citizens to play a more active role in sustainable development. Through collaboration with scientists in organised research projects citizens can contribute valuable information that can be used to develop and deliver policies, improve understanding and respond to many of the challenges facing society today.

THE FUTURE OF CITIZEN SCIENCE WATER MONITORING

New technology for smartphones, sensors, and imagery is changing the way volunteers can monitor water and contribute to public information. Additionally, more researchers are using big data and visualization tools to analyze the vast quantities of information. Now the Secchi disc may be enhanced or substituted by a phone app, which has the advantages of being easy to use with little instruction, with built-in quality control, geolocation for both finding and reporting a site, and automatic uploading to a common information store.

The percent of U.S. adults who have a smartphone is 68% (Pew Research Center, 2015). Technology add-ons to smartphones have become increasingly sophisticated and affordable. Research labs have created phone add-ons to measure standard chemical and nutrient water parameters (e.g., acidity, Nitrates) at a fraction of the cost of non-phone-based portable sensors (Hossain, et al., 2015; Carnegie Mellon Create Lab, 2016, Colorimetrix, 2016, Gunda, et al., 2014). The future for phone technology used for water monitoring is bright.

CONCLUSION

The Great Lakes of Region 5 supplies drinking water to 51 million people. This drinking water is increasingly threatened by pollution. The enormity of the problem is such that no government agency can monitor all of the streams and lakes. The Federal Government and states are using volunteer citizen scientists to help monitor the water quality of the lakes and streams in Region 5 so that water recovery plans can be made and executed. More citizens must become involved in watershed monitoring to cover more lakes and streams at more locations. Technology can help make better volunteers and can automate monitoring processes. To encourage more volunteers, governments and organizations should incorporate social networking into their recruitment strategies.

REFERENCES

Au, J., Bagchi, P., Chen, B., Martinez, R., Dudley, S. A., & Sorger, G. J. (2000). Methodology for public monitoring of total coliforms, *Escherichia Coli*, and toxicity in waterways by Canadian high school students. *Journal of Environmental Management*, *58*(3), 213–230. doi:10.1006/jema.2000.0323

Canfield, D. E. Jr, Brown, C. D., Bachmann, R. W., & Hoyer, M. V. (2002). Volunteer lake monitoring: Testing the reliability of data collected by the Florida LAKEWATCH program. *Lake and Reservoir Management*, *18*(1), 1–9. doi:10.1080/07438140209353924

Carnegie Mellon Create Lab. (2016). *Flamingo*. Retrieved from http://www.cmucreatelab.org/projects/Water_Quality_Monitoring/pages/Flamingo

Castilla, E. P., Cunha, D. G. F., Lee, F. W. F., Loiselle, S., Ho, K. C., & Hall, C. (2015). Quantification of phytoplankton bloom dynamics by citizen scientists in urban and peri-urban environments. *Environmental Monitoring and Assessment*, *690*, 187–198. doi:10.1007/s10661-015-4912-9

Ceccaroni, L., Wernand, M., Subirats, L., Piera, J., Farrés, R., Price, I., & Steblin, A. (2014). *Extending historic water-quality data sets, using old-fashioned techniques, citizen science and smartphones*. Poster presented at the Second International Ocean Research Conference, Barcelona, Spain. Retrieved from www.1000001labs.org/we-content/uploads/2014

Challenge.gov. (2016). *Visualize Your Water*. Retrieved from https://www.challenge.gov/challenge/visualize-your-water

Citclops. (2016a). *Citizen's Observatory for Coast and Ocean Optical Monitoring*. Retrieved from http://Citclops.eu

Citclops. (2016b). *Instructions how to use*. Retrieved from Citclops.eu/participate/instructions-how-to-use

Colorimetrix. (2016). *Environment*. Retrieved from http://www.colorimetrix.com

EarthEcho International. (2015). *World Water Monitoring Challenge*. Retrieved from http://monitor-water.org

European Citizen Science Association. (2016). *About citizen science and ECSA*. Retrieved from http://ecsa.citizen-science.net/about

Fisheries and Oceans Canada. (2015, August 24). *Community Aquatic monitoring program (CAMP)*. Retrieved from http://inter.dfo-mpo-gc.ca/Gulf/CAMP

Fore, L. S., Paulsen, K., & O'Laughlin, K. (2001). Assessing the performance of volunteers in monitoring streams. *Freshwater Biology*, *46*(1), 109–123. doi:10.1111/j.1365-2427.2001.00640.x

Ganda, N. S. K., Naicker, S., Shinde, S., Kimbahune, S., Shrivastava, S., & Mitra, S. (2014). Mobile water kit (MWK): A smartphone compatible low-cost water monitoring system for rapid detection of total coil form and E. Coli. *Analytical Methods*, *6*(16), 6236–6346. doi:10.1039/C4AY01245C

Garaba, P. S., Friedrichs, A., Voß, D., & Zielinski, O. (2015). Classifying natural waters with the Forel-Ule Colour Index System: Results, applications, correlations and crowdsourcing. *International Journal of Environmental Research and Public Health, 12*(12), 16096–16109. doi:10.3390/ijerph121215044

Hossain, M. A., Canning, J., Ast, S., Rutledge, P. J., Yen, T. L., & Jamalipour, A. (2015). *Lab-in-phone:* Smartphone-based portable fluorometer for pH field measurements of environmental water. *IEEE Sensors Journal, 15*(9), 5095–5102. doi:10.1109/JSEN.2014.2361651

Hunter, J., Alabri, A., & van Ingen, C. (2013). Assessing the quality and trustworthiness of citizen science data. *Concurrency and Computation, 25*(4), 454–466. doi:10.1002/cpe.2923

Illinois Environmental Protection Agency. (2016a). *Volunteer Lake Monitoring Program.* Retrieved from http://epa.illinois.gov/topics/water-quality/monitoring/vlmp/index

Illinois Environmental Protection Agency. (2016b). *Inland Lake Monitoring Programs.* Retrieved from epa.illinois.gov/topics/water-quality/monitoring/inland-lakes/index

Indiana Clean Lakes Program. (2016). *Volunteer Monitoring.* Retrieved from http://www.indiana.edu/~clp/volunteermonitoring.php

Michigan Department of Environmental Quality. (2016). *Cooperative Lake Monitoring Program.* Retrieved from http://www.michigan.gov/deq/0,4561,7-135-3313_3681_3686_3731-195536--,00.html

Minnesota Pollution Control Agency. (2008). *Citizen Lake-Monitoring Program: Instruction Manual.* Retrieved from https://www.pca.state.mn.us/sites/default/files/wq-s1-13.pdf

Minnesota Pollution Control Agency. (2009). *Citizen Monitoring of Surface Water Quality: 2009 Report to the Legislature.* Author.

Minnesota Pollution Control Agency. (2011). *Citizen Monitoring of Surface Water: 2011 Report to the Legislature.* Retrieved from https://www.pca.state.mn.us/sites/default/files/lrwq-s-1sy11.pdf

Minnesota Pollution Control Agency. (2012, October). *Water Quality Assessments of Select Lakes within the Mississippi River (St. Cloud) Watershed.* Retrieved from https://www.pca.state.mn.us/sites/default/files/wq-ws3-07010203.pdf

Minnesota Pollution Control Agency. (2016a). *Transparency trends for Upper Prior Lake.* Retrieved from http://cf.pca.state.mn.us/water/watershed web/whip/wawterwatc.cfm?wid=70-0072-00

Minnesota Pollution Control Agency. (2016b). *Lakes and water quality.* Retrieved from https://www.pca.state.mn.us/water/lakes-and-water-quality

Minnesota pollution Control Agency. (2016a). *Surface water data.* Retrieved from http://cf.pca.state.mn.us/water/watershedweb/wdip/index.cfm

Minnesota pollution Control Agency. (2016b). *Citizen Monitoring Program: Annual summary.* Retrieved from https://www.pca.state.mn.us/water/citizen-monitoring-program-annual-summary

National Water Quality Council. (2016). *Water Quality Portal National Results Coverage.* Retrieved from www.waterqualitydata.us/coverage

Obrecht, D. V., Milanik, M., Perkins, B. D., Ready, D., & Jones, J. R. (1998). Evaluation of data generated from lake samples collected by volunteers. *Journal of Lake and Reservoir Management*, *14*(1), 21–27. doi:10.1080/07438149809354106

Office of Science and Technology Policy. (2014). *National Plan for Civil Earth Observations*. Retrieved from https://www.whitehouse.gov/sites/default/files/microsites/ostp/NSTC/2014_national_plan_for_civil_earth_observations.pdf

Office of Science and Technology Policy. (2015, September 30). *Memorandum to the heads of executive departments and agencies*. Retrieved from https://www.whitehouse.gov/sites/default/files/microsites/ostp/holdren_citizen_science_memo_092915_0.pdf

Ohio State University Extension. (2016). *Ohio Watershed Network*. Retrieved from http://production.cfaes.ohio-state.edu/apps/courses/cfaes/course/view.php

Olmanson, L. G., Bauer, M. E., & Brezonik, P. L. (2002). *Use of Landsat imagery to develop a water quality atlas of Minnesota's 10,000 lakes*. Retrieved from http://www.isprs.org/proceedings/XXXIV/part1/paper/00102.pdf

Open Government Partnerships. (2016). *Levels of Credible Data*. Retrieved on March 16, 2016 from Ohio Environmental Protection Agency: http://www.epa.state.oh.us/dsw/credibledata/levels_of_data.aspx

Ottinger, G. (2010). Buckets of Resistance: Standards and the Effectiveness of Citizen Science. *Science, Technology & Human Values*, *35*(2), 244–270. doi:10.1177/0162243909337121

Pew Research Center. (2015, July). *Device ownership over time*. Retrieved from http://www.pewinternet.org/data-trend/mobile/device-ownership/

Scientific American. (2013, September 3). *Citizen Science: World Water Monitoring Challenge*. Retrieved March 10, 2016, from http://www.scientificamerican.com/citizen-science/world-water-monitoring-challenge

Shelton, A. M. (2013). *The accuracy of water quality monitoring data: A comparison between citizen scientists and professionals*. (Unpublished master's thesis). Saint Mary's University, Halifax, Nova Scotia, Canada.

United States Environmental Protection Agency. (2015). *Nonpoint source: Volunteer monitoring*. Retrieved from www.epa.gov/polluted-runoff-nonpoint-source-pollution/nonpoint-source-volunteer-monitoring

United States Environmental Protection Agency. (2016a). *EPA's Budget and Spending*. Retrieved from www.epa.gov/planandbudget

United States Environmental Protection Agency. (2016b). *STOrage and RETrieval and water quality eXchange*. Retrieved from http http://www.epa.gov/water data/storage-and-RETrival-and-water-quality-exchange

United States Environmental Protection Agency. (2016c). *Nutrient policy and data: Region 5*. Retrieved from https://www.epa.gov/nutrient-policy-data/region-5

United States Geological Survey. (2016, March 19). *WaterWatch*. Retrieved from http://waterwatch.usgs.gov

United States Geological Survey. (2016). *Monitoring our Rivers and Streams*. Retrieved from http://pubs.usgs.gov/fs/fs-077-02_ver1/

University of Wisconsin Extension-Lakes. (2016). *Citizen Lake Monitoring Network*. Retrieved at http://www.uwsp.edu/cnr-ap/UWEXLakes/Pages/programs/clmn/default.aspx

Wernand, M., & van der Woerd, H. (2010). Spectral analysis of the Forel-Ule Ocean colour comparator scale. *Journal Of The European Optical Society*, *5*, 10014s. doi:10.2971/jeos.2010.10014s

Wisconsin Department of Natural Resources. (2016). *Citizen Lake Monitoring Network*. Retrieved from http://dnr.wi.gov/lakes/clmn/

KEY TERMS AND DEFINITION

Clean Water Act: A United States law passed in 1972 that governs water pollution.

Environmental Protection Agency: The United States Environmental Protection Agency (USEPA or EPA) was created to protect citizens health and protect the environment by enforcing federal laws.

Great Lakes: A series of large interconnected freshwater lakes in the northeastern United States bordering Canada that drain into the Atlantic Ocean.

Region 5: EPA Region 5 consists of six states in the upper Midwest of the United States: Ohio, Michigan, Indiana, Illinois, Wisconsin, and Minnesota. EPA has 10 regions.

Secchi Disc: The Secchi Disc was created in 1865 by Angelo Secchi to measure the transparency of bodies of water. The circular disc is lowered into water by a calibrated line and a measurement (Secchi Depth) is taken at the point the disc is no longer visible.

Surface Water: Surface water is lake, stream, wetland, and ocean water.

United States Geological Survey: The United States Geological Survey (USGS) is a scientific agency that studies land and natural resources. USGS is concerned with natural hazard such as floods and droughts.

Watershed: The land and everything upon it on which water drains to a common point. Larger watersheds are made from smaller watersheds just as streams flow into rivers.

Compilation of References

Aguilar, P. L. (2010). *Development Discourse and Social Policies: Challenges to Theory and Contemporary Debates.* Retrieved April 27, 2016 from: https://www.scribd.com/doc/283054476/Aguilera-2010-Development-Discourse-and-Social-Policies-Challenges-to-Theory-and-Contemporary-Debates

Aji, A., Wang, F., Vo, H., Lee, R., Liu, Q., Zhang, X., & Saltz, J. (2013). Hadoop GIS: A high performance spatial data warehousing system over mapreduce. *Proceedings of the VLDB Endowment, 6*(11), 1009–1020. doi:10.14778/2536222.2536227 PMID:24187650

Ajzen, I. (1985). From Intentions to Actions: A Theory of Planned Behavior. In J. Kuhl & J. Beckmann (Eds.), *Action Control* (pp. 11–39). Springer Berlin Heidelberg. doi:10.1007/978-3-642-69746-3_2

Ajzen, I. (1991). The theory of planned behavior. *Organizational Behavior and Human Decision Processes, 50*(2), 179–211. doi:10.1016/0749-5978(91)90020-T

Ajzen, I. (2002). Perceived Behavioral Control, Self-Efficacy, Locus of Control, and the Theory of Planned Behavior. *Journal of Applied Social Psychology, 32*(4), 665–683. doi:10.1111/j.1559-1816.2002.tb00236.x

Ajzen, I., & Madden, T. J. (1986). Prediction of goal-directed behavior: Attitudes, intentions, and perceived behavioral control. *Journal of Experimental Social Psychology, 22*(5), 453–474. doi:10.1016/0022-1031(86)90045-4

Alabri, A. (2010). Enhancing the quality and trust of citizen science data. IEEE eScience 2010.

Alabri, A., & Hunter, J. (2010). Enhancing the quality and trust of citizen science data. In *e-Science (e-Science), 2010 IEEE Sixth International Conference on* (pp. 81-88). IEEE. doi:10.1109/eScience.2010.33

Alguezaui, S., & Filieri, R. (2010). Investigating the role of social capital in innovation: Sparse versus dense network. *Journal of Knowledge Management, 14*(6), 891–909. doi:10.1108/13673271011084925

Almond, G. A., & Verba, S. (1963). Organizational membership and civic competence, from The Civic Culture In *The Civic Society Reader.* Academic Press.

Altman, D. G. (1995). Sustaining interventions in community systems: On the relationship between researchers and communities. *Health Psychology, 14*(6), 526–536. doi:10.1037/0278-6133.14.6.526 PMID:8565927

Amano, T., & Sutherland, W. J. (2013). Four barriers to the global understanding of biodiversity conservation: Wealth, language, geographical location and security. *Proceedings. Biological Sciences, 280*(1756), 20122649. doi:10.1098/rspb.2012.2649 PMID:23390102

Anderson, L. M., Scrimshaw, S. C., Fullilove, M. T., Fielding, J. E., & Normand, J. (2003). Task Force on Community Preventive Services. Culturally competent healthcare systems: A systematic review. *American Journal of Preventive Medicine, 24*(3), 68–79. doi:10.1016/S0749-3797(02)00657-8 PMID:12668199

Andrianandrasana, H. T., Randriamahefasoa, J., Durbin, J., Lewis, R. E., & Ratsimbazafy, J. H. (2005). Participatory ecological monitoring of the Alaotra wetlands in Madagascar. *Biodiversity and Conservation*, *14*(11), 2757–2774. doi:10.1007/s10531-005-8413-y

Annenberg Public Policy Center. (2014). *Americans know surprisingly little about their government, survey finds.* Retrieved from http://www.annenbergpublicpolicycenter.org/americans-know-surprisingly-little-about-their-government-survey-finds/

Ansell, A., & Gash, A. (2008). *Collaborative Governance in Theory and Practice. J Pblic Adm Res Theory*, *18*(4), 543–571. doi:10.1093/jopart/mum032

Antelio, M., Esteves, M. G. P., Schneider, D., & Souza, J. M. (2012), Qualitocracy: A data quality collaborative framework applied to citizen science. In *Proceedings of the International Conference on Systems, Man and Cybernetics (IEEE SMC 2012).* doi:10.1109/ICSMC.2012.6377847

Armitage, C. J., & Conner, M. (1999). Distinguishing Perceptions of Control From Self-Efficacy: Predicting Consumption of a Low-Fat Diet Using the Theory of Planned Behavior. *Journal of Applied Social Psychology*, *29*(1), 72–90. doi:10.1111/j.1559-1816.1999.tb01375.x

Arnstein, S. R. (1969). A ladder of citizen participation. *Journal of the American Institute of Planners*, *35*(4), 216–224. doi:10.1080/01944366908977225

Auer, S., Bizer, C., Kobilarov, G., Lehmann, J., Cyganiak, R., & Ives, Z. (2007). *Dbpedia: A nucleus for a web of open data*. Springer Berlin Heidelberg.

Au, J., Bagchi, P., Chen, B., Martinez, R., Dudley, S. A., & Sorger, G. J. (2000). Methodology for public monitoring of total coliforms, *Escherichia Coli*, and toxicity in waterways by Canadian high school students. *Journal of Environmental Management*, *58*(3), 213–230. doi:10.1006/jema.2000.0323

Azzurro, E., Broglio, E., Maynou, F., & Bariche, M. (2013). Citizen science detects the undetected: The case of Abudefduf saxatilis from the Mediterranean Sea. *Management of Biological Invasions*, *4*(2), 167–170. doi:10.3391/mbi.2013.4.2.10

Ballard, H. L., Trettevick, J. A., & Collins, D. (2008). Comparing participatory ecological research in two contexts: an immigrant community and a Native American community on Olympic Peninsula, Washington. In *Partnerships for empowerment: participatory research for community-based natural resource management*. London, UK: Earthscan.

Bandura, A. (1991). Social cognitive theory of self-regulation. *Organizational Behavior and Human Decision Processes Organizational Behavior and Human Decision Processes*, *50*(2), 248–287. doi:10.1016/0749-5978(91)90022-L

Barber, B. (1983). *The logic and limits of trust* (Vol. 96). Rutgers University Press New Brunswick.

Barcelona City Council, Citizen Science. (2016). *20 projects to promote the city*. Retrieved from: https://issuu.com/bcnlabcienciaciudadana/docs/llibret_icub__v.eng_

Bardaji, R., Sánchez, A. M., Simon, C., Wernand, M. R., & Piera, J. (2016). Estimating the Underwater Diffuse Attenuation Coefficient with a Low-Cost Instrument: The KdUINO DIY Buoy. *Sensors (Basel, Switzerland)*, *16*(3), 373. doi:10.3390/s16030373 PMID:26999132

Barron, B. (2006). Interest and self-sustained learning as catalysts of development: A learning ecologies perspective. *Human Development*, *49*(4), 193–224. doi:10.1159/000094368

Barron, B., & Darling Hammond, L. (2010). Prospects and challenges for inquiry-based approaches to learning. In H. Dumont (Ed.), *The Nature of Learning. Using research to inspire practice*. Paris: OECD. doi:10.1787/9789264086487-11-en

Bastiaensen, J., DeHerdt, T., & D'Exelle, B. (2005). Poverty reduction as a local institutional process. *World Development, 33*(6), 979–993. doi:10.1016/j.worlddev.2004.09.019

Baylis, J., & Smith, S. (2001). *The Globalization of World Politics: An introduction to international relations.* Oxford University Press.

Beaman, R. S., & Cellinese, N. (2012). Mass digitization of scientific collections: New opportunities to transform the use of biological specimens and underwrite biodiversity science. *ZooKeys*, (209): 7. PMID:22859875

Belbin, L., & Williams, K. J. (2016). Towards a national bio-environmental data facility: Experiences from the Atlas of Living Australia. *International Journal of Geographical Information Science, 30*(1), 108–125. doi:10.1080/13658816.2015.1077962

Bell, S., Marzano, M., Cent, J., Kobierska, H., Podjed, D., Vandzinskaite, D., … Mursic, R. (n.d.). What counts? Volunteers and their organisations in the recording and monitoring of biodiversity. *Biodiversity and Conservation, 17*(14), 3443–3454. http://doi.org/10.1007/s10531-008-9357-9

Bell, S., Cornford, D., & Bastin, L. (2013). The state of automated amateur weather observations. *Weather, 68*(2), 36–41. doi:10.1002/wea.1980

Benz, S., Miller-Rushing, A., Domroese, M., Ballard, H. L., Bonney, R., DeFalco, T., & Young, A. et al. (2013). Workshop 1: Conference on public participation in scientific research 2012; An international, interdisciplinary conference. *Bulletin of the Ecological Society of America, 194*(1), 112–117.

Berger, J. (1972). *Ways of Seeing*. London: Penguin.

Bermeo, N. G. (2004). Authoritarian legacies and democracy in Latin America and Southern Europe. *University of Notre Dame Press.*

Bird, T. J., Bates, A. E., Lefcheck, J. S., Hill, N. A., Thomson, R. J., Edgar, G. J., & Frusher, S. et al. (2013). Statistical solutions for error and bias in global citizen science datasets. *Biological Conservation, 173*, 144–154. doi:10.1016/j.biocon.2013.07.037

Björk, B. C., Laakso, M., Welling, P., & Paetau, P. (2014). Anatomy of green open access. *Journal of the Association for Information Science and Technology, 65*(2), 237–250. doi:10.1002/asi.22963

Bocciolesi, E., & González, S. J. G. (2015). Educar a leer entre realidad y complejidad. Las variables de la contemporaneidad entre México, España e Italia. *Memorias del Encuentro Internacional de Educación a Distancia.*

Bogers, M., & West, J. (2012). Managing distributed innovation: Strategic utilization of open and user innovation. *Creativity and Innovation Management, 21*(1), 61–75. doi:10.1111/j.1467-8691.2011.00622.x

Bolivariana de Venezuela, C. R., & Oficial, G. (2005). *Law of Community Service, Gaceta Oficial de la República de Venezuela, Nº 38272, 14 de Septiembre, 2005.* Retrieved April 15, 2016 from: http://www.ciens.ula.ve/scciens/organizacion/ley.php

Bonaparte, N. (2004). His Rise and Fall (3rd ed.). Oxford, UK: Blackwell.

Bonney, R., Ballard, H., Jordan, R., McCallie, E., Phillips, T., Shirk, J., & Wilderman, C. (2009). Public participation in scientific research: Defining the field and assessing its potential for informal science education. Washington, DC: Center for Advancement of Informal Science Education (CAISE).

Bonney, R., Ballard, H., Jordan, R., McCallie, E., Phillips, T., Shirk, J., & Wilderman, C. C. (2009). Public participation in scientific research: Defining the field and assessing its potential for informal science education. A CAISE Inquiry Group Report. Washington, DC: Center for Advancement of Informal Science Education (CAISE).

Bonney, R. (1996). Citizen Science: A lab tradition. *Living Bird*, *15*(4), 7–15.

Bonney, R. I. C. K., & Dickinson, J. L. (2012). *Overview of citizen science. Citizen Science: Public Participation in Environmental Research*. New York: Cornell University Press.

Bonney, R., Ballard, H., Jordan, R., McCallie, E., Phillips, T., Shirk, J., & Wilderman, C. C. (2009). *Public Participation in Scientific Research: Defining the Field and Assessing Its Potential for Informal Science Education. A CAISE Inquiry Group Report*. Online Submission.

Bonney, R., Cooper, C. B., Dickinson, J., Kelling, S., Phillips, T., Rosenberg, K. V., & Shirk, J. (2009). Citizen science: A developing tool for expanding science knowledge and scientific literacy. *Bioscience*, *59*(11), 977–984. doi:10.1525/bio.2009.59.11.9

Bonney, R., Phillips, T. B., Ballard, H. L., & Enck, J. W. (2015). Can citizen science enhance public understanding of science? *Public Understanding of Science (Bristol, England)*. PMID:26445860

Bonney, R., Shirk, J., Phillips, T., Wiggins, A., Ballard, H., Miller-Rushing, A., & Parrish, J. (2014). Next steps for citizen science. *Science*, *343*(6178), 1436–1437. doi:10.1126/science.1251554 PMID:24675940

Bonter, D. N., & Cooper, C. B. (2012). Data validation in citizen science: A case study from Project FeederWatch. *Frontiers in Ecology and the Environment*, *10*(6), 305–307. doi:10.1890/110273

Bowser, A., McMonagle, R., & Tyson, E. (2015). Citizen Science and Crowdsourcing Metadata Workshop Summary. Washington, DC: Woodrow Wilson International Center for Scholars. Retrieved from https://www.wilsoncenter.org/publication/citizen-science-and-crowdsourcing-metadata-workshop-summary

Bowser, A., Schade, S., Newman, G., Göbel, C., Brenton, P., & Cavalier, D. (2016, April 14). *Reflections on the Federal Catalog of Crowdsourcing and Citizen Science Projects* [Blog Post]. Retrieved May 16, 2016, from https://wilsoncommonslab.org/2016/ 04/14/ reflections-on-the-federal-catalog-of-citizen-science-and-crowdsourcing-projects/

Bowser, A., Hansen, D., & Preece, J. (2013, April). Gamifying citizen science: Lessons and future directions. In *Workshop on Designing Gamification: Creating Gameful and Playful Experiences*.

Bowser, A., & Shanley, L. (2013). *New Visions in Citizen Science*. Washington, DC: Woodrow Wilson International Center for Scholars.

Bowser, A., Wiggins, A., Shanley, L., Preece, J., & Henderson, S. (2014). Sharing data while protecting privacy in citizen science. *Interaction*, *21*(1), 70–73. doi:10.1145/2540032

Brabham, D. (2013). *Using Crowdsourcing in Government. Collaboration Across Boundaries*. IBM Center for The Business of Government. Available at: http://www.businessofgovernment.org/report/using-crowdsourcing-government

Brabham, D. C. (2009). Crowdsourcing the public participation process for planning projects. *Plan Theory*, *8*(3), 242–262. doi:10.1177/1473095209104824

Bradshaw, B. (2003). Questioning the credibility and capacity of community-based resource management. *The Canadian Geographer*, *47*(2), 137–150. doi:10.1111/1541-0064.t01-1-00001

Bramston, P., Pretty, G., & Zammit, C. (2011). Assessing environmental stewardship motivations. *Environment and Behavior*, *43*(6), 776–788. doi:10.1177/0013916510382875

Brandon, A., Spyreas, G., Molano-Flores, B., Carrol, C., & Ellis, J. (2003). Can volunteers provide reliable data for forest vegetation surveys? *Natural Areas Journal*, *23*, 254–261.

Bransford, J., Vye, N., Stevens, R., Kuhl, P., Schwartz, D., Bell, P., & Sabelli, N. et al. (2006). Learning Theories and education: Toward a Decade of synergy. In P. A. Alexander & P. H. Winne (Eds.), *Handbook of Educational Psychology* (2nd ed.; pp. 209–244). Mahwah, NJ: Lawrence Erlbaum Associates.

Bray, G. S., & Schramm, H. L. Jr. (2001). Evaluation of a statewide volunteer angler diary program for use as a fishery tool. *North American Journal of Fisheries Management*, *21*(3), 606–615. doi:10.1577/1548-8675(2001)021<0606:EOASVA>2.0.CO;2

Brody, S. (2013). Impact factor: Imperfect but not yet replaceable. *Scientometrics*, *96*(1), 255–257. doi:10.1007/s11192-012-0863-x

Brossard, D., Lewenstein, B., & Bonney, R. (2005). Scientific knowledge and attitude change: The impact of a citizen science project. *International Journal of Science Education*, *27*(9), 1099–1121. doi:10.1080/09500690500069483

Brovelli, M. A., & Raghavan, V. (2015). Reflections on FOSS4G Seoul 2015. *Directions Magazine*. Retrieved from: http://www.directionsmag.com/entry/reflections-on-foss4g-seoul-2015/455554

Brovelli, M. A., Dotti, L., Minghini, M., Pancaldi, M., & Zamboni, G. (2014). Volunteered Geographic Information For Water Management: A Prototype Architecture.*International Conference on Hydroinformatics*.

Brovelli, M. A., Minghini, M., & Zamboni, G. (2015). Public participation in GIS via mobile applications. *ISPRS Journal of Photogrammetry and Remote Sensing*.

Brown, W. T., Krasny, M. E., & Schoch, N. (2001). Volunteer monitoring of nonindigenous invasive plant species in the Adirondack Park, New York, USA. *Natural Areas Journal*, *21*, 189–196.

Bruyere, B., & Rappe, S. (2007). Identifying the motivations of environmental volunteers. *Journal of Environmental Planning and Management*, *50*(4), 503–516. doi:10.1080/09640560701402034

Buesching, C. D., & Newman, C. (2005). Volunteers in ecological research: Amateur ecological monitors – the benefits and challenges of using volunteers. *Bulletin of the Ecological Society of America*, *36*, 20–22.

Bueshing, C. D., & Slade, E. (2012). Citizen science in Wytham Woods: collecting scientific data using amateur volunteers. In D. W. Macdonald & R. E. Feber (Eds.), *Wildlife Conservation on Farmland*. Oxford, UK: Oxford University Press.

Burke, J. A., Estrin, D., Hansen, M., Parker, A., Ramanathan, N., Reddy, S., & Srivastava, M. B. (2006) Participatory sensing. WSW'06 at SenSys '06.

Bushdid, C., Magnasco, M. O., Vosshall, L. B., & Keller, A. (2014). Humans Can Discriminate More than 1 Trillion Olfactory Stimuli. *Science*, *343*(6177), 1370–1372. doi:10.1126/science.1249168

Bushway, J., Dickinson, L., Stedman, C., Wagenet, P., & Weinstein, A. (2011). Benefits, motivations, and barriers related to environmental volunteerism for older adults: Developing a research agenda. *International Journal of Aging & Human Development*, *72*(3), 189–206. doi:10.2190/AG.72.3.b PMID:21834387

Buytaert, W., Dewulf, A., De Bièvre, B., Clark, J., & Hannah, D. (2016). Citizen Science for Water Resources Management: Toward Polycentric Monitoring and Governance? *Journal of Water Resources Planning and Management*, *142*(4), 01816002. doi:10.1061/(ASCE)WR.1943-5452.0000641

Buytaert, W., Zulkafli, Z., Grainger, S., Acosta, L., Alemie, T. C., & Bastiaensen, J. et al. (2014). Citizen science in hydrology and water resources: Opportunities for knowledge generation, ecosystem service management, and sustainable development. *Frontiers of Earth Science*, *2*, 1–21.

Cairn, R., & Kielsmeier, J. (1991). Growing Hope: A Sourcebook on Integrating Youth Service into the School Curriculum. National Youth Leadership Council.

Caldwell Johnson, E. (2005). *Communities of Practice for Local Capacity in Central Asia: The Community Empowerment Network*. Retrieved April 16, 2016 from http://learningforsustainability.net/capacity-building-empowerment/

California Energy Commission. (2012). *Community-based climate adaptation planning: case study of Oakland*. Author.

Cameron, A., Salisbury, C., Lart, R., Stewart, K., Peckham, S., Calnan, M., & Thorp, H. et al. (2011). Policy makers' perceptions on the use of evidence from evaluations. *Evidence & Policy*, *7*(4), 429–447. doi:10.1332/174426411X603443

Canfield, D. E. Jr, Brown, C. D., Bachmann, R. W., & Hoyer, M. V. (2002). Volunteer lake monitoring: Testing the reliability of data collected by the Florida LAKEWATCH program. *Lake and Reservoir Management*, *18*(1), 1–9. doi:10.1080/07438140209353924

Carnegie Mellon Create Lab. (2016). *Flamingo*. Retrieved from http://www.cmucreatelab.org/projects/Water_Quality_Monitoring/pages/Flamingo

Carolan, M. S. (2006). Science, expertise and the democratization of the decision-making process. *Society & Natural Resources*, *19*(7), 661–668. doi:10.1080/08941920600742443

Carr, A. J. L. (2004). Why do we all need community science? *Society & Natural Resources*, *17*(9), 841–849. doi:10.1080/08941920490493846

Castell, N., Kobernus, M., Liu, H. Y., Schneider, P., Lahoz, W., Berre, A. J., & Noll, J. (2014). *Mobile technologies and services for environmental monitoring: The Citi-Sense-MOB approach*. Urban Climate.

Castells, M. (2000). *The rise of the Network Society*. Malden, MA: Blackwell Publishers.

Castilla, E. P., Cunha, D. G. F., Lee, F. W. F., Loiselle, S., Ho, K. C., & Hall, C. (2015). Quantification of phytoplankton bloom dynamics by citizen scientists in urban and peri-urban environments. *Environmental Monitoring and Assessment*, *690*, 187–198. doi:10.1007/s10661-015-4912-9

Catlin-Groves, C. L. (2012). The citizen science landscape: from volunteers to citizen sensors and beyond. *International Journal of Zoology,* 1–14. http://doi.org/10.1155/2012/349630

Cavalier, D., Cheetham, R., Emanuele, R., Frankl, J., Johnson, M., Manik-Perlman, R., . . . Redeschi, M. (2014). *Citizen Science Data Factory*. Retrieved from http://scistarter.com/research

Ceccaroni, L., Wernand, M., Subirats, L., Piera, J., Farrés, R., Price, I., & Steblin, A. (2014). *Extending historic water-quality data sets, using old-fashioned techniques, citizen science and smartphones*. Poster presented at the Second International Ocean Research Conference, Barcelona, Spain. Retrieved from www.1000001labs.org/we-content/uploads/2014

Ceccaroni, L., Velickovski, F., Steblin, A., & Subirats, L. (2015) Citclops: data interpretation and knowledge-based systems integration.*Environmental Information Infrastructures and Platforms Workshop*.

Chalkley, S., Brendan, E., & Gowland, K. (1999). *Who is watching our water? Waterwatch Australia. In The Challenge of Rehabilitating Australia's Streams* (pp. 163–168). Melbourne: Co-operative Research Centre for Catchment Hydrology.

Challenge.gov. (2016). *Visualize Your Water*. Retrieved from https://www.challenge.gov/challenge/visualize-your-water

Chen, C.-J., & Hung, S.-W. (2010). To give or to receive? Factors influencing members knowledge sharing and community promotion in professional virtual communities. *Information & Management*, *47*(4), 226–236. doi:10.1016/j.im.2010.03.001

Chiu, C.-M., Hsu, M.-H., & Wang, E. T. G. (2006). Understanding knowledge sharing in virtual communities: An integration of social capital and social cognitive theories. *Decision Support Systems*, *42*(3), 1872–1888. doi:10.1016/j.dss.2006.04.001

Cho, H., Chen, M., & Chung, S. (2010). Testing an integrative theoretical model of knowledge-sharing behavior in the context of Wikipedia. *ASI Journal of the American Society for Information Science and Technology*, *61*(6), 1198–1212.

Cho, M. K., McGee, G., & Magnus, D. (2006). Lessons of the stem cell scandal. *Science*, *311*(5761), 614–615. doi:10.1126/science.1124948

Chow, W. S., & Chan, L. S. (2008). Social network, social trust and shared goals in organizational knowledge sharing. *Information & Management*, *45*(7), 458–465. doi:10.1016/j.im.2008.06.007

Christakis, N. A., & Fowler, J. H. (2009). *Connected: the surprising power of our social networks and how they shape our lives*. Little Brown.

Chu, M., Leonard, P., & Stevenson, F. (2012). Growing the base for citizen science: Recruiting and Engaging Participants. In J. L. Dickinson & R. Bonney (Eds.), *Citizen Science: Public Participation in Environmental Research* (pp. 69–81). Ithaca, NY: Cornell University Press. doi:10.7591/cornell/9780801449116.003.0005

Cima, O. (2015). Expanding spaces of participation: Insights from an infrastructural project in rural Nepal. *Geographica Helvetica*, *70*(4), 353–361. doi:10.5194/gh-70-353-2015

Ciravegna, F., Huwald, H., Lanfranchi, V., & Wehn de Montalvo, U. (2013). Citizen observatories: the WeSenseIt Vision. In Proceeding of the Infrastructure for Spatial Information in the European Community (INSPIRE 2013). Florence, Italy: INSPIRE 2013.

Citclops. (2016a). *Citizen's Observatory for Coast and Ocean Optical Monitoring*. Retrieved from http://Citclops.eu

Citclops. (2016b). *Instructions how to use*. Retrieved from Citclops.eu/participate/instructions-how-to-use

Citizen Science. (n.d.a). Retrieved April 16, 2016, from https://en.wikipedia.org/wiki/Citizen_science

Citizen Science. (n.d.b). In *National Geographic*. Retrieved April 16, 2016 from http://education.nationalgeographic.org/encyclopedia/citizen-science/

Citizen Scientist. (n.d.). In *Open Scientist*. Retrieved April 16, 2016 from http://www.openscientist.org/2011/09/finalizing-definition-of-citizen.html

Citizen's Observatory. (n.d.). *swe4citizenscience*. Retrieved from: http://github.com/opengeospatial/swe4citizenscience

Citizens' Observatory. (n.d.). *About the Projects*. Retrieved March 6, 2016, from http://www.citizen-obs.eu/About.aspx

Clery, D. (2011). Galaxy Zoo volunteers share pain and glory of research. *Science*, *333*(6039), 173–175. doi:10.1126/science.333.6039.173 PMID:21737731

Clow, D., & Makriyannis, E. (2011). iSpot Analysed: Participatory Learning and Reputation. In *Proceedings of 1st International Conference on Learning Analytics and Knowledge*. doi:10.1145/2090116.2090121

Cobweb project. (2016). *JRC's Citizen Science: data and service infrastructure meeting - Ispra (Italy)*. Retrieved June 1, 2016, from https://cobwebproject.eu/jrcs-citizen-science-data-and-service-infrastructure-meeting-ispra-italy-jan-2016

Coghlan, A. (2005). *Towards an Understanding of the Volunteer Tourism Experience.* (Unpublished doctoral dissertation). School of Business, James Cook University, Townsville, Australia.

Cohn, J. P. (2008). Citizen science: Can volunteers do real research? *Bioscience*, *58*(3), 192–197. doi:10.1641/B580303

Coll, C. (2016). La educación formal en la nueva ecología del aprendizaje: tendencias, retos y agenda de investigación. In J.M. Mominó & C. Sigalés (Eds.), El impacto de las TIC en educación. Más allá de las promesas. Barcelona: Editorial UOC.

Colorimetrix. (2016). *Environment*. Retrieved from http://www.colorimetrix.com

Conrad, C. C., & Hilchey, K. G. (2011). A review of citizen science and community-based environmental monitoring: Issues and opportunities. *Environmental Monitoring and Assessment*, *176*(1-4), 273–291. doi:10.1007/s10661-010-1582-5 PMID:20640506

Conrad, T. C., & Daoust, T. (2008). Community-based monitoring frameworks: Increasing the effectiveness of environmental stewardship. *Environmental Management*, *41*(3), 358–366. doi:10.1007/s00267-007-9042-x PMID:18026783

Cooper, C. B. (2012). Links and Distinctions Among Citizenship, Science, and Citizen Science. A Response to "The Future of Citizen Science". *Democracy & Education*, *20*(2), 13.

Cooper, C. B., Dickinson, J., Phillips, T., & Bonney, R. (2007). Citizen science as a tool for conservation in residential ecosystems. *Ecology and Society*, *12*(2), 11.

Cooper, C. B., Shirk, J., & Zuckerberg, B. (2014). The Invisible Prevalence of Citizen Science in Global Research: Migratory Birds and Climate Change. *PLoS ONE*, *9*(9), e106508. doi:10.1371/journal.pone.0106508 PMID:25184755

Cooper, S., Khatib, F., Treuille, A., Barbero, J., Lee, J., Beenen, M., & Players, F. et al. (2010). Predicting protein structures with a multiplayer online game. *Nature*, *466*(7307), 756–760. doi:10.1038/nature09304 PMID:20686574

Correia, M. G., & Bleicher, R. E. (2008). Making connections to teach reflection. *Michigan Journal of Community Service Learning*, *14*, 41–49.

Craglia, M., de Bie, K., Jackson, D., Pesaresi, M., Remetey-Fülöpp, G., Wang, C., & Woodgate, P. et al. (2012). Digital Earth 2020: Towards the vision for the next decade. *International Journal of Digital Earth*, *5*(1), 4–21. doi:10.1080/17538947.2011.638500

Craglia, M., Goodchild, M. F., Annoni, A., Camara, G., Gould, M., Kuhn, W., & Parsons, E. (2008). Next-generation digital earth: A position paper from the vespucci initiative for the advancement of geographic information science. *International Journal of Spatial Data Infrastructures Research*, *3*(6), 146–167.

Crain, R., Cooper, C., & Dickinson, J. L. (2014). Citizen science: A tool for integrating studies of human and natural systems. *Annual Review of Environment and Resources*, *39*(1), 641–665. doi:10.1146/annurev-environ-030713-154609

Crall, A.W. (2013). *CitizenScience.org Beta Test: Findings and Recommendations.* Report prepared for the Cornell Lab of Ornithology.

Crall, A. W., Newman, G. J., Jarnevich, C. S., Stohlgren, T. J., Waller, D. M., & Graham, J. (2010). Improving and integrating data on invasive species collected by citizen scientists. *Biological Invasions*, *12*(10), 3419–3428. doi:10.1007/s10530-010-9740-9

Crall, A. W., Newman, G. J., Stohlgren, T. J., Holfelder, K. A., Graham, J., & Waller, D. M. (2011). Assessing citizen science data quality: An invasive species case study. *Conservation Letters*, *4*(6), 433–442. doi:10.1111/j.1755-263X.2011.00196.x

Crowston, K., & Prestopnik, N. R. (2011). Gaming for (Citizen) Science: Exploring Motivation and Data Quality in the Context of Crowdsourced Science through the Design and Evaluation of a Social-Computational System. In *IEEE Seventh International Conference on e-Science Workshops (eScienceW).*

Crowston, K., & Prestopnik, N. R. (2013). Motivation and data quality in a citizen science game: A design science evaluation. In *Forty-sixth Hawai'i International Conference on System Sciences (HICSS-46).* doi:10.1109/HICSS.2013.413

Cuban, L. (2012). *Dilemes polítics i docents de l'ús de les TIC a l'aula. El cas dels Estats Units.* Barcelona: Fundació Jaume Bofill. Retrieved from http://www.debats.cat/sites/default/files/debats/pdf/dilemes-politics-docents.pdf

Cuban, L. (1986). *Teachers and machines. The classroom use of technology since 1920.* Teachers College Press.

Cuban, L. (2001). *Oversold & Underused. Computers in the classroom.* Cambridge, MA: Harvard University Press.

Cuff, B. D., Hansen, M., & Kang, J. (2008). Urban Sensing: Out of the Woods. *Communications of the ACM, 51*(3), 1–33. doi:10.1145/1325555.1325562

Dale, R. (2005). Globalisation, knowledge economy and comparative education. *Comparative Education, 41*(2), 117–149. doi:10.1080/03050060500150906

Damon, W. (1984). Peer education: The untapped potential. *Journal of Applied Developmental Psychology, 5*(4), 331–343. doi:10.1016/0193-3973(84)90006-6

Danielsen, F., Burgess, N. D., & Balmford, A. (2005). Monitoring matters: Examining the potential of locally-based approaches. *Biodiversity and Conservation, 14*(11), 2507–2542. doi:10.1007/s10531-005-8375-0

Danielsen, F., Burgess, N. D., Balmford, A., Donald, P. F., Funder, M., Jones, J. P., & Yonten, D. et al. (2009). Local participation in natural resource monitoring: A characterization of approaches. *Conservation Biology, 23*(1), 31–42. doi:10.1111/j.1523-1739.2008.01063.x PMID:18798859

Darwall, W. R. T., & Dulvy, N. K. (1996). An evaluation of the suitability of non-specialist volunteer researches for coral reef surveys. Mafia Island, Tanzania – A case study. *Biological Conservation, 78*(3), 223–231. doi:10.1016/0006-3207(95)00147-6

David, J. S., McCarthy, W. E., & Sommer, B. S. (2003). Agility---: The key to survival of the fittest in the software market. *Communications of the ACM, 46*(5), 65–69. doi:10.1145/769800.769803

Davidson, C. N., & Goldberg, D. T. (2010). *The future of thinking. Learning institutions in a digital age.* Cambridge, MA: The MIT Press. Retrieved from https://mitpress.mit.edu/books/future-thinking

Davies, I. (2004). Science and citizenship education. *International Journal of Science Education, 26*(14), 1751–1763. doi:10.1080/0950069042000230785

Davis, P. (1996). *Museums and the natural environment: the role of natural history museums in biological conservation.* Leicester, UK: Leicester University Press.

Davis, R., Shrobe, H., & Szolovits, P. (1993). What is a Knowledge Representation? *AI Magazine, 14*(1), 17–33.

de Jong, S. P., Smit, J., & van Drooge, L. (2015). Scientists' response to societal impact policies: A policy paradox. *Science & Public Policy.*

De Smith, M. J., Goodchild, M. F., & Longley, P. (2007). *Geospatial analysis: a comprehensive guide to principles, techniques and software tools.* Leicester, UK: Troubador Publishing Ltd.

Defining Citizen Science. (n.d.). In *The Cornell Lab of Ornithology.* Retrieved April 16, 2016, from http://www.birds.cornell.edu/citscitoolkit/about/definition

DEFRA. (2006). *Evidenced Based Policy Making.* Available at: https://www.wbginvestmentclimate.org/uploads/modgov.pdf

DEFRA. (2014). *Making the most of our evidence: A strategy for Defra and its network.* Available: https://www.gov.uk/government/uploads/system/uploads/attachment_data/file/318610/evidence-strategy-defra.pdf

Delaney, D., Sperling, C. D., Adams, C. S., & Leung, B. (2008). Marine invasive species: Validation of citizen science and implications for national monitoring networks. *Biological Invasions*, *10*(1), 117–128. doi:10.1007/s10530-007-9114-0

Delors, J. (Ed.). (1996). Educació: hi ha un tresor amagat a dins. Informe de la UNESCO de la comissió internacional sobre Educació per al s. XXI. Barcelona: Centre UNESCO de Catalunya.

Devictor, V., Whittaker, R. J., & Beltrame, C. (2010). Beyond scarcity: Citizen science programs as useful tools for conservation biogeography. *Diversity & Distributions*, *16*(3), 354–362. doi:10.1111/j.1472-4642.2009.00615.x

Dewey, J. (1916). Democracy and Education. In The middle works of John Dewey (vol 9). Academic Press.

D'Hondt, E., Zaman, J., Philips, E., Boix, E. G., & De Meuter, W. (2014). Orchestration Support for Participatory Sensing Campaigns. In *Proceedings of the 2014 ACM International Joint Conference on Pervasive and Ubiquitous Computing (UbiComp 2014)*. ACM.

Dickinson, J. L., & Bonney, R. (Eds.). (2012). *Citizen Science: Public participation in environmental research.* Ithaca, NY: Cornell University Press.

Dickinson, J. L., Crain, R. L., Reeve, H. K., & Schuldt, J. P. (2013). Can evolutionary design of social networks make it easier to be 'green'? *Trends in Ecology & Evolution*, *28*(9), 561–569. doi:10.1016/j.tree.2013.05.011 PMID:23787089

Dickinson, J. L., Shirk, J., Bonter, D., Bonney, R., Crain, R. L., Martin, J., & Purcell, K. et al. (2012). The current state of citizen science as a tool for ecological research and public engagement. *Frontiers in Ecology and the Environment*, *10*(6), 291–297. doi:10.1890/110236

Dickinson, J. L., Zuckerberg, B., & Bonter, D. N. (2010). Citizen science as an ecological research tool: Challenges and benefits. *Annual Review of Ecology Evolution and Systematics*, *41*(1), 149–172. doi:10.1146/annurev-ecolsys-102209-144636

Dierkes, M., & von Grote, C. (Eds.). (2000). *Between understanding and trust: the public, science and technology.* Amsterdam: Harwood Academic.

Dillon, R. (2014). *Engage, empower, energize – Leading tomorrow's schools today.* London: Rowwan & Littlefield.

DiMaggio, P., & Hargittai, E. (2001). *From the "digital divide" to "digital inequality": Studying Internet use as penetration increases.* Working Paper 15. Center for Arts and Cultural Policy Studies. Retrieved from http://www.princeton.edu/~artspol/workpap15.html

DiMaggio, P., Hargittai, E., Celeste, C., & Shafer, S. (2004). From unequal access to differentiated use: A literature review and agenda for research on digital inequality. In K. M. Neckerman (Ed.), *Social inequality* (pp. 355–400). New York: Russell Sage Foundation.

Dimock, M., Kiley, J., Keeter, S., Doherty, C., & Tyson, A. (2014). Beyond Red vs. Blue: The Political Typology. *The Pew Research Center for the People & the Press.* Retrieved from http://www.people-press.org/files/2014/06/6-26-14-Political-Typology-release1.pdf

du Toit, A. (2012). *Making sense of "evidence": Notes on the discursive politics of research and pro-poor policy making', Working Paper 21.* PLAAS, UWC.

Dua, A., Bulusu, N., Feng, W., & Hu, W. (2009). Towards trustworthy participatory sensing. In *Proceedings of the 4th USENIX conference on hot topics in security (HotSec'09)*. Berkeley, CA: USENIX Association.

Dumont, H., Istance, D., & Benavides, F. (2010). *The Nature of Learning. Using research to inspire practice*. Paris: OECD.

Dunkley, R. A., (2016). Learning at Eco-attractions: Exploring the Bifurcation of Nature and Culture through Experiential Environmental Education. *The Journal of Environmental Education*, 1-9.

Dunn, W. N. (2015). *Public Policy Analysis*. Routledge.

Earth, F. (2013). *Future Earth Initial Design: Report of the Transition Team*. Paris: International Council for Science (ICSU). Available at: http://www.icsu.org/news-centre/future-earth/media-centre/relevant_publications/future-earth-initial-design-report

EarthEcho International. (2015). *World Water Monitoring Challenge*. Retrieved from http://monitorwater.org

Earthwatch Institute Australia. (2013a, October 14). Call for the establishment of a national research and coordination network to advance public participation in Australian scientific studies 'Citizen Science Network (Australia)'. *Public Discussion Paper*. [Blog]. Retrieved May 30, 2016, from http://inspiringaustralia.net.au/citizen-science-network-australia/

Earthwatch Institute Australia. (2013b). *Summary of responses to Earthwatch's public discussion paper on the establishment of a National Citizen Science Network for Australia*. Discussion Paper.

ECSA. (2016). *About Citizen Science and ECSA*. Retrieved June 1, 2016, from http://ecsa.citizen-science.net/about

Eisnor, D. (2006). *What is neogeography anyway?* Retrieved from: http://platial.typepad.com/news/2006/05/what_is_neogeog.html

Ellwood, E. R., Dunckel, B. A., Flemons, P., Guralnick, R., Nelson, G., Newman, G., & Mast, A. R. et al. (2015). Accelerating the digitization of biodiversity research specimens through online public participation. *Bioscience*, *65*(4), 383–396. doi:10.1093/biosci/biv005

Ely, E. (2008a). Getting volunteer data into science journals. *The Volunteer Monitor*, *19*, 8–11.

Ely, E. (2008b). Volunteer monitoring and the democratization of science. *The Volunteer Monitor*, *19*, 1–5.

Endfield, G. H., & Morris, C. (2012). Well weather is not a girl thing is it? Contemporary amateur meteorology, gender relations and the shaping of domestic masculinity. *Social & Cultural Geography*, *13*(3), 233–253. doi:10.1080/14649365.2012.677470

Engelken-Jorge, M., Moreno, J., Keune, H., Verheyden, W., & Bartonova, A., & the CITISENSE consortium. (2014). Developing citizens' observatories for environmental monitoring and citizen empowerment: challenges and future scenarios. In *Proceedings of the Conference for E-Democracy and Open Government (CeDEM14)*. Danube University.

Engel, S. R., & Voshell, J. R. (2002). Volunteer biological monitoring: Can it accurately assess the ecological condition of streams? *American Entomologist*, *48*(3), 164–177. doi:10.1093/ae/48.3.164

EPA - United States Environmental Protection Agency. (2012). *Starting Out in Volunteer Monitoring*. EPA 941-F002-004.

Esman, M. J. (2003). *Social capital and empowerment*. Cornell University. Retrieved April 12, 2016 from: http://siteresources.worldbank.org/INTEMPOWERMENT/Resources/486312-1097679640919/think_pieces_esman.pdf

España, L. P. (2004). La cultura y las causas de la pobreza en Venezuela. In Luís Ugalde, Luis Pedro España y otros: Detrás de la Pobreza. Percepciones, Creencias, Apreciaciones. Editan: Asociación Civil para la Promoción de Estudios Sociales y la Universidad Católica Andrés Bello.

Estrin, D. (2010). Participatory sensing: Applications and architecture. *IEEE Internet Computing*, *14*(1), 12–42. doi:10.1109/MIC.2010.12

European Citizen Science Association. (2015a). *10 Principles of Citizen Science*. Retrieved January 10, 2016, from http://ecsa.citizen-science.net

European Citizen Science Association. (2015b). *ECSA draft policy paper: Citizen science as part of EU policy delivery*. Retrieved March 6, 2016, from http://ecsa.citizen-science.net

European Citizen Science Association. (2015c). *ECSA draft policy paper: Endorsing the white paper on citizen science for Europe*. Retrieved March 6, 2016, from http://ecsa.citizen-science.net

European Citizen Science Association. (2015d). *ECSA policy paper: Endorsing the German green paper on citizen science*. Retrieved January 10, 2016, from http://ecsa.citizen-science.net

European Citizen Science Association. (2015e). *ECSA strategy*. Retrieved January 10, 2016, from http://ecsa.citizen-science.net

European Citizen Science Association. (2016). *About citizen science and ECSA*. Retrieved from http://ecsa.citizen-science.net/about

European Citizen Science Association. Citizen Science Association, & Australian Citizen Science Association. (2015). *Memorandum of Understanding between CSA, ECSA, CSNA*. Retrieved March 6, 2016, from http://ecsa.citizen-science.net

European Commission. (2012). *Responsible research and innovation: Europe's ability to respond to societal challenges*. Retrieved January 10, 2016, from https://ec.europa.eu/research/swafs/pdf/pub_public_engagement/responsible-research-and-innovation-leaflet_en.pdf

European Commission. (2013a). *Digital science in Horizon 2020*. Retrieved January 10, 2016, from http://ec.europa.eu/information_society/newsroom/cf/dae/document.cfm?doc_id=2124

European Commission. (2013b). *Factsheet: Horizon 2020 budget*. Retrieved January 10, 2016, from http://ec.europa.eu/research/horizon2020/pdf/press/fact_sheet_on_horizon2020_budget.pdf

European Commission. (2016). *Open Science*. Retrieved March 6, 2016, from https://ec.europa.eu/digital-agenda/en/open-science

European Commission. (n.d.). *Responsible research & innovation*. Retrieved March 6, 2016, from http://ec.europa.eu/programmes/horizon2020/en/h2020-section/responsible-research-innovation

Eveleigh, A., Jennett, C., Lynn, S., & Cox, A. L. (2013). 'I Want to Be a Captain! I Want to Be a Captain!': Gamification in the Old Weather Citizen Science Project. In *Proceedings of the First International Conference on Gameful Design, Research, and Applications*. doi:10.1145/2583008.2583019

Evidence. (2013). In Oxford Dictionaries. Available at http://www.oxforddictionaries.com/definition/english/evidence

Eye on Earth. (2015). *Summary of outcomes Eye on Earth summit 2015*. Retrieved January 10, 2016, from http://www.eoesummit.org/news/eye-on-earth-summary-of-outcomes/

Fairweather, J. (2008). *Linking evidence and promising practices in science, technology, engineering, and mathematics (STEM) undergraduate education*. Washington, DC: Board of Science Education, National Research Council, The National Academies.

Falk, J. H., & Dierking, L. D. (2010). The 95 Percent Solution School is not where most Americans learn most of their science. *American Scientist*, *98*(6), 486–493. doi:10.1511/2010.87.486

Fantini, A., & Tirmizi, A. (2006). *Exploring and assessing intercultural competence*. Academic Press.

Fernandez-Gimenez, M. E., Ballard, H. L., & Sturtevant, V. E. (2008). Adaptive management and social learning in collaborative and community-based monitoring: A study of five community-based forestry organizations in the western USA. *Ecology and Society*, *13*, 4.

Firehock, K., & West, J. (1995). A brief history of volunteer biological water monitoring using macroinvertebrates. *Journal of the North American Benthological Society*, *14*(1), 197–202. doi:10.2307/1467734

Fisheries and Oceans Canada. (2015, August 24). *Community Aquatic monitoring program (CAMP)*. Retrieved from http://inter.dfo-mpo-gc.ca/Gulf/CAMP

Fitzpatrick, M. C., Preisser, E. L., Elison, A. M., & Elkinton, J. S. (2009). Observer bias and detection of low-density populations. *Ecological Applications*, *19*(7), 1673–1679. doi:10.1890/09-0265.1 PMID:19831062

Flanagin, A. J., & Metzger, M. J. (2008). The credibility of volunteered geographic information. *GeoJounal*, *72*(3-4), 137–148. doi:10.1007/s10708-008-9188-y

Florida, R. (2004). *The Rise of The Creative Class*. New York: Perseus Books Group.

Fore, L. S., Paulsen, K., & O'Laughlin, K. (2001). Assessing the performance of volunteers in monitoring streams. *Freshwater Biology*, *46*(1), 109–123. doi:10.1111/j.1365-2427.2001.00640.x

Forte, A., & Lampe, C. (2013). Defining, understanding and supporting open collaboration: Lessons from the literature. *The American Behavioral Scientist*, *57*(5), 535–547. doi:10.1177/0002764212469362

Foursquare. (n.d.). Retrieved from: https://foursquare.com/

Frank, M. R., Mitchell, L., Dodds, P. S., & Danforth, C. M. (2013). Happiness and the patterns of life: A study of geolocated tweets. *Scientific Reports*, *3*(2625), 1–9.

Franzoni, C., & Sauermann, H. (2014). Crowd science: The organization of scientific research in open collaborative projects. *RESPOL Research Policy*, *43*(1), 1–20. doi:10.1016/j.respol.2013.07.005

Freire, P. (2005). *Pedagogía del oprimido*. Siglo XXI.

Friedman, A. J. (Ed.). (2008). *Framework for evaluating impacts of informal science education projects*. The National Science Foundation. Available from: http://insci.org/resources/Eval Framework.pdf

Fritz, S., McCallum, I., Schill, C., Perger, C., See, L., Schepaschenko, D., & Obersteiner, M. et al. (2012). Geo-Wiki: An online platform for improving global land cover. *Environmental Modelling & Software*, *31*, 110–123. doi:10.1016/j.envsoft.2011.11.015

Fróes, B. T. (2002). Análise Contrastiva: memória da construção de uma metodologia para investigar a tradução de conhecimento científico em conhecimento público. *DataGramaZero - Revista de Ciência da Informação, 3*, 16. Retrieved April 20, 2016 from: http://repositorio.ufba.br/ri/handle/ri/2985

Furco, A. (2010). The community as a resource for learning: an analysis of academic service-learning in primary and secondary education. In H. Dumont (Ed.), *The Nature of Learning. Using research to inspire practice*. Paris: OECD. doi:10.1787/9789264086487-12-en

Ganapati, S. (2011). Use of Public Participation Geographical Information Systems. *Applications in e-Government, 71*(3), 425-434.

Ganda, N. S. K., Naicker, S., Shinde, S., Kimbahune, S., Shrivastava, S., & Mitra, S. (2014). Mobile water kit (MWK): A smartphone compatible low-cost water monitoring system for rapid detection of total coil form and E. Coli. *Analytical Methods*, *6*(16), 6236–6346. doi:10.1039/C4AY01245C

Ganzevoort, W., & van den Born, R. J. G. (2016). Beyond 'data drones': Citizen scientists' concerns and motivations. Paper presented at the Citizen Observatories for Water Management Conference, Venice, Italy.

Garaba, P. S., Friedrichs, A., Voß, D., & Zielinski, O. (2015). Classifying natural waters with the Forel-Ule Colour Index System: Results, applications, correlations and crowdsourcing. *International Journal of Environmental Research and Public Health*, *12*(12), 16096–16109. doi:10.3390/ijerph121215044

García Hamilton, J. I. (2002). *El autoritarismo y la improductividad en Hispanoamérica*. Buenos Aires.

García, L., Rodríguez, D., Wijnen, M., & Pakulski, I. (2016). *Earth Observation for Water Resources Management: Current Use and Future Opportunities for the Water Sector*. World Bank Publications. doi:10.1596/978-1-4648-0475-5

Gaventa, J. (Interviewee). (2015). *Professor John Gaventa for AURA Research 1*. Retrieved May 10 from https://www.youtube.com/watch?v=HEy4YTnY4II

Geiger, D., Rosemann, M., & Fielt, E. (2011b). *Crowdsourcing Information Systems– A Systems Theory Perspective*. Presented at the 22nd Australasian Conference on Information Systems, Sydney, Australia.

Geiger, D., Seedorf, S., Schulze, T., Nickerson, R., & Schader, M. (2011a). Managing the crowd: Towards a taxonomy of crowdsourcing processes.*Proceedings of the Seventeenth Americas Conference on Information Systems*.

Gellman, R. (2015). *Crowdsourcing, Citizen Science, and the Law: Legal Issues Affecting Federal Agencies*. Washington, DC: Woodrow Wilson International Center for Scholars.

Genet, K. S., & Sargent, L. G. (2003). Evaluation of methods and data quality from a volunteer-based amphibian call survey. *Wildlife Society Bulletin*, *31*, 703–714.

Gharesifard, M., & Wehn, U. (2016). To share or not to share: Drivers and barriers for sharing data via online amateur weather networks. *Journal of Hydrology (Amsterdam)*, *535*, 181–190. doi:10.1016/j.jhydrol.2016.01.036

Giddings, B., Hopwood, B., & O'Brien, G. (2002). Environment, Economy and Society: Fitting Them Together into Sustainable Development. *Sustainable Development*, *10*(4), 187–196. doi:10.1002/sd.199

Global Community Monitor. (2006). Retrieved from www.gcmonitor.org

Goeschl, T., & Jürgens, O. (2012). Environmental quality and welfare effects of improving the reporting capability of citizen monitoring schemes. *Journal of Regulatory Economics*, *42*(3), 264–286. doi:10.1007/s11149-012-9191-6

Goldman, J., Shilton, K., Burke, J., Estrin, D., Hansen, M., Ramanathan, N., . . . West, R. (2009). *Participatory Sensing: A citizen-powered approach to illuminating the patterns that shape our world*. Foresight & Governance Project, White Paper.

Goldsmith, A. (2005). Police reform and the problem of trust. *Theoretical Criminology*, *9*(4), 443–470. doi:10.1177/1362480605057727

Gollan, J., De Bruyn, L. L., Reid, N., & Wilkie, L. (2012). Can volunteers collect data that are comparable to professional scientists? A study of variables used in monitoring the outcomes of ecosystem rehabilitation. *Environmental Management*, *50*(5), 969–978. doi:10.1007/s00267-012-9924-4

Goodchild, M. F. (2007). Citizens as sensors: The world of volunteered geography. *GeoJournal*, *69*(4), 211–221. doi:10.1007/s10708-007-9111-y

Goodchild, M. F., & Li, L. (2012). Assuring the quality of volunteered geographic information. *Spatial Statistics*, *1*, 110–120. doi:10.1016/j.spasta.2012.03.002

Goode, D. (2011). Planning for nature in towns and cities. In I. Douglas, D. Goode, M. C. Houck, & R. Wang (Eds.), *The Routledge Handbook of Urban Ecology* (pp. 84–92). Abingdon, UK: Routledge.

Goode, D. (2014). *Nature in Towns and Cities*. London: HarperCollins.

Google Crisis Response. (2012). *Superstorm Sandy: NYC*. Retrieved from: http://google.org/crisismap/2012-sandy-nyc

Gore, A. (1998). The digital earth: understanding our planet in the 21st century. *Australian Surveyor, 43*(2), 89-91.

Graesser, A. C., & King, B. (2008). Technology-based training. *Human Behaviour in Military Contexts*, 127-149.

Graesser, A. C., Chipman, P., & King, B. G. (2008). Computer-mediated technologies. In Handbook of research on educational communications and technology, (pp. 211-224). Academic Press.

Great Barrier Reef Marine Park Authority. (n.d.). *Eye on the Reef program*. Retrieved March 6, 2016, from http://www.gbrmpa.gov.au/managing-the-reef/how-the-reefs-managed/eye-on-the-reef

Greenwood, J. J. D. (2007). Citizens, science and bird conservation. *Journal für Ornithologie*, *148*(S1), S77–S124. doi:10.1007/s10336-007-0239-9

Groom, Q. J., O'Reilly, C., & Humphrey, T. (2014). Herbarium specimens reveal the exchange network of British and Irish botanists, 1856–1932. *Nordic Journal of Botany*, *4*(2), 95–103. doi:10.1179/2042349714Y.0000000041

Grossberndt, S., & Liu, H.-Y. (2016 in press). Citizen participation approaches in environmental health. In J. Pacyna & E. Pacyna (Eds.), *Environmental Determinants of Human Health* (pp. 1–25). London, UK: Springer.

Gruber, K. (2015). Deep influence of soil microbes. *Nature Plants*, *1*(12), 15194. doi:10.1038/nplants.2015.194 PMID:27251726

Gruber, T. R. (1993). A translation approach to portable ontology specifications. *Knowledge Acquisition*, *5*(2), 199–220. doi:10.1006/knac.1993.1008

Guerra, F. X. (1999). *El soberano y su reino. Reflexiones sobre la génesis del ciudadano en América Latina. In Ciudadanía política y formación de naciones: perspectivas históricas de América Latina*. México: Fondo de Cultura Económica.

Gurin, P. (1999). The compelling need for diversity in education. University of. Massachusetts School of Education Journal, 32, 36-62.

Güss, C. D., Tuason, M. T., & Gerhard, C. (2010). Cross-National Comparisons of Complex Problem-Solving Strategies in Two Microworlds. *Cognitive Science*, *34*(3), 489–520. doi:10.1111/j.1551-6709.2009.01087.x PMID:21564221

Habermas, J. (2006). Political communication in media society: Does democracy still enjoy an epistemic dimension? the impact of normative theory on empirical research1. *Communication Theory*, *16*(4), 411–426. doi:10.1111/j.1468-2885.2006.00280.x

Hagel, J., Brown, J. S., & Davison, L. (2010). *From Do It Yourself to Do It Together*. Retrieved February 3, 2016, from https://hbr.org/2010/02/from-do-it-yourself-to-do-it-t.html

Haklay, M. (2011). *Citizen Science as Participatory Science*. Retrieved February 3, 2016, from https://povesham.wordpress.com/2011/11/27/citizen-science-as-participatory-science/

Haklay, M. (2015). *Citizen Science and Policy: A European Perspective*. Retrieved from https://www.wilsoncenter.org/publication/citizen-science-and-policy-european-perspective

Haklay, M. (2015). Citizen Science and Policy: A European Perspective. Washington, DC: Woodrow Wilson International Center for Scholars. Retrieved from https://www.wilsoncenter.org/sites/default/files/Citizen_Science_Policy_European_Perspective_Haklay.pdf

Haklay. (2013). *Citizen Science and Policy: A European Perspective*. Commons Lab. Retrieved May 16, 2016, from: https://www.wilsoncenter.org/sites/default/files/Citizen_Science_Policy_European_Perspective_Haklay.pdf

Haklay, M. (2010). How good is volunteered geographical information? A comparative study of OpenStreetMap and Ordnance Survey datasets. *Environment and Planning. B, Planning & Design*, *37*(4), 682–703. doi:10.1068/b35097

Haklay, M. (2013). Citizen science and volunteered geographic information: Overview and typology of participation. In *Crowdsourcing geographic knowledge* (pp. 105–122). Springer Netherlands. doi:10.1007/978-94-007-4587-2_7

Haklay, M. (2015). *Citizen science and policy: a European perspective*. Washington, DC: Woodrow Wilson International Center for Scholars.

Haklay, M. (2015). *Citizen Science and Policy: A European Perspective*. Washington, DC: Woodrow Wilson International Center for Scholars.

Haklay, M., Antoniou, V., Basiouka, S., Soden, R., & Mooney, P. (2014). *Crowdsourced geographic information use in government Report to GFDRR*. London: World Bank.

Haklay, M., Antoniou, V., Basiouka, S., Soden, R., & Mooney, P. (2014). *Crowdsourced geographic information use in government, Report to GFDRR*. London: World Bank.

Hamari, J. (2013). Transforming homo economicus into homo ludens: A field experiment on gamification in a utilitarian peer-to-peer trading service. *Electronic Commerce Research and Applications*, *12*(4), 236–245. doi:10.1016/j.elerap.2013.01.004

Hamari, J., Koivisto, J., & Sarsa, H. (2014). Does gamification work? - a literature review of empirical studies on gamification. In *47th Hawaii International Conference on System Sciences (HICSS)*, 2014 (pp. 3025-3034). IEEE. doi:10.1109/HICSS.2014.377

Hand, E. (2010). Citizen science: People power. *Nature*, *466*(7307), 685–687. doi:10.1038/466685a PMID:20686547

Hargreaves, A. (2000). Four ages of professionalism and professional learning. *Teachers and Teaching: History and Practice*, *6*(2), 151-182.

Harjanne, A., Ervasti, T., Karhu, J. A., & Tuomenvirta, H. (2015). Combining science education with citizen science - Experiences from a research institute led science education project. *LUMAT*, *3*(7), 948–959.

Harnad, S., & Brody, T. (2004). Comparing the impact of open access (OA) vs. non-OA articles in the same journals. *D-Lib Magazine*, *10*(6).

Harrison, T. M., & Stephen, T. (Eds.). (1996). *Computer networking and scholarly communication in the twenty-first-century university*. SUNY Press.

Hart, A., & Wolff, D. (2006). Developing local communities of practice through local community–university partnerships. *Planning Practice and Research*, *21*(1), 121–138. doi:10.1080/02697450600901616

Harvey, F., Kuhn, W., Pundt, H., Bishr, Y., & Riedemann, C. (1999). Semantic interoperability: A central issue for sharing geographic information. *The Annals of Regional Science*, *33*(2), 213–232. doi:10.1007/s001680050102

Haslauer, E., Delmelle, E. C., Keul, A., Blaschke, T., & Prinz, T. (2015). Comparing Subjective and Objective Quality of Life Criteria: A Case Study of Green Space and Public Transport in Vienna, Austria. *Social Indicators Research*, *124*(3), 911–927. doi:10.1007/s11205-014-0810-8

Hawelka, B., Sitko, I., Beinat, E., Sobolevsky, S., Kazakopoulos, P., & Ratti, C. (2014). Geo-located Twitter as proxy for global mobility patterns. *Cartography and Geographic Information Science*, *41*(3), 260–271. doi:10.1080/15230406.2014.890072

Heath, T., & Bizer, C. (2011). Linked data: Evolving the web into a global data space. *Synthesis Lectures on the Semantic Web: Theory and Technology, 1*(1), 1-136.

Heimlich, J. E. (2012). *Public participation in scientific research: 2012 conference evaluation*. Columbus, OH: Ohio State University Extension.

Henshaw, V. (2013). *Urban smellscapes: understanding and designing city smell environments*. Oxford, UK: Taylor & Francis Group Ltd.

Hepburn, L., Tegart, P., Roetman, P., von Gavel, S., Niedra, S., & Roger, E. ... Lambkin, C. (2015). *The Australian Guide to Running a BioBlitz*. Retrieved January 10, 2016, from http://csna.gaiaresources.com.au/wordpress/wp-content/uploads/2015/07/BIOBLITZ_Guidelines_WEB+-ACSA-logo.pdf

Hess, D. (2010). *Social movements, publics, and scientists. Invited Plenary Lecture*. Tokyo: Japanese Society for Science and Technology Studies.

Hew, K. F., & Hara, N. (2007). Knowledge sharing in online environments: A qualitative case study. *ASI Journal of the American Society for Information Science and Technology*, *58*(14), 2310–2324. doi:10.1002/asi.20698

Higgins, C., Williams, J., Leibovici, D., Simonis, I., Davis, M. J., Muldoon, C., & O'Grady, M. (2015). Citizen OBservatory WEB (COBWEB): A Generic Infrastructure Platform to Facilitate the Collection of Citizen Science Data for Environmental Monitoring. In *Proceedings of the Workshop "Environmental Infrastructures and Platforms 2015 - Infrastructures and Platforms for Environmental Crowd Sensing and Big Data" for the European Citizen Science Association General Assembly 2015* (ECSA GA'2015).

Higgins, C. I., Williams, J., Leibovici, D. G., Simonis, I., Davis, M. J., Muldoon, C., & O'Hare, G. et al. (Manuscript submitted for publication). Citizen OBservatory WEB (COBWEB): A Generic Infrastructure Platform to Facilitate the Collection of Citizen Science data for Environmental Monitoring. *International Journal of Spatial Data Infrastructures Research*.

Hochachka, W. M., Fink, D., Hutchinson, R. A., Sheldon, D., Wong, W. K., & Kelling, S. (2012). Data-intensive science applied to broad-scale citizen science. *Trends in Ecology & Evolution*, *27*(2), 130–137. doi:10.1016/j.tree.2011.11.006 PMID:22192976

Höchtl, J., & Lampoltshammer, T. J. (2016) ADEQUATe - Analytics and Data Enrichment to Improve the Quality of Open Data. In Parycek, P.; Edelmann, N. (Eds.), *Proceedings of the International Conference for E-Democracy and Open Government CeDEM16*, (pp. 27-32). Krems: Edition Donau-Universität Krems.

Hodges, C. J., Leibovici, D. G., & Ties, S. F. (2014). Building confidence in crowdsourced data for biological monitoring and policy making as part of FP7 Project COBWEB. Role of Volunteered Geographic Information in Advancing Science: Effective Utilization (Workshop). In *Proceedings of GIScience 2014 Eight International Conference on Geographical Information Science*.

Hofer, B., Lampoltshammer, T. J., & Belgiu, M. (2015). Demography of Twitter Users in the City of London: An Exploratory Spatial Data Analysis Approach. In J. Brus, A. Vondrakova, & V. Vozenilek (Eds.), *Modern Trends in Cartography* (pp. 199–211). Cham: Springer International Publishing. doi:10.1007/978-3-319-07926-4_16

Hoffmann, A., Penner, J., Vohland, K., Cramer, W., Doubleday, R., Henle, K., & Häuser, C. et al. (2014). Improved access to integrated biodiversity data for science, practice, and policy - the European Biodiversity Observation Network (EU BON). *Natureza & Conservação*, *8*, 49–65. doi:10.3897/natureconservation.6.6498

Holdren, J. P. (2013). *Memorandum for the heads of executive departments and agencies: Increasing access to the results of federally funded scientific research.* Washington, DC: Office of Science and Technology Policy.

Hopkins, A. D. (1918). Periodical events and natural law as guides to agricultural research and practice. *U.S. Department of Agriculture, Weather Bureau, Monthly weather review, Suppl. 9.*

Hopkins, G. W., & Freckleton, R. P. (2002). Declines in the numbers of amateur and professional taxonomists: Implications for conservation. *Animal Conservation*, *5*(3), 245–249. doi:10.1017/S1367943002002299

Hosmer, L. T. (1995). Trust: The connecting link between organizational theory and philosophical ethics. *Academy of Management Review*, *20*(2), 379–403.

Hossain, M. A., Canning, J., Ast, S., Rutledge, P. J., Yen, T. L., & Jamalipour, A. (2015). *Lab-in-phone:* Smartphone-based portable fluorometer for pH field measurements of environmental water. *IEEE Sensors Journal*, *15*(9), 5095–5102. doi:10.1109/JSEN.2014.2361651

Hotho, A., Jäschke, R., Schmitz, C., & Stumme, G. (2006). *Information retrieval in folksonomies: Search and ranking.* Springer Berlin Heidelberg.

Howe, J. (2009). *Crowdsourcing: Why the Power of the Crowd Is Driving the Future of Business.* Crown Publishing Group.

Hsu, C.-L., & Lin, J. C.-C. (2008). Acceptance of blog usage: The roles of technology acceptance, social influence and knowledge sharing motivation. *INFMAN Information & Management*, *45*(1), 65–74.

Hsu, M.-H., Ju, T. L., Yen, C.-H., & Chang, C.-M. (2007). Knowledge sharing behavior in virtual communities: The relationship between trust, self-efficacy, and outcome expectations. *International Journal of Human-Computer Studies*, *65*(2), 153–169. doi:10.1016/j.ijhcs.2006.09.003

Hunter, J., Alabri, A., & van Ingen, C. (2013). Assessing the quality and trustworthiness of citizen science data. *Concurrency and Computation*, *25*(4), 454–466. doi:10.1002/cpe.2923

Huotari, K., & Hamari, J. (2012). Defining gamification: a service marketing perspective. In *Proceeding of the 16th International Academic MindTrek Conference* (pp. 17-22). ACM. doi:10.1145/2393132.2393137

Hutt, H., Everson, R., Grant, M., Love, J., & Littlejohn, G. (2013). How clumpy is my image? Evaluating crowdsourced annotation tasks. In *UK Workshop on Computational Intelligence (UKCI).*

Hyatt, S. (2001). From citizen to volunteer. In The new poverty studies: The ethnography of power, politics, and impoverished people in the United States. Academic Press.

Ibercivis. (2016). *Ibercivis starts to draw the map of citizen science in Spain.* Retrieved June 2, 2016, from http://www.ibercivis.com/ibercivis-starts-draw-map-citizen-science-spain/

Illinois Environmental Protection Agency. (2016a). *Volunteer Lake Monitoring Program.* Retrieved from http://epa.illinois.gov/topics/water-quality/monitoring/vlmp/index

Illinois Environmental Protection Agency. (2016b). *Inland Lake Monitoring Programs*. Retrieved from epa.illinois.gov/topics/water-quality/monitoring/inland-lakes/index

Indiana Clean Lakes Program. (2016). *Volunteer Monitoring*. Retrieved from http://www.indiana.edu/~clp/volunteer-monitoring.php

International Organization for Standardization. (2003). *ISO 19115:2003 - Geographic information - Metadata*. Retrieved from: http://www.iso.org/iso/catalogue_detail.htm?csnumber=26020

Irvin, R. A., & Stansbury, J. (2004). Citizen participation in Decision Making: Is it Worth the Effort? *Public Administration Review*, *64*(1), 55–65. doi:10.1111/j.1540-6210.2004.00346.x

Irwin, A., & Horst, M. (2015). Engaging in a decentred world. *Remaking Participation: Science, Environment and Emergent Publics*, 64.

Irwin, A. (1995). *Citizen science: a study of people, expertise and sustainable development*. London: Routledge.

Irwin, A. (1995). *Citizen science: A study of people, expertise and sustainable development*. Psychology Press.

Jacobson, S. K., Carlton, J. S., & Monroe, M. C. (2012). Motivation and satisfaction of volunteers at a Florida natural resource agency. *Journal of Park and Recreation Administration*, *30*(1).

Jahn, T., Bergmann, M., & Keil, F. (2012). Transdisciplinarity: Between mainstreaming and marginalization. *Ecological Economics*, *79*, 1–10. doi:10.1016/j.ecolecon.2012.04.017

Jakus, G., Milutinović, V., Omerović, S., & Tomažić, S. (2013). *Concepts, Ontologies, and Knowledge Representation*. New York: Springer. doi:10.1007/978-1-4614-7822-5

Janowicz, K. (2009). Semantic Interoperability. In B. Warf (Ed.), *Encyclopedia of Geography*. SAGE Publications.

Janzen, D. H. (2004). Setting up tropical biodiversity for conservation through non-damaging use: Participation by parataxonomists. *Journal of Applied Ecology*, *41*(1), 181–187. doi:10.1111/j.1365-2664.2004.00879.x

Jarnevich, C. S., Graham, J. J., Newman, G. J., Crall, A. W., & Stohlgren, T. J. (2007). Balancing data sharing requirements for analysis with data sensitivity. *Biological Invasions*, *9*(5), 597–599. doi:10.1007/s10530-006-9042-4

Jekel, T., Koller, A., & Strobl, J. (2011). Research – education cooperations for GI in secondary education. In *International perspectives on GIS in secondary education* (pp. 27-36). New York: Springer.

Jiang, B., & Thill, J. C. (2015). Volunteered Geographic Information: Towards the establishment *of a new paradigm. Computers, Environment and Urban Systems*, *53*(September), 1–3. doi:10.1016/j.compenvurbsys.2015.09.011

Johnson, M., Hannah, C., Acton, L., Popovici, R., Karanth, K., & Weinthal, E. (2014). Network environmentalism: Citizen scientists as agents for environmental advocacy. *Global Environmental Change*, *29*, 235–245. doi:10.1016/j.gloenvcha.2014.10.006

Joint Research Center. (2016). *Citizen Science: data and service infrastructure meeting: JRC-Ispra, 26th-27th January 2016* [Meeting Report]. Retrieved May 16, 2016, from https://ec.europa.eu/jrc/sites/default/files/Citizen_science_27022016_next_steps.pdf

Jones, M., Riddell, K., & Morrow, A. (2013). *The impact of citizen science activities on participant behavior and attitudes: Project Report 2013: The Conservation Volunteers*. Retrieved May 16th 2016, from: http://www.environment.scotland.gov.uk/media/58143/Impact_Of_Citizen_Science_Activities_On_Participant_Behaviour_And_Attitude.pdf

Jones, R. C. (2014). Creating new technical universities in emerging countries. In *Interactive Collaborative Learning (ICL), 2014 International Conference on* (pp. 493-498). IEEE. doi:10.1109/ICL.2014.7017822

Joost, S., Baumann, R., Ertz, O., Ingensand, J., Widmer, I., & Rappo, D. (2014). *A participatory WebGIS platform to support biodiversity inventory in the Geneva cross-border area.* Third Open Source Geospatial Research & Education Symposium (OGRS), Espoo, Finland.

Jordan, R. C., Ballard, H. L., & Phillips, T. (2012). Key issues and new approaches for evaluating citizen-science learning outcomes. *Frontiers in Ecology and the Environment*, *10*(6), 307–309. doi:10.1890/110280

Jordan, R. C., Gray, S. A., Howe, D. V., Brooks, W. R., & Ehrenfeld, J. G. (2011). Knowledge gain and behavioral change in citizen-science programs. *Conservation Biology*, *25*(6), 1148–1154. doi:10.1111/j.1523-1739.2011.01745.x PMID:21967292

Jørgensen, M. S., Hall, I., Hall, D., Gnaiger, A., Schroffenegger, G., Brodersen, S., et al. (2004). *Democratic Governance Through Interaction between NGOs, Universities and Science Shops: Experiences, Expectations, Recommendations.* Final Report of INTERACTS. Retrieved from http://www.wilawien.ac.at/interacts/interacts_report_final2.pdf

Kahn, R. (2010). *Critical pedagogy, ecoliteracy, & planetary crisis: The ecopedagogy movement.* New York, NY: Peter Lang.

Kamar, E., Hacker, S., & Horvitz, E. (2012). Combining human and machine intelligence in large-scale crowdsourcing. In *Proceedings of the 11th International Conference on Autonomous Agents and Multiagent Systems (AAMAS 2012).* Valencia, Spain: International Foundation for Autonomous Agents and Multiagent Systems.

Kampen, H., Medlock, J. M., Vaux, A. G. C., Koenraadt, C. J. M., van Vliet, A. J. H., Bartumeus, F., & Werner, D. et al. (2015). Approaches to passive mosquito surveillance in the EU. *Parasit Vectors*, *8*(1), 9. doi:10.1186/s13071-014-0604-5 PMID:25567671

Katzenbach, J., Steffen, I., & Kronley, C. (2012). Cultural Change That Sticks. *Harvard Business Review*, *90*(7/8), 110–117. PMID:22852451

Kaufmann, N., Schulze, T., & Veit, D. (2011, August 4 - 7). *More than fun and money. Worker Motivation in Crowdsourcing-A Study on Mechanical Turk.* Paper presented at the AMCIS, Detroit, MI.

Keddem, S., Barg, F. K., Glanz, K., Jackson, T., Green, S., & George, M. (2015). Mapping the urban asthma experience: Using qualitative GIS to understand contextual factors affecting asthma control. *Social Science & Medicine*, *140*(September), 9–17. doi:10.1016/j.socscimed.2015.06.039

Kehrer, J., Ladstadter, F., Muigg, P., Doleisch, H., Steiner, A., & Hauser, H. (2008). Hypothesis generation in climate research with interactive visual data exploration. *Visualization and Computer Graphics. IEEE Transactions on*, *14*(6), 1579–1586.

Kennedy, E. B. (2016). When Citizen Science Meets Science Policy. In D. Cavalier & E. B. Kennedy (Eds.), *The Rightful Place of Science: Citizen Science.* Tempe, AZ: Arizona State University.

Kight, C. (2012). *A Brief History of Citizen Science.* Retrieved June 2, 2016, from http://www.science20.com/anthrophysis/brief_history_citizen_science-93317

King, K., & McGrath, S. (2002). *Globalisation, enterprise and knowledge: Education, training and development in Africa.Symposium Books Ltd.*

Kingston, R., Carver, S., Evans, A., & Turton, I. (2000). Web-based public participation geographic information systems: An aid to local environmental decision-making. *Computers, Environment and Urban Systems*, *24*(2), 109–125. doi:10.1016/S0198-9715(99)00049-6

Klettner, S., Huang, H., & Schmidt, M. (2011). EmoMap – considering emotional responses to space for enhancing LBS. In *Advances in Location-Based Services Proceedings 8th International Symposium on Location-Based Services*. Springer.

Klien, E., Lutz, M., & Kuhn, W. (2006). Ontology-based discovery of geographic information services—An application in disaster management. *Computers, Environment and Urban Systems*, *30*(1), 102–123. doi:10.1016/j.compenvurbsys.2005.04.002

Kliskberg, B. (2001). *El Capital Social*. Caracas: Editorial PANAPO.

Kobernus, M., Berre, A. J., Gonzalez, M., Liu, H.-Y., Fredriksen, M., Rombouts, R., & Bartonova, A. (2015). A practical approach to an integrated citizens' observatory: The CITI-SENSE framework. In *Proceedings of the Workshop 'Environmental Information Systems and Services - Infrastructures and Platforms 2013' (ENVIP 2013)*. Neusiedl am See, Austria: CEUR Workshop Proceedings.

Kolb, D. A. (1984). *Experiential Learning: Experience as the source of learning and development*. Prentice-Hall.

Koplan, J. P., Liverman, C. T., & Kraak, V. I. (Eds.). (2005). *Preventing Childhood Obesity: Health in the Balance*. Washington, DC: National Academies Press.

Korpela, E. J., Anderson, D. P., Bankay, R., Cobb, J., Howard, A., & Lebofsky, M., … Werthimer, D. (2011). Status of the UC-Berkeley SETI efforts. In *Proceedings of SPIE conference 8152, Instruments, Methods, and Missions for Astrobiology XIV, 815212*. San Diego, CA: Cornell University Library.

Kotler, P., Wong, V., Saunders, J., & Armstrong, G. (2005). *Principles of Marketing* (European Edition). Pearson Education Limited.

Kozma, R. B. (2012). *Les TIC i la transformació de l'educació en l'economia del coneixement*. Barcelona: Fundació Jaume Bofill. Retrieved from http://www.debats.cat/sites/default/files/debats/pdf/kozma.pdf

Kruger, L. E., & Shannon, M. A. (2000). Getting to know ourselves and our places through participation in civic social assessment. *Society & Natural Resources*, *13*(5), 461–478. doi:10.1080/089419200403866

Kullenberg, C., & Kasperowski, D. (2016). What Is Citizen Science?–A Scientometric Meta-Analysis. *PLoS ONE*, *11*(1), e0147152. doi:10.1371/journal.pone.0147152 PMID:26766577

Lakhani, K. R., & Panetta, J. A. (2007). The principles of distributed innovation. *Innovations*, *2*(3), 97–112. doi:10.1162/itgg.2007.2.3.97

Lakshminarayanan, S. (2007). Using citizens to do science versus citizens as scientists. *Ecology and Society, 12. Response*, *2*, 1.

Lall, N. (2011). Estructuras de investigación colaborativa comunidad universidad: aproximación a su posible impacto. *Rizoma-Freireaano, 9*. Retrieved April 22, 2016 from: http://www.rizoma-freireano.org/index.php/estructuras-de-investigacion-colaborativa-comunidad-universidad-aproximacion-a-su-posible-impacto--nirmala-lall

Lampoltshammer, T. J., & Heistracher, T. (2014). Ontology evaluation with Protégé using OWLET. *Infocommunications Journal*, *6*(2), 12–17.

Lampoltshammer, T. J., Kounadi, O., Sitko, I., & Hawelka, B. (2014). Sensing the public's reaction to crime news using the 'Links Correspondence Method'. *Applied Geography (Sevenoaks, England)*, *52*, 57–66. doi:10.1016/j.apgeog.2014.04.016

Lampoltshammer, T. J., & Wiegand, S. (2015). Improving the Computational Performance of Ontology-Based Classification Using Graph Databases. *Remote Sensing*, *7*(7), 9473–9491. doi:10.3390/rs70709473

Lanfranchi, V., Wrigley, S. N., Ireson, N., Ciravegna, F., & Wehn, U. (2013). Citizens' observatories for situation awareness in flooding. In *Proceedings of the 11th International ISCRAM Conference*. University Park, PA: ISCRAM.

Lanfranchi, V., Wrigley, S., Ireson, N., Ciravegna, F., & Wehn, U. (2014). *Citizens' Observatories for Situation Awareness in Flooding*. Paper presented at the 11th International ISCRAM Conference (Information Systems for Crisis and Response Management), University Park, PA.

Lanthaler, M., Sporny, M., & Kellogg, G. (2014). *JSON-LD 1.0*. W3C Recommendation. Retrieved from https://www.w3.org/TR/json-ld/

Lawrence, A. (2006). No personal motive? Volunteers, biodiversity, and the false dichotomies of participation. *Ethics Place and Environment*, *9*(3), 279–298. doi:10.1080/13668790600893319

Lawrence, R. L., & Deagan, D. A. (2001). Choosing public participation methods for natural resources: A context-specific guide. *Society & Natural Resources*, *14*, 857–872. doi:10.1080/089419201753242779

Lawrence, S. (2001). Online or invisible. *Nature*, *411*(6837), 521–523. doi:10.1038/35079151

Lee, Y. W., Strong, D. M., Kahn, B. K., & Wang, R. Y. (2002). AIMQ: A methodology for information quality assessment. *Information & Management*, *40*(2), 133–146. doi:10.1016/S0378-7206(02)00043-5

Leibovici, D. G., Jackson, M., & Higgins, C. (2015). *Geospatial Data Curation and Interoperability in the COBWEB project: citizen science & crowdsourcing for environmental policy*. Paper presented at RDA (Research Data alliance) symposium Plenary 5, joint meeting of IG Geospatial & IG Big Data, San Diego, CA.

Lenhart, A., & Horrigan, J. B. (2003). Re-visualizating the digital divide as a digital spectrum. *IT & Society*, *1*(5), 23–39.

Leopold, A. (2001). *A Sand County Almanac: Outdoor Essays and Reflections*. New York: Ballantine.

Lepczyk, C. A., Boyle, O. D., Vargo, T. L., Gould, P., Jordan, R., Liebenberg, L., & Vaughan, H. et al. (2009). … Vaughan, H. (2009). Citizen science in ecology: The intersection of research and education. *Bulletin of the Ecological Society of America*, *90*(3), 308–317. doi:10.1890/0012-9623-90.3.308

Lerner, J., & Tirole, J. (2002). Some simple economics of open source. *The Journal of Industrial Economics*, *50*(2), 197–234. doi:10.1111/1467-6451.00174

Levrel, H., Fontaine, B., Henry, P. Y., Jiguet, F., Julliard, R., Kerbiriou, C., & Couvet, D. (2010). Balancing state and volunteer investment in biodiversity monitoring for the implementation of CBD indicators: A French example. *Ecological Economics*, *69*(7), 1580–1586. doi:10.1016/j.ecolecon.2010.03.001

Lewenstein, B. V. (2004). *What does citizen science accomplish?* Cornell University. Retrieved from https://ecommons.cornell.edu/bitstream/handle/1813/37362/Lewenstein.2004.What%20does%20citizen%20science%20accomplish.pdf

Lin, H. F. (2008). Determinants of successful virtual communities: Contributions from system characteristics and social factors. *Information & Management*, *45*(8), 522–527. doi:10.1016/j.im.2008.08.002

Lin, M.-J. J., Hung, S.-W., & Chen, C.-J. (2009). Fostering the determinants of knowledge sharing in professional virtual communities. *Computers in Human Behavior*, *25*(4), 929–939. doi:10.1016/j.chb.2009.03.008

Lintott, C. J., Schawinski, K., Slosar, A., Land, K., Bamford, S., Thomas, D., & Vandenberg, J. et al. (2008). Galaxy Zoo: Morphologies derived from visual inspection of galaxies from the Sloan Digital SkySurvey. *Monthly Notices of the Royal Astronomical Society*, *389*(3), 1179–1189. doi:10.1111/j.1365-2966.2008.13689.x

Litke, S., & Day, J. S. (1998). Building local capacity for stewardship and sustainability: The role of community-based watershed management in Chilliwack, British Colombia. *Environments*, *25*, 91–110.

Liu, S. B., Iacucci, A. A., & Meier, P. (2010). Ushahidi Haiti and Chile: next generation crisis mapping. *ACSM Bulletin*, *246*.

Liu, H.-Y., Bartonova, A., Neofytou, P., Yang, A., Kobernus, M. J., Negrenti, E., & Housiadas, C. (2012a). Facilitating knowledge transfer: Decision support tools in environment and health. *Environmental Health*, *11*(Suppl 1), S17. doi:10.1186/1476-069X-11-S1-S17 PMID:22759499

Liu, H.-Y., Bartonova, A., Pascal, M., Smolders, R., Skjetne, E., & Dusinska, M. (2012b). Approaches to integrated monitoring for environment and health impact assessment. *Environmental Health*, *11*(1), 88. doi:10.1186/1476-069X-11-88 PMID:23171406

Liu, H.-Y., Kobernus, M., Broday, D., & Bartonova, A. (2014). A conceptual approach to a citizens' observatory – supporting community-based environmental governance. *Environmental Health*, *13*(1), 107. doi:10.1186/1476-069X-13-107 PMID:25495204

Liu, H.-Y., Skjetne, E., & Kobernus, M. (2013). Mobile phone tracking: In support of modelling traffic-related air pollution contribution to individual exposure and its implications for public health impact assessment. *Environmental Health*, *12*(1), 93. doi:10.1186/1476-069X-12-93 PMID:24188173

Lloyd, A., Lipu, S., & Kennan, M. A. (2010). On becoming citizens: Examining social inclusion from an information perspective. *Australian Academic & Research Libraries*, *41*(1), 42–53. doi:10.1080/00048623.2010.10721433

Lodge, D. M., Williams, S., MacIsaac, H. J., Hayes, K. R., Leung, B., Reichard, S., & McMichael, A. (2006). Biological invasions: Recommendations for US policy and management. *Ecological Applications*, *16*(6), 2035–2054. doi:10.1890/1051-0761(2006)016[2035:BIRFUP]2.0.CO;2 PMID:17205888

Longley, P. A., Goodchild, M. F., Maguire, D. J., & Rhind, D. W. (2005). Geographic Information Systems and Science. John Wiley & Sons.

Loukides, M., Warden, P., Watters, A., & Croll, A. (2013). *Big Data Now - Current Perspectives from O'Reilly Radar* (A. Noren, Ed.). O'Reilly Media.

Louv, R. (2005). *Last Child in the Woods: Saving our Children from Nature Deficit Disorder*. Chapel Hill, NC: Algonquin Books.

Lukyanenko, R., Parsons, J., & Wiersma, Y. F. (2016). Emerging problems of data quality in citizen science. *Conservation Biology*, *30*(3), 447–449. doi:10.1111/cobi.12706 PMID:26892841

Lundh, P., Stanford, T., & Shear, L. (2014). *Nano and Society: Case Study of a Research-to Practice Partnership between University Scientists and Museum Professionals*. Menlo Park, CA: SRI International.

Lupia, A. (2015). *Uninformed: Why People Seem to Know So Little about Politics and What We Can Do about It*. Oxford University Press.

Lynam, T., Jong, W., Sheil, D., Kusumanto, T., & Evans, K. (2007). A review of tools for incorporating community knowledge, preferences and values into decision making in natural resources management. *Ecology and Society*, *12*, 5–18.

Macdonald, L., Cummins, S., & Macintyre, S. (2007). Neighbourhood fast food environment and area deprivation—substitution or concentration? *Appetite*, *49*(1), 251–254. doi:10.1016/j.appet.2006.11.004

Mackney, P., & Spring, N. (2000). *The Earthwatch Perspective on Volunteers and Data Validity*. Oxford, UK: Earthwatch Institute.

Macnaghten, P. (2003). Embodying the environment in everyday life practices. *The Sociological Review*, *51*(1), 63–84. doi:10.1111/1467-954X.00408

Mahr, D. (2014). *Citizen Science: Partizipative Wissenschaft im späten 19. und frühen 20. Jahrhundert*. Baden-Baden, Germany: Nomos; doi:10.5771/9783845253732

Maisonneuve, N., Stevens, M., Niessen, M. E., Hanappe, P., & Steels, L. (2009). Citizen Noise Pollution Monitoring. In *Proceedings of the 10th International Digital Government Research Conference*.

Manley, G. (1952). The Weather and Diseases: Some Eighteenth-Century Contributions to Observational Meteorology. *Notes and Records of the Royal Society of London*, *9*(2), 300–307. doi:10.1098/rsnr.1952.0017

Manstead, A. S. R., & Van Eekelen, S. A. M. (1998). Distinguishing Between Perceived Behavioral Control and Self-Efficacy in the Domain of Academic Achievement Intentions and Behaviors. *Journal of Applied Social Psychology*, *28*(15), 1375–1392. doi:10.1111/j.1559-1816.1998.tb01682.x

Marsh, L., & Onof, C. (2008). Stigmergic epistemology, stigmergic cognition. *Cognitive Systems Research*, *9*(1), 136–149. doi:10.1016/j.cogsys.2007.06.009

Martella, R., Kray, C., & Clementini, E. (2015). A Gamification Framework for Volunteered Geographic Information. In F. Bacao, M. Y. Santos, & M. Painho (Eds.), *AGILE 2015* (pp. 73–89). Cham: Springer International Publishing. doi:10.1007/978-3-319-16787-9_5

McGrath, S., King, K., Leach, F., & Carr-Hill, R. (1994). *Education and Training for the Informal Sector.* UK Department for International Development, Education Research Paper, 11.

McKinley, D.C., Miller-Rushing, A.J., Ballard, H.L, Bonney, R., Brown, H., Evans, D.M., … Soukup, M.A. (2015). Investing in Citizen Science Can Improve Natural Resource Management and Environmental Protection. *Issues in Ecology,* 19.

McLean, K. (2016). Smellmap: Amsterdam—Olfactory Art & Smell Visualisation. *Leonardo*.

McLure Wasko, M., & Faraj, S. (2000). It is what one does: Why people participate and help others in electronic communities of practice. *The Journal of Strategic Information Systems*, *9*(2-3), 155–173. doi:10.1016/S0963-8687(00)00045-7

McQuillan, D. (2014). The Countercultural Potential of Citizen Science. *M/C Journal, 17*(6), 1-9.

Meek, S., Jackson, M., & Leibovici, D. G. (2014). *A flexible framework for assessing the quality of crowdsourced data.* Paper presented at the AGILE conference, Castellon, Spain.

Meek, S., Jackson, M., & Leibovici, D. G. (2015). Addressing the quality assurance challenge for location-based crowdsourced data through workflow composition of OGC web services. *Computers & Geosciences*, (Jan): 2015.

Meier, P., & Brodock, K. (2008, October 23). Crisis mapping Kenya's election violence: Comparing mainstream news, citizen journalism and Ushahidi. *iRevolution Blog*.

Meneses, J. & Mominó, J.M. (2010). Putting Digital Literacy in Practice: How Schools Contribute to Digital Inclusion in the Network Society. *The Information Society: An International Journal, 26*(3), 197-208. 10.1080/01972241003712231

Michigan Department of Environmental Quality. (2016). *Cooperative Lake Monitoring Program*. Retrieved from http://www.michigan.gov/deq/0,4561,7-135-3313_3681_3686_3731-195536--,00.html

Michinel, J. L. (2011). Un proyecto paraguas para la realización del servicio comunitario por estudiantes de la Facultad de Ciencias de la UCV: mejoramiento de la educación en escuelas públicas (y privadas subsidiadas) a través del SC. *Revist@ serviciocomunitario, 1*. Retrieved March 22, 2016 from: http://saber.ucv.ve/ojs/index.php/rev_scfc/article/view/631/578

Miller, H. J., & Goodchild, M. F. (2015). Data-driven geography. *GeoJournal*, *80*(4), 449–461. doi:10.1007/s10708-014-9602-6

Miller, J. D. (2004). Public understanding of, and attitudes toward, scientific research: What we know and what we need to know. *Public Understanding of Science (Bristol, England)*, *13*(3), 273–294. doi:10.1177/0963662504044908

Miller-Rushing, A., Primack, R., & Bonney, R. (2012). The history of public participation in ecological research. *Frontiers in Ecology and the Environment*, *10*(6), 285–290. doi:10.1890/110278

Minghini, M. (2014). *Multi-dimensional geoweb platforms for citizen science and civic engagement applications*. PhD thesis.

Minkman, E. (2015). Citizen Science in Water Quality Monitoring: Developing Guidelines for Dutch Water Authorities for Contributory Mobile Crowd Sensing. TU Delft, Delft University of Technology.

Minnesota Pollution Control Agency. (2008). *Citizen Lake-Monitoring Program: Instruction Manual*. Retrieved from https://www.pca.state.mn.us/sites/default/files/wq-s1-13.pdf

Minnesota Pollution Control Agency. (2009). *Citizen Monitoring of Surface Water Quality: 2009 Report to the Legislature*. Author.

Minnesota Pollution Control Agency. (2011). *Citizen Monitoring of Surface Water: 2011 Report to the Legislature*. Retrieved from https://www.pca.state.mn.us/sites/default/files/lrwq-s-1sy11.pdf

Minnesota Pollution Control Agency. (2012, October). *Water Quality Assessments of Select Lakes within the Mississippi River (St. Cloud) Watershed*. Retrieved from https://www.pca.state.mn.us/sites/default/files/wq-ws3-07010203.pdf

Minnesota pollution Control Agency. (2016a). *Surface water data*. Retrieved from http://cf.pca.state.mn.us/water/watershedweb/wdip/index.cfm

Minnesota Pollution Control Agency. (2016a). *Transparency trends for Upper Prior Lake*. Retrieved from http://cf.pca.state.mn.us/water/watershed web/whip/wawterwatc.cfm?wid=70-0072-00

Minnesota pollution Control Agency. (2016b). *Citizen Monitoring Program: Annual summary*. Retrieved from https://www.pca.state.mn.us/water/citizen-monitoring-program-annual-summary

Minnesota Pollution Control Agency. (2016b). *Lakes and water quality*. Retrieved from https://www.pca.state.mn.us/water/lakes-and-water-quality

Mitchell, T. (2008). Traditional vs. Critical Service-Learning: Engaging the Literature to Differentiate Two Models. *Michigan Journal of Community Service Learning*, *2*, 50–65.

Morgan, N., Pritchard, A., & Pride, R. (2003). Marketing to the Welsh diaspora: The appeal to hiraeth and homecoming. *Journal of Vacation Marketing*, *9*(1), 69–80. doi:10.1177/135676670200900105

Mullen, M. W., & Allison, B. E. (1999). Stakeholder involvement and social capital: Keys to watershed management success in Alabama. *Journal of the American Water Resources Association*, *35*(3), 655–662. doi:10.1111/j.1752-1688.1999.tb03621.x

Müller, R. M., Thoring, K., & Oostinga, R. (2010). Crowdsourcing with semantic differentials: A game to investigate the meaning of form. *AMCIS Proceedings*.

Muller, C. L., Chapman, L., Johnston, S., Kidd, C., Illingworth, S., Foody, G., & Leigh, R. R. et al. (2015). Crowdsourcing for climate and atmospheric sciences: Current status and future potential. *JOC International Journal of Climatology*, *35*(11), 3185–3203. doi:10.1002/joc.4210

Musick, M., & Wilson, J. (2007). *Volunteers: A Social Profile*. Bloomington, IN: Indiana Univ. Press.

Narayan-Parker, D. (Ed.). (2002). *Empowerment and poverty reduction: A sourcebook*. World Bank Publications.

Nascimento, S., Guimarães Pereira, Â., & Ghezzi, A. (2014). *From citizen science to do it yourself science: An annotated account of an on-going movement (Science and Policy Report)*. Joint Research Centre of the European Commission.

National Science Board. (2008). *Science and Technology: Public Attitudes and Understanding. Science Engineering Indicators 2008*. National Science Board.

National Water Quality Council. (2016). *Water Quality Portal National Results Coverage*. Retrieved from www.waterqualitydata.us/coverage

Natural Research Council. (1996). *The National Science Education Standards*. Washington, DC: National Academy Press.

Nature. (2015). Rise of the citizen scientist. *Nature, 524*(7565), 265-265. doi:10.1038/524265a

Nerbonne, J. F., & Nelson, K. C. (2008). Volunteer macroinvertebrate monitoring: Tensions among group goals, data quality, and outcomes. *Environmental Management, 42*(3), 470–479. PubMeddoi:10.1007/s00267-008-9103-9

Neteler, M., & Mitasova, H. (2008). Open Source GIS: A GRASS GIS Approach (3rd ed.). Springer. doi:10.1007/978-0-387-68574-8

Neuhaus, F., Vizedom, A., Baclawski, K., Bennett, M., Dean, M., Denny, M., …, , & Obrst, L. (2013). Towards ontology evaluation across the life cycle. The Ontology Summit 2013. *Applied Ontology*, *8*(3), 179–194.

New South Wales Department of Primary Industries. (n.d.). *NSW DPI Game Fish Tagging Program*. Retrieved March 6, 2016, from http://www.dpi.nsw.gov.au/fisheries/recreational/saltwater/gamefish-tagging

New South Wales Office of Environment and Heritage. (2015a). *NSW Office of Environment and Heritage (OEH) Citizen Science Position Statement*. Retrieved May 18, 2016, from http://www.environment.nsw.gov.au/research/citizenscience.htm

New South Wales Office of Environment and Heritage. (2015b). *OEH Citizen Science Strategy 2016–18. Driving a new era of public participation in science to support OEH decision-making*. Retrieved May 18, 2016, from http://www.environment.nsw.gov.au/research/citizenscience.htm

Newman, G. (2014). Citizen CyberScience: New directions and opportunities for human computation[Editorial]. *Human Computation*, *1*(2), 103–109. doi:10.15346/hc.v1i2.2

Newman, G., Wiggins, A., Crall, A., Graham, E., Newman, S., & Crowston, K. (2012). The future of Citizen Science: Emerging technologies and shifting paradigms. *Frontiers in Ecology and the Environment*, *10*(6), 298–304. doi:10.1890/110294

Newman, G., Zimmerman, D., Crall, A., Laituri, M., Graham, J., & Stapel, L. (2010). User-friendly web mapping: Lessons from a citizen science website. *International Journal of Geographical Information Systems*, *24*(12), 1851–1869. doi:10.1080/13658816.2010.490532

Ngo Thu, H., & Wehn, U. (2016). Data sharing in international transboundary contexts: The Vietnamese perspective on data sharing in the Lower Mekong Basin. *Journal of Hydrology Journal of Hydrology*, *536*(2), 351–364.

Nichols, J. D., & Williams, B. K. (2006). Monitoring for conservation. *Trends in Ecology & Evolution*, *21*(12), 668–673. doi:10.1016/j.tree.2006.08.007

Nicholson, E., Ryan, J., & Hodgkins, D. (2002). Community data - where does the value lie? Assessing confidence limits of community collected water quality data. *Water Science and Technology*, *45*, 193–200.

Nielsen, K. A., & Nielsen, B. S. (2006). Methodologies in action research: Action research and critical theory. In K. A. Nielsen & L. Svensson (Eds.), *Action research and interactive research: Beyond practice and theory* (pp. 63–87). Maastricht, The Netherlands: Shaker Publishing.

No somos suecos… ¡somos latinoamericanos!. (n.d.). In *Interconectados in Interconectados*. Retrieved April 18, 2016 from: http://interconectados.org/no-somos-suecos-somos-latinoamericanos/)

NOAA, National Weather Service. (2014). Retrieved from http://www.nws.noaa.gov/os/coop/what-is-coop.html

Nov, O., Arazy, O., & Anderson, D. (2010). Crowdsourcing for science: understanding and enhancing SciSourcing contribution. In *ACM CSCW 2010 Workshop on the Changing Dynamics of Scientific Collaborations*.

Nov, O. (2007). What Motivates Wikipedians? *Communications of the ACM*, *50*(11), 60–64. doi:10.1145/1297797.1297798

Nov, O., Arazy, O., & Anderson, D. (2011). Dusting for science: motivation and participation of digital citizen science volunteers.*Proceedings of the 2011 iConference*. New York: ACM doi:10.1145/1940761.1940771

Nov, O., Arazy, O., & Anderson, D. (2011). Technology-Mediated Citizen Science Participation: A Motivational Model. In *Proceedings of the AAAI International Conference on Weblogs and Social Media (ICWSM 2011)*.

Nov, O., Arazy, O., & Anderson, D. (2014). Scientists@ Home: What drives the quantity and quality of online citizen science participation? *PLoS ONE*, *9*(4), e90375. doi:10.1371/journal.pone.0090375 PMID:24690612

Nov, O., Arazy, O., Lotts, K., & Naberhaus, T. (2013). Motivation-targeted personalized UI design: a novel approach to enhancing citizen science participation.*ECSCW 2013: Proceedings of the 13th European Conference on Computer Supported Cooperative Work*. London: Springer. doi:10.1007/978-1-4471-5346-7_15

Nutley, S., Powell, A., & Davies, H. (2013). *What counts as good evidence? Provocation paper for the Alliance of Useful Evidence*. Retrieved from http://www.alliance4usefulevidence.org/assets/What-Counts-as-Good-Evidence-WEB.pdf

Obrecht, D. V., Milanik, M., Perkins, B. D., Ready, D., & Jones, J. R. (1998). Evaluation of data generated from lake samples collected by volunteers. *Journal of Lake and Reservoir Management*, *14*(1), 21–27. doi:10.1080/07438149809354106

OECD (2012). *Better Skills, Better Jobs, Better Lives: A Strategic Approach to Skills Policies*. OECD Publishing. 10.1787/9789264177338-en

OECD. (2009). *21st Century Skills and Competences for New Millennium Learners in OECD Countries*. OECD Education Working Papers, No. 41. OECD Publishing.

OECD. (2010). *The nature of learning. Using research to inspire practice*. OECD Publishing. Retrieved from http://www.oecd.org/edu/ceri/thenatureoflearningusingresearchtoinspirepractice.htm-3

OECD. (2015). *Innovative Learning Environments: Implementation and change*. OECD Publishing. Retrieved from http://www.oecd.org/edu/ceri/The ILE project.pdf

Office of Science and Technology Policy. (2014). *National Plan for Civil Earth Observations*. Retrieved from https://www.whitehouse.gov/sites/default/files/microsites/ostp/NSTC/2014_national_plan_for_civil_earth_observations.pdf

Office of Science and Technology Policy. (2015, September 30). *Memorandum to the heads of executive departments and agencies*. Retrieved from https://www.whitehouse.gov/sites/default/files/microsites/ostp/holdren_citizen_science_memo_092915_0.pdf

Office of the Chief Scientist. (2013). *Science, technology, engineering and mathematics in the national interest: A strategic approach*. Canberra, Australia: Australian Government.

Ohio State University Extension. (2016). *Ohio Watershed Network*. Retrieved from http://production.cfaes.ohio-state.edu/apps/courses/cfaes/course/view.php

Okolloh, O. (2009). Ushahidi, or 'testimony': Web 2.0 tools for crowdsourcing crisis information. *Participatory Learning and Action, 59*(1), 65-70.

Okolloh, O. (2009). Ushahidi or 'testimony': Web 2.0 tools for crowdsourcing crisis information. Participatory Learning and Action. *IIED Publications Database, 59*, 65–70.

Olmanson, L. G., Bauer, M. E., & Brezonik, P. L. (2002). *Use of Landsat imagery to develop a water quality atlas of Minnesota's 10,000 lakes*. Retrieved from http://www.isprs.org/proceedings/XXXIV/part1/paper/00102.pdf

Olson, J. S., & Olson, G. M. (2003). Culture surprises in remote software development teams. *Queue, 1*(9), 52. doi:10.1145/966789.966804

Open Geospatial Consortium. (n.d.a). *Open Geospatial Consortium*. Retrieved from: http://www.opengeospatial.org/

Open Geospatial Consortium. (n.d.b). *Sensor Web Enablement (SWE)*. Retrieved from: http://www.opengeospatial.org/ogc/markets-technologies/swe

Open Government Partnerships. (2016). *Levels of Credible Data*. Retrieved on March 16, 2016 from Ohio Environmental Protection Agency: http://www.epa.state.oh.us/dsw/credibledata/levels_of_data.aspx

OpenScientist. (2011). *Finalizing a Definition of "Citizen Science" and "Citizen Scientists*. Retrieved from http://www.openscientist.org/2011/09/finalizing-definition-of-citizen.html

OpenStreetMap. (n.d.). *Copyright and License*. Retrieved from: http://www.openstreetmap.org/copyright

Ordóñez Vela, L. A., Álvarez, G. C. L., Bocciolesi, E., Lombardi L, G., Martínez, A. K., & Said-Hung, E. (in press). Cultura, conectividad y colaboración: las tres "C" que marcan el futuro de la ciencia en el siglo XXI. In E. Said-Hung (Ed.), Educación, participación y escenarios digitales. Debates sobre la mediación digital en el siglo XXI. Barranquilla: Ediciones Uninorte.

Osborne, J., & Dillon, J. (2008). *Science education in Europe: Critical reflections* (Vol. 13). London: The Nuffield Foundation.

Ostermann, F. O., & Spinsanti, L. (2011). A conceptual workflow for automatically assessing the quality of volunteered geographic information for crisis management. In *Proceedings of 14thAGILE International Conference on Geographic Information Science*.

Ostrom, E. (2007). A diagnostic approach for going beyond panaceas. *Proceedings of the National Academy of Sciences of the United States of America, 104*(39), 15181–15187. doi:10.1073/pnas.0702288104 PMID:17881578

Ostrom, E., Janssen, M. A., & Anderies, J. M. (2007). Going beyond panaceas. *Proceedings of the National Academy of Sciences of the United States of America, 104*(39), 15176–15178. doi:10.1073/pnas.0701886104 PMID:17881583

Ottinger, G. (2010). Buckets of resistance: Standards and the effectiveness of citizen science. *Science, Technology & Human Values, 35*(2), 244–270. doi:10.1177/0162243909337121

Overdevest, C., & Mayer, B. (2008). Harnessing the power of information through community monitoring: Insights from social science. *Texas Law Review, 86*, 1493–1526.

Oxford Dictionaries. (2014). Oxford University Press. Retrieved from http://www.oxforddictionaries.com/definition/american_english/citizen-science

Oxford English Dictionary. (2014). Oxford University Press.

Pandya, R. E. (2012). A framework for engaging diverse communities in citizen science in the US. *Frontiers in Ecology and the Environment*, *10*(6), 314–317. doi:10.1890/120007

Pant, R. B. (2009). *The Social Media Marketing Mix (The 4 "P's")*. Retrieved January, 26, 2012 from http://ritubpant.com/social-media-marketing-mix/

Park, P. (2014). Realizing the potential of citizen science. In *UNEP Year Book 2014* (pp. 36–41). UNEP.

Parliamentary Office of Science and Technology. (2014). *Environmental citizen science*. Parliamentary Office of Science and Technology (POST), Houses of Parliament, POSTNOTE 476, August 2014. Retrieved May 18· 2016, from http://researchbriefings.parliament.uk/ResearchBriefing/Summary/POST-PN-476

Participatory Development. (n.d.). In *Wikipedia*. Retrieved April 16, 2016 from https://en.wikipedia.org/wiki/Participatory_development

Partnership for the 21st Century Skills. (2005). *A Report on the Landscape of 21st Century Assessment*. Washington, DC: Partnership for the 21st Century Skills.

Pavlou, P. A., & Fygenson, M. (2006). Understanding and predicting electronic commerce adoption: An extension of the theory of planned behavior. *Management Information Systems Quarterly*, 115–143.

Pecl, G. T., Gillies, C., Sbrocchi, C., & Roetman, P. (2015). Building Australia through citizen science. *Australian Government Office of the Chief Scientist: Occasional Paper Series, 11*.

Pedró, F. (2006). *The New Millennium learners: Challenging our Views on ICT and Learning*. Paris: OECD – CERI.

Pekrun, R., Renzel, A. C., Goetz, T., & Perry, R. P. (2007). Theoretical perspectives on emotion in education. In P. Schutz, R. Pekrun, & G. Phye (Eds.), Emotion in Education. Academic Press.

Penrose, D., & Call, S. M. (1995). Volunteer Monitoring of Benthic Macroinvertebrates – Regulatory Biologists Perspectives. *Journal of the North American Benthological Society*, *14*(1), 301–306. doi:10.2307/1467735

Pergams, O. R., & Zaradic, P. A. (2006). Is love of nature in the US becoming love of electronic media? 16-year downtrend in national park visits explained by watching movies, playing video games, internet use, and oil prices. *Journal of Environmental Management*, *80*(4), 387–393. doi:10.1016/j.jenvman.2006.02.001 PMID:16580127

Pettibone, L. (2015). *Constructing "citizens" in the citizen science literature: Who are they and what do they do?* Paper presented at Interpretive Policy Analysis Conference, Lille, France.

Pettibone, L., Ziegler, D., Bonn, A., & Vohland, K. (Eds.). (2015). *GEWISS Dialogforum: Forschungsförderung für Citizen Science*. GEWISS Bericht Nr. 7. Deutsches Zentrum für Integrative Biodiversitätsforschung (iDiv) Halle-Jena-Leipzig, Helmholtz-Zentrum für Umweltforschung – UFZ, Leipzig; Berlin-Brandenburgisches Institut für Biodiversitätsforschung (BBIB), Museum für Naturkunde, Leibniz-Institut für Evolutions- und Biodiversitätsforschung – MfN, Berlin in Kooperation mit der Leopoldina – Nationale Akademie der Wissenschaften. Retrieved January 10, 2016, from http://www.buergerschaffenwissen.de/sites/default/files/assets/dokumente/gewiss_7_foerderbericht.pdf

Pew Research Center. (2015, July). *Device ownership over time*. Retrieved from http://www.pewinternet.org/data-trend/mobile/device-ownership/

Pfeffer, M. J., & Wagenet, L. P. (2007). Volunteer Environmental Monitoring, Knowledge Creation and Citizen-Science Interaction. In J. Pretty, A. Ball, T. Benton, J. Guivant, D. R. Lee, D. Orr, & P. H. Ward et al. (Eds.), *The SAGE Handbook of Environment and Society*. SAGE. doi:10.4135/9781848607873.n16

Phillips, T., Ferguson, M., Minarchek, M., Porticella, N., & Bonney, R. (2014). *User's Guide for Evaluating Learning Outcomes in Citizen Science*. Ithaca, NY: Cornell Lab of Ornithology.

Pickard, G., Pan, W., Rahwan, I., Cebrian, M., Crane, R., Madan, A., & Pentland, A. (2011). Time-Critical Social Mobilization. *Science*, *334*(6055), 509–512. doi:10.1126/science.1205869 PMID:22034432

Pimentel, D. (Ed.). (2011). *Biological invasions: economic and environmental costs of alien plant, animal, and microbe species*. CRC Press. doi:10.1201/b10938

Pimm, S. L., Alibhai, S., Bergl, R., Dehgan, A., Giri, C., Jewell, Z., & Loarie, S. et al. (2015). Emerging technologies to conserve biodiversity. *Trends in Ecology & Evolution*, *30*(11), 685–696. doi:10.1016/j.tree.2015.08.008 PMID:26437636

Pipino, L. L., Lee, Y. W., & Wang, R. Y. (2002). Data quality assessment. *Communications of the ACM*, *45*(4), 211–218. doi:10.1145/505248.506010

Plengsaeng, B., Wehn, U., & van der Zaag, P. (2014). Data-sharing bottlenecks in transboundary integrated water resources management: A case study of the Mekong River Commissions procedures for data sharing in the Thai context. *Water International*, *39*(7), 933–951. doi:10.1080/02508060.2015.981783

Pocock, M. J. O., Chapman, D. S., Sheppard, L. J., & Roy, H. E. (2014). A Strategic Framework to Support the Implementation of Citizen Science for Environmental Monitoring. Final Report to SEPA. Wallingford, UK: Centre for Ecology & Hydrology. Retrieved from http://www.ceh.ac.uk/products/publications/documents/hp1114final_5_complete.pdf

Pocock, M. J. O., Chapman, D. S., Sheppard, L. J., & Roy, H. E. (2014). *Choosing and using citizen science: A guide to when and how to use citizen science to monitor biodiversity and the environment*. Centre for Ecology & Hydrology. Retrieved January 10, 2016, from http://www.ceh.ac.uk/citizen-science-best-practice-guide

Pocock, M. J. O., Chapman, D. S., Sheppard, L. J., & Roy, H. E. (2014). *Choosing and Using Citizen Science: a guide to when and how to use citizen science to monitor biodiversity and the environment*. London: Centre for Ecology & Hydrology.

Pocock, M. J. O., Roy, H. E., Preston, C. D., & Roy, D. B. (2015). The Biological Records Centre: A pioneer of citizen science. *Biological Journal of the Linnean Society. Linnean Society of London*, *115*(3), 475–493. doi:10.1111/bij.12548

Pollock, R. M., & Whitelaw, G. S. (2005). Community based monitoring in support of local sustainability. *Local Environment*, *10*(3), 211–228. doi:10.1080/13549839.2005.9684248

Ponciano, L., Brasileiro, F., Simpson, R., & Smith, A. (2014). Volunteers' Engagement in Human Computation Astronomy Projects. *Computing in Science Engineering*.

Pont, B., Nusche, D., & Moorman, H. (2008). *Improving School Leadership: Policy and Practice*. OECD. Retrieved from http://www.oecd.org/edu/school/44374889.pdf

POSTnote. (2014). *Environmental Citizen Science*. POSTnote number 476. Houses of Parliament. Retrieved from http://researchbriefings.parliament.uk/ResearchBriefing/Summary/POST-PN-476

Preece, J. (2002). Supporting Community as Building Social Capital. *Communications of the ACM*, *45*(4), 37–39. doi:10.1145/505248.505269

Prestopnik, N. R., & Crowston, K. (2012). Citizen science system assemblages: understanding the technologies that support crowdsourced science. In *Proceedings of the 2012 iConference* (pp. 168-176). ACM. doi:10.1145/2132176.2132198

Pretty, J. (2003). Social capital and the collective management of resources. *Science*, *302*(5652), 142–148. doi:10.1126/science.1090847 PMID:14671287

Price, C. A., & Lee, H. S. (2013). Changes in participants scientific attitudes and epistemological beliefs during an astronomical citizen science project. *Journal of Research in Science Teaching, 50*(7), 773–801. doi:10.1002/tea.21090

Primack, R. B., Higuchi, H., & Miller-Rushing, A. J. (2009). The impact of climate change on cherry trees and other species in Japan. *Biological Conservation, 142*(9), 1943–1949. doi:10.1016/j.biocon.2009.03.016

Process, B., & Communiqué, D. L. (2007). *Towards the European Higher Education Area: responding to challenges in a globalised world.* Retrieved from http://www.ehea.info/Uploads/Declarations/London_Communique18May2007.pdf

Pycsw. (n.d.). Retrieved from: http://pycsw.org/

Quercia, D., Schifanella, R., Aiello, L. M., & McLean, K. (2015). *Smelly maps: the digital life of urban smellscapes.* arXiv preprint arXiv:1505.06851

Quigley, K., & Roy, J. (2012). Cyber-Security and Risk Management in an Interoperable World An Examination of Governmental Action in North America. *Social Science Computer Review, 30*(1), 83–94. doi:10.1177/0894439310392197

Quinn, A. J., & Bederson, B. B. (2009). *A Taxonomy of Distributed Human Computation.* University of Maryland.

Raby, P. (2001). *Alfred Russel Wallace: A Life.* Chatto & Windus.

Raddick, M. J., Bracey, G., Carney, K., Gyuk, G., Borne, K., Wallin, J., & Jacoby, S. (2009). *Citizen science: status and research directions for the coming decade.* Astro2010: The Astronomy and Astrophysics Decadal Survey, Position Papers, no. 46.

Raddick, M. J., Bracey, G., Carney, K., Gyuk, G., Borne, K., Wallin, J., & Jacoby, S. (2010). Citizen Science: Status and Research Directions for the Coming Decade. In astro2010: The Astronomy and Astrophysics Decadal Survey.

Raddick, M. J., Bracey, G., Gay, P. L., Lintott, C. J., Murray, P., Schawinski, K., & Vandenberg, J. et al. (2010). Galaxy Zoo: Exploring the motivations of citizen science volunteers. *Astronomy Education Review, 9*(1), 18. doi:10.3847/AER2009036

Rademacher, P. (2005). *HousingMaps.* Retrieved from: http://www.housingmaps.com/

Ramsey, P. (2007). *PostGIS case studies.* Retrieved from: http://www.refractions.net/expertise/whitepapers/postgis-case-studies/

Ratcliff, J. (2008). *The Transit of Venus Enterprise in Victorian Britain.* London: Pickering & Chatto.

Raven, C. E. (1942). *John Ray, Naturalist: his life and work.* Cambridge, UK: Cambridge University Press.

Reed, M., Evely, A. C., Cundill, G., Fazey, I. R. A., Glass, J., Laing, A., et al. (2010). What is social learning? *Ecology and Society.* Retrieved April 25, 2016 from https://research-repository.st-andrews.ac.uk/bitstream/handle/10023/1624/Fazey2010EcologySociety15WhatIsSocialLearning.pdf?sequence=1&isAllowed=y

Reed, S. L., & Lenat, D. B. (2002). Mapping ontologies into Cyc. In *AAAI 2002 Conference Workshop on Ontologies For The Semantic Web* (pp. 1-6).

Reed, J., Raddick, M. J., Lardner, A., & Carney, K. (2013) An Exploratory Factor Analysis of Motivations for Participating in Zooniverse, a Collection of Virtual Citizen Science Projects. In *Proceedings of the Hawaii International Conference on Systems Sciences* (pp. 610–619). doi:10.1109/HICSS.2013.85

Reed, M. S. (2008). Stakeholder participation for environmental management: A literature review. *Biological Conservation, 141*(10), 2417–2431. doi:10.1016/j.biocon.2008.07.014

Reid, C., Marshall, J., Logan, D., Kleine, D., & Dean, A. (2012). *Coral Reefs and Climate Change. The guide for education and awareness. Published by CoralWatch*. The University of Queensland.

Riddell, K. (n.d.). Retrieved April 16, 2016, from http://www.wcmt.org.uk/sites/default/files/report-documents/Riddell%20K%20Report%202014.pdf

Riessman, C. K. (1993). *Narrative Analysis*. London: Sage.

Robinson, L. (2014): *Presentation at ECSA General Assembly April 2014*. Retrieved January 10, 2016, from http://ecsa.biodiv.naturkundemuseum-berlin.de/sites/ecsa.biodiv.naturkundemuseum-berlin.de/files/ECSA-GA-2014-04-14-Robinson.pdf

Robinson, L. D., Tweddle, J., Postles, M. C., West, S. E., & Sewell, J. (2013). *Guide to Running a BioBlitz 2.0*. Retrieved May 30, 2016, from http://www.nhm.ac.uk/content/dam/nhmwww/take-part/Citizenscience/bioblitz-guide.pdf

Robinson, J. P., DiMaggio, P., & Hargittai, E. (2003). New social survey perspectives on the digital divide. *IT & Society*, *1*(5), 1–22.

Robinson, L. D., Tweddle, J. C., Postles, M. C., West, S. E., & Sewell, J. (2013). *Guide to running a BioBlitz*. London: Natural History Museum, Bristol Natural History Consortium, Stockholm Environment Institute York and Marine Biological Association.

Robson, C., Hearst, M., Kau, C., & Pierce, J. (2013). Comparing the use of social networking and traditional media channels for promoting citizen science. In *Proceedings of the 2013 conference on Computer Supported Cooperative Work Companion* (pp. 1463-1468). doi:10.1145/2441776.2441941

Rocha Menocal, A., & Sharma, B. (2008). *Joint evaluation of citizens' voice and accountability: synthesis report*. London: DFID.

Rodriguez, M. A. (2012). *Proyectos de Servicio Comunitario en Facultad de Ciencias Económicas y Sociales de la UCV*. Comunicación Personal.

Roger, E., & Klistorner, S. (2016). BioBlitzes help science communicators engage local communities in environmental research. *Journal of Science Communication*, *15*(3), 1–18.

Rotman, D., Hammock, J., Preece, J., Hansen, D., Boston, C., Bowser, A., & He, Y. (2014). Motivations affecting initial and long-term participation in citizen science projects in three countries. In iConference 2014 Proceedings. London: ACM.

Rotman, D., Hammock, J., Preece, J., Hansen, D., Boston, C., Bowser, A., & He, Y. (2014). Motivations affecting initial and long-term participation in citizen science projects in three countries. In Proceedings of iConference (pp. 110–124). http://doi.org/ doi:<ALIGNMENT.qj></ALIGNMENT>10.9776/14054

Rotman, D., Preece, J., Hammock, J., Procita, K., Hansen, D., Parr, C., & Jacobs, D. et al. (2012). Dynamic changes in motivation in collaborative citizen-science projects. In *Proceedings of the ACM 2012 conference on Computer Supported Cooperative Work*. London: ACM doi:10.1145/2145204.2145238

Roy, H. (2012, November 23). *Review highlights role of citizen science projects*. Retrieved June 1, 2016, from http://www.bbc.com/news/science-environment-20445296

Roy, H. E., Pocock, M. J. O., Preston, C. D., Roy, D., Savage, J., Tweddle, J. C., & Robinson, L. D. (2012). *Understanding Citizen Science and Environmental Monitoring*. Final Report on behalf of UK-EOF. NERC Centre for Ecology & Hydrology and Natural History Museum. Available at: http://www.ceh.ac.uk/news/news_archive/documents/understandingcitizenscienceenvironmental monitoring_report_final.pdf

Roy, H. E., Pocock, M. J. O., Preston, C. D., Roy, D. B., Savage, J., Tweddle, J. C., & Robinson, L. D. (2012). *Understanding Citizen Science & Environmental Monitoring*. London: NERC Centre for Ecology & Hydrology and Natural History Museum.

Roy, H., Pocock, M. J. O., Preston, C. D., Roy, D. B., Savage, J., Tweeddle, J. C., & Robinson, L. D. (2012). *Understanding citizen science & environmental monitoring*. NERC Centre for Ecology & Hydrology and Natural History Museum.

Russo, J. E., Schoemaker, P. J., & Russo, E. J. (1990). *Decision traps: Ten barriers to brilliant decision-making and how to overcome them*. New York, NY: Simon & Schuster.

Ryan, R. L., Kaplan, R., & Grese, R. E. (2001). Predicting Volunteer Commitment in Environmental Stewardship Programmes. *Journal of Environmental Planning and Management*, *44*(5), 629–648. doi:10.1080/09640560120079948

Sabini, J. (1995). *Social Psychology* (2nd ed.). New York: W. W. Norton & Company, Inc.

Sagl, G., Delmelle, E., & Delmelle, E. (2014). Mapping Collective Human Activity in an Urban Environment based on Mobile Phone Data. *Cartography and Geographic Information Science*, *41*(3), 272–285. doi:10.1080/15230406.2014.888958

Salonen, A., & Åhlberg, M. (2013). Towards sustainable society – From materialism to post-materialism. *International Journal of Sustainable Society*, *5*(4), 374–393. doi:10.1504/IJSSOC.2013.056846

Saltmarsh, J. (2005). The Civic Promise of Service Learning. *Liberal Education*, *91*(2), 50–55.

Sampson, S. D. (2012). The topophilia hypothesis: Ecopsychology meets evolutionary psychology. In P.H. Kahn & P.H. Hasbach (Eds.), Ecopsychology: Science, totems, and the technological species (pp. 23-53). Cambridge, MA: MIT Press.

Samy, G., Chavan, V., Ariño, A. H., Otegui, J. O., Hobern, D., Sood, R., & Robles, E. (2013). Content assessment of the primary biodiversity data published through GBIF network: Status, challenges and potentials. *Biodiversity Informatics*, *8*(2), 1546–9735. doi:10.17161/bi.v8i2.4124

Sanderson, I. (2009). Intelligent Policy Making for a Complex World: Pragmatism, Evidence and Learning. *Political Studies*, *57*(4), 699–719. doi:10.1111/j.1467-9248.2009.00791.x

Scassa, T., & Chung, H. (2015). *Typology of citizen science projects from an intellectual property perspective*. Washington, DC: Woodrow Wilson International Center for Scholars —Commons Lab.

Scassa, T., & Sattler, A. (2011). Location-based services and privacy. *Canadian Journal of Law and Technology*, *9*(1 & 2).

Science Communication Unit, University of the West of England, Bristol. (2013). *Science for Environment Policy In-depth Report: Environmental Citizen Science*. Report produced for the European Commission DG Environment. Retrieved January 10, 2016, from http://ec.europa.eu/environment/integration/research/newsalert/pdf/IR9_en.pdf

Science Communication Unit, University of the West of England, Bristol. (2013). *Science for Environment Policy In-depth Report: Environmental Citizen Science*. Retrieved June 1, 2016, from http://ec.europa.eu/science-environment-policy

Scientific American. (2013, September 3). *Citizen Science: World Water Monitoring Challenge*. Retrieved March 10, 2016, from http://www.scientificamerican.com/citizen-science/world-water-monitoring-challenge

SciStarter. (2016). Retrieved from http://scistarter.com/page/Citizen%20Science.html

Sclove, R. (2010). *Reinventing technology assessment: A 21st Century model*. Washington, DC: Woodrow Wilson International Center for Scholars.

SCU. (2013). *Science Communication Unit, University of the West of England, Bristol (2013). Science for Environment Policy In-depth Report: Environmental Citizen Science*. Report produced for the European Commission DG Environment. Available at: http://ec.europa.eu/environment/integration/research/newsalert/pdf/IR9_en.pdf

Scyphers, S. B., Powers, S. P., Akins, J. L., Drymon, J. M., Martin, C. W., & Schobernd, Z. H. et al. (2015). The Role of Citizens in Detecting and Responding to a Rapid Marine Invasion. *Conservation Letters*, *8*(4), 242–250. doi:10.1111/conl.12127

Senge, P. (2006). *The Fifth Discipline. The Art and Practice of the Learning Organization*. New York: Currency Doubleday.

Serrano Sanz, F., Holocher-Ertl, T., Kieslinger, B., Sanz Garcia, F., & Silva, C. (2014). *White Paper on Citizen Science for Europe*. Socientize Consortium. Retrieved January 10, 2016, from http://www.socientize.eu/?q=eu/content/download-socientize-white-paper

Serrano Sanz, F., Holocher-Ertl, T., Kieslinger, B., Sanz García, F., & Silva, C. G. (2014). *White paper on citizen science for Europe*. European Commission. Retrieved from http://ec.europa.eu/newsroom/dae/document.cfm?doc_id=6913

Shaxson, L., Bielak, A., Ahmed, I., Brien, D., Conant, B., Fisher, C., & Phipps, D. (2012). Expanding our understanding of K*(KT, KE, KTT, KMb, KB, KM, etc.). A concept paper emerging from the K* conference held in UNU-INWEH, Hamilton, ON.

Shelton, A. M. (2013). *The accuracy of water quality monitoring data: A comparison between citizen scientists and professionals*. (Unpublished master's thesis). Saint Mary's University, Halifax, Nova Scotia, Canada.

Sheppard, S. A., & Terveen, L. (2011). Quality is a Verb: The Operationalization of Data Quality in a Citizen Science Community. In *Proceedings of the 7th International Symposium on Wikis and Open Collaboration*. doi:10.1145/2038558.2038565

Shirk, J. L., Ballard, H. L., Wilderman, C. C., Phillips, T., Wiggins, A., Jordan, R., & Bonney, R. et al. (2012). Public participation in scientific research: A framework for deliberate design. *Ecology and Society*, *17*(2), 29. doi:10.5751/ES-04705-170229

Sieber, R., & Haklay, M. (2015). The epistemology(s) of volunteered geographic information: A critique. *Geography and Environment*, *2*(2), 122–136. doi:10.1002/geo2.10

Silvertown, J., Buesching, C. D., Jacobson, S. K., & Rebelo, T. (2013). Citizen science and nature conservation. In Key Topics in Conservation Biology (vol. 2, pp. 127-142). John Wiley & Sons, Ltd. doi:10.1002/9781118520178.ch8

Silvertown, J. (2009). A new dawn for citizen science. *Trends in Ecology & Evolution*, *24*(9), 467–471. doi:10.1016/j.tree.2009.03.017 PMID:19586682

Simonis, I. (2015). *Citizen Observatories: A Standards Based Architecture*. EGU General Assembly 2015, Vienna, Austria.

Simonis, I. (2015). Standardised Information Models to Optimise Exchange, Reusability and Comparability of Citizen Science Data. A Specialised Approach. In *Proceedings of the Workshop "Environmental Infrastructures and Platforms 2015 - Infrastructures and Platforms for Environmental Crowd Sensing and Big Data" for the European Citizen Science Association General Assembly 2015* (ECSA GA'2015).

Simpson, R. (2013). *Explainer: what is citizen science?* The Conversation. Retrieved from http://theconversation.com/explainer-what-is-citizen-science-16487

Sklar, E., & Richards, D. (2010). Agent-based systems for human learners. *The Knowledge Engineering Review*, *25*(02), 111–135. doi:10.1017/S0269888910000044

Slavin, R. E. (2010). Co-*operative* learning: what makes group-work work? In H. Dumont (Ed.), *The Nature of Learning. Using research to inspire practice*. Paris: OECD. doi:10.1787/9789264086487-9-en

Snow, J. (1854). *On the Mode of Communication of Cholera* (2nd ed.). Retrieved from http://matrix.msu.edu/~johnsnow/images/online_companion/chapter_images/fig12-5.jpg

Soares, M. D. (2011). *Employing Citizen Science to Label Polygons of Segmented Images*. (doctoral dissertation), National Institute For Space Research.

Socientize. (2014). *White Paper on Citizen Science for Europe*. European Commission.

Socientize. (n.d.). *Socientize*. Retrieved March 6, 2016, from http://www.socientize.eu/?q=eu/content/socientize-0

Solesbury, W. (2001). *ESRC UK Centre for Evidence Based Policy and Practice: Working Paper 1*. Retrieved from http://eur-lex.europa.eu/legal-content/en/TXT/?uri=celex:52001DC0428

Sowa, J. F. (1999). *Knowledge representation: logical, philosophical, and computational foundations*. Brooks/Cole.

SparklingScience . (2016). Retrieved May 09, 2016, from http://www.sparklingscience.at/en

Spindler, G. D. (1959). *Learning in Culture: Anthropological Perspective*. Educational.

Stack, B., & Donkin, M. (2013). Teaching botanical identification to adults: Experiences of the UK participatory science project Open Air Laboratories. *Journal of Biological Education*, *47*(2), 104–110. doi:10.1080/00219266.2013.764341

Stafford, R., Hart, A. G., Collins, L., Kirkhope, C. L., Williams, R. L., Rees, S. G., & Goodenough, A. E. et al. (2010). Eu-Social Science: The Role of Internet Social Networks in the Collection of Bee Biodiversity Data. *PLoS ONE*, *5*(12), e14381. doi:10.1371/journal.pone.0014381 PMID:21179423

Stallman, R. M., & Gay, J. (2009). *Free Software, Free Society: Selected Essays of Richard M. Stallman*. Paramount, CA: CreateSpace.

Steiniger, S., & Bocher, E. (2009). An overview on current free and open source desktop GIS developments. *International Journal of Geographical Information Science*, *23*(10), 1345–1370. doi:10.1080/13658810802634956

Steiniger, S., & Hay, G. J. (2009). Free and open source geographic information tools for landscape ecology. *Ecological Informatics*, *4*(4), 183–195. doi:10.1016/j.ecoinf.2009.07.004

Steiniger, S., & Hunter, A. J. S. (2010). Teaching GIScience with Free and Open Source Software? A first assessment. In *6th International Conference of GIScience*.

Steyaert, J. (2002). Inequality and the digital divide: Myths and realities. In S. F. Hick & J. G. McNutt (Eds.), *Advocacy, activism and the Internet* (pp. 199–211). Chicago: Lyceum Press.

Stöckli, B., Wiesmann, U., & Lys, J. A. (2012). A guide for transboundary research partnerships: 11 principles. Berne: Swiss Commission for Research Partnership with Developing Countries (KFPE).

Storksdieck, M., Shirk, J. L., Cappadonna, J. L., Domroese, M., Göbel, C., Haklay, M., & Vohland, K. (2016). Associations for citizen science: Regional knowledge; global collaboration. *Citizen Science: Theory and Practice*, *2*(1), 1.

Subkowski, P. (2006). On the psychodynamics of collecting. *The International Journal of Psycho-Analysis*, *87*(2), 383–401. doi:10.1516/4UMF-YF9G-FVFR-JM09 PMID:16581582

Sui, D., Elwood, S., & Goodchild, M. (Eds.). (2013). *Crowdsourcing geographic knowledge: volunteered geographic information (VGI) in theory and practice*. Springer Science & Business Media. doi:10.1007/978-94-007-4587-2

Sullivan, B. L., Aycrigg, J. L., Barry, J. H., Bonney, R. E., Bruns, N., Cooper, C. B., & Kelling, S. et al. (2014). The eBird enterprise: An integrated approach to development and application of citizen science. *Biological Conservation, 169*, 31–40. doi:10.1016/j.biocon.2013.11.003

Sullivan, B., Wood, C., Iliff, M., Bonney, R., Fink, D., & Kelling, S. (2009). eBird: A citizen-based bird observation network in the biological sciences. *Biological Conservation, 142*(10), 2282–2292. doi:10.1016/j.biocon.2009.05.006

Sultana, P., & Abeyasekera, S. (2008). Effectiveness of participatory planning for community management of fisheries in Bangladesh. *Journal of Environmental Management, 86*(1), 201–213. doi:10.1016/j.jenvman.2006.12.027 PMID:17291678

Tennessee Water Resources Research Center, University of Tennessee. (2004). *A comparative analysis of water quality monitoring programs in the southeast: lessons for Tennessee*. Retrieved June 2, 2016, from http://isse.utk.edu/wrrc/programsprojects/pdfs/mainbook.pdf

Terhune, K. (2013). Retired Justice Sandra Day O'Connor, in Boise, laments 'alarming degree of public ignorance'. *Idaho Statesman*. Retrieved from http://www.mcclatchydc.com/news/nation-world/national/article24755302.html

The Royal Society. (n.d.). *Knowledge, Networks and Nations: Global scientific collaboration in the 21st century*. Author.

The WNYC Data News Team. (2015). *Know Your Evacuation Zone*. Retrieved from: http://www.wnyc.org/story/know-your-evacuation-zone/

Theobald, E. J., Ettinger, A. K., Burgess, H. K., DeBey, L. B., Schmidt, N. R., Froehlich, H. E., & Parrish, J. K. et al. (2015). Global change and local solutions: Tapping the unrealized potential of citizen science for biodiversity research. *Biological Conservation, 181*, 236–244. doi:10.1016/j.biocon.2014.10.021

Thompson, S. (2010, September 24). *Citizen Science Toolkit Project Evaluation*. Retrieved March 6, 2016, from http://www.informalscience.org/citizen-science-toolkit-project

Tian, H., Stige, L. C., Cazelles, B., Kausrud, K. L., Svarverud, R., Stenseth, N. C., & Zhang, Z. (2011). Reconstruction of a 1910-y-long locust series reveals consistent associations with climate fluctuations in China. *Proceedings of the National Academy of Sciences of the United States of America, 108*(35), 14521–14526. doi:10.1073/pnas.1100189108 PMID:21876131

Tinati, R., Van Kleek, M., Simperl, E., Luczak-Rösch, M., Simpson, R., & Shadbolt, N. (2015). *Designing for citizen data analysis: a cross-sectional case study of a multi-domain citizen science platform*. Paper presented at the Association for Computing Machinery's Human Computer Interaction Conference on Human Factors in Computing Systems, Seoul, South Korea. doi:10.1145/2702123.2702420

Torney-Purta, J., Lehmann, R., Oswald, H., & Schulz, W. (2001). Citizenship and education in twenty-eight countries: Civic knowledge and engagement at age fourteen. *Herengracht, 487*, 1017.

Tsai, W., & Ghoshal, S. (1998). Social capital and value creation: The role of intrafirm networks. *Academy of Management Journal, 41*(4), 464–476. doi:10.2307/257085

Tuan, Y. F. (1974). *Topophilia*. Englewood Cliffs, NJ: Prentice-Hall.

Tulloch, D. L. (2008). Is VGI Participation? From Vernal Pools to Video Games. *GeoJournal, 72*(3–4), 161–171. doi:10.1007/s10708-008-9185-1

Tweddle, J. C., Robinson, L. D., Pocock, M. J. O., & Roy, H. E. (2012). *Guide to citizen Science: Developing, implementing and evaluating citizen science to study biodiversity and the environment in the UK*. Natural History Museum and NERC Centre for Ecology & Hydrology for UK-EOF.

Tweddle, J.C., Robinson, L.D., Pocock, M.J.O., & Roy, H.E. (2012). *Guide to citizen science: developing, implementing and evaluating citizen science to study biodiversity and the environment in the UK*. Natural History Museum and NERC Centre for Ecology & Hydrology for UK-EOF.

Tweddle, J. C., Robinson, L. D., Pocock, M. J. O., & Roy, H. E. (2012). *Guide to Citizen Science: Developing, Implementing and Evaluating Citizen Science to Study Biodiversity and the Environment in the UK*. London, UK: Natural History Museum and Centre for Ecology & Hydrology for UK-EOF.

Tweddle, J. C., Robinson, L. D., Pocock, M. J. O., & Roy, H. E. (2012). *Guide to citizen science: developing, implementing and evaluating citizen science to study biodiversity and the environment in the UK*. London: Natural History Museum and NERC Centre for Ecology & Hydrology for UK-EOF.

Tyack, D., & Tobin, W. (1994). The grammar of Schooling: Why has it been so hard to change? *American Educational Research Journal*, *31*(3), 453–480. doi:10.3102/00028312031003453

Type III Error. (n.d.). In *Wikipedia*. Retrieved April 16, 2016, from https://en.wikipedia.org/wiki/Type_III_error

Uchoa, A. P., Esteves, M. G. P., & Souza, J. M. (2013). Mix4Crowds - Toward a framework to design crowd collaboration with science. In *2013 IEEE 17th International Conference on Computer Supported Cooperative Work in Design (CSCWD)*.

Ugalde, L., España, L.P., Lacruz, T., De Viana, M., González, L., Luengo, N., & Ponce, M. (2004). *Detrás de la pobreza*. Universidad Católica Andrés Bello.

Uhrner, U., Grosso, G., Romain, A. C., Hutsemekers, V., Delva, J., Kunz, W., . . . Ledent, Ph. (2013). Development of an Environmental Information System for Odour using Citizen and Technology Innovative Sensors and Advanced Modelling. *CEUR Workshop Proceedings*.

Ulrich, C., & Nadkarni, N. M. (2009). Sustainability research and practices in enforced residential institutions: Collaborations of ecologists and prisoners. *Environment, Development and Sustainability*, *11*(4), 815–832. doi:10.1007/s10668-008-9145-4

Umbrich, J., Neumaier, S., & Polleres, A. (2015). Quality Assessment and Evolution of Open Data Portals. In *Future Internet of Things and Cloud (FiCloud), 2015 3rd International Conference on* (pp. 404-411). IEEE. doi:10.1109/FiCloud.2015.82

UNEP. (2009). *Environmental governance*. Retrieved May 29, 2016, from http://www.unep.org/pdf/brochures/EnvironmentalGovernance.pdf

UNESCO, Leeds-Hurwitz, W., & Stenou, K. (2013). *Intercultural Competences: Conceptual and Operational Framework*. UNESCO.

UNESCO. (2015). *Rethinking Education - Towards a global common good?* París: UNESCO.

United States Environmental Protection Agency. (2015). *Nonpoint source: Volunteer monitoring*. Retrieved from www.epa.gov/polluted-runoff-nonpoint-source-pollution/nonpoint-source-volunteer-monitoring

United States Environmental Protection Agency. (2016a). *EPA's Budget and Spending*. Retrieved from www.epa.gov/planandbudget

United States Environmental Protection Agency. (2016b). *STOrage and RETrieval and water quality eXchange*. Retrieved from http http://www.epa.gov/water data/storage-and-RETrival-and-water-quality-exchange

United States Environmental Protection Agency. (2016c). *Nutrient policy and data: Region 5*. Retrieved from https://www.epa.gov/nutrient-policy-data/region-5

United States Geological Survey. (2016). *Monitoring our Rivers and Streams*. Retrieved from http://pubs.usgs.gov/fs/fs-077-02_ver1/

United States Geological Survey. (2016, March 19). *WaterWatch*. Retrieved from http://waterwatch.usgs.gov

University of Wisconsin Extension-Lakes. (2016). *Citizen Lake Monitoring Network*. Retrieved at http://www.uwsp.edu/cnr-ap/UWEXLakes/Pages/programs/clmn/default.aspx

Vallaeys, F. (2003). Orientaciones para la enseñanza de la ética, el capital social y el desarrollo en las universidades latinoamericanas. *Revista venezolana de Gerencia, 21,* 38-73.

Vallaeys, F. (2007). *La Responsabilidad Social Universitaria: ¿Cómo entenderla para querer practicarla?* Retrieved April 16, 2016 from: https://www.academia.edu/21125213/LA_RESPONSABILIDAD_SOCIAL_UNIVERSITARIA_C%C3%93MO_ENTENDERLA_PARA_QUERERLA_Y_PRACTICARLA

Van Den Berg, H. A., Dann, S. L., & Dirk, J. M. (2009). Motivations of adults for non-formal conservation education and volunteerism: Implications for Programming. *Applied Environmental Education and Communication*, *8*(1), 6–17. doi:10.1080/15330150902847328

van Dijk, J. A. G. M. (2005). *The deepening divide: Inequality in the information society*. Thousand Oaks, CA: Sage.

van Dijk, J. A., & Hacker, K. (2003). The digital divide as a complex and dynamic phenomenon. *The Information Society*, *19*(4), 315–326. doi:10.1080/01972240309487

Van Rijsoort, J., & Jinfeng, Z. (2005). Participatory resource monitoring as a means for promoting social change in Yunnan, China. *Biodiversity and Conservation*, *14*(11), 2543–2573. doi:10.1007/s10531-005-8377-y

Venkatraman, V. (2010). Conventions of scientific authorship. *Science*. Retrieved May 18, 2916, from http://www.sciencemag.org/careers/2010/04/conventions-scientific-authorship

Vetter, J. (2011). Introduction: Lay participation in the history of scientific observations. *Science in Context*, *24*, 259–280. doi:10.1017/S0269889711000093 PMID:21797076

Vigoda, E. (2002). From responsiveness to collaboration: Governance, citizens, and the next generation of public administration. *Public Administration Review*, *62*(5), 527–540. doi:10.1111/1540-6210.00235

Villa, S. A., & Villa, L. O. (2007). El aprendizaje basado en competencias y el desarrollo de la dimensión social en las universidades. *Educar*, *40*, 15–48.

Vyver, J. (2014). Maratus harrisi: the tiny peacock spider discovered by Canberra man Stuart Harris in Namadgi National Park. *ABC News, Canberra, Australia.* Retrieved from http://www.abc.net.au/news/2014-08-14/discovering-maratus-harrisi/5670424

Waissbluth, M. (2003). La insularidad en la gestión pública latinoamericana. *Revista del CLAD Reforma y Democracia*, *27*, 75–88.

Wals, A. E., Brody, M., Dillon, J., & Stevenson, R. B. (2014). Convergence between science and environmental education. *Science*, *344*(6184), 583–584. doi:10.1126/science.1250515 PMID:24812386

Walter-Herrmann, J., & Büching, C. (Eds.). (2013). *FabLab: Of machines, makers and inventors*. Bielefeld, Germany: Transcript; doi:10.14361/transcript.9783839423820

Warschauer, M. (2003). *Technology and social inclusion: Rethinking the digital divide*. Cambridge, MA: MIT Press.

Waxman, O. B. (2016, April 4). Pigeons That Track Pollution. *Time,* 23.

Webb, N. (2008). Co-operative Learning. In T.L. Good (Ed.), 21st Century Education: A Reference Handbook. Sage.

WEF. (2015). *Global Risks 2015, Insight Report* (10th ed.). World Economic Forum.

WEF. (2016). *Global Risks 2016, Insight Report* (11th ed.). World Economic Forum.

Wehn de Montalvo, U. (2003a). In search of rigorous models for policy-oriented research: A behavioural approach to spatial data sharing. *Journal of the Urban and Regional Information Systems Association*, *15*(1), 19–28.

Wehn de Montalvo, U. (2003b). *Mapping the determinants of spatial data sharing*. Ashgate Pub Ltd.

Wehn, U., Rusca, M., Evers, J., & Lanfranchi, V. (2015). Participation in flood risk management and the potential of citizen observatories: A governance analysis. *Environmental Science & Policy*, *48*(0), 225–236. doi:10.1016/j.envsci.2014.12.017

Weiling, F. (1991). Historical study: Johann Gregor Mendel 1822–1884. *American Journal of Medical Genetics*, *40*(1), 1–25. doi:10.1002/ajmg.1320400103 PMID:1887835

Wenger, E. (1998). Communities of practice: Learning, meaning, and identity. New York: Cambridge. doi:10.1017/CBO9780511803932

Wenger, E., Mcdermott, R., & Snyder, W. M. (2002). *Seven Principles for Cultivating Communities of Practice*. HB-SWK Pub.

Wernand, M. R., Ceccaroni, L., Piera, J., & Zielinski, O. (2012). Crowdsourcing technologies for the monitoring of the colour, transparency and fluorescence of the sea. In Proceedings of Ocean Optics XXI (pp. 8-12).

Wernand, M. R., Ceccaroni, L., Piera, J., & Zielinski, O. (2012). Crowdsourcing technologies for the monitoring of the colour, transparency and fluorescence of the sea. *Proceedings of Ocean Optics XXI*.

Wernand, M., & van der Woerd, H. (2010). Spectral analysis of the Forel-Ule Ocean colour comparator scale. *Journal Of The European Optical Society*, *5*, 10014s. doi:10.2971/jeos.2010.10014s

Westbrooks, R. G. (2004). New approaches for early detection and rapid response to invasive plants in the United States. *Weed Technology*, *18*(sp1), 1468–1471. doi:10.1614/0890-037X(2004)018[1468:NAFEDA]2.0.CO;2

Whitelaw, G., Vaughan, H., Craig, B., & Atkinson, B. (2003). Establishing the Canadian community monitoring network. *Environmental Monitoring and Assessment*, *88*(1/3), 409–418. doi:10.1023/A:1025545813057 PMID:14570426

Wiemann, S., Meek, S., Chapman, C., Leibovici, D. G., Jackson, M., & Lars, B. (2015). *Service-based combination of quality assurance and fusion processes for the validation of crowdsourced observations*. Paper presented at AGILE 2015 conference, Lisbon, Portugal.

Wiersma, Y. F. (2010). Birding 2.0: citizen science and effective monitoring in the Web 2.0 world. *Avian Conservation and Ecology*, *5*(2), 13. Available at: http://www.ace-eco.org/vol5/iss2/art13/

Wiggins, A., & Crowston, K. (2011). From conservation to crowdsourcing: A typology of citizen science. In *System Sciences (HICSS), 2011 44th Hawaii international conference on* (pp. 1-10). IEEE. doi:10.1109/HICSS.2011.207

Wiggins, A., & Crowston, K. (2012). Goals and Tasks: Two Typologies of Citizen Science Projects. In *Proceedings of the Fourth-fifth Hawaii International Conference on Systems Sciences* (HICSS-45). doi:10.1109/HICSS.2012.295

Wiggins, A., & Crowston, K. (2010). Developing a Conceptual Model of Virtual Organizations for Citizen Science. *International Journal of Organizational Design and Engineering*, *1*(1/2), 148–162. doi:10.1504/IJODE.2010.035191

Wikipedia. (2016). *List of citizen science projects*. Retrieved May 30, 2016, from https://en.wikipedia.org/wiki/List_of_citizen_science_projects

Wikipedia. (n.d.). *List of deadliest Atlantic hurricanes*. Retrieved from: https://en.wikipedia.org/wiki/List_of_deadliest_Atlantic_hurricanes

Williams, A. C., Wallin, J. F., Yu, H., Perale, M., Carroll, H. D., Lamblin, A., & Brusuelas, J. H. (2014). A Computational Pipeline for Crowdsourced Transcriptions of Ancient Greek Papyrus Fragments. In *IEEE International Conference on Big Data*. doi:10.1109/BigData.2014.7004460

Wilson Center. (n.d.). *The Commons Lab Inventory*. Retrieved March 6, 2016, from http://wilsoncenter.org/the-commons-lab-inventory

Wisconsin Department of Natural Resources. (2016). *Citizen Lake Monitoring Network*. Retrieved from http://dnr.wi.gov/lakes/clmn/

Wooley, A. W., Chabris, C. F., Pentland, A., Hashmi, N., & Malone, T. W. (2010). Evidence for a collective intelligence factor in the performance of human groups. *Science*, *330*(6004), 686–688. doi:10.1126/science.1193147 PMID:20929725

World Commission on Environment and Development. (1987). *Our Common Future*. Oxford: Oxford University Press.

Wright, M. T., Gardner, B., Roche, B., von Unger, H., & Ainlay, C. (2010). Building an international collaboration on participatory health research. *Progress in Community Health Partnerships – Research. Education, and Action*, *4*(1), 31–36.

Xu, C., & Yang, C. (2014). Introduction to big geospatial data research. *Annals of GIS*, *20*(4), 227–232. doi:10.1080/19475683.2014.938775

Yang, C., Raskin, R., Goodchild, M., & Gahegan, M. (2010). Geospatial cyberinfrastructure: Past, present and future. *Computers, Environment and Urban Systems*, *34*(4), 264–277. doi:10.1016/j.compenvurbsys.2010.04.001

Yasuhiko, A. (2014). Safecast-Radiation Risks. *The Asia-Pacific Journal*, 11.

Zaman, J., D'Hondt, E., Boix, E. G., Philips, E., Kambona, K., & De Meuter, W. (2014). Citizen-Friendly Participatory Campaign Support. In *Proceedings of 2014 IEEE International Conference on Pervasive Computing and Communications Work in Progress (PerCom WiP'14)*. Budapest, Hungary: PerCom WiP'14.

Zhang, J., & Huang, X. (2015). Environment: China needs more monitoring apps. *Nature*, *520*(7548), 436. PubMeddoi:10.1038/520436d

Zhang, J., Chen, S., Chen, B., Du, Y., Huang, X., Pan, X., & Zhang, Q. (2013). Citizen science: Integrating scientific research, ecological conservation and public participation. *Biodiversity Science*, *21*(6), 738–749.

Zikopoulos, I. B. M. P., Eaton, C., & Zikopoulos, P. (2011). Understanding Big Data: Analytics for Enterprise Class Hadoop and Streaming Data. New York: McGraw-Hill Professional.

Zoellick, B., Nelson, S. J., & Schauffler, M. (2012). Participatory science and education: Bringing both views into focus. *Frontiers in Ecology and the Environment*, *10*(6), 291–297. doi:10.1890/110277

About the Contributors

Luigi Ceccaroni is founder and research lead at 1000001 Labs [http://www.1000001labs.org], and a member of the Knowledge Engineering and Machine Learning group (KEMLg) at the Universitat Politècnica de Catalunya - BarcelonaTech (UPC) (Spain) [http://upc.edu]. Luigi obtained a BSc degree in Environmental Sciences (first class honors) from the University of Bologna (Ravenna, Italy); an MSc degree in Information-Technology Languages and Systems from UPC; a PhD in Artificial Intelligence (first class honors), with a thesis entitled "OntoWEDSS - An Ontology-based Environmental Decision-Support System for the management of Wastewater treatment plants", from UPC in December 2001; and completed an Executive MBA at EAE Business School in March 2016. His main research interests combine: hazards forecasting in various environments; citizen science; ontologies; recommendation systems; semantics; the semantic Web; personalization; and application of artificial intelligence to healthcare and environmental sciences. In 2012-2015, Luigi was the principal investigator and coordinator of the Citclops project, an FP7 5 M€ European project about a citizens' observatory for coast and ocean monitoring. In 2010-2014, he has worked as a senior member of research staff at Eurecat (BDigital) (Spain) [http://eurecat.org/en/], where he founded and coordinated the Personalized Computational Medicine research group and wrote the winning proposal for the Synergy-COPD project (2011-2014), an FP7 5 M€ European project about modeling and simulation for systems medicine, of which he was the principal investigator and coordinator. In 2011 he wrote the winning proposal for the BackHome project (2012-2015), an FP7 4 M€ European project about brain-neural computer interfaces. In 2003-2011, Luigi was a senior member of research staff of the Software Department (LSI) at UPC, where he developed an advanced technology demonstration network for education and cultural applications in Europe and Latin America; and led the activities in the Laboranova European IP project, which developed a collaboration environment for strategic innovation. In 2003-2010, he was director of research at TMT Factory (Barcelona, Spain). In TMT Factory, he developed interactive television for blind people; a service of personalized and accessible orientation for tourism; and combined coordination and organization mechanisms with model driven design to create new software-development tools and services. In 2005-2009, Luigi was adjunct professor of Artificial Intelligence at UPC. In 2001-2003, he has worked as a member of research staff at Fujitsu Laboratories of America, in the their Silicon Valley labs. He took part in the evaluation of European research proposals submitted in response to the call FP7-ENV-2007-1 for Theme 6 "Environment (including climate change)", specifically on forecasting methods and assessment tools for sustainable development taking into account differing scales of observation. Luigi is the author of more than 70 publications peer-reviewed and managed a total of more than 14 M€ as lead researcher in the period 2007-2015. Since 2014, he is a member of the Board of Directors of the European Citizen

Science Association. In 2014-2016 he was a member of the Board of Directors and the vice-president of Greenpeace Spain.

Jaume Piera is a Tenured Scientist of the Institute of Marine Sciences (ICM-CSIC) and Scientific Collaborator at the Ecological and Forestry Applications Research Centre (CREAF). He is Ph.D. in Environmental Sciences (University of Girona), M.S. in Biology (University of Barcelona) and B.S. in Telecommunications Engineering (Technical University of Catalonia). His current research focuses on designing and evaluating scientific and technological infrastructures for environmental monitoring, particularly those based on citizen science approaches (known as Citizen Observatories). His research interests include designing and evaluating technologies to facilitate citizen participation and engagement in scientific research, promoting the use of local and traditional knowledge in conservation and education programs and developing new methods for analyzing citizen science data. As examples of his current research, he was leading the design if Do-It-Yourself instruments for water quality monitoring in the European project CITCLOPS (Citizens' Observatory for Coast and Ocean Optical Monitoring), and now is collaborating in the design of new technologies for the MOSQUITO ALERT project and the environmental citizen science platform NATUSFERA. Since 2014, he is a member of the Steering Committee of the European Citizen Science Association, chairing the Working Group "Projects, Data, Tools and Technology".

* * *

Marcio Antelio is an Associate Professor at the Federal Center for Technological Education Celso Suckow da Fonseca and is currently a PhD candidate at the Federal University of Rio de Janeiro. His research interests include Data Quality, Crowdsourcing, Citizen Science and Big Data.

Enrico Bocciolesi is an Assistant Professor of General and Social Pedagogy at eCampus University (Como-Italy), PhD. Vice director of CERISUS - Center for International Research in the Humans and Social Sciences. Past lecturer of Pedagogy of Socials and Interculturals Phenomena, past collaborator with the chair of Philosophy of Education, at the Faculty of Education of the University of Perugia (Italy). Past collaborator at Complutense University of Madrid and at European University of Madrid. Member of research group 125 of the UNED - Spain and the University of Florence at CSL - Communications Strategies Lab. Expert invited speaker in Brazil, Colombia, Hong Kong, Italy, Mexico, Peru, Republic of Latvia and Spain. First official Italian translator of the Learning Styles, CHAEA-Alonso and Gallego, is already in the Board of Directors of SRI - Strategic Research Initiative. Editor-in-Chief of the peer-reviewed publication "VEGA Journal".

Anne Bowser is a Senior Program Associate with the Science and Technology Innovation Program (STIP) at the Woodrow Wilson International Center for Scholars. Within STIP she co-directs the Commons Lab, a program and research agenda dedicated to supporting public participation in science, technology, and policy [http://www.commonslab.wilsoncenter.org]. Major Commons Lab activities include managing citizenscience.gov with the General Services Administration (GSA), and leading an international initiative on citizen science data and metadata standardization and interoperability. Anne also leads the Wilson Center's participation in a H2020 project on exploring bilateral EU/US cooperation in Science and Technology Innovation, and supports the Wilson Center's Serious Games Initiative. In 2016 Anne received her PhD from the University of Maryland College of Information Studies. Her dissertation, "Cooperative Design, Cooperative Science: Designing Floracaching to Motivate New Citizen Science Volunteers" explored how design-based research can become tool to create better citizen science technologies while engaging new populations in civic education. In addition to further exploring the role of participatory technology design in citizen science, Anne is interested in supporting research agendas including the ethical implications of citizen science and Internet research, and the potential for further mobilizing the public to drive research and policy dialogues. She is a member of the Board of Directors of the Citizen Science Association (CSA) since 2016.

Peter Brenton, as the manager of software development and support for field data capture and citizen science with the Atlas of Living Australia, works with citizen science and other communities to translate their requirements into functional and design specifications for software development and guides the development of open source software solutions to support these communities. Peter has an operational background including 20 years as a practitioner in many aspects of natural resource management in Australia and more than 15 years as a software project manager and business analyst, specializing in browser-based business applications.

Jessica L. Cappadonna has, for the past ten years, been working as an ecologist and a science communicator. Her current research relates to engaging citizen scientists with ecological acoustics using participatory design strategies. She is also investigating the state of citizen science across Australia, and is very involved in the development of the Australian Citizen Science Association.

Darlene Cavalier is a Professor of Practice at Arizona State University's Center for Engagement and Training in Science and Society and the School for the Future of Innovation in Society. She is the founder of SciStarter and serves on the Board of the Citizen Science Association.

Colin Chapman is Senior Data Manager for the Strategic Evidence & Assessment Branch of the Welsh Government.

Caren Cooper, PhD, is an Associate Professor in Forestry and Environmental Resources at North Carolina State University in the Chancellor's Faculty Excellence Program in Leadership in Public Science, jointly appointed as Assistant Head of the Biodiversity Research Lab at the North Carolina Museum of Natural Sciences. Cooper is co-editor-in-chief of Citizen Science: Theory & Practice, a journal of the Citizen Science Association in cooperation with the European Citizen Science Association and the Australian Citizen Science Association. Cooper is founder and moderator of #CitSciChat, a Twitter discussion about citizen science, sponsored by SciStarter. She can be followed on Twitter @CoopSciScoop.

Ria Dunkley's research interests focus upon human-nature encounters in various contexts, which translates into a number of empirical projects. Firstly, she is particularly interested in how and why transitions towards sustainability are formed through environmental education (especially in informal, experiential environments) and community-led action (including through the practices of grassroots and artistic communities). Secondly, she examines how changing perceptions of nature influence the motivations of individuals and communities to participate in environmental sustainability-related initiatives. She is also interested in alternative means of knowledge creation and understanding, beyond scientific and how these influence the transmission and circulation of notions of environmental sustainability. These interests have led to a number of projects such as, Homing In: Sensing, Sense-Making and Sustainable Place-Making (an arts and social sciences collaborative network) (AHRC 2013-2014); Community Action for Climate Change: a learning review (Welsh Government 2012-13) and Building Bridges for Education for Sustainability (University of Warwick and Monash University 2012-2013). Her recent publications address community involvement in environmental sustainability initiatives, as well as the effects of environmental learning experiences. She is also currently investigating the meaning of citizen science experiences to involved individuals and the role of participatory science in sustainable development. Ria also has a research interest in the meanings of heritage tourism to both host communities and audiences in the present day. She is currently involved in a number of networks through which she explores the relationship of this interest to sustainable development (including the GW4 environmental humanities group).

Mahmud Farooque is the Associate Director of Arizona State University Consortium for Science, Policy and Outcomes (CSPO) in Washington, DC where he focuses on linking science and innovation policy to improved decision-making and better societal outcomes. Mahmud coordinates CSPO's New Tools for Science Policy Breakfast Seminar, CSPO Conversations and Science Program Managers Network. As a principal coordinator of Expert and Citizen Assessment of Science and Technology (ECAST), Mahmud co-led WWViews on Biodiversity (2012), Participatory Technology Assessment of NASA's Asteroid Initiative (2014), and WWViews on Climate and Energy (2015) projects. Mahmud is a Co-PI in the NOAA grant titled "Science Center Public Forums: Community Engagement for Environmental Literacy, Improved Resilience, and Decision-Making" and PI in the "Participatory Engagement for Energy Policy Planning and Decision-making" project funded by the U.S. Department of Energy (DOE). Since 2004, Mahmud has been working with faculty and senior administration at large American Universities developing, coordinating and facilitating collaborative and trans-disciplinary research in the physical, natural, social sciences, engineering and public policy. Mahmud holds a Ph.D. in science, technology and public policy from George Mason University and an MPA in technology and information policy from the Maxwell School, Syracuse University.

Mohammad Gharesifard is a PhD research fellow working at UNESCO-IHE Institute for Water Education in Delft, the Netherlands. He has seven years of combined study/design and construction supervision work experience in Iran's water sector. During this period he worked for water and environmental consulting companies and was involved in several socio-technical water and sanitation projects. In 2014 he enrolled in Water Resources Management MSc. programme at UNESCO-IHE and graduated with distinction in April 2015. During this programme, he developed a keen interest in studying innovative water management paradigms and especially the role of Information Communication Technologies (ICTs) in citizen science initiatives. His MSc. thesis research focused on understanding the drivers and barriers for citizen engagement in ICT-based citizen observatories of the environment. He employed a social psychology framework from behavioural sciences to map influential factors on willingness of citizens to share their Personal Weather Station (PWS) data. He is currently working on two EU-funded projects in the area of environmental monitoring utilizing Citizen Observatories; the WeSenseIt (an EU-FP7 project, funded 2012–2016), and Ground Truth 2.0 (a Horizon 2020 project, funded 2016–2019). His PhD research focuses on the dynamic link between mass participation in ICT-enabled citizen observatories of the environment and the level of impact of the citizens/communities on decision making processes (the institutional impact of citizen observatories).

Mark Gillingham is an avid boater who enjoys sailing Lake Michigan. Mark has decades of experience with technologies that improve people's lives as they work in places as varied as elementary classrooms to foundation boardrooms. Now, he is putting this experience to use to monitor our local water quality for us and future generations.

Claudia Göbel works at the ECSA Secretariat currently hosted by the Museum für Naturkunde Berlin, Leibniz-Institute for Evolution and Biodiversity Research. She is responsible for community management and networking, analysis and review of citizen science projects, communication and PR as well as conference organisation and workshop facilitation. She has a background in History, Philosophy and Sociology of Science and Political Sciences. Her work and research focuses on social and epistemological orders in participatory research, research policy and Science 2.0.

Crona Hodges has been working on the FP7 Project COBWEB (Citizens Observatories) since October 2013 as part of the Earth Observation Group in Aberystwyth University. She has worked closely with COBWEB colleagues at Welsh Government and members of volunteer organisations in the Dyfi Biosphere, Wales, to further our understanding of how citizen science and mobile data collection programmes can enhance and contribute to the evidence-base of environmental data that can effectively contribute to policy development. Crona's research interests are specifically in European land cover monitoring programmes and she has been working with the EIONET's EAGLE group to explore ways in which that work can be enhanced by citizen science groups and voluntary groups on the ground. Aside from her work in research Crona has been Director of her own company Geo Smart Decisions Ltd since August 2011 which specialises in the management and analysis of spatial data for environmental applications. In particular, much of Crona's work has been in the area of land cover mapping using satellite imagery particularly for the purposes of social and environmental impact analyses. After gaining a First Class Honours Degree in Environmental Science at University of Stirling in Scotland (1998 – 2002), Crona went on to her PhD studies (2002 – 2006) which explored the relationship between vegetation assemblages and spectral reflectance using both in situ data and airborne CASI imagery. Crona is a member of the Association for Geographic Information and the Remote Sensing and Photogrammetric Society in the UK.

Catherine Hoffman works at the interface between citizen science and informal education. She has experience in environmental, STEM-based education including implementing citizen science into informal centers and for community-based non-profits. She is currently the Managing Director for SciStarter and serves as the Project Manager for the development of SciStarter 2.0. She holds an M.Sc. in Zoology from the University of British Columbia.

Elena Jurado combines passion for coordinating projects in the interface of education, science and technology at 1000001 Labs and for teaching at La Miranda School. She obtained a BSc degree in Industrial Engineering, an MSc degree in Technology, Innovation and Education; and completed her PhD in Environmental Modelling at CSIC-UPC. Her research on climate change effects on contaminants' dispersion led her to work in different countries such as Spain (IDAEA-CSIC), Netherlands (University of Utrecht), and Italy (Joint Research Centre European Commission) and to write a number of scientific articles in international journals. A profound interest in education has led her to teach and disseminate science and technology in different educational institutions. Her main research interests include: environmental modelling; artificial intelligence; sustainability; educational robotics; citizen science; and education.

Eric Kennedy is a PhD student in the Human and Social Dimensions of Science and Technology program at Arizona State University. His work examines the interface between expertise, public engagement, and decision-making. Kennedy is a 2014 Breakthrough Generation Fellow, and publishes in environmental science, innovation systems, and the social studies of science and knowledge. His most recent book is "The Rightful Place of Science: Citizen Science" (2016), co-edited with Darlene Cavalier.

Mike Kobernus (M) has twenty years of experience in large-scale software development projects. In the last ten years, he has focused on the design and implementation of Environmental Information systems. He has extensive knowledge on web based tools and web communication protocols.

Thomas J. Lampoltshammer holds a doctoral degree in Applied Geoinformatics in addition to his Master's degrees in the fields of Information and Communication Technology, Embedded and Intelligent Systems, as well as in Adult Education. He currently works as a Senior Researcher (postdoc) at the Department for E-Governance and Administration at the Danube University Krems/Austria. His research experience covers national and EU-funded projects in ICT-related topics, such as Geoinformatics, Semantics, Social Media, Legal Informatics, and E-Health. Furthermore, he acts as reviewer for several SCI-indexed journals and international conferences.

Hai-Ying Liu is a senior scientist at Department of Environmental Impact and Economics, NILU. She is expert on environmental health impact assessment, and integrative approaches to environmental issues. Dr. Liu has five years R&D experience on large EU projects at NILU and five years working experience from environmental regulatory authorities in China (Ministry of Environmental Protection). Her current research interests include innovative approaches for citizens' involvement in urban air pollution including micro sensors and mobile phone tracking systems. She has published more than 20 papers within ecology, air pollution research and integrated environmental monitoring, including three recent "highly accessed" papers in Environmental Health.

Giovanna Lombardi has a Bachelor's degree in Chemistry, Postgraduate Specialization in Instructional Psychology, Doctorate in Science Teaching. Professor and researcher at the Central University of Venezuela (UCV). Area of study: Learning the language of a discipline and the development of reading strategies for scientific texts as a way to improve the academic performance of poorly performing students, who usually come from public schools. Experience in the instructional design of Community Service Projects at the UCV.

Jano Moreira de Souza holds a B.Sc. in Mechanical Engineering from the Federal University of Rio de Janeiro (1974), a M.Sc. in Systems Engineering and Computer Science from the Federal University of Rio de Janeiro (1978) and a PhD in Information Systems from the University of East Anglia (1986). He is currently Professor at the Federal University of Rio de Janeiro. He has experience in computer science with emphasis on database and is currently working on the following topics: database, knowledge management, negotiation support systems, autonomic computing and computer-supported cooperative work.

Josep M. Mominó has a PhD in Education and Masters in Educational Research from the Ramon Llull University (URL), Barcelona. Professor on the Department of Psychology and Education in the Open University of Catalonia (UOC) and on the Information and Knowledge Society PhD programme. He's also a teacher and has experience in primary education. His research focus stresses the evolution of educational institutions in the Network Society.

Gregory J. Newman is a research scientist, ecologist, and informatics specialist at the Natural Resource Ecology Laboratory (NREL) at Colorado State University (CSU). He received his PhD from CSU in citizen science, community-based monitoring, and ecological informatics. His current research focuses on designing and evaluating the effectiveness of cyber-infrastructure support systems for citizen science programs. His research interests include evaluating various citizen science program models, understanding the socio-ecological benefits of engaging the public in scientific research, designing and evaluating data management systems for socio-ecological research, assessing the value of local and traditional ecological knowledge for conservation and education outcomes, and developing spatial-temporal decision support systems.

Luis Arnoldo Ordóñez Vela was born in Venezuela, graduated from UCV (Caracas, Venezuela), with a Ph.D. in Biochemistry from MIT (Cambridge, Massachusetts, USA), completed the Advanced Management Program (PAG) at IESA (Caracas, Venezuela). In the past, Secretary General of the Venezuelan Association for the Advancement of Science (AsoVAC), President of the Venezuelan Foundation for the Advancement of Science (FUNDAVAC) and Executive Secretary of the Permanent Commission for Science and Technology of the House of Representatives at the them National Congress. Currently is President of Interconectados and teaches at several universities in graduate courses, among which are Doctorate of Social Sciences and Humanities, USB (Caracas, Venezuela, Seminar on Participation); Doctoral Program in Education, FacE, UC (Valencia, Venezuela, Research Seminar); and Doctoral Program in Management, FACES, UC (Valencia, Venezuela, Organizations with Social Relevance). Other courses (MOOC): Citizen Participation, Structuring Complex Problems, and Collaborative Research. Current research interests focus on the determinants of participation in society and the impact of digital technologies, especially the role of information in decision-making mechanisms in Latin American environment and mobilization of knowledge between universities and Latin American society.

Carla Viana Pereira has a B.Sc in Engineering at the State University of Rio de Janeiro, and currently works as business analyst at DATAPREV, a public company that provides technology solutions in information and communication aimed at the improvement and implementation of social policies for the Brazilian state. She holds a MSc degree in computer science from the Federal University of Rio de Janeiro, where she conducts research on promoting behavioral change using an app called We4Fit, implemented to improve eating habits. Her main research interests are the study of gamification, persuasion, behavioral changes, collaboration and medicine 3.0.

Maria Gilda P. Esteves has a B.Sc in Biological Sciences (1983) and a B.Sc. in Marine Biology (1984). She also holds a M.Sc. in Marine Geology and Geophysics (1996) from Fluminense Federal

University and a doctoral degree in Systems Engineering and Computer Science at the Federal University of Rio de Janeiro - COPPE/UFRJ (2016). Her current research focuses on design of sociotechnical systems to support crowd collaboration with science.

Alexandre Prestes Uchoa is a project manager with extensive experience in design and deployment of information management processes and analytical methods for very large research intensive organizations. Work as a customer consultant for the world's biggest research management technology and data provider assisting decision and policy makers of South American governmental and academic organizations to define and assess research initiatives and strategies. Holds a MSc in computer science and is a PhD candidate in Universidade Federal do Rio de Janeiro (Federal University of Rio de Janeiro), one of the top 5 Brazilian institutions. Ongoing studies focus on knowledge representation and mediation in smart services.

Johannes Scholz started as Research Assistant at University of Applied Sciences, School of Geoinformation in 2004 until 2006. From 2007 until 2010 he was Project Assistant at the Institute of Geoinformation at Graz University of Technology. In 2010 he defended his Ph.D. thesis entitled "Real-time spatial optimization" with distinction. From 2010 until 2011 Johannes Scholz was University Assistant (PostDoc) at Vienna University of Technology, Institute of Geoinformation and Cartography. From 2012 until 2015 he was Senior Researcher at Research Studios Austria-Studio iSPACE a non-profit research organization. Currently, he is Assistant Professor for Geoinformation at Graz University of Technology, Institute of Geodesy. Besides, he teaches at the University of Salzburg and Alps-Adria University Klagenfurt and he is Associated Faculty at the Doctoral College 'GIScience' of University of Salzburg. His research interests focus on space and time, spatial-temporal events & processes, combinatorial optimization in GIS with focus on transport planning and decision support systems in general.

Joana Simoes is a software engineer and data scientist, with a strong background on geospatial technologies and algorithms. Throughout her career, she has mostly focused on customizing and applying Free and Open Source technologies to solve problems in a wide range of domains such as Smart Cities, Fisheries Stock Management or Disease Outbreak Modelling; she also enjoys coding low-level, generic, algorithms, and during her PhD she created a specific evolution of a quadtree for indexing vector data. Although Joana has accumulated a large experience with object oriented languages and relational databases, she is always interested in picking up new technologies. Lately she has been involved in the challenges of dealing with unstructured information, real-time or near-real time streams of data, and scaling systems to store and manage large volumes of information (Big Data).

Alexander Steblin holds a Degree in Telecommunications Engineering from the Pompeu Fabra University. He is a researcher in the R&D eHealth group in Eurecat - Technology Center of Catalonia, where is part of the Integrated Continuous Care research line. He has developed his experience in healthcare and citizen science sectors being involved in several national and European projects. In the European project REWIRE, he has coordinated a work package which consists of the development and integration of an application for planning, monitoring and customizing rehabilitation therapy after a stroke. Currently, he is working on the national project MyOSA, where he is leading the work package on analysis and data mining to provide statistics, recommendations and decision support to different agents involved in the treatment of obstructive sleep apnea. A part of the healthcare sector he was also involved in the

European citizen science project CITCLOPS where he participated in the development and analysis of methods that collect quality digital samples of seawater by citizens. His interests are related on the one hand with web-based technology solutions, using the techniques of artificial intelligence and machine learning and on the other hand with the techniques of usability and user experience.

Laia Subirats is a researcher of eHealth R&D group at Eurecat. She holds a PhD in Computer Science from Autonomous University of Barcelona, a Degree in Telecommunications Engineering from Pompeu Fabra University (first class honors) and a M.Sc. in Telematics Engineering from the Technical University of Catalonia and Carlos III University of Madrid. She has been working in R&D projects for 8 years in disciplines such as machine learning, citizen science, social networks, web services, ontologies, reputation and smart cities; both in national and international centers (Telefónica I+D and European Organization of Nuclear Research in Geneva, Switzerland) and within initiatives such as Google Summer of Code. She is co-/author of 20 technical and scientific publications and 1 patent in the area of artificial intelligence.

Patricia Tiago is a biologist, born in Lisbon. Obtained her MSc in Land and Environmental Planning from Universidade Nova de Lisboa in 2005. Research interests include involvement of people with science, specifically with biodiversity and biodiversity changes. In her PhD is working with citizen science databases to increase global biodiversity knowledge and to improve the data collected in those projects. From 2002 until 2013 was working with environmental education and science dissemination for different kinds of audiences and, in 2009, was part of a team that created a Portuguese biodiversity database - http://www.biodiversity4all.org.

Robin M. Urquhart was educated at Glasgow University, Scotland and Central University, Venezuela (UCV). Founded 3 schools in Venezuela, covering Kindergarten to 12th grade. Principal of K-12 school in Caracas, Asociación para una Nueva Educación for 30 years. Teacher in the UCV Masters program for Instructional Psychology for 15 years. Research interests: reading, writing, argument and critical thinking, third age education, project design. Independent consultant in Venezuela for 20 years. Founder and Instructional Director since 2011 of K-5 charter school in New York, NYC Montessori Charter School.

Katrin Vohland is director of the research department "Public Engagement with Science" at the Museum für Naturkunde, Leibniz Institute for Evolution and Biodiversity Research. Germany. Her main research interest is in the interface between science - especially biodiversity science - and different public audiences. She is active in the German Network for Biodiversity Research (NeFo) and investigates how scientific knowledge can be mainstreamed into policies as in the case of IPBES (Intergovernmental science-policy Platform on Biodiversity and Ecosystem Services). As another pathway to intensify science--society interactions, Katrin is very active in developing citizen science as an integrative approach in Germany and Europe. She initiated the German citizen science platform including the development of quality criteria, she is chair of the European scientific network of the COST Action "Citizen Science to promote creativity, scientific literacy, and innovation throughout Europe", and she is vice-chair of the European Citizen Science Association (ECSA) which promotes and conducts citizen science.

Uta Wehn is Associate Professor of Water Innovation Studies in the Integrated Water Systems and Governance Department at UNESCO-IHE. She holds a BSc in Computer Science and an MSc and PhD in Science and Technology Policy (Innovation Studies). She draws on almost 20 years of combined industrial and research experience. Her research focuses on ICT-enabled citizen science and the social, policy and governance issues arising from innovations in the water sector and beyond. She has over 60 publications in peer-reviewed journals, international conference proceedings and book chapters in the areas of citizen science, data sharing, knowledge transfer, capacity development and innovation. She currently leads the H2020 citizen science project 'Ground Truth 2.0 – Environmental knowledge discovery' (2016-2019) and she has a leading role in several other citizen science projects. She is also the Project Director of AfriAlliance: Africa-EU Innovation Alliance on Water and Climate (H2020, 2016-2021) and a member of several high level international initiatives, incl. the OECD Water Governance Initiative and the EIP Water Steering Group.

Jian Zhang is a Professor of Ecology and Biogeography at East China Normal University, Shanghai, China. His research interests span a broad range of topics, including macroecology, conservation biology, community ecology, climate change, forest dynamics, and spatial statistics. His research involves a wide range of spatial and temporal scales and a variety of taxa.

Index

T

U

V

W

www.ingramcontent.com/pod-product-compliance
Lightning Source LLC
Chambersburg PA
CBHW081215180926
51613CB00003B/244
9781522509622